B+T
NO 25 '81
$45.00

THE BOREAL
ECOSYSTEM

PHYSIOLOGICAL ECOLOGY
A Series of Monographs, Texts, and Treatises

EDITED BY

T. T. KOZLOWSKI
University of Wisconsin
Madison, Wisconsin

T. T. KOZLOWSKI. Growth and Development of Trees, Volumes I and II – 1971

DANIEL HILLEL. Soil and Water: Physical Principles and Processes, 1971

J. LEVITT. Responses of Plants to Environmental Stresses, 1972

V. B. YOUNGNER AND C. M. MCKELL (Eds.). The Biology and Utilization of Grasses, 1972

T. T. KOZLOWSKI (Ed.). Seed Biology, Volumes I, II, and III – 1972

YOAV WAISEL. Biology of Halophytes, 1972

G. C. MARKS AND T. T. KOZLOWSKI (Eds.). Ectomycorrhizae: Their Ecology and Physiology, 1973

T. T. KOZLOWSKI (Ed.). Shedding of Plant Parts, 1973

ELROY L. RICE. Allelopathy, 1974

T. T. KOZLOWSKI AND C. E. AHLGREN (Eds.). Fire and Ecosystems, 1974

J. BRIAN MUDD AND T. T. KOZLOWSKI (Eds.). Responses of Plants to Air Pollution, 1975

REXFORD DAUBENMIRE. Plant Geography, 1978

JOHN G. SCANDALIOS (Ed.), Physiological Genetics, 1979

BERTRAM G. MURRAY, JR. Population Dynamics: Alternative Models, 1979

J. LEVITT. Responses of Plants to Environmental Stresses, 2nd Edition.
Volume I: Chilling, Freezing, and High Temperature Stresses, 1980
Volume II: Water, Radiation, Salt, and Other Stresses, 1980

JAMES A. LARSEN. The Boreal Ecosystem, 1980

In preparation
SIDNEY A. GAUTHREAUX, JR. (Ed.), Animal Migration, Orientation, and Navigation
F. JOHN VERNBERG AND WINONA B. VERNBERG (Eds.), Functional Adaptations of Marine Organisms

THE BOREAL ECOSYSTEM

JAMES A. LARSEN
University–Industry Research Program
The University of Wisconsin
Madison, Wisconsin

1980

ACADEMIC PRESS
A Subsidiary of Harcourt Brace Jovanovich, Publishers

New York   London   Toronto   Sydney   San Francisco

COPYRIGHT © 1980, BY ACADEMIC PRESS, INC.
ALL RIGHTS RESERVED.
NO PART OF THIS PUBLICATION MAY BE REPRODUCED OR
TRANSMITTED IN ANY FORM OR BY ANY MEANS, ELECTRONIC
OR MECHANICAL, INCLUDING PHOTOCOPY, RECORDING, OR ANY
INFORMATION STORAGE AND RETRIEVAL SYSTEM, WITHOUT
PERMISSION IN WRITING FROM THE PUBLISHER.

ACADEMIC PRESS, INC.
111 Fifth Avenue, New York, New York 10003

United Kingdom Edition published by
ACADEMIC PRESS, INC. (LONDON) LTD.
24/28 Oval Road, London NW1 7DX

Library of Congress Cataloging in Publication Data

Larsen, James Arthur.
  The boreal ecosystem.

  (Physiological ecology series)
  Bibliography: p.
  Includes index.
  1. Taiga ecology--Canada. 2. Taiga ecology.
I. Title.
QH106.L28   574.5'2642'091813   79-8533
ISBN 0-12-436880-8

PRINTED IN THE UNITED STATES OF AMERICA

80 81 82 83   9 8 7 6 5 4 3 2 1

Contents

Preface ... xi
A Note on Nomenclature ... xv

1. Introduction: Boreal Ecology and Ecosystems Analysis
North American and Eurasian Similarities ... 5
Boreal Ecosystematics ... 6
Community Structure: Continuum Theory ... 6
Investigating Native Communities ... 10
The Basic Elements ... 12
Problems of Modeling ... 13
Summary ... 16
References ... 16

2. History of the Boreal Vegetation
Species and Environment ... 19
Dispersal ... 20
Origins of the Flora ... 21
The End of the Tertiary ... 23
Survival in Refugia ... 24
Postglacial Vegetation ... 25
The Palynological Record ... 26
Postglacial History ... 28
The Southern Boreal Border ... 31
The Northern Border ... 33
Forest-Tundra Communities ... 36
Ecological Significance of Historical Events ... 40
References ... 41

3. Climate of the Boreal Forest
Boreal Climate: General Discussion ... 49
Climatic Parameters ... 50

Local Boreal Climates	62
Forest Systems Climatology	67
Atmospheric Subsystem	69
Radiation and Temperature	70
Local Energy Budget	71
Climate and Permafrost	74
Climate and Species Distribution	76
References	84

4. The Boreal Soils Subsystem

Genesis of Podzol Soils	92
The Chemical Processes	96
The Geography of Podzols	97
Soil Classification: Boreal Regions	102
Genetic Variations and Plant Communities	106
Moisture and Permafrost	113
Soil Organisms	117
Soil Microorganisms	119
Soil Animals	121
Summary: Climate and Soils	122
References	123

5. Boreal Communities and Ecosystems: The Broad View

Plants and Environment	130
The Circumpolar Boreal Forest	132
The North American Boreal Continuum	134
Alaska	137
The Cordillera	139
Northwestern Mackenzie–Yukon Region	140
Southwestern Mackenzie and Northern Alberta	149
The Canadian Shield Region	161
Shield Communities; Species Composition	165
Boreal Forests of Eastern Canada	176
Gaspé-Maritime Region	178
The Forests of Balsam Fir	180
Cape Breton and Newfoundland	180
Labrador–Ungava	182
Northern Central Quebec	184
Black Spruce Communities	188
Altitudinal Tree Line	192
The Northern Forest Border	195
Lichens and Mosses of the Forest Floor	208
Mosses in the Communities	212
The Appalachian Extension	224
Extent of the Eurasian Boreal Forest	225

Summary	227
References	228

6. Relationships of Canadian Boreal Plant Communities

Procedure	240
Statistical Methods	244
Results	251
Discussion	256
Broad Climatic Relationships	258
The Study	258
Species Behavior	261
Regional Environmental Analysis	266
Future Possibilities	274
Summary	275
References	277

7. Boreal Communities and Ecosystems: Local Variation

Environmental Gradients	285
Environmental Gradients: Moisture	287
Environmental Gradients: Light and Shade	295
Environmental Gradients: Nutrients	298
The Dynamics of Competition: Adaptation	307
Succession	314
Neoclassic Succession Concepts	321
Permafrost and Succession	324
Succession in Peatlands	327
The Ecology of Fire	329
Sapling Establishment	335
Conclusion	339
References	341

8. Nutrient Cycling and Productivity

Nutrient Cycling and the Boreal Ecosystem	353
The Nature of Cycling Processes	353
Chemical Changes during Decomposition	355
Nutrient Cycles Through Soil Populations	355
The Carbon Cycle	357
The Nitrogen Cycle	357
The Nitrogen Cycle in Forest Stands	361
Other Elements	364
Effects of Fire on Cycling	366
Productivity: Community Comparisons	368
The Lichen Associates of the Communities	373

viii *Contents*

Ecosystem Nutrients and Management	376
References	377

9. The Trophic Pyramid: Animal Populations

Plants and Animals	382
Wildlife: A Biotic Factor	390
Biomass of the Avifauna	390
The Trophic Levels	396
The Plant–Moose–Wolf Subsystem	397
The Wolverine and Wolf Compared	399
Role of Insects	401
Effects of Fire on Animal Populations	403
Population Cycles	404
References	408

10. Boreal Ecology and the Forest Economy

Regional Variations	414
Succession	416
The Forest Border	418
Soils and Podzolization	421
Community Structure	422
The Forest Economy: Resources	423
Managing the Forest Ecosystem	426
Exploitation and Management	427
The Bioregenerative System	428
Management in Northern Regions	429
Wildlife Management	430
Production and Management	433
Forest Management	433
The Outlook for the Future	435
References	436

11. Epilogue

Text	440
References	443

Appendix I Analysis of Boreal Soils	445
Appendix II Broad Geographical Species Relationships	449
Appendix III Community Composition	452
Appendix IV Frequency of Occurrence of Lichens	457
Appendix V Species Frequently in Black Spruce Communities	463

Appendix VI Species in Boreal Forest Literature **468**
Appendix VII Special Definitions **480**

Subject Index 483
Species Index 495

Preface

There are, as yet, no single great works of synthesis in ecology such as Darwin and Wallace accomplished in the development of evolutionary biology. Ecology possesses a wide-ranging subject matter, and it may never be possible to organize ecological knowledge in quite the same way that the subject matter of evolutionary biology, physics, chemistry, and many other fields of science is organized.

Ecologists undertake the study of ecosystems—in this case the boreal ecosystem—by describing and systematizing knowledge available on how organisms live together in communities. The work of ecologists in the boreal biome includes the description of communities, tabulation of species composition of communities, and the study of behavior of species populations—all topics representing interrelated facets of the vegetational ecology of boreal regions.

Initial sections of this book deal with aspects of the floristic composition and evolutionary history of the boreal vegetation. These introduce subsequent discussion on the *processes* at work in vegetation, soils, and the atmosphere—in short, with the boreal forest as an ecosystem, the sum total of the influences of many closely interlaced biotic and physical factors. These include not only plant species that make up the visible vegetation but also nutrients, soil, temperature, rainfall, progression of the seasons, soil microflora, arthropods, insects, and larger animals such as marten, otter, beaver, moose, caribou, bear, and wolf, and man. All are closely linked strands in the web of life, a web apart from, yet dependent on and influencing, the raw physical environment.

This is not to say that the boreal forest is well understood as an ecosystem. It is a biome in which scientific investigation is still in the early stages, but it is obviously worthwhile now to summarize existing knowledge and to encourage research needed to eliminate gaps in the knowledge of this major ecosystem. It is a topic on which almost everything remains to be done.

The science of ecology has advanced to the point where it is now possible to systematize study of the interaction of biotic communities with environment. By this interaction is meant the way in which raw materials such as mineral nutrients, oxygen, nitrogen, carbon, and, equally important, energy are trapped, utilized, moved from lower to higher trophic levels (from plants to herbivorous animals to carnivorous), and recycled through microorganisms back to plants again, with gains or losses involved in each step.

The technique known as systems analysis has been devised to handle, with the aid of high-speed computers, complicated webs of cause and effect such as those found in chemical and industrial engineering and in ecosystems. The procedures employed in both engineering and ecology systematize knowledge. Ecosystem analysis is a tool with great potential for assisting in the effort to conserve and manage effectively the world's renewable resources.

This volume deals with boreal ecology, and it is probably reasonable to state that ecology is a branch of the sciences that has not as yet advanced to the state of strict precision and rigor that characterizes the basic sciences, but it does draw heavily on physiology and biochemistry, on applied climatology and soil science, and on microbiology, plant nutrition, geology, hydrology, as well as on other disciplines. Ecology and studies of environmental relationships are rapidly gaining in prestige, strength, and importance, and it is with potential future significance in the protection and management of boreal lands in mind that I undertook the preparation of this work.

The quantitative description of communities found in various regions of the boreal forest in Canada necessarily involves a method. I believe that ecological relationships are the important facets to be considered, and selection of the method chosen to study them should rest with the individual researcher.

This work is a brief summary of the knowledge available on the boreal forest region of North America, with extensive reference to the boreal regions of Europe and Asia. It should serve as an introduction and reference source to its audience: undergraduate and graduate students in the biological and ecological disciplines, research workers in these fields as well as in related areas such as soil science, agronomy, genetics, and climatology; in short, everyone with an interest in boreal ecology.

References are included at the end of each chapter. Some of the citations have been referred to in the text; others are included for individuals interested in perusing the list for more material on particular subjects. The point of view is that of ecosystem analysis—a subject basic to forestry and resource management, protection of the environment, and to all activities involving preservation of natural resources on which human beings depend for sustenance.

Although the title page asserts that the book has been written by one person, no book is ever the sole product of its author. Many associations go into every literary and scientific work. In this case they are too numerous to thank individually. Without the help others have given, this work would not have achieved whatever modest level of accomplishment it may represent.

James A. Larsen

A Note on Nomenclature

In most of this volume, the scientific names of trees are used without reference to the common names. In several chapters, however, there are selections quoted from other authors, and these do contain common names of the species without reference to the scientific names. To avoid editorial modification of the original text, common names and scientific names are used, often interchangeably and without consistency. To assist those readers who are unfamiliar with the names of these tree species, a short list of the most frequently mentioned species is given in the tabulation below. I have also used the common names of trees in sections in which the names of only a few tree species are used, but these are used frequently and repetition of scientific names in this instance seemed both unnecessary and, from the point of view of literary style, somewhat pretentious. A complete list of trees and other plant species mentioned in the text is provided in Appendix VI. Scientific and common names (where they exist unambiguously) are both given. The botanical manuals that can be employed to help identify the species in the field are also given in this Appendix, and those most often followed in selecting the species names to be used in cases in which more than one name is in common usage for a given species are identified.

Scientific and Common Names of the More Frequently Mentioned Trees

Genus and species	Common name
Abies balsamea	Balsam fir
Betula lutea	Yellow birch
Betula papyrifera	White birch, paper birch
Larix laricina	Tamarack, American larch
Picea engelmannii	Engelmann spruce
Picea glauca	White spruce
Picea mariana	Black spruce

(continued)

Genus and Species	Common name
Picea rubens	Red spruce
Pinus banksiana	Jack pine
Pinus contorta	Lodgepole pine
Pinus flexilis	Limber pine
Pinus resinosa	Norway pine, red pine
Pinus strobus	Eastern white pine
Populus balsamifera	Poplar, balsam poplar
Populus grandidentata	Bigtooth aspen
Populus tremuloides	Aspen, trembling aspen
Pseudotsuga menziesii	Douglas fir
Quercus rubra	Northern red oak
Thuja occidentalis	Northern white cedar
Tsuga mertensiana	Mountain hemlock

1   Introduction: Boreal Ecology and Ecosystems Analysis

The great northern coniferous biome is known as the *boreal forest* or, from the Russian, *taiga*. Its northern limit—where it meets the treeless lands of the arctic tundra—ranges in North America from roughly lat. 68°N in the Brooks Range of Alaska southeastward across Canada to about lat. 58°N on the west coast of Hudson Bay. It crosses the Labrador peninsula at roughly lat. 58°N and, across the Atlantic, follows the northern coasts of Sweden and Finland. Except for a fringe of coast tundra along northern Russia, here also it follows the Arctic coast. In Siberia, however, the northern tundra should perhaps not be considered exclusively coastal, since the treeless Siberian barrens extend 500 or more kilometers inland from the coasts of the Laptev and East Siberian seas.

The southern limit of the boreal forest is less easily defined, since the boundaries here are not usually as sharp as the forest–tundra ecotone. The transition, furthermore, takes place to a wide variety of vegetational types—so that rather than grading into one biome exclusively, as into tundra in the north, the transition in the south is to broadleaf deciduous forest, to parkland, and to grassland, depending apparently upon the existing regional climate.

The boreal forest is, thus, circumpolar in extent, occupying a belt as wide as 1000 km in certain regions of both North America and Eurasia. Although it is remarkably uniform in general appearance throughout this entire range, important regional differences are, nevertheless, clearly discernible upon close inspection. There may be only minor differences among regions in the general appearance of the dominant trees making up the forest canopy, but similarities in appearance obscure the fact that there are distinct differences in the species and varieties of plants found therein. At the northern forest border, for example, white spruce, Siberian larch, cedar, and balsam fir, as well as pine, black

Fig. 1. Sunlight penetrates to the ground in this small opening in a closed-canopy black spruce forest located in the southern portion of the boreal forest of central Canada.

spruce, and aspen, all attain positions of dominance at least in certain restricted portions of the circumpolar boreal range (Hustich, 1966; Hare and Ritchie, 1972).

Regional differences are apparent not only along the northern border but also throughout the forest. George H. La Roi (1967), for example, in a study of the composition of the North American boreal forest, has found that species composition in the southern boreal forest changes continually as one travels east or west, with only small changes apparent over short distances but with communities in Newfoundland and Alaska

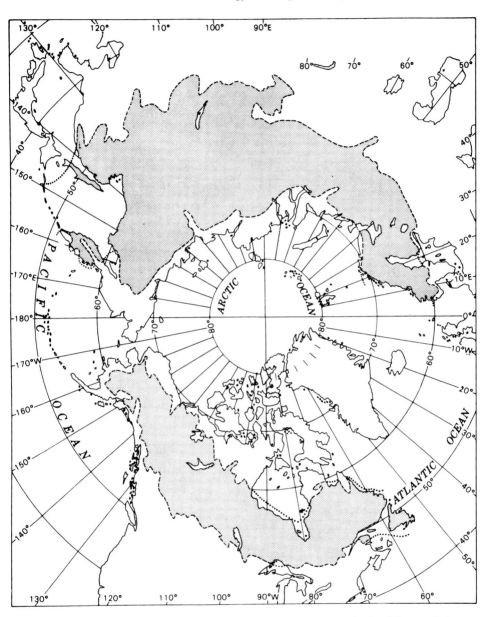

Fig. 2. The circumpolar range of the boreal forest. About two-thirds of the area is in Eurasia. The sector in eastern Canada lies farthest from the North Pole. The map is from Hare and Ritchie (1972) and is based on a number of sources, among them Sjörs (see References). The map is reprinted here by permission of the authors and of the American Geographical Society of New York.

TABLE 1
Chief Boreal Tree Species by Longitudinal Sector[a,b]

Genus	North America, 55°W–160°W	Northern Europe, 5°E–40°E	Western Siberia, 40°E–120°E	Eastern Siberia, 120°E–170°W
Conifers				
Picea (spruce)	*glauca* *mariana*	*excelsa*	*obovata*	*obovata* *jezoënsis*
Abies (fir)	*balsamea*		*sibirica*	*nephrolepis* *sachalinensis*
Pinus (pine)	*banksiana*	*silvestris*	*sibirica* *silvestris*	*silvestris* *pumila* *cembra*
Larix (larch)	*laricina*		*sibirica* *sukachzewski*	*dahurica*
Hardwoods				
Populus (popular)	*tremuloides* *balsamifera*	*tremula*	*tremula*	*tremula* *suaveolens*
Betula (birch)	*papyrifera* *kenaica*	*pubescens* *verrucosa* *kusmisscheffi*	*verrucosa* *pubescens*	*ermani*
Alnus (alder)	*tenuifolia* *crispa* *rugosa*	*incana*	*fruticosa*	*fruticosa*
Salix (willow)	*Salix species*	*Salix species*	*Salix species*	*Salix species*

[a] From Hare and Ritchie (1972), with permission
[b] Certain species of nonboreal type range into the southern edge of the boreal forest. In North America, for example, these include the white and red pines (*Pinus strobus*, *P. resinosa*), *Thuja occidentalis*, and species of *Ulmus*, *Fraxinus*, and *Acer*. In Europe and Asia, species of *Quercus*, *Tilia*, and *Fraxinus* also have a limited boreal range. The taxonomies of *Betula*, *Alnus*, and *Salix* are highly confused, and this table is only a rough guide.

quite dissimilar in terms of the species present. Just as there are differences from east to west, so are there differences from north to south at any given longitude, and in magnitude these are the equal of the east–west differences; in fact, in most regions greater differences over shorter distances are found along the north–south axis than along the east–west axis. There are differences also in the composition of the alpine variants of the boreal forest, as found, for example, in the Rocky Mountains. Some of the similarities and differences in the boreal forest communities of the various regions, in terms of the more abundant tree species, are shown in Table 1.

NORTH AMERICAN AND EURASIAN SIMILARITIES

Botanists familiar with the circumpolar boreal coniferous vegetation of northern regions of both Eurasia and North America attest that in appearance the forests are remarkably alike, and it has long been recognized that the floras of both regions have many genera and even species in common. In appearance, the North American and Eurasian forests are similar because they are virtually identical in general physical structure. The canopy is one-layered and composed of coniferous evergreen trees. Large shrubs are normally widely scattered and sparse in number. The ground layer of vegetation is dominated by low shrubs, there is a scattering of herbaceous plants, and all are underlain by mosses and lichens which may attain the density and thickness of a rich, deep carpet. By reason of its permanent nature, the northern coniferous evergreen forest canopy creates light conditions that, in at least one aspect, differ greatly from light conditions in temperate deciduous forests. Evergreen coniferous forests lack seasonal periods—such as occur in spring and fall in deciduous forests—when direct solar rays penetrate to the forest floor unobstructed by the canopy of trees. There is, as a consequence, no burst of flowering of early spring ephemeral species such as occurs in temperate deciduous forests, and there is little to correspond to the temperate forest's midsummer lull in activity.

Many boreal plant species have extensive, often circumpolar, distributions (Hultén, 1968). The tree species, however, are each confined to a single continent. The principle seems to hold that species making up the lower strata are more wide-ranging: the lower the layer the broader the range of the species commonly found therein. There are, thus, no circumboreal trees, but the mosses and lichens are almost all circumboreal or cosmopolitan. The reasons for this difference between the trees and other species of boreal communities are at present not at all clear (Kornás, 1972).

BOREAL ECOSYSTEMATICS

In considering the boreal forest or any other aggregation of native communities on land or in lakes and oceans, we first must recognize the nature of the raw materials with which we deal. Essentially these fall into two categories, living organisms and the physical environment. Living organisms are identified by means of species names. The environment is described in terms of the physical conditions that prevail and the chemical composition of the soil and atmosphere. Basic to an understanding of ecosystems all over the world is knowledge of the *physiological and biochemical processes* at work within living organisms. Also, one must understand the characteristic *individual responses to the environment* that lend each species an ecological identity as recognizable as morphology (and the two are related). Finally, the dynamics of the *relationships among species* living together in an environment must be understood. We thus have virtually the whole range of sciences playing a role in the study of boreal ecology.

We must, for obvious reasons, however, here limit our scope to studies of boreal species in their interactions with the environment. We are limited, further, to the study of relationships between communities and the environment (synecology) rather than to the study of individual species in the environment (autecology). We have in one sense simplified the task, but what remains are the aspects of community ecology that can only be considered the least precise and the least known, often involving familiar but vague concepts such as *succession, climax, ecocline,* and *continuum,* all of which are at best loosely defined, at worst controversial, or even fundamentally meaningless. Topics in ecosystem dynamics such as *productivity, trophic levels,* and *nutrient cycling* are, on the other hand, all more accessible to study than those listed above, but these subjects are largely in initial stages of investigation. It is with these reservations in mind that we attempt our review of knowledge available on the ecology of boreal vegetation.

COMMUNITY STRUCTURE: CONTINUUM THEORY

It is becoming generally accepted that a continuum of communities exists in natural environments. The existence of a continuum can be demonstrated in the case of plant communities by counting the individual plants of each species in numerous study plots located at random throughout a given area. The communities in these plots do not fall into a few distinct types or categories but are arrayed along various ecoclines, changing gradually in the number of individuals of each species from wet to dry sites, for example, and from sites rich to poor in nutrients.

These relationships are illustrated in a study conducted by Jeglum (1971, 1972, 1973) on boreal peatland vegetation in central Saskatchewan. In a sample of 119 study plots, the number of individuals of each species changed gradually along the ecoclines of soil moisture and nutrient status. These relationships can be seen in greater detail in Tables 25 and 26, but a graphic summary is given in Fig. 3, which shows the arrangement of communities by dominant species along the moisture and nutrient gradients. What is perhaps not obvious here, but shown clearly in Tables 25 and 26, is the gradual nature of the transition in species composition along the gradients. Species that are dominant in a community at a point along the continuum are present, but in lesser abundance, in many other communities as well. They are demonstrated to be increasingly important in environments more closely resembling the one in which they attain the greatest importance.

It is not to be inferred, however, that the remarkably close fit between plant communities and the environment is a static relationship. It is a

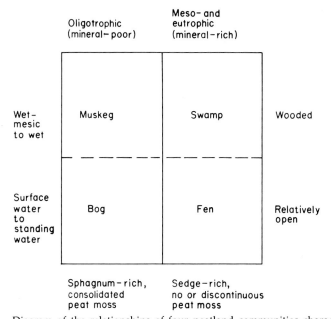

Fig. 3. Diagram of the relationships of four peatland communities characteristic of northern environments; shown are the relationships of the communities to physical aspects of the environment. The diagram is from Jeglum *et al.* (1973), with permission.

dynamic one, continually in the process of change as a consequence of *competition* and *succession*. By evolution and adaptation, plant species have developed ways to coexist with individuals of some species and to compete with others. The result, in terms of community dynamics, is succession. *Primary succession* is the process by which a plant community progressively changes the character of the environment, such as occurs during the filling in of a shallow lake with peat and the eventual establishment of a forest through the sequence:

fen → swamp → bog → muskeg → treed bog → forest

Secondary succession is the process by which any one of these communities is reestablished after having been destroyed by some agency such as flooding or fire: There will then be a series of changes from year to year in which initial invaders of the newly opened land, *the pioneer species*, are replaced by *secondary invaders* and finally a community of the type initially present is reestablished. This kind of succession can be represented by the sequence:

bare surface → annual herbs → perennial herbs and shrubs → treed bog → forest

The *climax community* carries the implication that it is the final stage succession can reach; it is the community which, barring fire or other disturbance, will be perpetuated for a very long period of time. Climax is essentially a misleading term, at least in boreal regions, since rarely is any community free from disturbance for more than a few decades. Many areas support *subclimax communities* which, even though undisturbed, appear not to undergo successional processes but remain stable for long periods of time. This poses the question of how long a community must persist to be considered climax.

Perhaps it is now obvious that at least some of the traditionally accepted concepts of ecological dynamics do not apply, at least without modification, to boreal regions. Many relationships between plant communities and environment in boreal regions cannot be described adequately by means of the traditional concepts of succession, competition, and climax. In some instances, for example, competition among species may be largely nonexistent because the raw physical forces in the environment are so harsh that any plant capable of gaining a foothold on a plot of ground has few or no competitors. If this is the case—or, we should say, in instances where this is the case—succession and climax become meaningless concepts, ecoclines and continuua are strictly determined by topography and surface geology, and all species are

Fig. 4. From the air, the northern boreal forest in northern central Canada (Kasba Lake, southern Keewatin, Northwest Territories, in the background) is a continuous vegetational cover over the landscape. The light-colored ground lichens show between trees of the lichen woodland on the low hills in the foreground, and areas of open muskeg fill the low sites at the edges of lakes and streams.

pioneers. The farther north one travels the more frequently one encounters situations of this kind (Fig. 4).

While many traditionally useful ecological concepts for the present must remain vague in their applicability to communities in northern lands, other concepts are more readily adaptable to application in studies of the boreal forest. *Productivity* is the term employed to designate the mass of organic material produced by a community, measured in terms of weight per unit of time. *Trophic levels* can be defined as the hierarchy established by the *food chain* and *energy cycle:*

  soil → plant→ herbivore → predator → top predator → decomposer and so back to the soil in a recurring cycle now considered the classical approach to studies of ecosystem structure (Dobben and Lowe-McConnel, 1975). *Nutrient cycling* can be described in terms of amounts of the various nutrient elements—nitrogen, carbon, potassium, and so on—moving through the food chain and tightly linked to the energy cycle. It will be shown that these concepts have ready and useful application to studies of the northern coniferous forest.

INVESTIGATING NATIVE COMMUNITIES

The central problem encountered in studies of native plant and animal communities is that the methods employed must be simple and at the same time capable of measuring with reasonable accuracy the various components of a complex system. Under conditions of fieldwork that are seldom physically comfortable and which, at worst, can be exceedingly arduous, only simple methods can be successfully employed; they should, however, minimize biasing influences and utilize replication.

The development of statistical methods for the study of plant communities opens a whole new range of possibilities for investigation. Many of the techniques, however, require long and tedious calculations, and their potential cannot be fully realized without using electronic computers. Even with computers it is still not possible, moreover, to handle with precision more than a few factors in the complex of environmental influences found in native communities, and even today we are often restricted to the use of relatively simple field methods and unsophisticated statistical treatments in characterizing the composition of communities in terms of the proportion in which each species is represented—numbers of individuals, percentage of the total population represented by each species, weight, space occupied, and so on.

Communities often include many individuals capable of attaining longer average life spans than those of a human investigator. Studies of forests, for example, must be based on inferences as to what has happened to the plant communities within the forest in years past and what presumably will happen in the future. For these inferences to be an accurate assessment of reality, exceptionally clever techniques must be employed. It is a credit to the modern science of ecology that progress has been made in studies of such processes as, for example, succession, competition, and evolution in native plant and animal communities.

The questions to be asked initially in a study of a native community are

1. What is its structure, i.e., what species are present, what size are they, what age are they, in what density do they occur, and is there pattern in the way they are arranged on the surface of the ground?
2. What are the relationships between the species and the environmental factors?
3. What role, if any, does each species play in maintaining and perpetuating the community?
4. Is the system self-perpetuating in its present state or will it continue to change, through the processes of succession, until it attains a more static condition?
5. What evolutionary development has given each species its present form and function; are any of the species evolving rapidly, as evidenced by ecotypic variation, at the present time?

It is probable that, in most instances, basic knowledge needed to answer these questions is still almost totally lacking. To be sure, beginnings have been made in studies of communities found in the world's major biomes, but it is not always certain that the techniques employed accurately characterize the communities, and so few quantitative studies may have been conducted that even community structure has not yet been adequately described.

For many biomes, descriptions of native communities will never be forthcoming, because the communities will have been destroyed long before the needed research is completed—or perhaps even begun. For many biomes we have already lost the opportunity to obtain even such data as can be obtained by means of quadrat counts of species frequency, density, and dominance and the other simple parameters commonly employed in studies of community structure (see "Methods" in Appendix VII).

It is fortunate that in the more remote boreal regions the disturbance of native communities has still been minimal, and human influence with cow, plow, and axe (not to mention steam shovel, jackhammer, and cement mixer) has not destroyed the cover of native vegetation. In these regions, we perhaps have time for the studies needed to understand the role of vegetation in the global ecosystem, knowledge that may one day be needed to rebuild an ecologically devastated planet. At the rate the human race is currently remaking the face of the earth, this knowledge will one day be needed to reestablish a balanced ecosystem in which atmosphere and soil, land surfaces, oceans, and lakes are capable of sustaining a stable global human population—one not so large that it continually destroys renewable resources more rapidly than they can be replenished.

THE BASIC ELEMENTS

Living organisms are the basic elements of ecosystems, continually undergoing interactions with one another and with the environmental complex. Fundamentally, there are two ways of measuring the organisms present in an ecosytem—*counting* and *weighing* by individual species or by such categories as (in the case of plants) herbs, shrubs, trees, or (in the case of animals) herbivores, carnivores, and other categories.

Communities can then be compared on the basis of the number of individuals of each species present, using *indexes of similarity* of which an example is given in Chapter 6. The changes that occur over time as a result of succession, or differences existing among communities because of differences in prevailing environmental conditions, can be studied using such indexes. The objective of much ecological research is to relate changes in biomass or numbers of organisms to environmental changes and to the consequences of competition and succession. This objective can be described in mathematical equations. A matrix of species and environmental factors can, for example, be expressed as a set of differential equations in which the terms are defined as follows: dN/dt is the rate of change in the number of individuals N_i of any species, and dE/dt is the rate of change in the intensity of the environmental factors E, a response proportional to the sum of products representing various interactions. Such a matrix is simply a mathematical way of saying that the rates of change in the densities of species representing the community depend upon the initial density values and the kind of response each species makes to a changing environment.

An example might be the set of differential equations (Margalef, 1968):

	E	N_1	N_2	N_3
$dE/dt =$		$-aEN_1$	$-bEN_2$	
$dN_1/dt =$	$+eEN_1$	$-fN_1^2$	$-qN_1N_2$	$-hN_1N_3$
$dN_2/dt =$		$+iN_1N_2$	$-jN_2$	$-kN_2N_3$
$dN_3/dt =$		$+lN_1N_3$	$-mN_2N_3$	$-nN_3^2$

giving the matrix of possible cross-products:

E^2	EN_1	EN_2	EN_3
EN_1	N_1^2	N_1N_2	N_1N_3
EN_2	N_1N_2	N_2^2	N_2N_3
EN_3	N_1N_3	N_2N_3	N_3^2

and a matrix of coefficients:

0	−a	−b	0
e	−f	−q	−h
0	i	−j	−k
0	l	−m	−n

expressing the strength of interactions.

Numerical methods utilizing computers can be employed to predict the steady state the system will approach after the introduction of appropriate feedback controls, time lags, and other effects. Externally induced changes in E and N represent environmental changes as well as changes in species densities caused by external forces such as grazing and natural disturbance. Often, species interact feebly with a great number of other species; strong interactions are found in systems having small numbers of species, and these are often subject to marked fluctuations in the populations of the species present.

There are many aspects of modeling that, at present, are understood only insufficiently, and the result is that modeling must still be considered an inexact art. With the present state of knowledge, many factors have an undetermined influence upon community behavior, and their effects can at best be only estimated. In addition, the response of plants to environmental factors may change greatly under certain conditions. Competing individuals of the same or other species will often have an effect upon behavior. Normal responses under some conditions may be quite different from responses to identical stimuli under other conditions. To be an accurate representation of the plant community, a model must take all such variations into account.

PROBLEMS OF MODELING

In our effort to describe the boreal ecosystem, we will not plunge into the more sophisticated stages of ecosystem analysis—the stages in which all relationships are reduced to differential equations for the elucidation of working arrangements within and among subunits. Basic outlines of many aspects of the boreal ecosystem are only now emerging and, while efforts have begun at various research centers to construct models of this ecosystem, it is generally acknowledged that these are laying the foundations for a task that will take many years to complete. There is now, however, sufficient knowledge available to begin fitting together a picture of the boreal forest as an ecosystem, and this must suffice for our purposes here.

A descriptive understanding of the boreal ecosystem is needed as a foundation upon which to build conceptual systems models. It is the initial step in any effort to construct, ultimately, a valid computerized model, one that, for example, can be used to predict future development of a vegetational community from knowledge of its present condition. Such models become useful for developing forest management practices and, equally important, for providing knowledge of the principles that make ecosystem behavior comprehensible. The initial step, however, is simply to describe the community to be analyzed and to inventory species and discern relative abundance, major community relationships, and interactions with the environment.

To state the matter in bold terms, expectations that mathematical representation of plant communities would possess a verisimilitude comparable to that obtained with systems analysis in, for example, the chemical industries, have proved overly optimistic. The reasons for this merit some discussion. First of all, most studies of ecosystems to date have necessarily been devoted to measurement of matter and energy transformations—the essential first step in analysis. These are basic to understanding, but they do not represent the entire system. Thus, some species, comparable in energy content and elemental composition to others, may produce allelopathic chemicals and generate distinctive effects in terms of behavior of the system. Examples of such differences in species behavior are many. They involve differing responses to sunlight and temperature, variations in palatability and nutrient status for herbivores, resistance to infection, toxic mineral tolerance and, more recently, susceptibility to damage by pollutants, herbicides, and pesticides. Organisms found in ecosystems are not interchangeable; a shift in species populations results in changes in the behavior of the system.

Matter and energy cycling is characteristic of natural systems, but the characteristics of the cycling are as much a response to the kinds of organisms present as to the raw physical aspects of the environment. The only way to obtain adequate knowledge of interspecies relationships is to undertake adequate sampling of plants and animals in the wild state.

There are, as an additional hazard in model building, simplifying assumptions that can seriously endanger the accuracy of the model if they are adopted uncritically. Pielou (1977) has listed a few of the more prevalent, Kellman (1975) has noted others, and additional major assumptions are mentioned in every book on systems modeling. Some of the more insidious dangers include sequential sampling of a system when it is in a state of rapid flux, assumptions that a system is closed when in fact it is not, and assumptions that responses of populations to changes in the environment are instantaneous and that they are not related to the age structure of the population. The behavior of species may be greatly influenced by the composition of the community, but in many studies the composition of the whole community is ignored. Competitive success may be enhanced or restricted according to population structure, and in many cases this, too, has been ignored.

There is, thus, much opportunity for error in models of natural systems. When errors in modeling are present, the predictions made by the model regarding the nature of steady states to be achieved at some future date will bear no relationship to the steady state attained by the natural community. Models are constructed to emulate a system that, it is assumed, is at or near a state of equilibrium, and it is also assumed that small changes in the system will result from small changes in the environmental factors. But, in nature, large fluctuations in environmental factors are probably common. They occur at times that are often distributed randomly, elicit responses that are nonlinear in intensity, and induce rates of change that occur independently of the rates of change in the inducing factor(s).

It is apparent from this discussion that the frontiers of community ecology are no longer to be found in descriptions of vegetation—whatever the method of sampling and data presentation. Today, the focus is on behavior of species populations, behavior of individual species in physiological terms, dynamics of interspecies as well as intraspecies relationships, competition, commensalism, parasitism, allelopathy, and so on. Realistic ecosystem analysis and modeling must allow for the influence of many complicating factors.

To build and test a model requires that populations of plant and animal species be adequately sampled. The model, in short, will not be simple; it must allow for all relevant complications. It must not, moreover, be based purely on guesses, however educated they may be. On the other hand, nothing would be gained by making the model as complex as the system; it would then actually no longer be a model but would be itself a system. Moreover, models that satisfy a mathematician's sense of elegance are not necessarily models of the real world. There is much to be gained in explorations of mathematical theory, but mathematicians are quick to admit that what is theoretically interesting may not bear much relationship to reality. It may, however, lead to pertinent questions and, ultimately, clearer insights into operating principles. There may be—there undoubtedly is—much still not known about systems dynamics. For example, do "waves" of action and reaction sweep through natural systems, the result of chance or unpredictable oscillations in the components of the systems? This and similar questions are basic to ecosystem analysis today. Continuing improvement in the performance of systems models is dependent upon enhanced knowledge and understanding.

SUMMARY

It is the purpose here to describe what is known and what still must be learned before it will be possible to begin to understand in mathematical terms the simplest boreal communities. Once this knowledge has been attained, it will be time to apply more sophisticated methods.

Three essential observations can be made in summary concerning the boreal forest ecosystem. First, in terms of areal extent, the forest is immense. Second, the boreal forest is surprisingly uniform in appearance over the entire expanse of subarctic circumpolar territory it occupies. Third, there exist readily discernable relationships between boreal vegetational communities and the environment in which they are found. The initial step in ecosystem analysis is to break the subject into manageable subunits and then to describe each subunit in as great detail as can be undertaken with the time and labor available. Finally, the subunits are conceptually reassembled to reveal how they work together in the natural integrated ecosystem.

Let us turn, then, to the subject with which we are to be concerned throughout the pages of this book, the boreal forest, and initially, with what is known of its history throughout the last million or so years, a necessary preliminary to an understanding of the plant and animal communities found there today.

REFERENCES

Bell, R. (1885). Report on part of the basin of the Athabaska River, Northwest Territory. *Geol. Surv. Can. Rep. Progr. 1882–83–84, Part CC* pp. 3–37.
Bell, R. (1900). Preliminary report of explorations about Great Slave Lake in 1899. *Geol. Surv. Can. Summ. Rep. 1899, Part A,* pp. 103–110.
Bostock, H. S. (1970). Physiographic subdivisions of Canada. In "Geology and Economic Minerals of Canada." Dept. Energy, Mines and Resources, Ottawa, Canada.
Grime, J. P. (1979). "Plant Strategies and Vegetation Processes." Wiley, New York.
Hare, F. K., and Ritchie, J. C. (1972). The boreal bioclimates. *Geogr. Rev.* **62,** 334–365.
Hearne, S. (1775). "A Journey from Prince of Wales Fort, in Hudson's Bay, to the Northern Ocean, etc." (J. B. Tyrrell, ed.). Champlain Society, Toronto.
Hosie, R. C. (ed.). (1979). "Native Trees of Canada." Can. Gov. Publ. Centre, Ottawa.
Hultén, E. (1968). "Flora of Alaska and Neighboring Territories: A Manual of Vascular Plants." Stanford Univ. Press, Stanford, California.

Hustich, I. (1966). On the forest–tundra and the northern tree-lines. *Ann. Univ. Turku (Rep. Kevo Subartic Stn). Part A.II:* **36,** 7–47.
Hustich, I. (1979). Ecological concepts and biographical zonation in the North: the need for a generally accepted terminology. *Holarctic Ecol.* **2,** 208–217.
Innis, G. S. (ed.). (1979). "New directions in the analysis of ecological systems." Soc. for Computer Simulation, La Jolla, California.
Jeglum, J. K. (1971). Plant indicators of pH and water level in peatlands at Candle Lake, Saskatchewan. *Can. J. Bot.* **49,** 1661–1676.
Jeglum, J. K. (1972). Boreal forest wetlands near Candle Lake, central Saskatchewan. I. Vegetation. *Musk-Ox* **11,** 41–58.
Jeglum, J. K. (1973). Boreal forest wetlands near Candle Lake, central Saskatchewan. II. Relationships of vegetational variation to major environmental gradients. *Musk-Ox* **12,** 32–48.
Jeglum, J. K., Boissinneau, A. N., and Haavisto, V. F. (1973). "Wetland classification by vegetational physiognomy and dominance." *Can. Bot. Assoc. Meet. London, Ontario, 1973.*
Kellman, M. C. (1975). "Plant Geography." St. Martin's, New York.
Kendeigh, S. C. (1954). History and evaluation of various concepts of plant and animal communities in North America. *Ecology* **35,** 152–171.
Kornás, J. (1972). Corresponding taxa and their ecological background in the forests of temperate Eurasia and North America. *In* "Taxonomy, Phytogeography, and Evolution" (D. H. Valentine, ed.), pp. 37–59. Academic Press, New York.
Krylov, G. V. (1973). "Prioroda taiga zapadnoi Bibiri" ("Nature of the taiga of western Siberia"). Novosibirsk.
La Roi, G. H. (1967). Ecological studies in the boreal spruce–fir forests of the North American taiga. I. Analysis of the vascular flora. *Ecol. Monogr.* **37,** 229–253.
Larsen, J. A. (1965). The Vegetation of the Ennadai Lake area, N.W.T.: Studies in arctic and subarctic bioclimatology. *Ecol. Monogr.* **35** 37–59.
Larsen, J. A. (1972). The vegetation of northern Keewatin. *Can. Field Nat.* **86,** 45–72.
Larsen, J. A. (1973). Plant communities north of the forest border, Keewatin, Northwest Territories. *Can. Field Nat.* **87,** 241–248.
Larsen, J. A. (1974). Ecology of the northern continental forest Border. *In* "Arctic and Alpine Environments" (J. D. Ives and R. G. Barry ed.), pp. 341–369. Methuen, London.
Low, A. P. (1889). Report on exploration in James Bay and country east of Hudson Bay, drained by the Big, Great Whale, and Clearwater rivers in 1887 and 1888. *Geol. Nat. Hist. Surv. Can. Annu. Rep.* **3,** Rept. J. pp. 6–80.
Low, A. P. (1896). Report on explorations in the Labrador peninsula, along the East Main, Koksoak, Hamilton, Manicuagan and portions of other rivers, in 1892–93-94-95. *Geol. Surv. of Can. 1896 Annu. Rep.* **8,** Rep. L, pp. 5–387.
Low, A. P. (1897). Report on a traverse of the northern part of the Labrador peninsula from Richmond Gulf to Ungava Bay. *Geol. Surv. Can. Annu. Rep.* **9,** Rept. L, pp. 6–80.
MacArthur, R. H. (1972). "Geographical Ecology." Harper, New York.
Margalef, R. (1968). "Perspectives in Ecological Theory." Univ. Chicago Press, Chicago, Illinois.
Pielou, E. C. (1969). "An Introduction to Mathematical Ecology." Wiley, New York.
Pielou, E. C. (1977). "Mathematical Ecology." Wiley, New York.
Rosswall, T., ed. (1971). Systems analysis in northern coniferous forests. Ecol. Res. Comm., Bull. 14. Nat. Sci. Res. Council, Sweden.

Shugart, H. H., and O'Neill, R. V. (eds.). (1979). "Systems Ecology." Dowden, Hutchinson and Ross, Inc., Stroudsburg, Penn. (Distributed by Academic Press, New York).
Sjörs, H. (1963). Amphi-Atlantic zonation, nemoral to arctic. In "North atlantic biota and their history" (A. Löve and D. Löve, eds.), pp. 109–125. Oxford Univ. Press, London and New York.
Tyrrell, J. B. (1897). Report on the Doobaunt, Kazan, and Ferguson rivers and the northwest coast of Hudson Bay, and on two overland routes from Hudson Bay to Lake Winnipeg. Geol. Surv. Can. Annu. Rep. **9**, Rept. "F", pp. 5–218.
van Dobben, W. H., and Lowe-McConnel, R. H., eds. (1975). "Unifying Concepts in Ecology." D. W. Junk, The Hague.
Walter, H., (1973). "Vegetation of the Earth, in Relation to Climate and the Eco-Physiological Conditions." Springer-Verlag, Berlin and New York.

2 History of the Boreal Vegetation

For an understanding of the plant species making up the vegetation of boreal regions, we must know something of the history of these species—their history in terms of the geological past, at least during the Tertiary and the Quaternary periods. But to understand this history, we must first have a general appreciation of the relationships that exist between living, growing, evolving plants and their environment.

SPECIES AND ENVIRONMENT

Each plant species, as a consequence of its evolutionary history, inhabits a geographical area possessing environmental characteristics to which the species is adapted. These environmental characteristics include not only those features that can be classed as physical but also include biological influences—other plants, animals, fungi, bacteria, viruses, all of which are capable of modifying the raw abiotic environment or of influencing plants through various processes and activities.

Native plant species found in a given region live together in communities, each species taking advantage of its unique adaptations to occupy an environment available in one or more of the various sites found over the landscape—woodland, meadow, bog, and so on. As one moves from one geographical area to another, climatic conditions, soils, and topography often change, so that species inhabiting a characteristic kind of site in one region are found on a somewhat different site in another. As one travels farther and farther from any reference region, more and more changes in plant communities occur, until there are only a few species held in common between the two regions or eventually none at all.

The distribution of boreal plant species seems to be basically a response to climate, with edaphic factors playing a secondary role. Climatic factors determine whether a given species is a potential occupant of a given area, and a broad parallel between temperature belts and vegetation can usually be discerned. Available moisture is a secondary influence in the control of plant distribution and is allied to temperature as a controlling factor. Many genera restricted to particular conditions of temperature range completely across a continent, inhabiting both humid and arid subdivisions of the climatic zones. In contrast, no species restricted to particular conditions of rainfall or humidity ranges across all the temperature zones.

While many of these relationships appear self-evident, detailed knowledge of the interaction between native plant species and the climate is still not available. There is today a wide gap indeed between the climatic interpretation of vegetational distribution and the description of the ranges of individual plant species in terms of their physiological response to meteorological factors.

DISPERSAL

Effective dispersal of seeds has obvious competitive advantages for plant species. From the wide variety of methods by which dispersal is accomplished, it can be assumed that evolutionary development of the dispersal methods of most species has undergone much field testing, so to speak, and has resulted in selection for dispersal mechanisms reasonably effective for colonizing or recolonizing available environments. Additionally, speed of dispersal has been of considerable significance in establishing the present-day geographical distribution of species. Most plant species growing in central Canada must have extended their range 1000 km or more since the end of the last continental glaciation 5000–7000 years ago (Savile, 1956). The present distribution of many plant species clearly shows that they must have traversed the Bering Strait while the sea level was depressed during the last glaciation and that they have traveled hundreds of kilometers east and west from the strait since the ice retreated (Hultén, 1968; Hopkins, 1967; Wolfe and Leopold, 1967; Wolfe, 1972, 1978; Colinvaux, 1967). It seems that 100–200 km or more per 1000 years is a minimum rate of dispersal for many plants under favorable conditions. During interglacial ages of the Pleistocene epoch, many plant species could easily have dispersed throughout the world. Plants invading unoccupied or sparsely occupied country can travel rapidly because they are not forced to compete with existing vegetation; the migration rates given for plants reoccupying deglaciated country cannot, thus, be applied to plants that must pass through densely vegetated country. It is true, however, that many plants are not good pioneers and are only capable of invading deglaciated country after it has first become adequately vegetated with pioneer species.

Wind, water, and animals can account for surprisingly rapid long-distance migrations of plant species. One study, for example, has shown that viable seeds of several common genera were regurgitated from the digestive tracts of birds after 144–340 hr or more, long enough to be transported several thousands of kilometers (Proctor, 1968).

Whatever may be the fastest and slowest rates of dispersal among boreal plant species, it must be assumed that differences do exist among species in the alacrity with which seeds are disseminated. Consequently, the rate of dispersal for the entire complement of species capable of colonizing newly available regions is slower than the rate at which the species with the fastest dispersal methods can become established. The maximum period required for the vegetation to reach equilibrium with the climate following a period of climatic amelioration is the time needed for the slowest migrant to reach the limit of its possible range, taking into consideration the possibility that more rapidly migrating species may constitute competitors for slower migrants by seizing most of the readily available habitats. It is also necessary to take into consideration the need for sufficient time for natural succession to establish an environment suitable for species occupying only climax communities. But on a geological time scale, periods of this magnitude are obviously available.

Rates of reaction to a deterioration of climate will of course be much more rapid; an unfavorable climate may manifest itself in changes in plant communities practically without a time lag, and particularly sensitive species can be exterminated by a single unfavorable season. For many species, the present limits of geographical range cannot be very far removed from their climatic limit. This appears to be the case with spruce in the northern central portions of Canada, at the northern edge of the forest, where deteriorating climatic conditions forced retreat to a more southern position about 1000 years ago.

ORIGINS OF THE FLORA

Fossil evidence has failed to provide a lucid record of the origins of the arctic and boreal floras, although it is inferred that they developed throughout the Tertiary period, beginning about 70 million years ago, and probably earlier (Fig. 5). Interpretations of the fossil evidence from

Years ago Millions	Duration	PERIOD	Era
2	2	QUARTERNARY Recent Pleistocene	CENOZOIC
65	63	TERTIARY ⎧ Pliocene ⎪ Miocene ⎨ Oligocene ⎪ Eocene ⎩ Palaeocene	
135	70	CRETACEOUS	MESOZOIC
190	55	JURASSIC	
225	35	TRIASSIC	
280	55	PERMIAN	PALEOZOIC
345	65	CARBONIFEROUS	
395	50	DEVONIAN	
430	35	SILURIAN	
500	70	ORDOVICIAN	
570	70	CAMBRIAN	

Fig. 5. The paleontological record.

the Tertiary hold that conditions then prevailing favored development of a tropical flora in the southern portions of North America, which graded northward into an "Arcto-Tertiary flora" in which deciduous trees and conifers were predominant. Wolfe (1971, 1972, 1978) has demonstrated, however, that the concept of an Arcto-Tertiary geoflora does not apparently apply to the Alaskan region; this further opens the interesting question whether it prevailed elsewhere as widely as has been generally accepted. The Tertiary genera are readily identified, in any case, through similarities to genera common in forests, prairies, and savannas today, and it appears that Tertiary vegetational zones in North America were displaced latitudinally about 20° north of the comparable present-day vegetational zones. In the early Tertiary, for example, the poleward limit of palms lay in Alaska and Germany. Sequoia was found in Spitzbergen. Species of a rich Tertiary flora known from Greenland include willows, poplars, birch, hazel, *Liquidambar*, elm, plane, sassafras, ash, *Liriodendron*, maple, and grape.

Tertiary fossils indicate that the climate of Banks Island, in the western Canadian Arctic, and also of Iceland, supported tree and smaller plant species now characteristic of the boreal and even the hardwood zone—spruce, pine, fir, hemlock, and small but significant amounts of elm, hazel, hornbeam, birch, willow, alder, and temperate-zone herbaceous plant species. Upper strata, corresponding to late Tertiary, have yielded a pollen assemblage dominated by spruce and pine and lacking the hardwoods and hemlocks found in the lower, earlier strata. The upper formation appears to represent a late stage in the transition from the warm climate of the early middle Tertiary to the cooler conditions of the subsequent Pleistocene epoch. General climatic conditions in the world had been fairly constant throughout the Cretaceous period, during which the angiosperms may have originated. Even if angiosperms arose earlier, it was in the Cretaceous that rapid evolution must have occurred, since a rather large number of angiosperm genera first make an appearance in the fossil record of this period. For some unknown reason, no fossils have yet been found that are indisputably transitional forms between angiosperms and "proto-angiosperms" from which they are derived; some evidence can be interpreted as showing that angiosperms evolved from forms at least resembling extinct seed ferns (Andrews, 1963) although this view is by no means universally accepted. In any case, the ancestral primitive angiosperms are believed to be completely extinct, unrepresented in modern flora by any living order. Conjectures are that the early angiosperms were some form of low-growing shrub, possessing simple leaves, a single vascular cylinder in a woody stem, and with flowers at the ends of branches or in a loose cyme. The time of angiosperm origin is uncertain, but the most likely geologic periods are the Triassic or Jurassic (Stebbins, 1974).

THE END OF THE TERTIARY

The Tertiary period closed with a change in climate occurring most notably in the Pliocene epoch, although a general cooling trend can be noted much earlier in the Miocene. Particularly in the high latitudes, temperatures declined rapidly in the Pliocene; tropical or warm-temperate conditions shifted markedly southward, and for the first time in millions of years glacial ice began to accumulate over the northern portions of North America and Eurasia, culminating in the widespread glaciations of the Pleistocene (Larsen and Barry, 1974; Beaty, 1976; Schneider and Thompson, 1979). With the steadily declining temperatures of the late Tertiary, it is probable that the modern species now characteristic of both the arctic and the boreal flora evolved from the species of the Tertiary flora of the northern portions of the hemisphere. By the beginning of the Quaternary period, about 3 or 4 million years ago, a predominantly herbaceous perennial flora similar in composition to modern floras existed throughout the North.

The history of the arctic tundra flora is interesting because in many respects it helps us to explain the history of the boreal flora. Although no fossil evidence yet indicates the existence of an actual tundra flora prior to the Quaternary, physiological characteristics indicate that both the boreal and the arctic flora are old ones, with well-developed adaptations to their environments. Many endemic genera even today have not moved beyond the confines of the Arctic regions. During the times of maximum warmth in the Tertiary (the late Eocene and early Oligocene), neither the polar ice caps nor large areas of tundra apparently existed, although at high altitudes there may have been small, isolated tundras along the edges of mountain glaciers. Strips of land conceivably only a few meters wide could have existed in places that were perennially chilled by ice, but in the middle and late Tertiary these small tundra strips with their specialized floras merged with each other on the lowlands and developed into circumpolar tundra belts. Species such as *Empetrum nigrum* and *Dryas octopetala* are at the latest preglacial in origin. Dwarf willows and birches probably appeared during the late Cretaceous or early Tertiary, as did the northern pines and spruces of the sub-Arctic, but knowledge of the pre-Pleistocene history of even the most typical species is limited because of the scant fossil record of late Tertiary vegetation in the Arctic. The fossil evidence that does exist is contradictory and difficult to interpret. For example, fossil coniferous wood extends to Meighen Island (lat. 80°N), an area that today possesses the most severe sea level climate of any ice-free area in the Arctic, much more severe than the climate at 1000 m altitude at lat. 82°N in northern interior Ellesmere Island. If the area in Tertiary times bore the same climatic relationship to the remainder of the Arctic that it does today, then it is hardly conceivable that conditions sufficiently cold for an arctic flora existed anywhere in the Northern Hemisphere (Savile, 1961, 1972; Löve and Löve, 1963).

SURVIVAL IN REFUGIA

To explain survival of the arctic and boreal vegetation through the Pleistocene glaciations it is important to consider the probable existence of unglaciated refugia in which plants could survive the rigors of each glacial stage. Evidence for the existence of such refugia is convincing.

There were areas in the Far North that evidently remained free of both glacial ice and summer-long snow cover, where conditions were not sufficiently severe to make survival of the arctic species impossible. The boreal flora, on the other hand, must have been driven southward before the advancing glaciers. It seems that the arctic plants now found along the cliffs of the north shore of Lake Superior must have migrated into the region along the edge of the last glacier as it withdrew in late-glacial times. Some species, it is true, may have survived the glacial period to the south and east of the ice in the western and eastern United States, but for much of the arctic flora occupying at least the Western Hemisphere survival must have occurred in the Far North. This may not have been the case, however, in Europe and Asia. Available evidence shows that the present-day tundra associations of Eurasia formed relatively recently. Tundra species were formerly interspersed with various types of Eurasian steppe vegetation. Thus, it appears that the arctic flora in North America survived north of the ice, with a few species perhaps waiting out the Pleistocene in small, favorable mountain habitats south of it. In Europe, the arctic species must have been south of the ice in somewhat greater proportions. That the boreal flora survived largely south of the ice on both continents seems a reasonable assumption. Evidence indicates that the boreal flora in North America survived in areas where northern coniferous, mixed, and southern deciduous forests existed relatively close to the ice border in what is now the central and eastern United States. A reserve of northern conifers and deciduous trees existed close to the foot of the Rocky Mountains, and this was at times joined to a refugium in Yukon–Alaska by a corridor through Alberta.

POSTGLACIAL VEGETATION

The close of the Pleistocene, which has been set rather arbitrarily at about 10,500 years ago, was characterized by temperatures that had ameliorated to the point where glacial ice was fast disintegrating, and it can be assumed that summer temperatures were well above what they had been in previous centuries. The behavior of plants in periods of ameliorating climate and glacial retreat must have been characterized principally by a very rapid occupation of areas left open by the disintegrating ice margin. During these periods, migration of species occurred as rapidly as means of dispersal would permit and for as great a distance as physiological tolerance allowed. During at least one postglacial episode, the climatic optimum (hypsithermal, altithermal), temperatures were generally warmer than at present, and we can assume that, in the case of many species, migration into deglaciated areas followed very closely the melting of the ice. With conditions even warmer than today, a major readjustment of relationships among species even in unglaciated areas must have occurred. As the ice retreated, species characteristic of the boreal forest moved northward in its wake.

When the glacial ice disintegrated, species that had retreated south before advancing ice in the central and eastern portions of the North American continent now formed an unusual mosaic of representatives of both the boreal and deciduous forests. Far to the west, species that had found refuge in the mountains also now marched northward. Each species advanced as far in all directions as the limits of its environmental tolerance or competition from better adapted species permitted. Many boreal species attained a northern range as far as 200–400 km beyond their present-day limit because of favorable conditions prevailing during the postglacial climatic optimum. Eventually, after the period of migration and readjustment, communities resembling in essential respects those existing today began to emerge. Subsequent deterioration of the climate following the climatic optimum forced many of these species southward once again, and only pollen evidence in peat bogs and other sediments remain to show that they once occupied regions farther north.

THE PALYNOLOGICAL RECORD

The sequence outlined above has, in general, been derived from studies of fossil pollen in peat deposits and lake sediments, in which the various strata have been dated by carbon-14 methods. Windborne pollen and bits of leaves and stems found in these natural archives of postglacial vegetational history are used to identify plant species occupying the landscape in past millenia. The quantity of pollen from each species is used to obtain some indication of the abundance of each species relative to the others. In the case of insect-pollinated species, it is unlikely that pollen grains in sufficient numbers have been preserved to furnish a basis for inferences as to importance in postglacial plant communities. So, when this is considered, there are difficulties obviously involved in interpreting pollen and macrofossil data in terms of actual composition of plant communities, since entomophilous species are largely absent from the record. In analyzing the data, moreover, the numbers indicating pollen abundance are usually tabulated on charts, along with carbon-14 chronology, and the imposing solidity of the resulting array of numbers tends to conceal the uncertainties involved. Additional difficulties in interpretation arise when two or more species are closely related, especially if one or more of the species is northern and the others southern in affinity, and the difficulties become quite intractable when the pollen of one species is indistinguishable from that of another. Thus, in the case of birch (*Betula*), for example, one species is a tree (*Betula papyrifera*) reaching greatest abundance in the southern boreal forest, another is a northern shrub (*Betula glandulosa*), and the two species overlap in geographical range. Similarities in the pollen of the two species make use of birch as an indicator of either the northern or southern environment unreliable at best (Ives, 1977). Similar situations are encountered with a number of herbaceous genera. It is, likewise, not possible to distinguish between black spruce and white spruce on the basis of pollen, hence the habitat preferences of spruces cannot be made to yield inferences as to the nature of the plant community. This is not the place to discuss in detail the other possibilities for error in interpreting pollen data; they are described in the various references at the end of the chapter. Despite the disadvantages of the method, however, it is nevertheless likely that knowledge obtained of boreal postglacial history will necessarily be derived from pollen and macrofossil analysis, for lack of other sources of information. It can be said, moreover, that knowledge of the limitations of palynological analysis has progressed rapidly, and it is possible to place a degree of confidence in at least the more recent findings and the more cautious conclusions of skilled palynologists. There is now considerable agreement as to the gross structure and floristic composition of many communities existing in past millenia. Thus, in Far Northern regions, peat or sediment samples lacking abundant pollen of spruce and pine, but containing pollen and macrofossils of sedges, grasses, and perhaps arctic herbaceous or shrubby species, are assumed to represent a treeless tundra community of herbs and shrubs. An abundance of spruce, poplar, and pine pollen represents a southern boreal community. Spruce alone or with a scattering of larch and possibly small amounts of pine represents the northern boreal forest. Final attainment of a detailed and completely accurate synthesis of postglacial vegetational history in North America and Eurasia may be, in the final analysis, impossible, but it seems safe to say that at least a reasonably reliable outline of the history of the postglacial vegetational communities in northern regions has been—or can be—achieved. In many areas the chronology of glacial retreat is as yet imperfectly known, and peat deposits or lake sediments are rare or nonexistent. For such areas, it may only be possible to make conjectures as to the nature of the postglacial plant communities on the basis of knowledge obtained from areas where stratigraphy and chronology can be accurately discerned.

POSTGLACIAL HISTORY

In summarizing the evidence available at the time, Ritchie (1976) points out that the earliest late-glacial pollen samples show a high degree of uniformity over a wide area of northern central North America, including Manitoba, Saskatchewan, Minnesota, and Michigan, as well as Keewatin, Mackenzie, the Yukon, and Alaska. Samples obtained at suitable sites throughout the region indicate that the vegetation must have consisted of a variety of plant species that now possess geographical affinities primarily North American, wide-ranging, boreal, temperate and, in a few cases, arctic. Ritchie also states: "It is unlikely that any vegetation exists today on a regional scale which is similar to the late-glacial communities. One suspects that the late-glacial spruce communities were youthful and primeval...."

As the glacial ice disintegrated, there was an initial rapid immigration from the south of a mixed flora that later differentiated into low-arctic, boreal, and temperate groups depending upon the particular reaction of each species to the changed environment. The regional plant communities prior to 6500 B.P. (before the present) were, according to studies by Ritchie (1977), different in composition and structure from any modern examples. Then, slowly, these communities developed into those we recognize today, and they have remained relatively stable throughout the past four or five millenia. The major changes seen in the pollen record, however, as Ritchie adds, are due to relatively few species, and changes in the composition of communities such as the lichen–heath mat are not discernable. In general, however, the pollen spectra throughout North America tend to corroborate the observations of West (1964) that the effect of glaciation (in Europe) was to produce plant communities with "no long history in the Quaternary... merely temporary aggregations under given conditions of climate, other environmental factors, and historical factors." Invasion from adjacent refugia, Ritchie states, was followed by adjustments of the early postglacial communities as a result of competition from other species, developing soils, and stabilizing topographical features.

During the numerous glacial advances of the Pleistocene, ice covered large portions of the North American continent and, as a consequence, when the ice disappeared, vegetation reoccupied almost the entire area we know now as Canada. The ice began to disintegrate along the southern border of the continental ice sheet about 12,000 years ago, releasing northern parts of the United States and southern parts of Canada from its grip, and then, about 7000 years ago, the final vestiges disappeared from an area of northern central Canada known as the Keewatin Ice Divide, inland from the west coast of Hudson Bay. The plant species involved in recolonization had survived the glacial episodes in areas south of the maximum limit of glaciation, as well as in refugia on the continental shelf, the Arctic Islands, mountainous areas along the east and west coasts, and unglaciated parts of the Yukon and Alaska. Regions where many species survived the glaciation are well-known from palynological studies. Both spruce and pine, for example, survived in areas of the United States. But a satisfactory explanation for the distribution during glaciation, as well as patterns of postglacial migration, of many plants, especially certain arctic species, is as yet not available. Most striking is the distribution of the arctic species now found along the coast of James Bay and on rocky cliffs along the northeastern shores of Lake Superior. These species, as noted by Soper and Maycock (1963) include *Woodsia alpina, Trisetum spicatum* var. *pilosiglume, Polygonum viviparum, Sagina nodosa,* and *Saxifraga aizoön* var. *neogaea.* The explanation put forth is that as the ice retreated the arctic species invaded from refugia in the northwest, following lakeshore and riverbank habitats. Since there is no evidence of a long-lasting tundra community anywhere in the area, it is assumed that boreal forest species soon crowded arctic invaders out. Colonizing boreal vegetation moved northward rapidly from the midwestern plains regions, south of the maximum position of the ice sheet, where grassland occupied a much less extensive area during glaciation than is now the case. The oldest pollen assemblages deposited in the Great Lakes region resemble those deposited today in tundra regions near the northern tree limit, and so it is inferred that, although short-lived, tundra occupied the land for a brief period immediately following disintegration of the glacial ice.

In northern Minnesota, spruce pollen dominates the next lowest sediments. Other tree genera found include tamarack, ash, and poplar. The same assemblage is found in southern Saskatchewan as well as in Wisconsin and Michigan. In northern Minnesota there were two major periods of dominance by the spruce–fir–birch assemblage; the first occurred as the initial postglacial invasion by vegetation, and the second continues to the present day. They are separated, however, by a period in which pine attained dominance. Most of the pollen diagrams can be fitted to a scheme of climatic change that postulates a hypsithermal interval followed by a decrease in temperature, an increase in precipitation, or both, which continues to the present day (Cushing, 1965; Amundson and Wright, 1979). In some areas, the lowest pollen zone is characterized by a high frequency of the pollen of the sedges, *Dryas integrifolia*, *Salix herbacea*, *Vaccinium uliginosum*, *Juncus balticus*, and *Betula*, the latter probably *B. glandulosa* since it occurs conspicuously at the transition from tundra to the boreal forest pollen assemblage, the community in which it is most abundant today. Spruce is the most abundant tree pollen type. This zone also contains unidentified fragments of mosses and fungi. The next zone above it has spruce as the most abundant pollen type, and *Larix*, *Juniperus/Thuja*, *Populus balsamifera* type/ *Fraxinus nigra* type, sedges, *Artemisia*, *Ambrosia*, and *Urtica* type; all are important.

In southern Minnesota, the boreal forest was present by 13,000 years B.P., and before roughly 10,000 years ago it was replaced by a forest in which species most heavily represented by pollen include *Betula*, *Abies* (probably *A. balsamea*), and alder (probably *Alnus rugosa*). In northeastern Minnesota, the beginning of the boreal forest pollen assemblage dates to about 10,500 B.P., indicating that the boreal forest was present in southern Minnesota as long as 3000 years before it reached the northeast (Watts, 1967). In the northeast, tundra vegetation was present until about 10,500 B.P., when it was invaded by the boreal forest, at the margins of which *Betula glandulosa* was a pioneer species. The boreal forest then lasted for a few centuries before it was displaced by fir–birch–alder–pine, from which the present-day Great Lakes forest evolved. In eastern North America, late-glacial pollen assemblages resemble those being deposited today in regions of tundra and boreal forest, the oldest (late Wisconsin) resembling those deposited today in tundra areas far from any forest. In southern New England, the tundra was replaced by plant communities resembling present-day taiga or forest–tundra border vegetation with a lichen–heath ground cover and widely spaced spruce trees (Terasmae, 1967, 1973; Davis, 1967; Cushing, 1967; Watts, 1967). In the western and central interior of Canada, the earliest pollen spectra suggest that forest followed closely behind the retreating ice, since evidence of tundra in many areas is nonexistent (Ritchie and Yarranton, 1978a,b). As the boreal communities expanded northward, they were replaced along their southern margin by grassland. Spruce appeared in the period 14,000–10,000 B.P. on southern sites and 8000–6500 B.P. on northern sites. Pine and birch arrived thereafter (Ritchie, 1976). It is interesting that, according to studies on central Alberta (Lichti-Federovich, 1970), poplar forest preceded the spruce forest. According to Ritchie (1976) it is conceivable, however, that a late-glacial tundra may have existed in these areas, briefly as was the case in the Lake Superior region, and that evidence for its existence may eventually be obtained. The possibility that a tundra existed in a narrow band between the receding ice front and the encroaching forest in western areas has long been the subject of speculation among palynologists, and final resolution of the question has not been achieved. In some areas, notably the Far Northwest (northern Mackenzie), the vegetation was clearly tundra but, according to Ritchie, the pollen and macrofossil evidence so far does not permit reconstruction of tundra community composition in the west central region.

THE SOUTHERN BOREAL BORDER

The initial interest in studies of the history of development of the postglacial boreal plant communities in North America derives to a large extent from pollen analyses of samples obtained in central and more southern areas of what is now the boreal region—northern Minnesota, southern Manitoba, Saskatchewan—where grassland, open parkland, and the southern boreal transition zone moved north or south at intervals during postglacial time in evident response to shifting climatic regimes. The zonation of these three major vegetational formations today is roughly as follows: across the northern portion, the boreal forest; in the south, grassland; and between them a transitional region occupied by aspen parkland and what is known as the transitional boreal forest (Zoltai, 1975; Ritchie, 1976). The aspen parkland consists of patches of poplar grove interspersed with low shrub and herbaceous communities. There is, as Ritchie (1976) points out, some possibility that aspen parkland is a convenient category for a number of poorly described forest communities dominated by poplar, oak, birch, elm, and ash. Much of the area, as he states, was repeatedly burned both before and after the colonial period. The transitional boreal forest is distinguished from aspen parkland by the presence of coniferous tree species. It is distinguished from the boreal forest proper by the absence of continuous stands of spruce–birch–pine–aspen forests characteristic of the southern edge of the boreal biome. In some representative areas of the transitional forest there is often found an open community of white spruce, with a ground cover of *Juniperus* and *Arctostaphylos uva-ursi*, with many shrubs typical of grasslands. The southern boreal forest adjoining the transition zone consists of forest communities in which the major arboreal species are *Picea glauca, P. mariana, Larix laricina, Abies balsamea, Pinus banksiana, Betula papyrifera, Populus tremuloides,* and *P. balsamifera,* and with *Quercus macrocarpa, Ulmus americana, Fraxinus pennsylvanica,* and *Acer negundo* forming local stands in areas with poorly developed soils (Rowe, 1972; Ritchie, 1976). The northern boreal forest is distinguished by the great preponderance of conifers and diminished importance of deciduous angiosperm tree species.

Following in the wake of the disintegrating ice, these vegetational belts became established as soon as the migrational capacities of the species permitted them to invade. In early postglacial time, the transitional forest and the grassland reached a point farther north than they occupy today. There is, however, no pollen-stratigraphic evidence of changes in the position of the southern boundary of the boreal forest in central Canada during the past 2000 years (Zoltai, 1975; Ritchie and Yarranton, 1978a,b). Ritchie and Hadden (1975) have shown that an episode of spruce–grassland or –parkland at Grand Rapids, Manitoba, from 7300 to 6200 B.P., was contemporary with the culmination of the prairie period in the entire area of southwestern Manitoba, Saskatchewan, and the adjacent United States west of the Red and Missouri rivers. Grasslands and parklands probably did not extend further northeast in Manitoba than Grand Rapids, while in Saskatchewan the northern limit of aspen parklands was in the area near Prince Albert. To the north, east, and southeast of Grand Rapids, the land areas exposed before 7200 B.P. by the diminishing ice sheet were probably occupied by a spruce-dominated boreal forest (Nichols, 1969; Wright, 1970; Mott, 1973; Ritchie, 1976).

It is difficult to make accurate inferences concerning the chronology of climatic shifts in the southern regions, because some species representative of each of the major zones—grassland, aspen parkland, transitional forest, and southern boreal forest—migrate more slowly than others into areas that have become climatically suitable. A lag between the time when an area becomes available and the time it is occupied often results in assignment of much later dates to a shift in climate than was actually the case; the influx of more slowly migrating species will corroborate the climatic shift, but the inference should, of course, not be made that the shift occurred synchronously with the time of influx. Since knowledge concerning migration rates of most species is fragmentary, there are obvious difficulties in making accurate climatic interpretations from pollen abundance in peat and sediment samples. It is now being recognized that puzzling discontinuities among areas in chronology of vegetational and, by inference, climatic change can often be accounted for by different migration rates. Such problems are encountered in greatest intensity in southern areas, where a greater number of species, as well as greater variety of vegetational zones, exist. In northern portions of the boreal biome, particularly in the Far North where forest merges with tundra, the picture is more clearly outlined, even though the amount of research conducted in the area has been relatively much less, because of difficulties of fieldwork and transport in the region as well as the limited time when fieldwork is possible.

THE NORTHERN BORDER

In northern Canada, the transition zone between forest and tundra, called variously the continental arctic tree line, the northern forest border, or the northern forest–tundra ecotone, lies roughly along a WNW–ESE axis extending from the Mackenzie Delta in the northwest to the western shore of Hudson Bay at the mouth of the Churchill River. Along this entire length, the forest border swings northward at some points and southward at others, forming interdigitations of forest into tundra along river valleys and tundra into forest along expanses of uplands, and between these extremes there are the ubiquitous lichen woodlands—forests of widely spaced spruces with a ground carpet of lichens, scattered northern shrubs, and herbaceous species.

This striking geographical zonation in the vegetation is coincident with the pathway of cyclonic storms along the average summer position of the arctic front, a climatic zone separating arctic air masses prevailing over the tundra of northern arctic Canada from the air masses of southern origin prevailing over the subarctic vegetation of spruces and treed muskeg (see Chapter 3). The coincidence of these two transition zones, one climatic and the other vegetational, is striking indeed, and not only does the northern edge of continuous forest coincide with this climatic frontal zone, but there are also marked changes across this zone in the frequency counts of species making up the ground or understory vegetation (Larsen, 1971, 1973, 1974). Climatic variation across the boundary between the two regions is sufficient to account, at least in large measure, for the equally marked vegetational differences, since geological and topographical characteristics are generally uniform throughout (Lee, 1959). An interesting point to be explored is that there exists a zone of relatively depauperate vegetational communities just north of the forest border in central and southwestern Keewatin and that these communities apparently identify a climatic zone possessing rather definite and identifiable characteristics.

Studies of the postglacial history of the northern forest border in Canada corroborate studies in the northern central United States and southern central Canada and demonstrate also that there is a good correlation between events in North America and Europe. The latest major climatic deterioration recorded in northern central Canada occurred simultaneously with worsening climatic conditions in northern Europe, and the limited evidence available tends to confirm the same general scheme of postglacial events in northern Russia.

It appears that in both North America and Eurasia today the northern limit of tree growth—the northern forest border—coincides roughly with the climatic frontal zone existing between the arctic air masses and the southern air masses in summer (see Chapter 3). Since this zone would move north or south in response to major climatic changes, it is to be expected that the forest border would follow. After a major climatic deterioration, forests overtaken by colder and drier regimes characteristic of arctic air masses would be subject to fires that would open the landscape for occupancy by tundra vegetation. Evidence derived from plant macrofossils, as well as studies on soil stratigraphy and palynology, at a few widely spaced locations have established the extent of northward movement of the Canadian boreal forest–tundra ecotone during postglacial time. In the area of Ennadai and Yathkyed lakes (lat. 61°–63°N, long. 98°–101°W) in Keewatin, Northwest Territories, for example, the northernmost limit of tree species (not trees, in an accurate sense, since the individuals are very dwarfed or decumbent) is represented by clumps of *Picea glauca* or *Picea mariana* in sheltered river valleys and along lakeshores (Larsen, 1965, 1973, 1974; Savile, 1972; Nichols, 1967a,b, 1970, 1974, 1975, 1976a,b; Elliott, 1978; 1979a,b). Spruce reproduction is primarily vegetative, by layering (Larsen, 1965, 1973, 1974; Elliott, 1978) and sexual reproduction has been shown by Elliott (1978, 1979a,b) to occur rarely if at all in spruce at the northern limit of spruce range in Keewatin. This lends support to the concept that the outlying spruce clumps in tundra in this region are relicts. The same evidently is not the case, however, in Labrador-Ungava, for here the spruce forest appears capable of northward migration if the topography were favorable for further range expansion; migration is blocked, however, by high elevations that evidently have so far prevented spruce from becoming established north of a limit that exists because of topographic rather than climatic conditions (Elliott and Short, 1979). The northernmost spruce clumps in Keewatin can be regarded as relicts of consistently warmer summers in years past (Larsen, 1965; Savile, 1972; Nichols, 1976a,b). The coincidence of the mean summer location of the arctic front with the forest–tundra boundary suggests that the cold, dry arctic air prevents northward penetration of forest into tundra. This, in turn, implies that the summer position of the arctic front must have previously been located sufficiently far north to permit reproduction of the spruce by seeds, at least to some extent, rather than by layering (Larsen, 1965; Nichols, 1976a,b).

From studies of fossil pollen and palaeosols in the areas where relict spruce is found, it is apparent that no northward forest displacement extending more than 100 mi (160 km) has occurred within the past 2000–3000 years, according to evidence accumulated by Nichols (1967a, 1972, 1975, 1976a,b) and Sorenson (1977; Sorenson and Knox, 1973). A general period of warmer climatic conditions about 1200–1000 B.P. permitted forest to grow for a brief period about 70 mi (110 km) north of the present-day limit in southern Keewatin (Bryson et. al., 1965). The most northerly spruce clones, according to the most generally accepted interpretation, could be relicts that have survived since 3500 B.P. or even from periods of maximum postglacial warmth recorded in the central Canadian sub-arctic about 5000 years ago and again about 4000 years ago.

Numerous past fluctuations in the position of the forest–tundra boundary are, as Larsen (1965) and Nichols (1976a,b) point out, the probable explanation for the existence of open tundra patches left isolated in the northern woodlands, as well as of the relict spruce clumps left in tundra. This, in addition to the naturally diffuse nature of an ecosystem boundary, accounts for the latitudinal extent of the present day forest–tundra communities mapped by Rowe (1972). Moreover, the compression of this zone in northwestern Canada is the result of more relatively restricted ecotonal movement (Larsen, 1967, 1971a,b, 1973; Ritchie and Hare, 1971). The effect of the Arctic Ocean in the northwest is evidently influential in stabilizing the climate, and this, coupled with the stabilizing effect of the northern Rocky Mountains prevented synoptic latitudinal displacement of the atmospheric circulation, as appears to be the case today.

This general scheme of postglacial vegetational events in northern central Canada is corroborated by pollen diagrams—spanning 1400 km from the west shore of Hudson Bay to Great Bear Lake and beyond—prepared by Nichols (1975) and demonstrating a striking synchrony in and movement of the forest–tundra ecotonal boundary in response to climatic changes over at least the past 6000 years. There were numerous fluctuations in the forest position up until about 3500 B.P., as an apparent consequence of climatic warming and cooling trends; the forest was able to recover from cold climatic episodes until about 3500 B.P., when widespread and broadly synchronous fires swept the ecotone from one end to the other over a period of about 200 years. By 3000 B.P., the tundra had again expanded southward during a prolonged episode of cold, dry summers, and there was another marked retreat of the forest in about 2500 B.P., again in 1400 B.P., and again about 800–900 years ago, each episode alternating with warming periods. It is noteworthy that the fires ranging across the ecotone in the period 3500–3300 B.P. correspond chronologically to evidence obtained in western Greenland that Canadian windblown charcoal settled there from 3510 to 3360 B.P. This event, according to Nichols, may prove to be synchronous throughout the entire Canadian forest–tundra ecotone, and possibly across Eurasia as well.

FOREST–TUNDRA COMMUNITIES

North of the forest–tundra ecotonal area is the arctic tundra, a region with a largely unknown vegetational history and with many questions awaiting the exploration of both ecologists and palynologists. Vegetational studies conducted in the forest–tundra ecotone in the regions around Ennadai and Artillery lakes, Keewatin and Mackenzie, Northwest Territories, central Canada, and in northern Keewatin (Larsen, 1965, 1971, 1973, 1974) reveal that the communities at the extreme southern edge of the tundra, just north of the forest border, are floristically depauperate, characterized by fewer plant species than communities either north or south of this transition zone. Moreover, species making up the transition zone communities are geographically wide-ranging, many occupying habitats throughout both the northern edge of the forest and the low-arctic tundra. While these species, in aggregation, may be said to characterize the tundra communities of the regions, many are so ubiquitous throughout both forest and tundra that their ecological relationships, particularly those involving climatic conditions, pose some interesting and perplexing problems.

There are some rather definite relationships between forest border and regional climatic characteristics. The zone of floristically depauperate tundra communities extends for some distance north of the forest border, and it appears that this distinct vegetational zone coincides with a correspondingly distinct climatic zone. Frontal conditions obviously prevail with greatest frequency along the mean frontal position, and they occur with decreasing frequency north and south of this position. Plant species adapted to survival in the region dominated by arctic masses would be expected to occur with increasing presence, frequency, or both, in vegetational communities along a traverse from forest into the region occupied nearly continuously by arctic air. Apparently it can be inferred from the remarkable width of the transition zone characterized by depauperate vegetational communities that the climatic frontal conditions in Keewatin occupy a broader north–south band than they do farther to the west.

One area in which the vegetational communities have been studied by Ritchie (1960) is located about 140 km northwest of Churchill, within the extreme northern edge of the forest–tundra transition zone. Because of its proximity to the coastal tundra along the west shore of Hudson Bay, where there are communities rich in arctic species, the area possesses arctic species as members of the plant communities as well as the ubiquitous species associated with the depauperate communities found to the west at, for example, Ennadai Lake. This area, according to Ritchie, also is occupied in the greatest area of upland sites by a sparse, treeless vegetation that might be called tundra, but many stumps and wind-felled trunks of spruce indicate the presence of forest in the past. At present, a depauperate heath community grows on the upland sites, made up of the following species: *Loiseleuria procumbens, Empetrum hermaphroditum (E. nigrum), Ledum decumbens, Vaccinium vitis-idaea* ssp. *minus, V. uliginosum, Cetraria nivalis, Cladonia rangiferina, C. mitis, Alectoria altaica, Carex glacialis, C. bigelowii,* and *Luzula confusa*.

Certain ridges are considered by Ritchie to be definitely potential bearers of *Picea mariana* forest, and a regeneration forest community dominated by trees of *Picea mariana* with an average height of 7 m, a diameter at breast height of 6.5 cm, and 50 rings in the wood at breast height was observed in which the chief associated species recorded were *Betula papyrifera* var. *neoalaskana, B. glandulosa, Ledum groenlandicum, Alnus crispa, Larix laricina, Vaccinium vitis-idaea* ssp. *minus, V. uliginosum,* and *Empetrum hermaphroditum,* all typically members of spruce forest in the forest–tundra transition zone. A minor variant of the heath–tundra type was distinguished by the presence of cushions of the arctic species *Diapensia lapponica,* growing in open situations where instability of the soil surface was apparent.

The flora of the area is generally impoverished and monotonous, but some of the species present are exclusively arctic (often low-arctic) species: *Colpodium fulva, Hierochloe alpina, Carex stans, C. bigelowii, C. glacialis, C. capillaris* ssp. *arctogena, Juncus castaneus, J. albescens, Luzula confusa, Stellaria ciliatosepala, Sagina caespitosa, Arabis arenicola, Potentilla nivea, Ledum decumbens, Arctostaphylos alpina, Phyllodoce caerulea,* and *Diapensia lapponica*. The other vascular plant species, making up about 85% of the flora, are wide-ranging boreal or boreal-arctic species. The flora corroborates the view that there is a striking floristic association of arctic species in the Hudson Bay lowlands. Of interest is the fact that *Phyllodoce caerulea,* often described as a rare species, was recorded in local abundance in several localities.

The physiological responses of plant species to the environment of this zone in which arctic species are found inland some distance from Hudson Bay, and the climatic characteristics of the forest–tundra ecotonal region farther inland, are discussed in more detail in Chapter 3, but the historical aspects of the vegetation of the Hudson Bay lowlands and the forest–tundra ecotone also have a bearing upon the nature and extent of these interesting vegetational zonations. Ritchie (1976) asks: "What was the origin of the species-rich tundras and forest-tundras of the west shores of Hudson Bay?" He adds further that the many low-arctic and subarctic tundra species adapted to boreal environments led Gorodkov (1952) to propose that low-latitude tundra in Russia resulted from elimination of trees from taiga as a consequence of climatic change, bringing about the spread and increasing abundance of such species as *Betula glandulosa, Vaccinium uliginosum, V. vitis-idaea, Ledum decumbens, Rhododendron lapponicum, Loiseleuria procumbens,* and *Empetrum nigrum* var. *hermaphroditum.* Fire associated with a shift to a colder, drier climate has, at the northern limit of the forest in central Canada, had such an effect as described in the previous pages, but this does not explain the origin of many exclusively arctic species found in certain areas nor does it offer an explanation of the obvious fact that the arctic species now found in adjacent areas to the north—perhaps only 150–200 km distant—have not moved into the southern fringe of depauperate vegetation even though ample time evidently was available for migration to have occurred over such a relatively short distance. This explanation for the origin of the fringe of depauperate plant communities along the southern border of the tundra is all the more inadequate when it is noted that, in the Hudson Bay coastal region around Churchill, Manitoba, many exclusively arctic species seem well-adapted and are abundant in plant communities (Ritchie, 1956), and evidence indicates that at least 7000 years has been available for occupancy of the land since emergence from inundation after the retreat of glacial ice from the region (Mills and Velduis, 1978). There is forest inland and near the coast at Churchill and here the arctic species cohabit with forest species; to the northwest 300–400 km in the continental interior no arctic species are found in forest communities, and only the depauperate northern boreal forest flora is found for some distance northward from the edge of the forest–tundra ecotone. The existence of arctic species at Churchill may be due to migrational avenues that exist along the coast, in addition to cold climatic conditions that result from presence of Hudson Bay ice cover long into the summer, but the reasons are not clearly evident, in any case, and pose interesting ecological and historical questions. To answer the questions that arise when these puzzling aspects of postglacial history are considered, it would be necessary to recover the earliest evidence of late-Pleistocene vegetation, and it is to be hoped that sites where this record of vegetational history has been preserved will one day be discovered. As Noble (1974) and Short (1978) point out, all such studies also are of interest to archeologists, since they provide information on environmental changes, particularly climatic changes, affecting human occupancy of these regions. Prehistoric occupation by humans of the tundra–taiga of northern central Canada probably commenced at least as early as 6000–7000 B.P., and evidence suggests that spruce forest was already established in the area southeast of Great Bear Lake, for example, by 7000 B.P. Faunal remains from a roasting hearth at a campsite in this area include caribou, black bear, beaver, hare, eagle, and fish. A much longer history of human occupancy is recorded from boreal regions of northwestern Canada, and Irving and Harington (1973) have obtained evidence indicating that prehistoric campsites in the Yukon date to a time thousands of years before this.

Excavations and radiocarbon estimates from an area in central Canada near Dubawnt Lake in the northern barrens indicate that full deglaciation of the Keewatin Ice Divide occurred about 7000 years ago (Gordon, 1975, 1976). The Cree Lake area in central Saskatchewan was invaded by vegetation by 8000–7000 B.C., and the caribou population was likely migrating between Cree Lake and Dubawnt Lake by 7500–6500 B.C., followed closely by Indian caribou hunters who occupied the barrens prior to 6000 B.C.

A prolonged cold period following 1500 B.C. brought about a cultural change, with pre-Dorset people moving into the area in the wake of the southward retreating Indians. Cold persisted until about 700 B.C., when conditions improved, and the pre-Dorset peoples presumably returned northward, to be replaced by Indians of the Taltheilei tradition, who continued to inhabit the area until a period of colder weather beginning in the early nineteenth century, during which there was an influx of caribou Eskimos from Hudson Bay. Their occupation of the interior barrens was short-lived, however; they were relocated to the Hudson Bay coast after a period of starvation in the 1950s.

In his studies of the close association between human tribes and caribou populations in the barrens and the forest–tundra transition zone, Gordon (1979) demonstrates an ecological relationship between man and caribou, with prehistoric Indian and inland pre-Dorset peoples aligning with discrete caribou populations or herds and developing distinctive regional cultural patterns. There is, thus, a long tradition of ecological dependence by man upon caribou on the northern edge of the boreal forest, hence between man and vegetation, since at least to some degree, the caribou undoubtedly influence plant life in turn.

ECOLOGICAL SIGNIFICANCE OF HISTORICAL EVENTS

In looking back over this long history of the boreal vegetation, from Cretaceous times to the present day, one conclusion stands out as being perhaps of more general biological significance than the individual climatic and vegetational events we have been discussing. This conclusion is that the plant species comprising the boreal flora pursued, from at least early Cretaceous to late Tertiary times, an evolutionary course unmodified and unaffected by rapid or major environmental changes. Climatic change during this time seems not to have been of the overwhelming magnitude characteristic of the change that began in the Pliocene, culminating in the major continental glaciations of the Pleistocene.

The fossil record of the Tertiary flora suggests that for a very long time there had been a remarkably uniform vegetation extending throughout the northern portions of the Northern Hemisphere, reaching even into high latitudes. And then, quite rapidly on the geological time scale, this flora was disrupted. The species of which it was composed were forced to withstand extreme conditions to survive. Arctic species were driven northward into several northern ice-free refugia or southward along the coasts and mountain ranges. Boreal species moved southward before the advancing ice, following paths determined by favorable conditions of temperature and rainfall but always into regions where seasonal patterns of day length and conditions relating to winter dormancy were much different than those to which evolution during previous millenia had fitted them.

The Pleistocene must have been a time of great change in the genetic structure of boreal species populations. Then, with the disappearance of the glaciers, there followed a relatively brief time during which conditions were even warmer than they are now, and these same populations were forced to migrate once again, this time northward into regions they had once occupied and where they again encountered conditions of day length and dormancy to which they had once been, but no longer were, adjusted. These events must have brought about changes in adaptive responses of boreal species which, lacking preglacial plants for comparison, we will never be able to trace. But perhaps in retrospect we can use these events to furnish an explanation, however vague and unsatisfactory, for certain puzzling and unaccountable characteristics of boreal species. Perhaps these events are responsible for the relative simplicity of the northern flora. In tropical forests, for example, there may be a hundred species of trees in a square mile; in the boreal forest there are only one or two or, at most, a half-dozen. In the tropics, species appear to have acquired, through evolutionary time, a specific set of adaptations that fit them closely into the environmental niche they now occupy; in boreal regions there often appears to be an absence of such tightly knit relationships among species in the communities, and, in fact, a given species will often occupy a bewildering array of habitats and otherwise behave in an unaccountable manner.

These puzzling aspects of northern history enhance the fascination the boreal forest holds for ecologists, however, and in Chapter 6 we will discuss some individual characteristics of the species that make up communities found in boreal regions. First, however, we will have a look at the environment—climate and soil—in the regions where boreal plant and animal species are found at the present time.

REFERENCES

Amundson, D. C., and Wright, H. E., Jr. (1979). Forest changes in Minnesota at the end of the pleistocene. *Ecol. Monogr.* **49,** 1–16.
Andrews, H. N. (1963). Early seed plants. *Science* **142,** 925–931.
Barry, R. G., Arundale, W. H., Andrews, J. T., Bradley, R. S., and Nichols, H. (1977). Environmental change and cultural change in the eastern Canadian Arctic during the last 5000 years. *Arct. Alp. Res.* **9,** 193–210.
Beaty, C. B. (1976). The causes of glaciation. *Am. Sci.* **66,** 452–459.
Birks, H. J. B. (1973). Modern pollen rain studies in some arctic and alpine environments. *In* "Quaternary Plant Ecology" (H. J. B. Birks and R. G. West, eds.), pp. 143–168 Cambridge Univ. Press, London and New York.
Black, R. A. (1977). Reproductive biology of *Picea mariana* (Mill.) B.S.P. at treeline. Doctoral thesis, University of Alberta, Edmonton, Alberta, Canada.
Bryson, R. A., Irving, W. M., and Larsen, J. A. (1965). Radiocarbon and soil evidence of former forest in the southern Canadian tundra. *Science* **147,** 46–48.
Colinvaux, P. A. (1967). Quaternary vegetational history of arctic Alaska. *In* "The Bering Land Bridge" (D. M. Hopkins, ed.), pp. 207–231. Stanford Univ. Press, Stanford, California.
Curtis, J. T. (1959). "The Vegetation of Wisconsin." Univ. of Wisconsin Press, Madison.
Cushing, E. J. (1965). Problems in the phytogeography of the Great Lakes region. *In* "The Quaternary of the United States" (H. E. Wright, Jr., and D. G. Frey, eds.), pp. 403–416. Princeton Univ. Press, Princeton, New Jersey.
Cushing, E. J. (1967). Late-Wisconsin pollen stratigraphy and the glacial sequence in Minnesota. *In* "Quaternary Paleoecology" (E. J. Cushing and H. E. Wright, Jr., eds.), pp. 59–88. Yale Univ. Press, New Haven, Connecticut.

Davis, M. B. (1967). Late-glacial climate in northern United States: A comparison of New England and the Great Lakes region. *In* "Quaternary Paleoecology" (E. J. Cushing and H. E. Wright, Jr., eds.), pp. 11–44. Yale Univ. Press, New Haven, Connecticut.

Davis, M. B. (1976). Pleistocene biogeography of temperate deciduous forests. *In* "Geoscience and Man, Vol. XIII, Ecology of the Pleistocene," (R. C. West and W. A. Haag, eds.), pp. 13–26. Louisiana State University, Baton Rouge.

Elliott, D. L. (1979a). The stability of the northern Canadian tree limit: Current regenerative capacity. Doctoral thesis, University of Colorado, Boulder, Colorado.

Elliott, D. L. (1979b). The current regenerative capacity of the northern Canadian trees, Keewatin, N.W.T., Canada. *Arct. Alp. Res.* **11,** 243–251.

Elliott, D. L., and Short, S. K. (1979). The northern limit of trees in Labrador. *Arctic* **32,** 201–206.

Gordon, B. C. (1975). "Of Men and Herds in Barrenland Prehistory." Archeological Survey of Canada, National Museum of Man, Mercury Ser. No. 28, Ottawa.

Gordon, B. A. C. (1976). "Migod—8000 Years of Barrenland Prehistory." Archeological Survey of Canada, National Museum of Man, Mercury Ser. No. 56, Ottawa.

Gordon, B. C. (1977). Prehistoric Chipewyan harvesting at a barrenland caribou water crossing. *West. Can. J. Anthropol.* **7,** 69–83.

Gordon, B. C. (1979). Man–environment relationships in barrenland history. *Am. Antiq.* (in press).

Gorodkov, B. N. (1952). On the origin of arctic deserts and tundras. *Tr. Bot. Inst. Nauk Acad. SSSR Ser.* **3.**

Greig-Smith, P. (1964). "Quantitative Plant Ecology." 2nd ed. Butterworth, London.

Hare, F. K. (1950). Climate and zonal divisions of the boreal forest formation in eastern Canada. *Geogr. Rev.* **40,** 615–635.

Hare, F. K. (1959). A Photo-reconnaissance survey of Labrador–Ungava. *Geogr. Surv. Can. Mem.* **6,** 1–83.

Hare, F. K. (1968). The Arctic. *Q. J. R. Meteorol. Soc.* **94,** 439–459.

Hare, F. K., and Hay, J. E. (1971). Climate of Canada and Alaska. *In* "World Survey of Climatology, Vol. 11, The Climates of North America" (R. A. Bryson and F. K. Hare, eds.), pp. 49–192. Elsevier, Amsterdam.

Harp, E. Jr. (1958). Prehistory in the Dismal Lake area, Northwest Territories, Canada. *Arctic* **11,** 219–249.

Harp, E. Jr. (1961). The archeology of the lower and middle Thelon, Northwest Territories. Arct. Inst. North Am. Tech. Pap. 8.

Hopkins, D. M. (1967). "The Bering Land Bridge." Stanford Univ. Press, Stanford, California.

Hultén, E. (1968). "Flora of Alaska and Neighboring Territories: A Manual of the Vascular Plants." Stanford Univ. Press, Stanford, California.

Hustich, I. (1966). On the forest–tundra and the northern tree-lines. *Ann. Univ. Turku.* **3,** 7–47.

Irving, W. N. (1968). The barren grounds. *In* "Science, History, and Hudson Bay" (C. S. Beals and A. Shenstone (eds.), pp. 26–54. Dept. Energy, Mines, and Resources, Ottawa, Canada.

Irving, W. N., and Harington, C. R. (1973). Upper Pleistocene radiocarbon-dated artefacts from the northern Yukon. *Science* **179,** 335–340.

Ives, J. D. (1974). Biological refugia and the Nunatak hypothesis. *In* "Arctic and Alpine Environments" (J. D. Ives and R. G. Barry, eds.), pp. 605–636. Methuen, London.

Ives, J. D. (1978). The maximum extent of the Laurentide ice sheet along the East Coast of North America during the last glaciation. *Arctic* **31,** 24–53.

Ives, J. W. (1977). Pollen separation of three North American Birches. *Arct. Alp. Res.* **9**, 73–80.

Kulp, J. L. (1961). Geological time scale. *Science* **133**, 1105–1114.

Larsen, J. A. (1965). The vegetation of the Ennadai Lake area, N.W.T.: Studies in subarctic and arctic bioclimatology. *Ecol. Monogr.* **35**, 37–59.

Larsen, J. A. (1967). Ecotonal plant communities north of the forest border, Keewatin, N.W.T., central Canada. Univ. Wisconsin Dep. Meteor. Tech. Rep. 32, Madison, Wisconsin.

Larsen, J. A. (1971a). Vegetational relationships with air mass frequencies: Boreal forest and tundra. *Arctic* **24**, 177–194.

Larsen, J. A. (1971b). The vegetation of Fort Reliance, Northwest Territories. *Can. Field. Nat.* **85**, 147–178.

Larsen, J. A. (1973). Plant communities north of the forest border, Keewatin, Northwest Territories. *Can. Field Nat.* **87**, 241–248.

Larsen, J. A. (1974). Ecology of the northern continental forest border. *In* "Arctic and Alpine Environments" (J. D. Ives and R. G. Barry, eds.), pp. 341–369. Methuen, London.

Larsen, J. A., and Barry, R. G. (1974). Palaeoclimatology. *In* "Arctic and Alpine Environments" (J. D. Ives and R. G. Barry, eds.), pp. 253–276. Methuen, London.

Lee, H. A. (1959). Surficial geology of southern district of Keewatin and the Keewatin Ice Divide. *Geol. Surv. Can. Bull.* **51**.

Levins, R. (1968). "Evolution in Changing Environments." Princeton Univ. Press, Princeton, New Jersey.

Lichti-Federovich, S. (1974). Palynology of two sections of Late-Quaternary sediments from the Porcupine River, Yukon Territory. *Geol. Surv. Can. Pap.* **74–23**.

Lichti-Federovich, S. (1970). The pollen stratigraphy of a dated section of Late-Pleistocene lake sediment from central alberta. *Can. J. Earth Sci.* **7**, 938–945.

Löve, A. and Löve, D., eds. (1963). "North Atlantic Biota and Their History." Pergamon, Oxford.

Mills, G. F., and Veldius, H. (1978). A buried paleosol in the Hudson Bay lowland, Manitoba: Age and characteristics. *Can. J. Soil Sci.* **58**, 259–269.

Mott, R. J. (1973). Palynological studies in central Saskatchewan. *Geol. Surv. Can. Pap.* **72–49**.

Nash, R. J. (1975). "Archeological investigations in the Transitional Forest Zone: Northern Manitoba, Southern Keewatin, N.W.T." Manitoba Museum of Man and Nature, Winnipeg, Canada.

Nichols, H. (1967a). Central Canadian palynology and its relevance to northwestern Europe in the Late Quaternary period. *Rev. Palaeobot. Palynol.* **2**, 231–243.

Nichols, H. (1967b). The postglacial history of vegetation and climate at Ennadai Lake, Keewatin and Lynn Lake, Manitoba. *Eiszeitalter G. W.* **18**, 176–197.

Nichols, H. (1969). Chronology of peat growth in Canada. *Palaeogeogr. Palaeoclimato. Palaeoecol.* **6**, 61–65.

Nichols, H. (1970). Late Quaternary pollen diagrams from the Canadian Arctic Barren Grounds at Pelly Lake, northern Keewatin, N.W.T. *Arct. Alp. Res.* **2**, 43–61.

Nichols, H. (1972). Summary of the palynological evidence for Late Quaternary vegetational and climatic change in the Central and Eastern Canadian Arctic. *In* "Climatic changes in arctic areas during the last ten thousand years." *Acta Univ. Oulu, Ser. A3, Geol.* **1**, 309–339.

Nichols, H. (1974). Arctic North American paleoecology: The recent history of vegetation and climate deduced from pollen analysis. *In* "Arctic and Alpine Environments" (J. D. Ives and R. G. Barry, eds.), pp. 637–667. Methuen, London.

Nichols, H. (1975). Palynological and paleoclimatic study of the Late Quaternary displacement of the boreal forest–tundra ecotone in Keewatin and Mackenzie, N.W.T., Canada. Occas. Pap. No. 15, Inst. Arct. Alp. Res. Univ. of Colorado, Boulder, Colorado.
Nichols, H. (1976a). Pollen diagrams from sub-Arctic central Canada. *Science* **155**, 1665–1668.
Nichols, H. (1976b). Historical aspects of the Canadian forest-tundra ecotone. *Arctic* **29**, 38–47.
Nichols, H., Kelly, P. M., and Andrews, J. T. (1978). Holocene palaeo-wind evidence from palynology in Baffin Island, *Nature (London)* **273**, 140–142.
Noble, W. C. (1971). Archeological surveys and sequences in the central district of Mackenzie, N.W.T. *Arct. Anthropol.* **3**, 102–135.
Noble, W. C. (1974). The tundra–taiga ecotone: Contributions from the Great Slave–Great Bear Lake region. *Int. Conf. Prehist. Paleoecol. West. North Am. Arct. Subarct.*, Calgary, pp. 153–171.
Patten, B. C., ed. (1971). "Systems Analysis and Simulation in Ecology." Academic Press, New York.
Pielou, E. C. (1979). "Biogeography." Wiley, New York.
Proctor, V. W. (1968). Long distance dispersal of seeds by retention in digestive tract of birds. *Science* **160**, 321–322.
Ritchie, J. C. (1956). The native plants of Churchill, Manitoba. *Can. J. Bot.* **34**, 269–320.
Ritchie, J. C. (1959). The Vegetation of northern Manitoba. III. Studies in the Subarctic. Arct. Inst. North Am. Tech. Pap. 3.
Ritchie, J. C. (1960). The vegetation of northern Manitoba. IV. The Caribou Lake region. *Can. J. Bot.* **38**, 185–197.
Ritchie, J. C. (1962). A geobotanical survey of northern Manitoba. Arct. Inst. North Am. Tech. Pap. 9.
Ritchie, J. C. (1969). Absolute pollen frequencies and carbon-14 age of a section of Holocene lake sediment from the Riding Mountain area of Manitoba. *Can. J. Bot.* **47**, 1345–1349.
Ritchie, J. C. (1974). Modern pollen assemblages near the Arctic tree line, Mackenzie Delta region, Northwest Territories. *Can. J. Bot.* **52**, 381–396.
Ritchie, J. C. (1976). The Late-Quaternary vegetational history of the western interior of Canada. *Can. J. Bot.* **54**, 1793–1818.
Ritchie, J. C. (1977). The Modern and Late Quaternary vegetation of the Campbell–Dolomite uplands near Inuvik, N.W.T. Canada. *Ecol. Monogr.* **47**, 401–423.
Ritchie, J. C., and Hadden, K. A. (1975). Pollen stratigraphy of Holocene sediments from the Grand Rapids area, Manitoba, Canada. *Rev. Palaeobot. Palynol.* **19**, 193–202.
Ritchie, J. C., and Hare, F. K. (1971). Late-Quaternary vegetation and climate near the Arctic tree line of northwestern North America. *Quat. Res.* **1**, 331–42.
Ritchie, J. C., and Lichti-Federovich, A. (1967). Pollen dispersal phenomena in Arctic–Subarctic Canada. *Rev. Paleobot. Palynol.* **3**, 255–266.
Ritchie, J. C., and Yarranton, G. A. (1978). An investigation of the Late-Quaternary history of the boreal forest in central Canada. *J. Ecol.* **66**, 199–212.
Ritchie, J. C., and Yarranton, G. A. (1978b). Patterns of change in Late-Quaternary vegetation of the western interior of Canada. *Can. J. Bot.* **56**, 2177–2183.
Rowe, J. S. (1972). Forest regions of Canada. *Can. For. Serv. Publ.* **1300**.
Savile, D. B. O. (1956). Known dispersal rates and migratory potentials as clues to the origin of the North American biota. *Am. Midl. Nat.* **56**, 434–53.
Savile, D. B. O. (1961). The botany of the northwestern Queen Elizabeth Islands. *Can. J. Bot.* **39**, 909–42.

Savile, D. B. O. (1972). Arctic adaptations in plants. *Can. Dep. Agric. Res. Branch Monogr.* **6**, 1-81.
Schneider, S. H., and Thompson, S. L. (1979). Ice ages and orbital variations: Some simple theory and modeling. *Quat. Res.* **12**, 188-203.
Sellers, W. D. (1965). "Physical Climatology." Univ. of Chicago Press, Chicago, Illinois.
Short, S. K. (1978). Palynology: A Holocene environmental perspective for archeology in Labrador-Ungava. *Arct. Anthropol.* **15**, 9-35.
Short, S. and Nichols, H. (1977). Holocene pollen diagrams from subarctic Labrador-Ungava: Vegetational history and climatic change. *Arct. Alp. Res.* **9**, 265-290.
Soper, J. H., and Maycock, P. F. (1963). A community of Arctic-alpine plants on the east shore of Lake Superior. *Can. J. Bot.* **41**, 183-198.
Sorenson, C. J. (1977). Reconstructed Holocene bioclimates. *Ann. Assoc. Geogr.* **67**, 214-22.
Sorenson, C. J., and Knox, J. C. (1974). Paleosols and paleoclimates related to Late Holocene forest/tundra border migrations: Mackenzie and Keewatin, N.W.T. Canada. *Int. Conf. Prehist. Paleoecol. West. North Am. Arct. SubArct.* Calgary, pp. 187-204.
Sorenson, C. J., Knox, J. C., Larsen, J. A., and Bryson, R. A. (1971). Paleosols and the forest border in Keewatin, N.W.T. *Quat. Res.* **1**, 468-473.
Stebbins, G. L. (1972). Ecological distribution of centers of major adaptive radiation in angiosperms. *In* "Taxonomy, Phytogeography, and Evolution" (D. H. Valentine, ed.), pp. 7-36. Academic Press, New York.
Stebbins, G. L. (1974). "Flowering Plants: Evolution above the Species Level." Harvard Univ. Press, Cambridge, Massachusetts.
Tedrow, J. C. F. (1977). "Soils of the Polar Landscapes." Rutgers Univ. Press, New Brunswick, New Jersey.
Terasmae, J. (1967a). A review of Quaternary palaeobotany and palynology in Canada. *Geol. Surv. Can. Pap.* **67–13**.
Terasmae, J. (1967b). Postglacial chronology and forest history in the northern Lake Huron and Lake Superior regions. *In* "Quaternary Paleoecology" (E. J. Cushing and H. E. Wright, Jr., eds.), pp. 45-56. Yale Univ. Press, New Haven, Connecticut.
Terasmae, J. (1973). Notes on late Wisconsin and early Holocene history of vegetation in Canada. *Arct. Alp. Res.* **5**, 201-222.
Turner, B. L. (1977). Fossil history and geography. *In* "The Biology and Chemistry of the Compositae" (V. H. Heywood, J. B. Harborne, and B. L. Turner, eds.), Vol. 1, pp. 21-41. Academic Press, New York.
Van Dyne, G. M. ed. (1969). "The Ecosystem Concept in Natural Resource Management." Academic Press, New York.
Vilks, G. and Mudie, P. J. (1978). Early deglaciation of the Labrador shelf. *Science* **202**, 1181-1183.
Watts, W. A. (1967). Late-glacial plant macrofossils from Minnesota. *In* "Quaternary Paleoecology" (E. J. Cushing and H. E. Wright, Jr., eds.), pp. 90-96. Yale Univ. Press, New Haven, Connecticut.
Webb, T. (1974). A vegetational history from northern Wisconsin: Evidence from modern and fossil pollen. *Am. Midl. Nat.* **92**, 12-34.
West, R. G. (1964). Inter-relations of ecology and Quaternary paleobotany. *J. Ecol.* **52**, (Suppl.), 47-57.
Wolfe, J. A. (1971). Tertiary climatic fluctuations and methods of analysis of Tertiary floras. *Palaeogeogr. Palaeoclimatol. Palaeoecol.* **9**, 27-57.
Wolfe, J. A. (1972). An interpretation of Alaskan Tertiary floras. *In* "Floristics and Paleofloristics of Asia and Eastern North America" (Alan Graham, ed.), pp. 225-233 Elsevier, New York.

Wolfe, J. A. (1978). A paleobotanical interpretation of Tertiary climates in the northern hemisphere. *Am. Sci.* **66,** 694–703.
Wolfe, J. A., and Leopold, E. B. (1967). Neogene and early Quaternary vegetation of northwestern North America and northeastern Asia. *In* "The Bering Land Bridge" (D. M. Hopkins, ed.), pp. 193–206. Stanford Univ. Press, Stanford, California.
Wright, H. E., Jr. (1970). Vegetation history of the central plains. *In* "Pleistocene and Recent Environments of the Central Plains" (W. Dort and J. K. Jones, eds.), pp. 157–172. Univ. of Kansas Press, Lawrence, Kansas.
Wright, H. E., Jr., and Watts, W. A. (1969). Glacial and vegetational history of northeastern Minnesota. Minn. Geol. Surv. Spec. Publ. SP-11, Minneapolis.
Wright, J. V. (1972). "The Shield Archaic." Publ. Archeol. No. 3. Natl. Mus. Man, Ottawa, Canada.
Wright, J. V. (1975). "The Prehistory of Lake Athabasca." Archeolog. Surv. of Can. Natl. Mus. of Man, Ottawa, Canada.
Wright, J. V. (1976). The Grant Lake site, Keewatin District, N.W.T. Archeolog. Surv. Can. Natl. Mus. of Man, Mercury Series No. **47,** Ottawa, Canada.
Zoltai, S. C. (1975). Southern limit of coniferous trees on the Canadian prairies. Info. Rep. NOR-X-128, Environment Canada, Ottawa.

3 Climate of the Boreal Forest

The influence of climate on vegetation can only be described as all-pervasive. The close relationship between climate and the major world vegetation zones is, in fact, implied in the names given to many of the more distinctive vegetation types—desert scrub, tropical rain forest, cloud forest—and the word *boreal* itself carries climatic implications since it is taken from a term which, in a sense, combines the meanings of cold and northern. The relationship between climate and natural vegetation was recognized centuries ago, but von Humboldt (1807) apparently was the first to give it broad scientific credibility when he discerned that certain isotherms could be shown to be coincident with the boundaries of the earth's major vegetational zones. Many authorities believe that under natural conditions the ranges of many or most plant species, and of animals and microorganisms, are climatically determined; in other words, species would be capable of spreading out over a much greater region than they now occupy only if favorable climatic conditions existed there. Species are, however, restricted by environmental tolerance to a circumscribed area where favorable conditions exist and where they are not competing for space with organisms that are better adapted.

It has been shown many times that native grasses, herbaceous species, shrubs, and trees grow and compete for living space best in the climatic regime to which they have become adapted through evolutionary processes. This is not to say, however, that we know precisely and in great detail why this is true in terms of biochemistry and physiology. There has been, in fact, no great amount of significant research relating these two aspects of biology. What might be termed the biochemistry and physiology of climatic adaptations remain today largely an unexplored scientific field, yet one that is crucial to an understanding of the ecological relationships between living things and the environment. In short, we simply do not know in biochemical terms why all plant species cannot grow everywhere; we know from experience that they cannot, but it remains for future experimentation to demonstrate the basic reasons why this is so. There are fundamental biochemical and physiological connections between climatic regimes and the distinctive geographical ranges of many boreal plant and animal species, but the nature of these connections will remain obscure until the physiology of temperature and moisture relationships, and of environmentally induced growth and reproductive hormonal responses, is better understood.

In ecological studies of plant distribution, the balance between photosynthesis and respiration, a relationship that is dependent upon the balance between light and temperature, has been invoked to explain the different ecological affinities of different species. When photosynthesis is limited by environmental conditions, carbohydrate reserves are depleted and, if this condition is prolonged, it leads ultimately to reproductive failure. The balance between photosynthesis and respiration, for example, may explain the occurrence of the arctic tree line. Boreal and arctic plants are necessarily adapted to cold temperatures as well as the light conditions in northern regions, and they attain maximum photosynthetic rates at lower temperatures than is characteristic of plants of more southern regions. While this enables boreal and arctic species to survive at high latitudes, they are confined to such regions, because elsewhere they encounter species more vigorous in response to the conditions to which they, in turn, are best adapted. Northern plants also show an annual periodicity in growth that correlates with local annual climatic cycles. Many plants of northern latitudes have efficient means of vegetative reproduction—such as the layering characteristic of *Picea mariana* toward the northern limits of its range. Frost resistance, too, is related to geographical distribution; thus, *Vaccinium uliginosum*, with a more northerly distribution than *V. myrtillus* or *V. vitis-idaea*, has the greatest resistance to frost (Bannister, 1970, 1971, 1976).

The classic study of climatically related regional physiological and morphological differences in a plant species is that of Mooney and Billings (1961) who studied the variations in *Oxyria digyna* populations at 16 sites, making possible comparisons between the plants of the Colorado mountains, Point Barrow, Alaska, and Thule, Greenland. They found, both in field and controlled laboratory studies, a marked cline in populations along latitudinal and altitudinal gradients; the more northern plants had fewer flowers and more rhizomes, a higher leaf chlorophyll content, a higher photosynthetic rate at lower temperatures, and a higher respiration rate at all temperatures. Low-latitude, high-elevation plants attained photosynthetic light saturation at a higher light intensity than low-elevation, high-latitude plants. These and other differences showed clearly that *Oxyria* had both morphological and metabolic variations along altitudinal and latitudinal gradients that accounted for its ability to exist throughout such a wide range of conditions. An essentially similar variation in photosynthetic and respiration rates has been demonstrated in different regional populations of lichens (MacFarlane and Kershaw, 1978), *Picea mariana* (Vowinckel, *et al.*, 1975), and other species (see above for references.). It is apparent that temperature, light intensity, photoperiod, and other conditions related to climate are all environmental variables to which species adapt on a regional basis. These are factors that appear to be of great significance in establishing the ability of plants to survive in northern latitudes, but this conclusion is based mostly, if not entirely, on inference rather than convincing physiological data. There has been, in short, no rigorous elucidation of the physiological factors that establish the geographical limits of the boreal vegetational zone. Ecologists have yet to reveal the causal relationships involved in both the northern and southern limits of boreal vegetation. The limits seem to be thermally established, but it is not yet possible to say precisely in what way. A number of authors (Hopkins, 1959; Budyko, 1958; Hare, 1968; Ritchie and Hare, 1971) have confirmed that the number of degree days above 50°F seems to be the most reasonable correlate, but warm-season temperature is related to net radiation, and it may be that the total energy available at the surface of the earth is the critical factor involved in limitation of tree growth in northern regions. There are perhaps other factors importantly involved, but it seems reasonable to infer that this energy relationship is, at least in the case of tree species, linked directly with the photosynthetic capacity of these plants to synthesize organic compounds in sufficient quantity to maintain the arboreal structure. This relationship is discussed more fully in Chapter 7.

BOREAL CLIMATE: GENERAL DISCUSSION

The objective of systems analysis is to describe quantitatively the factors influencing the environment under consideration—both the macro- and the microenvironmental factors—and to trace the ways in which the processes involved influence, and are influenced by, one another. In terms of macroclimate, the broad features of the general global atmospheric circulation are the basic framework, although much remains to be done in developing a satisfactory quantitative theory of climate. New observational methods utilizing satellites and numerical analysis, however, ultimately will be of great value in extending knowledge of the macroclimatic aspects of environment, as well as of weather events that are the local consequence of variations in the general circulation. For present purposes, the general macroenvironmental features of weather and climate in boreal regions, i.e., radiation, cloudiness, precipitation, wind, temperature, and humidity, will be described in general terms, and mention will be made only of data from a few areas that may be taken as representative of the climatic environment of boreal regions. The climatic region delineated in general by the boreal forest is known as the sub-Arctic. It is bounded on the north over most of Canada and Eurasia approximately by the position of the July 13°C (55°F) isotherm, with marked departures in regions possessing montane or oceanic climatic influences. The southern limit of the boreal region in central and eastern Canada is bounded roughly by the position of the July 18°C (65°F) isotherm, but in the western provinces, Saskatchewan and Alberta, where drier conditions prevail, the southern edge of the forest border lies to the north of this isotherm, trending into regions where annual precipitation is greater than that in the southern portions of the western Canadian prairie provinces. The same general relationships between the boreal forest and summer isotherms hold also in Eurasia.

Bordered on the north by the cold, dry arctic tundra, the northern forest border extends in Canada from the Yukon southeastward to Hudson Bay and then roughly eastward across northern Quebec; the southern boundary crosses northern Alberta and Saskatchewan, trends southward across Manitoba, follows the northern rim of the Great Lakes, and then runs eastward along the St. Lawrence. In Eurasia, both the climatic lines and the northern edge of the forest trend more uniformly east and west than is the case in Canada.

To the north, the lands are more sparsely treed than is the case in the central areas, with a transition into tundra roughly north of the July 13°C isotherm. To the south the lands are heavily forested with conifers and, at the southern edge of the boreal forest, the coniferous species are typically intermixed with a larger proportion of broadleaf deciduous species. Appreciable snow cover lasts for more than one-half the year in most parts of the region. Extremely low temperatures are characteristic of the winters, and very warm afternoons usually occur during a few days of the summer.

CLIMATIC PARAMETERS

The general climatology of Canada and Alaska has been summarized in comprehensive fashion by Hare and Hay (1974), and descriptions of the general climatology of Eurasia have been published by Borisov (1959,

1970) and Lydolph (1977). There is no need to repeat the material here. Some elements of the climate, however, appear to bear more directly than others upon the relationships between vegetation and the environment, and the discussion that follows will emphasize these. Many of the factors with ecological significance can presently be treated only by an examination in detail of weather data from select stations, an effort that the authors mentioned above necessarily could not make in more generalized treatises.

A living plant exists in a microclimatic environment and, while the broad characteristics of its habitat are a function of the restraints imposed by the limits of the mesoclimatic parameters directly at and for a short distance above the land surface, the actual topographical niche occupied by the plant—a muskeg, swale, hilltop, valley, or shoreline—may greatly differ from what might be inferred from climatic averages and other summary parameters as seen from air mass analysis, seasonal weather patterns, or synoptic maps. At times, in fact, it may seem that the macroclimate and the microclimate are barely related, since, for example, a plant leaf in full sunlight at high latitudes may attain temperatures as high—if not higher—than a tropical plant in shade. There are, of course, differences in day length, precipitation, humidity, winds, and other factors to differentiate the northern subarctic environment from other regions of the world. These are conditions to which plant species have devised physiological and morphological adaptations through millenia of evolutionary selection processes, rendering them as well adapted, in terms of survival capability, as plants of tropical and temperate regions are to their environment, even though, were one to look only at the broad-scale macroclimatic data, it would possibly be something of a mystery as to how this could be the case.

The macroclimate of the boreal region is established by the same physical factors responsible for climate everywhere on the globe, which, working together, result in heat and moisture regimes characteristic of the seasons and expressed conventionally in terms of precipitation, temperature, humidity, wind, and so on. The incoming radiation from the sun is the single source of energy for all the climatological events observed in the atmosphere, and probably the most important aspects of solar radiation in the boreal regions are related to latitude—to the fact that, for a large part of the year, the radiation balance is negative and, for the rest of the time, relatively low compared with that for temperate or tropical zones. In short, compared with all but arctic regions, winters are long and frigid and summers short and generally cool with only a few days in midsummer in which afternoon temperatures attain 80°F or, rarely, 90°F.

TABLE 2
Mean Temperature (°C) for Selected Boreal Stations in Relation to Continental Positions in Canada and the USSR

	Western		Central		Eastern	
	USSR (Loukhy)	Canada (Inuvik)	USSR (Dudinka)	Canada (Reliance)[a]	USSR (Olenek)	Canada (Ennadai)
Northern						
April	− 2	−13	−16	−10	−13	−13
May	4	− 1	− 6	1	− 1	− 3
June	11	10	4	8	11	7
July	14	14	12	13	14	12
Aug.	12	10	10	13	10	12
Sept.	7	3	3	6	3	4

	Western		Central		Eastern	
	USSR (Arkhangelsk Oblast)	Canada (Normal Wells)	USSR (Turukhansk)	Canada (Yellowknife)	USSR (Vilyuysk)	Canada (Brochet)
Central						
April	− 1	− 8	−10	− 8	− 8	− 6

	Western		Central		Eastern	
	USSR (Volgoda)	Canada (Simpson)[a]	USSR (Yeniseysk)	Canada (Smith)[a]	USSR (Kirensk)	Canada (The Pas)
May	5	5	−1	4	4	3
June	12	13	9	12	14	11
July	16	16	15	16	18	15
Aug.	13	13	13	14	14	14
Sept.	8	6	5	7	5	7
Southern						
April	2	−3	−2	−2	−2	0
May	10	8	6	7	7	8
June	14	14	14	13	15	14
July	17	17	18	16	19	18
Aug.	15	14	15	14	15	16
Sept.	9	8	8	8	7	10

[a] Fort Reliance, Fort Simpson, Fort Smith.

TABLE 3

The Average of Maximum Temperatures (°C) for the Month during the Period of Record: Selected Boreal Stations in Relation to Continental Positions in Canada and the USSR

	Western		Central		Eastern	
	USSR (Loukhy)	Canada (Inuvik)	USSR (Dudinka)	Canada (Reliance)[a]	USSR (Olenek)	Canada (Ennadai)
Northern						
April	21	14	9	15	12	11
May	29	24	16	23	27	17
June	31	31	28	29	34	29
July	31	31	30	32	36	32
Aug.	29	29	30	30	33	27
Sept.	24	25	24	27	25	24

	Western		Central		Eastern	
	USSR (Arkhangelsk Oblast)	Canada (Norman Wells)	USSR (Turukhansk)	Canada (Yellowknife)	USSR (Vilyuysk)	Canada (Brochet)
Central						
April	23	18	14	16	19	20

	Western		Central		Eastern	
	USSR (Volgada)	Canada (Simpson)[a]	USSR (Yeniseysk)	Canada (Smith)[a]	USSR (Kirensk)	Canada (The Pas)
May	30	31	28	26	32	24
June	32	31	32	30	36	27
July	34	31	34	32	37	33
Aug.	33	31	31	30	35	31
Sept.	28	26	24	26	28	24
Southern						
April	28	22	23	27	24	31
May	31	32	33	32	32	24
June	32	35	36	33	36	36
July	35	36	37	39	37	37
Aug.	35	35	34	34	36	35
Sept.	29	30	29	32	28	30

[a] Fort Reliance, Fort Simpson, Fort Smith.

TABLE 4

The Average of Minimum Temperatures (°C) for the Month during the Period of Record: Selected Boreal Stations in Relation to Continental Positions in Canada and the USSR

	Western		Central		Eastern	
	USSR (Loukhy)	Canada (Inuvik)	USSR (Dudinka)	Canada (Reliance)[a]	USSR (Olenek)	Canada (Ennadai)
Northern						
April	−36	−44	−42	−37	−44	−38
May	−14	−28	−36	−31	−29	−29
June	− 7	− 6	−15	− 6	−15	− 8
July	− 3	− 2	− 1	− 1	− 4	− 1
Aug.	− 6	− 3	− 2	− 1	−12	− 1
Sept.	−11	−15	−20	− 7	−24	−10

	Western		Central		Eastern	
	USSR (Arkhangelsk Oblast)	Canada (Norman Wells)	USSR (Turukhansk)	Canada (Yellowknife)	USSR (Vilyuysk)	Canada (Brochet)
Central						
April	−27	−37	−41	−39	−40	−37

	Western		Central		Eastern	
	USSR (Volgoda)	Canada (Simpson)[a]	USSR (Yeniseysk)	Canada (Smith)[a]	USSR (Kirensk)	Canada (The Pas)
May	-14	-17	-29	-23	-23	-23
June	-4	-3	-8	-2	-4	-6
July	1	-1	-8	-2	-4	-6
Aug.	0	-6	-6	1	-6	1
Sept.	-7	-14	-17	-8	-15	-8
Southern						
April	-24	-40	-35	-40	-35	-30
May	-11	-17	-17	-20	-15	-13
June	-4	-3	-4	-7	-4	-4
July	1	1	1	-4	0	0
Aug.	-2	-4	-3	-7	-5	-5
Sept.	-6	-13	-13	-15	-11	-9

[a] Fort Reliance, Fort Simpson, Fort Smith.

With the boreal forest extending in a broad band over both North America and Eurasia, it is not surprising that climatic conditions vary greatly from the southern edge of the forest to the northern forest border. Some indications of the range of climate are given in Tables 2–4, in which data are presented from selected meteorological stations in the interior of Canada and the USSR where topography is flat and there is relatively little climatic influence in terms of disruption of airflow patterns by such features as mountains or oceans. In other areas, where there are mountains or where the frequency of air masses of oceanic or other origin is high, there is a greater individuality in the annual climatic patterns and more variation in pattern from one station to another.

Over much of the boreal region in both Canada and Eurasia, the flat topography results in rather interesting possibilities for correlating air mass patterns and frequency of occurrence of specific air mass types with major vegetational features. In a model used in synoptic meteorological studies and operational forecasting, for example, the climate of central Canada is considered to be dominated by four or five air masses and three frontal zones, all of which individually sweep unhindered across the region without being diverted or otherwise influenced by major topographical features. Pacific air masses, for example, stream across the plains from the southern reaches of the Cordillera to Manitoba and Ontario, swinging north and south and forming tropospheric troughs and ridges. There is, as a consequence, a high frequency of cyclonic passages in summer, and frontal activity is almost constant; the direction of movement is often in an east-southeasterly direction. There is some uncertainty concerning the nature and the location of this so-called arctic front, with doubts expressed concerning its existence as a definable entity in the sense that individual frontal developments can be said to be arctic or not arctic. There can be no doubt, however, that empirical analysis reveals frequent frontal activity in the region and, while its modal position and genesis are subject to some uncertainty and dispute, it will serve at least as a hypothesis until further data become available to characterize the air masses, their origins, and their trajectories in more accurate detail (Stupart, 1928; Thomas, 1953; Kendrew and Currie, 1955; Reed, 1960; Reed and Kunkel, 1960; Bryson, 1966; Barry, 1967; Hare, 1968; Hare and Hay, 1974). A region, it must always be remembered, is simply a human artifice employed to delineate a relatively homogeneous area on a map. The dividing line between climatic regions can be drawn roughly along climatic and vegetational transition zones, but in reality the regions merge almost imperceptibly. Where such a climatic transition zone occurs, it is also probably neither fixed nor stable, oscillating from year to year and even, as we have seen in the previous chapter, changing position over distances measured in hundreds of kilometers during past geological time. For many purposes, it is more satisfactory to demonstrate coincidence or correlation between two or more natural parameters—climate, vegetation, soils, for example—than to attempt to delineate regions by means of lines on maps, which are at best an approximation of the location of a transition zone and difficult to visualize in the abstract or observe in the field.

The air mass climatology of Eurasia resembles that of Canada in many important respects. It is relatively complex, enormous land areas are involved and, as a consequence, many air masses originate in one region, invade another region, become modified in the process, and no longer possess any of their original characteristics. In both Canada and Eurasia, the air mass pattern is most complex in the western areas, affected by the Cordillera in the case of Canada and by wind and weather patterns of irregular European land and sea masses in the case of Eurasia. As the air moves eastward on each continent, however, it tends to be modified, and the air masses are then known as continental polar in Canada and as continental temperate in Russia. The latter term, as Lydolph (1977) points out, is probably more accurately descriptive, since the air has derived its distinctive character from the north temperate, or boreal, land mass rather than from polar regions. The somewhat confusing Canadian terminology will be clarified by Table 5; in each case, the air mass adjoins colder air to the north and more moist warmer air to the south. Along each of these frontal zones where air masses confront one another, there are characteristic airstream paths of frontal disturbances, arising when the air masses intrude upon one another, forming troughs, ridges, and giant eddies at the interface of one air mass with another. This, of course, greatly simplifies the actual events by describing them in terms of climatic averages rather than in terms of the range of actual weather variation and the departures from average. Detailed descriptions, however, are furnished for Eurasia by Lydolph (1977) and Borisov (1959, 1970), and for Canada by Hare and Hay (1974).

The fact that storm tracks and annual weather patterns are so variable from day to day, as well as from year to year, could lead to the conclusion that there is no real and readily discernible relationship between air mass characteristics and the vegetational zonation that marks the boreal forest. Analysis of climatic patterns, however, does tend to confirm the rather tenuous hypothesis that certain average values of at least some climatic parameters are, indeed, coincident with the boreal forest throughout its vast extent and that they must be indicative of a causal or at least some kind of reciprocal relationship between climate and vegetation. It may well be that it is not the pattern of air mass or frontal activity

TABLE 5

Air Mass Terminology in Canada and the USSR

Canada[a]		USSR[b]	
Air mass	Origin	Air mass	Origin
Continental arctic (arctic airstreams)	Arctic regions: arctic islands, arctic Canada	Continental arctic, Maritime arctic	Northern tundra; Barents Sea
	Continental arctic front		
Maritime arctic (Alaska–Yukon; cool Pacific airstreams)	Northern Pacific Ocean; Beaufort, Chukchi seas, via Alaska–Yukon; as well as air masses of local origin because of extent of land area involved	Continental temperate	European USSR and western Siberia; local origin in large land masses
	Maritime arctic front (Midwesterly front)		
Maritime polar (mild pacific airstreams; Pacific air masses)	Pacific coast of the United States	Maritime temperate	Bering Sea; Sea of Okhotsk; Baltic Sea
	Polar front (Pacific front)		
Continental tropical and/or Maritime tropical (tropical air masses; tropical or southern anticyclone airstreams; continental airstreams)	Southern central United States	Maritime tropical, continental tropical	Eastern Mediterranean; India and environs

[a] Canadian operational air mass usage with frequently employed synonyms in parentheses. Partly from Hare and Hay (1974).
[b] From Lydolph (1977) as modified from Borisov (1970).

that is directly significant, but rather the energy budget at the surface of the earth as encountered in the growing season roughly from May to October. This would be, to be sure, a function of the frontal and air mass conditions, but, also, and of more direct impact on the growth of plants, of total sunlight, humidity, temperature, precipitation, runoff, storage of moisture in upper soil layers, and other conditions that directly affect the environment to which plants are exposed at the interface between atmosphere and soil. The macroclimate is significant, in other words, to the extent that it affects each of these factors and, for each, establishes the limits and the average values that characterize the mesoclimate and microclimates in the boreal regions. The particular combination of atmospheric parameters by which the frontal and air mass activity creates an environment suitable for boreal vegetational communities is not as significant as the fact that it does so; some combination of factors, perhaps best described ultimately by a scalar expression, results in a boreal climatic environment, and the particular combination probably varies from area to area within the whole region.

It should also be possible to approach the subject of boreal climatic relationships from the alternate pole. It should be of considerable significance to establish the range of boreal mesoclimatic conditions by studies not on the physical parameters of the boreal climate as found over areas of land surface, but rather by studies on the tolerance limits of boreal plant species as established by laboratory growth chamber experiments, thus delineating the climatic parameters and the limits characteristic of boreal environments by observing the response of boreal plants to a range of conditions established in experimental greenhouses. Certainly one of the more interesting experiments would be determination of the climatic limits to normal growth and development for the more abundant plant species that range throughout the boreal forest. These could be compared with the climatic limits of plants demonstrating restricted ranges within the boreal region, as well as with other species that range to some extent within the boreal region but also well into other adjacent regions. It probably remains to be learned by experiment whether the physiological response of the plants can be correlated with the rather broad-scale macroclimatic parameters of air mass analysis or whether the response will be found to be significantly correlated only with the more detailed and more highly variable measurements characteristic of local mesoclimatic or microclimatic description.

In summary, it is reasonable to question the efficacy of air mass analysis in casting light on the real relationships between climate and vegetation, and to suspect that relationships involve the temperature, humidity, precipitation, and winds at mesoscale and microscale levels rather than the ebb and flow of air masses and frontal zones as expressed in the macroscale climatic terms of air mass analysis. The two are, of course, related; the macroclimate is fundamental in its influence upon conditions characteristic of the mesoclimate, but the mesoscale departures from the conditions described by macroclimatic parameters are, it seems reasonable to say, at least of equal significance in establishing the environmental conditions that tend to favor development of one kind of vegetational community rather than another. It is this that makes possible the existence of islands of anomalous vegetational communities located some distance from the main biome with which they are normally thought to be associated. It is, also, this relationship that makes it seem more productive to study vegetation in terms of its relationship to mesoclimate, or microclimate, rather than to the macroclimate characteristic of the area in which it is found.

LOCAL BOREAL CLIMATES

The local climate characteristic of an area is consequent principally upon the intensity of solar radiation (shortwave) reaching the ground (a function in turn of latitude and cloud cover), the radiative exchanges (longwave) between surface features, which include the vegetation, the absorptivity of the atmosphere, and, finally, the character and velocity of advected air coming in from adjacent areas. All these are influenced by physical characteristics of the local area: topography (relief, aerodynamic roughness, slope inclination), surface and subsurface soil moisture, and albedo and emissivity of the surface features, as well as the temperature of vegetational and atmospheric interfaces.

There are also many areas with innumerable small lakes, and in summer these tend to reduce the continentality of the boreal region; the existence, too, of a large lake such as Great Slave or Great Bear Lake, must also have an influence on the summer climate and upon the length of the growing season at least along shores that are ice-bound well into spring or even early summer. In winter the effect on the cloud cover is minimal, but in spring and summer the existence of low clouds and fog in the vicinity of large lakes is a climatic feature worthy of note. The character of different areas can, thus, vary over a wide range of possible combinations of atmospheric and surface features, with local climate varying accordingly. The nature of the relationship is relatively complex, but the local climate is tractable in the sense that analysis of the causes of differences among local climates can provide explanations in terms of physical laws, and it is possible to construct numerical models to simulate accurately the behavior of the local climates under natural conditions. This latter activity has been termed climatonomy in Lettau (1969) and Lettau and Lettau (1975), and the latter publication includes a regional climatonomic analysis of the tundra and boreal forest of central Canada. In this study, the authors predict annual means for and month-to-month variations in exchangeable soil moisture and surface temperature, calculated as a response to intensity of insolation, attenuation of incoming shortwave radiation by atmospheric scattering and absorption, the albedo at both the lower and upper boundaries of the atmosphere, precipitation, the moisture regime including runoff, and, finally, the length of the frost (winter) season. The calculated results compare accurately with direct climatic observations made at meteorological stations located in the study areas—tundra at Baker Lake (lat. 65°N, long. 95°W), open forest near Churchill (lat. 58°N, long. 94°W), and closed *Picea* forest southwest of Churchill (about lat. 55°N, long. 95°W).

While it is of considerable interest that such efforts at modeling climate are exceedingly useful for characterizing the climatic environment in which plants and animals exist, a description of the details of such efforts cannot be undertaken here. Rather, it will perhaps be of more value for most readers concerned with boreal ecology to furnish a verbal description employing climatic data available for a few representative local areas.

The continental climate of both northern North America and Eurasia is distinguished by moderately warm temperatures, relatively abundant precipitation in summer, and exceedingly cold, very dry winters. The average relative humidity in the warm months ranges between 50 and 70%. Precipitation maxima occur in July or August, and at this time the greatest evaporation occurs from the forest surface. Autumn rainfall is considerable, but there is a minimum of precipitation in winter, when skies are clear and the wind is moderate. In the subarctic regions of Alaska and Canada there are frequent encounters in summer between arctic air masses and air masses of temperate latitudes, with the consequence that cold northern air intrudes into warm southern air at the rear of cyclones with accompanying sharp changes in temperature and winds and with the occurrence of low clouds and precipitation. In winter, strong, semipermanent inversions are characteristic of all polar and subpolar environments. An outstanding feature of the monthly mean temperatures in winter is the marked variation from year to year (Table 6). To demonstrate the extremes experienced during a study period, 1931–1960, the coldest month on record at Tanana, Alaska, was December

TABLE 6

Percentage of Winter Months at Tanana with Monthly Mean Temperature above or below Specified Departures from the 30-Year (1931–1960) Normal[a,b]

Month	Normal mean temperature (°F)	Below normal departure (%)			Above normal departure (%)			Months within normal range
		−10°F	−15°F	−20°F	−10°F	−15°F	−20°F	
Dec.	− 9.9	14	8	3	15	3	2	55
Jan.	−10.5	20	11	2	12	5	0	50
Feb.	− 4.6	17	6	2	17	2	0	56

[a] From Streten (1969), reproduced with permission.
[b] Based on 66 years of record, 1902–1968.

1956, with a mean temperature of −30.6°F. Four years later the data for December showed a mean of +6.4°F, the warmest since a record of +14.3°F in 1914, a difference in mean temperature of 37° and 45°F, respectively (Streten, 1969). Tanana offers a somewhat unusual case, since it is located in a broad valley entirely encircled by mountains, and inversions are protected from weak low-level weather systems which would otherwise disrupt the extended period of outgoing radiative heat loss required to attain such low temperatures. Elsewhere in the sub-Arctic, extremes may not have as great a range, but, even so, marked variation from year to year is a distinct feature of the climate everywhere. The extremes in the summer temperature patterns are not as striking as the extremes in winter. Meteorological stations in interior Alaska show that, on the average, about 13% of the summer months demonstrate departures of ±4° F, compared with the winter months in which nearly one-half the years of record demonstrated departures of ±10°–20°F from the mean (Streten, 1969). Individual days with extremely high temperatures, however, are common, and an interval with high temperatures is to be expected every summer. At such times, temperatures of 90°F can be attained, exceedingly warm for areas in which the mean maximum temperature for most days, even in protected valleys, is 72°–76°F. The highest temperatures ever recorded for such areas have often reached well over 90°F, and there are records of 100°F having been attained at least once at many weather-recording stations.

Interior Alaska records very low annual precipitation, with nearly 50% of the annual total between June and August. Rainfall is derived largely from summer convective activity which varies greatly in time and space. Periods of more rainfall are the result of invasion of the interior by moist air originating in lower latitudes of the North Pacific and, when they occur, there may be flooding in some low areas along waterways.

This temperature and rainfall pattern is similar in the central interior of Canada; in central and northern Manitoba, for example, mean monthly surface air temperatures range from −22° to −30°C in January, from 0° to −10°C in April, from 12° to 18°C in July, and from 4° to −1°C in October. Spring warming is rapid, with the mean daily temperature rising quite sharply within 2–3 weeks. The mean date of rise in mean daily temperature to 0°C is from April 1 to April 15 for inland areas. Autumn freeze-up, defined as the date on which the mean daily temperature falls to 0°C, occurs between October 10 and November 1, the earlier date being representative of stations in the northeastern corner of the province and the later date representing the northern central area at the southern edge of the boreal zone. The region is relatively dry, with a recorded precipitation range of about 40–45 cm throughout. About one-third of the precipitation in the entire central Canadian region falls as snow. To the north, in the Northwest Territories, precipitation is lower, ranging usually between 20 and 35 cm.

The same general patterns of weather prevail in Eurasia as a consequence of the similarities in latitude and the frequent encounters between air of cold, dry northern origin and warmer, moist air from more southern regions. In the northwestern Eurasian boreal region, the weather is characteristically changeable, with frequent passages of frontal depressions and, as a result, wet, cool summers. Northwesterly winds predominate. Borisov (1959) notes the great variability in weather conditions from year to year, pointing out, for example, that the average date of the blossoming of gooseberry near Leningrad is April 27 but can be as late as May 7 or as early as April 4. Full autumn coloring occurs at the extreme southern edge of the boreal zone by October 1 on the average but may be two weeks early or late. Summer in the northwestern zone is short and cool, with a frost-free period of 90–120 days. The first fall frost occurs by November 1 in the southern part of the region and by mid-October in the north. The last frost is in mid-April and in the beginning of May in these areas. Autumn brings a rapid increase in cloudiness and an increase in wind and rain. The first snow in the north occurs at the end of September and in the south by early October. The depth of snow in the forests at the height of winter is 90–100 cm.

The boreal forest in the northeastern region of Eurasia is distinguished by a continentality of climate, with the annual range of average temperature in many areas approaching 40°C. Compared with the northwest, there are fewer frontal depressions, but frosts are possible at night any time during the summer. The western Siberian boreal region is characterized by even stronger continentality, severe winters, deep snow, and adequate but not large amounts of moisture for optimal forest growth. Northern winds prevail in summer, and cold waves occur frequently, although hot days are also experienced when advection of warm air from the deserts of central Asia occurs. In winter there is more than a month during which average temperatures are below −25°C. A minimum of −54°C has been recorded. Spring comes not earlier than April, and summer lasts from mid-July to early September. There are only 30 days in the year with temperatures above 20°C. The precipitation in the western Siberian boreal region varies from 350 mm to more than 600 mm, with as much as 80 mm falling in a single thunderstorm. About one-third of the annual precipitation falls during the warm season. In some winters the snow depth may be 2 m, but in other years very little may fall and the land is virtually bare. The snow cover usually lasts 210 days in the north and 130 days in the far south of the region. Borisov (1959) adds that the thick early snow in the western Siberian region deflects the southern permafrost boundary to the north. Turukhansk has no permafrost, although it has an average annual temperature of −8°C, while other places with an average temperature of about −2°C have permafrost as a result of lack of snow cover.

The climate of the far eastern Siberian boreal forest is one of the most severe in the world. Cold waves in summer alternate with periods of extreme heat. At Yakutsk, for example, the mean July temperature is 19°C, but cold waves can drop that by 10°–12°C in an hour or two. The temperature often falls to near freezing by sunset, and only the southern parts of the region are free from frost in July. Winter is dry, temperatures fall to −70°C, and the Oimyakon and Verkhoyansk areas are known as the cold pole of Eurasia if not the world. In some years the mean January temperature falls to −56°C, the periods of lowest temperature coinciding with times of little or no wind. Borisov (1959) mentions the phenomenon known as the "whisper of the stars" during such times, when persons out-of-doors on cold nights hear the freezing of exhaled vapor, soft rustling as ice crystals fall to the snow.

FOREST SYSTEMS CLIMATOLOGY

The position of the boreal forest border across northern Canada, Europe, and Asia is more or less coincident with a number of the parameters employed to delineate climate. These include average pressure patterns for certain months of the year, average temperature distributions as revealed by isotherms, mass distribution fields in the upper atmosphere, and regions of maximum frontal activity. There are also a number of climate-related soil and geomorphological characteristics, notably podzolization and permafrost, which demonstrate distributions coincident in one way or another with circumpolar distributions of certain features associated with boreal vegetation (Tedrow, 1977; Brown, 1960, 1970a,b).

These relationships long ago led climatologists to acceptance of the general conclusion that spatial correlations between northern coniferous forest and the characteristics of climate are more than coincidental. Such easy acceptance of this conclusion has often been a source of wonder and delight to ecologists, who were inescapably more conscious of inadequacies in knowledge concerning the physiological responses that render climate so important in establishing range limits of plant species

(Britton, 1966). Curve matching is, at best, a dubious method for establishing proof, but recent expansion of knowledge of physiological responses and genetic processes tends to support the climatologists' faith. Many responses shown by plants growing in a variety of environments have been described in detail (Evans, 1963; Treshow, 1970, for example), and outlines of the genetic processes that enable plant species to adapt to environmental conditions have been discerned (Stebbins, 1974).

In recent years, there has been developed a more detailed and finer degree of understanding of the species composition of the boreal vegetation communities, and certain modern techniques using computers have given climatologists the ability to delineate climatic regions on a more accurate basis than was possible with older methods; these now have been employed to discern correlations existing between climate and the forest regions established by botanists and foresters (Newnham, 1968; Nicholson and Bryant, 1972; Bellefleur and Auclair, 1972; Miller and Auclair, 1974). The changes in frequency of common plant species in boreal communities along climatic gradients have also been investigated (Larsen, 1971a,b, 1974a,b), and a close interaction between climate and species abundance must exist. It is also apparent that the relationships are difficult to analyze because of the large number of variables involved. Studies of climatic and vegetational factors reveal correlations in regional distributions, but there are also seasonal and even erratic and rare climatic variations or extremes (e.g., a June frost occurring once every 10 years or more) that may profoundly influence reproduction and survival of certain plant species. Cause-and-effect relationships are thus not always clear. It has been assumed, for example, that the average summer position of the hypothetical arctic front is a determining factor in establishing the position of the northern forest border. More recently it has been asked whether the forest border does not, instead, influence and perhaps guide the direction of movement of the summer cyclonic storms, and hence establish the average position of a frontal zone. It has also been suggested that the positions of both forest border and arctic front may be determined by some other overriding influence such as the sharp differential in the net radiation in forest and tundra (Hare and Ritchie, 1972). This must have an influence upon climate, of an intensity as yet undetermined, and it may bring about stabilization of the forest border by a self-perpetuating feedback process at work within the forest itself, which tends to restrict reproduction of trees and invasion of the tundra (Larsen, 1973). Other influences, too, may be at work here, such as severity and frequency of forest fires, abrasion by wind-driven snow in winter, insect infestation, disease, and perhaps such events as the weathering of parent materials and the rate of soil formation. The relationships are complex, and much remains to be done before they will be fully understood (Hare, 1968; Hare and Ritchie, 1972; Johnson and Rowe, 1974; Larsen, 1973).

ATMOSPHERIC SUBSYSTEM

The atmosphere is a relatively simple mechanical mixture of elements and compounds essential to life—oxygen, nitrogen, carbon dioxide, and water. It is, moreover, not only a source of essential materials but also the sink to which they are returned in the perpetual cycling that occurs—from atmosphere to living plants to soil and ultimately back to the atmosphere again. Atmospheric factors include incoming solar radiation, the energy regime at the boundary layer of the earth, the effects of cloudiness, precipitation, wind, and so on. Air temperature and the internal temperature of plants and animals directly affect rates of physiological processes. Wind and humidity, as well as temperature, determine rates of evapotranspiration. The carbon dioxide content of the air, cloudiness, temperature, and moisture all affect the rates of photosynthesis; the latter, in turn, control other processes requiring the energy captured and stored in photosynthesis. Under natural conditions most of these factors are interrelated, so that variations in one factor from place to place are accompanied by corresponding variations in other factors. Biologists traditionally have attempted to correlate regional differences in plant and animal associations with regional differences in rainfall, sunlight, temperature, and wind, as well as soil moisture, nutrient content, texture, structure, and so on—with varying but often encouraging measures of success. Failure to achieve a consistently high correlation between these factors and the composition of vegetational associations of different regions has been explained as due to the modifying influences of unmeasured factors upon factors measured—but even with multiple correlation techniques it usually has not been possible to elucidate causative relationships with any certainty. All influences ascribed to one or more factors in classic ecology are almost without exception taken more on faith than based on demonstrable relationships; only recently, moreover, has there been an effort to distinguish between macroenvironmental factors (those not originating with, or markedly influenced by, plant and animal associations) and what might be termed microenvironmental factors, many of which originate as a consequence of, or are dependent upon, the presence of plants. Many microenvironmental factors have their intensity of influence markedly dependent upon the type and density of the plant cover and on the animal species present.

It is in this framework of the physical environment that the boreal ecosystem functions, and while, in many instances, the direct influence of a factor—temperature, for example—is difficult to discern and would be even more difficult to measure, it is intuitively such characteristics that lend the boreal ecosystem its unique identity.

RADIATION AND TEMPERATURE

Since solar radiation is the primary factor establishing temperature and moisture regimes, it is of fundamental significance in the climatology of vegetational communities, and a considerable amount of research has been conducted in an effort to characterize the radiation budget of the boreal forest. The total of incoming solar energy at a point on earth is known as the global solar radiation, and the global solar radiation per annum in Canada increases southward from values near 90 kilolangley (kly) at the arctic tree line to about 110 kly at the boundary between open woodland and closed forest zones, i.e., the northern "forest line" (Hare and Ritchie, 1972). More significant is absorbed radiation, which in the same span ranges from 50 to 55 kly at the tree line to about 80 kly at the forest line.

Zonal divisions of vegetation appear to correlate closely with mean net radiation. Growing season net radiation is fairly constant within the forest region, just as it is within the tundra region, and between the forest and the tundra there is a sharp drop. The gradient of net radiative heating across this sharp transition zone is due mainly to albedo effects in spring. In much of the forest zone, snow is covered by the crowns of the coniferous trees, and strong absorption of radiation occurs in winter, in contrast to tundra where the snow-covered ground is intensely reflective. In open woodland and in the forest–tundra ecotone the albedo is intermediate between these extremes. With the forest strongly heated, and the tundra much less so, an intense air temperature gradient develops. The low albedo of dense forest permits more rapid warming in spring, hence a longer above-freezing season than in the tundra. It is about 50 days longer in the forest than on the tundra, even though these zones are separated by less than 40 km. The structure of the vegetation, in this case, markedly influences the physical climate. Across Canada, the zonal divisions of the boreal forest seem to be in close relation to the distribution of both annual and warm season net radiative heating. Hare and Ritchie (1972) also point out that standing phytomass increases from less than 5 tons/ha in arctic tundra to as high as 25 tons just north of the forest border. It then rises rapidly southward in the forest to values as high as 300–400 tons/ha. In relation to the annual net radiation, much more phytomass exists per unit of energy absorbed in the southern boreal forest than in the southern fringes of the tundra.

LOCAL ENERGY BUDGET

The basic climatic processes—transfer of heat and moisture—proceed vigorously in a stand of trees during the warm months of the growing season (Tibbals *et al.*, 1964). The solar energy available to *Picea mariana* has been shown to be largely independent of seasonal changes within the growing season; the albedo (proportion of energy reflected back toward the sky by the forest) remains between 6 and 8% from April to November in the central portions of the forest. For comparison, the albedo of a sphagnum–sedge bog markedly increases during the spring when bud growth and shoot development are maximal; *Picea mariana* stands do not exhibit this increase, the albedo remaining essentially constant throughout the growing period. There are, in addition, no significant albedo differences when direct, direct-plus-diffuse, and diffuse radiation conditions are compared (Berglund and Mace, 1972). In winter, despite a snow cover beneath the trees, the albedo of a dense *Picea* forest increases only slightly, to about 10%, a result of the fact that the tree canopy effectively covers the snow. In contrast, the albedo of an open bog increases to 82% from summer values ranging between 12 and 16%.

Thus it is evident that during summer the absorbed energy is 5–10% greater in a bog covered with *Picea mariana* than in an open sphagnum–sedge bog. The added energy should result in greater evapotranspiration losses from *Picea mariana* than from the open bog, but evapotranspiration is ultimately controlled by physiological rather than by purely physical processes, and the actual characteristics of evapotranspiration from the two communities are as yet imperfectly known. A solar radiation model for *Picea*-dominated boreal woodlands in winter was constructed by Wilson and Petzold (1973), in a study of snow melt, using such vegetational parameters as mean tree height, mean radius of branches, and mean distance between trees. Similar models employed to characterize the energy regime of boreal forest stands in summer should be of value in developing an understanding of the temperature and moisture regimes during the growing season.

The pathways of heat and moisture, both of which have initial sources and ultimate sinks outside the forest itself, are complex. There are many ways in which incoming radiation and moisture are moved from one place to another within the forest stand. A relatively small amount of microclimatic research on conifers has been conducted in an effort to describe these pathways—a consequence probably, as Jarvis et al. (1976) point out, of the fact that the coniferous canopy is difficult to describe mathematically and work with. Leaf area is difficult to measure, tissues are physiologically somewhat sluggish, and stomata are hard to find. Moreover, in coniferous forest it is often difficult to locate study sites that can reasonably be considered representative of the forest as a whole—if such a theoretical entity can be said to exist at all. It is evident, however, that coniferous forests are quite different from other forest communities, and for this reason alone it is important that expanded research be undertaken.

Variation in vegetation from one site to another is readily interpreted as a response to variations in microclimate, but efforts to correlate the two have been tenuous because identification of the separate effects of what are interrelated variables is difficult. In a more comprehensive study that might be considered classic, MacHattie and McCormack (1961) made measurements of air, soil, and surface temperatures, evaporation, wind, and other parameters on a ridgetop, north slope, and south slope of a forest in Ontario (lat. 45°56'N, long. 77°33'W), a section of the Great Lakes–St. Lawrence forest region (Rowe, 1972). The ridge top was dominated by *Quercus* and *Pinus*, the north slope by *Populus*, *Pinus*, *Betula*, and *Picea*, and the south slope by *Pinus* and deciduous species. Ground vegetation was dominated by a distinct set of species at each site. The work was sufficiently detailed and complex to admit no easy summary, but the authors point out that the amount of energy absorbed as latent heat—influenced at each site by differences in all climatic parameters—is the major determinant of soil-surface temperature and low-level air temperature. Flowering of understory plant species was earliest on the ridgetop, next on the south slope, and last on the north slope, with a difference of a week between the north slope and the ridgetop. Radial growth of the trees, however, showed little difference among sites, the result of canopy temperatures being similar at all sites even though average soil and ground level air temperatures were distinctly different. The radiant energy available to the forest—the basic determinant of temperature—is largely dependent upon the macroclimate, but the canopy and understory influence the exchange of energy by reflection and emission. The needles of conifers, with a large leaf area index, make coniferous forests quite different from deciduous forests in energy-exchange characteristics (Jarvis *et al.*, 1976; Tajchman, 1972).

The canopy of a dense, mature forest is a barrier to vertical air movement and an effective absorber of radiation during daylight. As a result, air temperatures within a canopy during the day are higher than at either ground level or above the trees. The opposite may be true on cold, clear nights, when radiative heat loss is greatest from the canopy and air temperatures may be cooler than at ground level. However, air movement is sufficient in each instance to prevent the development of large temperature gradients, and transpiration reduces the daytime canopy temperature to some extent. Maximum temperatures in the upper canopy are attained in early afternoon. There then may be a difference of several degrees between canopy and ground. In a stand of widely spaced trees such as in a lichen woodland, canopy temperatures are no different from those of open ground (Baumgartner, 1956; Bannister, 1976).

High winds, of course, prevent the development of extremes of temperature and of vertical gradients from the ground upward through a canopy. They also prevent the deposition of dew and the occurrence of ground frosts on cold nights. Winds within a forest with a dense canopy, however, are never very intense, unless the air movement is strong enough to bend the trees sharply. There is little variation in wind speed within a forest, even when the air above the canopy is moving at high velocity; for air speeds above the canopy ranging from 1 to 7 m/sec, the variation within the canopy is from about 0.5 to 1.4 m/sec. The maximum reduction in wind speed is near the upper surface of a dense canopy, and air movement within a canopy is usually light and turbulent (Baumgartner, 1956, 1970; Bannister, 1976; Munn, 1970; Lee, 1978). Such movement is, however, important, for continual mixing of the air in the canopy is metabolically the most important event taking place; photosynthesis depends upon a continual supply of carbon dioxide (Jarvis *et al.*, 1976).

Evapotranspiration probably can be considered physiologically significant only in a negative sense—it is a necessary evil from the point of view of leaf metabolism. Leaves must have a moist surface to permit absorption of carbon dioxide, but this entails continual loss of water by transpiration, hence the need for large supplies from root systems. As a consequence, both soil moisture and atmospheric humidity (high humidity retards evaporation) influence metabolic activity in leaves and needles. When water is in short supply, stomata close to prevent further loss, but this in turn stops photosynthesis for lack of a carbon dioxide supply to chloroplasts. Plants such as *Vaccinium vitis-idaea* and *V.*

uliginosum show an ecotypic adaptation to sites with different degrees of moisture availability; plants from wet sites close stomata even though water levels in tissues are relatively high compared to those in plants on dry sites at the point of stomatal closure. Most other ericaceous dwarf shrub species show similar adaptations to habitat, those growing in moist and shaded habitats demonstrating stomatal closure at higher tissue water levels and less resistance to desiccation than plants from dry sites. It is of interest, moreover, that the boles of conifer trees are less capable of rapid transport of water than those of deciduous trees, and it seems reasonable that the needles must be more resistant to desiccation than leaves at times when the evaporation of water from stomata is rapid. Water deficits in needles may also become severe when aboveground parts of the trees are warm and transpiring and the roots are still encased in cold or frozen soil. The species with widest distribution in the boreal forest are better adapted in this respect to severe winters than species with southern or otherwise more restricted ranges (Bannister, 1970, 1971, 1976).

CLIMATE AND PERMAFROST

In the far northern regions, permafrost must exert a profound influence upon the vegetation, and it may indeed be one of the major influences restricting northward expansion of the forest into areas that are now tundra. Permanently frozen subsoil at depths as shallow as a few inches in summer at the northern limit of trees renders the entire rooting zone cold throughout the growing season, and this must greatly impede water and nutrient uptake in roots. The role of permafrost in the distribution of vegetational communities was studied by Dingman and Koutz (1974) who found that, on the Yukon–Tanana uplands, the *Picea glauca–Betula papyrifera* forest is confined to permafrost-free areas of the basin, and that other vegetational communities are underlain by permafrost at shallow depths. The boundary of the permafrost in the area appears to coincide with the isopleth of 265 cal/cm/day average annual insolation. These authors point out that the thickness of the seasonally thawed zone above the permafrost is also correlated with the insolation, and that the vegetation itself influences the depth of this so-called active layer of soil. They point out that a close correspondence between vegetational communities and the presence or absence of near-surface permafrost has been noted by numerous investigators on the uplands of central Alaska. Péwé (1966), for example, has observed that *Picea mariana* scrub forest occupies permafrost areas and that *Picea glauca–Betula–Populus* is found on frost-free slopes. Similar coincident distributions of permafrost and vegetation were noted in the central Mackenzie River valley by Crampton (1974). On higher slopes and ridges the permafrost table is near the surface in summer, and the vegetation is dominated by lichens (mostly *Cladonia alpestris*), with such shrubs as *Ledum groenlandicum, Betula glandulosa, Vaccinium vitis-idaea,* and *Potentilla fruticosa.* Stunted *Picea mariana* rarely exceeds 15 ft (4.6 m) in height. Where the permafrost table lies at a greater depth, as deep as 5 ft (152 cm) in places, sphagnum is found in greater abundance and lichens in less abundance, and shrub growth is thicker and higher. *Larix laricina* and *Betula papyrifera* are found here with the *Picea mariana,* which attains heights of up to 30 ft (9.1 m). *Picea glauca* is also present on these sites.

How greatly subsurface temperatures can differ from those of aerial plant organs such as leaves, needles, and twigs is demonstrated by data obtained at Tabane Lake, located at the northern edge of the forest–tundra ecotone in Keewatin, Northwest Territories, at 3:00 P.M. on a day, July 20, without clouds (Larsen, unpublished). With the use of a hand-held bolometer, the temperatures tabulated below were obtained.

Ambient air temperature	29°C
Bark of dead, downed *Picea mariana* with black lichen (*Alectoria jubata*)	50
Cones of same	40
Black lichens on rocks	42
Bare *Picea mariana* bark in sun	38–40
Dry *Sphagnum* mat in sun	37
Empetrum nigrum in sun	36–40
Stereocaulon mat in sun	38–40
Loiseleuria procumbens in sun	36
Bare sand in sun	38
Moist *Sphagnum, Vaccinium vitis-idaea,* and *Ledum* in moderate shade	32
Same in deeper shade	28
Dense *Polytrichum* mat in sun	32
Picea mariana twigs in sun	32
Picea mariana twigs in shade	26–28
Vaccinium and *Empetrum* in shade of lower *Picea mariana* branches	24–26
Salix catkin	28
Rubus chamaemorus leaves in sun	28
Moist *Sphagnum* mat in partial sun	24
Moist *Sphagnum* mat in deep shade	20
Deep depression in shaded moss	16
Opening of rodent hole	12
Hole under *Picea mariana* roots, rock-lined	3

The ambient air temperature was unusually warm for the region, one of the rare days on which temperatures attained 30°C; the temperature maximum for the day had passed by the time the observations were made. Nevertheless, the temperature of the black lichen-covered downed *Picea mariana* twigs was still 50°C, one reason why the fire hazard can become intense in such areas. Temperatures steadily declined toward ground level, and the influence of permafrost not far below the surface (actual depth at the site unknown) can be seen in the 3°C temperature under the roots of a *Picea mariana* tree—a temperature at which physiological activity would obviously be slow. It is apparent, also, that this temperature is rarely if ever exceeded for roots of *Picea mariana* trees growing in the area. Not more than 30 mi to the north, *Picea mariana* trees no longer dominate the landscape, and the species is found only in protected valleys (Larsen, 1965, 1974a; Elliott, 1979). The low temperatures obviously also will have an influence upon the growth and development of the species making up the ground cover and upon the mortality of seedlings, although detailed information on such matters is as yet lacking.

Forest fires occur most often in areas such as this, where temperatures of aboveground plant parts and low atmospheric humidity lead to severe desiccation. Most fires are caused by a lightning flash that ignites both lichen mat and tree branches. The trees are almost always killed, and the lichen mat is consumed. If the fire is intense, or if reburning occurs, the organic layer of the soil will be destroyed. The effect on both vegetation and the physical environment is profound. Rouse (1976), for example, made microclimatic measurements at four sites, burned 0, 1, 2, and 24 years previously, and compared them with measurements for a control site of mature, open *Picea*-lichen woodland. The measurements showed a substantial change in radiation absorbed over the recently burned surfaces, and soil temperatures were much higher at the burn sites during the growing season. Most striking was the decrease in albedo, leading to increased absorption of solar radiation, creating very high temperatures over the burned surfaces. There was a long-term increase in summer surface temperatures of as much as 14°C. The surface temperatures rose as high as 65°C over fresh burns, and diurnal temperature changes were extreme—up to 46°C. Soils of unburned forest start the growing season wetter than the soils of burned areas.

CLIMATE AND SPECIES DISTRIBUTION

All the studies described above point to the fact that minor variations in topography and vegetational community composition and structure can, indeed, be correlated with variations in climatic parameters. It is apparent, furthermore, that studies of the relationships between climate and vegetational communities yield pertinent ecological information, not only in terms of plant ecology but in terms of the bioclimate of the animal inhabitants of the forest as well. Thus, Pruitt (1957, 1959, 1978), for example, has conducted much work on the bioclimate of the subnivean environment, as well as on the effects of low temperatures and snow cover on larger mammals in northern regions; he states that animals and plants can scarcely be considered separate entities in boreal regions, since changes in plant cover result in habitat modification as a result of the changes in soil temperature. Moreover, successional development of vegetation, especially of a thick moss layer, results in colder soils; frozen soil prevents water percolation, resulting in wetter soils, inhibition of tree growth, and markedly changed conditions for animal life. It seems obvious that temperature is the single most important factor limiting the growth and development of vegetation in boreal regions, and four temperature-dependent physiological processes are most important in this relationship (Warren Wilson, 1967): (1) translocation, which is not greatly affected by temperature under normal temperate-zone conditions but which is markedly checked when plants are chilled to $0°$-$5°C$; (2) water absorption, which is affected similarly; (3) rate of photosynthesis, which increases roughly 50% with an increase in temperature from $0°$ to $1°C$, while an equal increase in photosynthetic rate is achieved only by a $10°C$ increase in temperature from $20°C$; and (4) rate of respiration, which is slow at low temperatures and increases steadily with a rise in temperature. If the rate of respiration exceeds that of photosynthesis, a plant eventually will run out of its store of carbohydrate and will be incapable of further growth or of reproduction. When photosynthesis exceeds respiration, sugar and starches are manufactured, stored, and used in growth, flowering, and production of seed. Temperature must often establish upper limits to growth in northern regions. It is, ultimately, gene structure that establishes the tolerance limits of a plant species, and it seems likely that distribution of a plant species in northern regions is not simply a response to frost but is conditioned by many specific temperature-dependent requirements (Warren Wilson, 1967). Illustrative of the complexity of physiological response to the environment is the work of Hicklenton and Oechel (1976, 1977) with the moss *Dicranum fuscescens*, specifically in relation to the acclimation and acclimation potential of the carbon dioxide exchange mechanism in relation to habitat, light and temperature. This and other work indicate the importance of photosynthetic acclimation in the distribution of plant species in arctic and subarctic environments (Kershaw and Rouse, 1971; Larson and Kershaw, 1975; Büttner, 1971; Billings and Mooney, 1968).

Temperature tolerance differs markedly among species. For each there is a fairly definite range of possible temperatures, and above or below these the plants will perish. Positive correlations between averages of a single atmospheric phenomenon and the large-scale regional geographical habitat preferences of plant species, however, should not be taken to imply a simple, direct one-to-one causal relationship. Air temperature, soil temperature, dew point, mixing ratio, solar radiation, moisture, precipitation, evaporation, wind speed, and indexes describing the relationship between two or more of these factors, are ways to describe the complex regime that is the total environment. While climatic parameters may have little significance individually, each does reflect to some degree the total atmospheric environment and as such may at times be employed usefully as a climatic indicator.

Statistical methods can be used to reduce many climatic variables to a relatively small number of components (Newnham, 1968; Nicholson and Bryant, 1972; Bellefleur and Auclair, 1972; Miller and Auclair, 1974), and research of this kind makes it possible to discern the climatic factors that appear to be most significant in regional distributions of species. In one study, for example, three factors—mean annual temperature, length of growing season, and annual potential evapotranspiration—appeared to be significant in establishing the broad-scale vegetational distribution demonstrated in Table 7 (Hare, 1950, 1954). It should be possible to discern the physiological characteristics of individual species that de-

TABLE 7

Forest Divisions and Potential Evapotranspiration (P-E) in Labrador–Ungava[a]

Division	Typical value of P-E (cm) along boundaries	Dominant vegetative cover type
Tundra	30.5–31.7	Tundra
Forest–tundra ecotone	35.5–36.8	Tundra and lichen woodland intermingled
Open boreal woodland	41.8–43.1	Lichen woodland
Main boreal forest	47.0–48.2	Closed spruce forest; spruce–fir association
Boreal–mixed forest ecotone	50.6	Closed forest with white and red pine, yellow birch, and other nonboreal invaders
Great Lakes–St. Lawrence mixed forest		Mixed forest

[a] Data from Hare (1950, 1954). Reprinted from *Geographical Review* with the permission of the American Geographical Society.

termine environmental tolerance limits, but this will have to be accomplished with time-consuming experiments in growth chambers. When the research has been accomplished, however, it will be possible to say there has been a start in acquiring an understanding of the ecological relationships of communities in terms of the individual capacities of the species of which they are composed. On the other hand, there is also a need for long-term records of clearly defined climatic parameters for many stations throughout the boreal region, and for a conceptual grouping of stations possessing similar climatic characteristics. In this effort, a beginning has been made. Analysis of mean temperature and precipitation data for 111 Canadian meteorological stations has delineated, for example, the distribution of areas possessing similar regimes and has revealed interesting similarities among forest regions widely separated geographically, indicating that for forest management purposes the regions can be considered homologous. Growth patterns of seedlings and trees, forest regeneration characteristics, forest structure, and forest fire control measures will likely be similar throughout each region (Miller and Auclair, 1974).

Efforts to relate climatic zones to vegetational communities are of considerable importance and, in the past, vegetation types have been used as the basis for climatic classification. Such systems are cumbersome, however, since climates designated as "warm/moist in all seasons" and "snow forest/moist all seasons" span wide latitudinal belts within which both vegetation and climate vary a great deal. Furthermore, the classification is verbal rather than numerical and cannot be employed in other than descriptive correlation. Under this system, however, the boreal forest is characterized by a cold snow-forest climate with adequate rainfall and warm summers.

There is a correlation between the forest divisions and potential evapotranspiration, but no ecological evidence has been forthcoming to support any suggestion that distribution of the forest community is controlled in this region by the moisture supply; studies, however, of the Russian taiga have shown that zonal divisions comparable to those defined for Labrador–Ungava give almost identical relationships to climate. It is of interest, additionally, that the approximate southern limit of continuous permafrost in Canada coincides with the northern limit of the forest–tundra ecotone, although one wonders if the relationship is not more assumed than real. The significant aspect of all these approaches is that in each there is an appreciable degree of coincidence between the northern forest border and climatic parameters, but what is lacking is an adequate ecological explanation for the coincidence. Although the coincidence appears much too exact and too readily amenable to ecological interpretation to be fortitious, nevertheless the fundamental nature of the relationship remains obscure. It is not known, in physiological and microenvironmental terms, why the transition between forest and tundra should coincide with climatic parameters. The physiological failure of spruce north of the present-day forest border may conceivably be found in the proportionately higher respiratory loss of trees in comparison to other vegetation, presumably the result, as Warren Wilson (1967) has pointed out, of the maintenance costs of the trunk-and-branches system:

> Probably over half of a tree's budget is spent on defense expenditure related solely to competition.... The particular compromises reached by different species are responsible for their characteristic tolerance ranges and are of great interest to the ecologist as the basis for species distributions. Strong adaptation to one aspect of the environment usually leads to susceptibility to another aspect.... The tree species dominating the forests in northern regions have, thus, apparently sacrificed physiological efficiency for dominance; the lowered efficiency, however, has cost them the possibility of survival in arboreal form in the harsh environment northward beyond the forest border.

They do nevertheless survive there as decumbent or very dwarfed individuals which can be found ranging far north of the present-day forest border in areas where forest once existed in postglacial times and has since retreated.

Picea mariana of decumbent form, or existing as small groves of dwarfed individuals, is widely scattered throughout the region north of the forest border in at least the regions north of Ennadai Lake and north of Artillery Lake, two areas investigated rather thoroughly. For example, small groves of *Picea* are found throughout the area at the south end of Dubawnt Lake and in widely scattered clumps at Yathkyed Lake. All at Yathkyed and many (if not all) at Dubawnt give every appearance of being remnants of a former more extensive and denser forest cover. The same is true of the area north of Artillery Lake; small groves are found in a few protected hollows as far north in this area as Clinton-Colden and Aylmer lakes (Larsen, 1971a,b, 1974a,b; Elliott, 1979; see Chapter 2).

Ecologically, the most striking characteristic of the northern forest border is its abrupt nature, particularly in view of the virtually continuous and dominant *Picea* cover for hundreds of miles south of this boundary. Aerial observations throughout the Ennadai Lake area support Tyrrell's notes of 1894 (Tyrrell, 1897) in which he states that, along the edge of the south end of Ennadai Lake, "within a few miles the forest disappears, or becomes confined to the ravines." The same holds true for the Artillery Lake area, as revealed in the notes of early explorers (Larsen, 1971a,b). Very few of the understory species attaining dominance at one latitude or another in the boreal forest zone have an equally wide latitudinal range. Among the most interesting subjects for future studies will be those relationships that permit *Picea* to occupy an unusually wide range but which fail in such dramatic fashion at the northern limit, giving *Picea* rather complete dominance over nearly 10° of latitude and then instituting limitations effective over a distance somewhat less than 50 mi.

One observation, supported by nothing more than a subjective experience with ambient conditions, may suggest the nature of the comparative forest and tundra temperature ralationships. During sampling of forest vegetation at the south end of Ennadai Lake on a clear day with light winds in early July, temperatures within the forest were notably cold, and it came as a marked relief to emerge into tundra on a hill summit. It was apparent that conditions stayed uniformly cold within the forest for a longer period into spring and summer than was the case on the tundra; on the tundra temperatures might become colder at night than would be the case in the forest, but daytime surface temperatures obviously were warmer. Growth limitations for trees in the forest may be at least partially in response to this colder microclimate, affecting soil, forest floor, and at least tree boles and lower branches of trees even during hours of maximum insolation. It is a subject that would lend itself to study with simple instrumentation.

Benninghoff (1952; see also Larsen, 1965) suggests that in many areas the environmental conditions that permit *Picea* growth initially on some sites are eventually modified as the forest approaches maturity and develops a closed canopy. Under the latter conditions, the active layer in summer becomes shallower, frost action damages roots, and the stand becomes degenerate and dies, to be replaced eventually by a new stock of trees capable of colonizing the renewed deep active layer found under open stands. Conditions at least related to these must be at work at the forest border, but the precise involvement of soils and within-stand radiation characteristics remains to be disclosed. Aerial albedo measurements in the region show little difference between radiation characteristics of open tussock muskeg tundra and those of the forest–tundra ecotone, so it appears initially at least, that radiation balances between the two terrain features in summer are similar (McFadden and Ragotzkie, 1967); the subject, however, is one concerning which further studies would surely be desirable.

Climatic effects, however, may be sufficient to accomplish the remarkable transformation in the character of the vegetation that occurs over a relatively short distance. On the other hand, some self-perpetuating system that augments the influence of the climatic parameters may be at work within the forest or the forest–environment complex, although the presence of such a system is not at present recognized, or understood if it indeed exists. A comparison of the energy budgets of forest and tundra has revealed differences of sufficient magnitude to warrant further study of the possibility that energy relationships on a gross forest–atmosphere scale are involved (Hare and Ritchie, 1972).

It is possible, however, that energy transfer relationships within the forest itself effectively contribute to maintenance of the forest where it now exists. The forest canopy undoubtedly serves as both a windbreak and an energy trap, resulting in higher soil- and ground-level air temperatures in the forest in late summer than occur in tundra, even when air temperatures at some short distance above forest and tundra vegetation are identical. Hence, the closed forest canopy tends to create a ground-level late summer environment more ameliorated in terms of growing conditions for mesophytic plants than that to which tundra vegetation is exposed, even on adjacent and similar topographic sites. Species comprising the understory community of a *Picea* forest are obviously capable of survival and reproduction in the forest environment, but many are equally obviously unable to advance into areas occupied by tundra. Even very small environmental differences must be critical, and these must account for the abrupt nature of the forest border (Larsen, 1973).

The active layer of soil appears to be deeper, hence late summer soil-surface temperatures are higher under the more closed-canopied *Picea* forest at the south end of Ennadai than under the small, isolated dwarfed *Picea* clumps at the north end (Larsen, 1965). The persistence of permafrost at shallow depths appears to render the environment a marginal one for *Picea* in the latter clumps, and permafrost levels may account for the failure of *Picea* clumps to regenerate after disturbance. The traditional concepts of forest succession must be inapplicable at the forest border since, once the forest has been eliminated, there appear to be no species, or aggregations of species, capable of ultimately creating conditions that will again permit ecesis by *Picea* and the *Picea* community, unless a change in climate has ameliorated the total environmental complex (see Larsen, 1965). The high heat-exchange capacity of water results in deeper active layers along shorelines, and this may account for the persistence of *Picea* northward along waterways and rivulets in regions where upland and inland areas are devoid of even dwarfed trees, except perhaps the rare relicts that can be seen to persist at times on what appear to be most unlikely upland sites. A number of other factors also known to affect the growth of trees at the forest border are presented by Savile (1963).

Of interest is the floristically depauperate zone in the forest–tundra transition; here there are few arctic species with ranges sufficiently far south to occupy the area, and there are fewer boreal species; many of those typical of boreal communities to the south are not found in the transition zone (Larsen, 1973). Since many arctic species probably are incapable of persisting beneath a *Picea* canopy, even though (improbably) climatic conditions might otherwise be favorable, it is likely that these species would be absent from an area recently occupied by *Picea* forest until sufficient time has elapsed for migration and recolonization. The evidence (Savile, 1956, 1964, and personal communication) indicates, however, that arctic species are capable of very rapid migration into environmentally suitable areas, since many opportunities exist for dispersal by animals and by physical events in arctic and subarctic regions.

In the light of this evidence, it appears that the absence of species of more arctic affinities at Ennadai, for example, cannot be attributed to insufficient time for migration. It is of interest in this regard, also, that *Diapensia lapponica*, an arctic species, was found atop an unusually high hill (400–500 ft above lake level) in the Ennadai area. Thus, at least one arctic species occupies a rare site which, microclimatically, must resemble arctic areas farther north. Significantly, it has failed to become a generally frequent component of the rock field communities at Ennadai, an ecological role it plays in the vegetation farther northward. One must assume that this, presumably, is because *Diapensia* is ill-adapted to the even very slightly more subarctic environment found on lower hills in the Ennadai area.

Finally, there exists one series of climatological observations that may bear importantly on the nature of the frontal zone in the region under consideration. McFadden (1965) and McFadden and Ragotzkie (1967) report on observations of dates of freeze-up and breakup of lakes in central Canada, providing maps showing (for various periods in spring and fall) the position of the zone in which some lakes are frozen and some are not (i.e., the zone between the line north of which all lakes are frozen and the line south of which all lakes are open). It is of interest that this zone is notably wider in the Keewatin area, in which the above vegetational data were obtained, than in areas farther west.

The floristically depauperate zone is correspondingly wider in the Keewatin area than it is in the area, for example, around the eastern arm of Great Slave Lake and Artillery Lake. In the Keewatin area, *Pinus banksiana* is absent from the forest for a considerable distance (perhaps 100 mi) south of the forest–tundra ecotone at Ennadai Lake, while it is found in abundance at one point along the portage between the east arm of Great Slave Lake and Artillery Lake. Additionally, *Rhododendron lapponicum* and *Dryas* species are found at Fort Reliance within the *Picea* forest, but to the east along the Kazan River they are not found until one travels many miles north of the forest–tundra ecotone. The same is true of other species of arctic affinity. It is thus apparent that, in the west, where the frontal zone characteristically occupies a narrower belt than in the east, boreal and arctic species overlap ranges, while in the east, where the frontal zone is wide, there exists a wide gap between species typically arctic or boreal in affinity. Here there exists a wide belt in which only the more ubiquitous species are found in sufficient abundance to appear regularly with high frequencies in transects. It will be of interest and perhaps of considerable ecological significance to explore these relationships further and attempt to determine the physiological characteristics that account for the distinctive response to the climatic conditions that prevail on the part of the species involved. Analyses of the energy budget profiles of a variety of boreal communities are much needed at the present time. Land use planning in northern forested regions should take into account the effects of removal of the forest canopy upon soil moisture and permafrost (Haag and Bliss, 1974); for in many areas, removal of forest cover in road construction and for other purposes quickly results in conditions that make the land quite unsuitable for the intended use.

REFERENCES

Ahti, T., Hämet-Ahti, L., and Jalas, J. (1968). Vegetation zones and their sections in northwestern Europe. *Ann. Bot. Fenn.* **5,** 169–211.
Bannister, P. (1970). The annual course of drought and heat resistance in heath plants from an oceanic environment. *Flora (Jena)* **159,** 105–123.
Bannister, P. (1971). The water relations of heath plants from open and shaded habitats. *J. Ecol.* **59,** 61–64.
Bannister, P. (1976). "Introduction to Physiological Plant Ecology." Wiley, New York.
Barry, R. G. (1967). Seasonal location of the Arctic front over North America. *Geogr. Bull.* **9,** 79–95.
Barry, R. G., and Chorley, R. J. (1968). "Atmosphere, Weather and Climate." Methuen, London.
Barry, R. G., and Hare, F. K. (1974). Arctic climate. In "Arctic and Alpine Environments" (J. D. Ives and R. G. Barry, eds.), pp. 17–54. Methuen, London.
Baumgartner, A. (1956). Untersuchungen über den Wärme- und Wasserhaushalt eines Junges Waldes. *Ber. Dtsch. Wetterdienstes* **5,** 4–53.

Baumgartner, A. (1970). Water and energy balances of different vegetation covers. *World Water Balance: Int. Assoc. Sci. Hydrol. Proc. Reading Symp.* pp. 56–65.
Bellefleur, P., and Auclair, A. N. (1972). Comparative ecology of Quebec boreal forests: A numerical approach to modeling. *Can. J. Bot.* **50**, 2357–2379.
Bendell, J. F. (1974). Effects of fire on birds and mammals. *In* "Fire and Ecosystems" (T. T. Kozlowski and C. E. Ahlgren, eds.), pp. 73–138. Academic Press, New York.
Benninghoff, W. S. (1952). Interaction of vegetation and soil frost phenomena. *Arctic* **5**, 34–44.
Berglund, E. R., and Mace, Jr., A. C. (1972). Seasonal albedo variation in black spruce and sphagnum–sedge bog cover types. *J. Appl. Meteorol.* **11**, 806–812.
Billings, W. D., and Mooney, H. A. (1968). The ecology of arctic and alpine plants. *Biol. Rev.* **48**, 481–529.
Borisov, A. A. (1959). "Climates of the U.S.S.R." (C. A. Halstead, ed.) Oliver & Boyd, Edinburgh.
Borisov, A. A. (1970). "Klimatografiya Sovetskogo Soyuza" (Climatography of the Soviet Union). Leningrad Univ., Leningrad.
Boughner, C. C., and Potter, J. G. (1953). Snow cover in Canada. *Weatherwise* **6**, 155–159.
Britton, M. E. (1966). Vegetation of the arctic tundra. *In* "Arctic Biology" (H. P. Hansen, ed.), 2nd ed. Oregon State Univ. Press, Corvallis.
Brown, R. J. E. (1960). The distribution of permafrost and its relation to air temperature in Canada and the U.S.S.R. *Arctic* **13**, 163–177.
Brown, R. J. E. (1969). Factors influencing discontinuous permafrost in Canada. *In* "The Periglacial Environment" (T. L. Péwé, ed.), pp. 11–54. McGill-Queen's Univ. Press, Montreal.
Brown, R. J. E. (1970a). "Permafrost in Canada." Univ. Toronto Press, Toronto.
Brown, R. J. E. (1970b). Permafrost as an ecological factor in the Subarctic. *Ecol. Subarct. Reg. Proc. Helsinki Symp.* pp. 129–139.
Bryson, R. A. (1966). Airmasses, streamlines, and the boreal forest. *Geogr. Bull.* **8**, 228–269.
Budyko, M. I. (1958). "The Heat Balance of the Earth's Surface." U.S. Weather Bureau, Washington, D. C.
Büttner, R. (1971). Untersuchungen zur Ökologie und Physiologie des Gasstoffwechsels bei einigen Strauchflechter. *Flora (Jena)* **160**, 72–99.
Cayford, J. H., and Haig, R. A. (1961). Glaze damage in forest stands in southeastern Manitoba. Can. Dep. For. For. Res. Br. Tech. Note 102.
Cayford, J. H., Hildahl, V., Nairn, L. D., and Wheaton, M. P. H. (1959). Injury to trees from winter drying and frost in Manitoba and Saskatchewan in 1958. *For. Chron.* **35**, 282–90.
Chapman, L. J., and Thomas, M. K. (1968). The climate of northern Ontario. Climatological Studies No. 6, Dept. Transport, Meteorolog. Branch, Ottawa, Canada.
Cleary, B. D., and Waring, R. H. (1969). Temperature: Collection of data and its analysis for the interpretation of plant growth and distribution. *Can. J. Bot.* **47**, 167–173.
Crampton, C. B. (1974). Linear-patterned slopes in the discontinuous permafrost zone of the central Mackenzie River valley. *Arctic* **27**, 265–272.
Dingman, S. L., and Koutz, F. R. (1974). Relations among vegetation, permafrost, and potential insolation in central Alaska. *Arct. Alp. Res.* **6**, 37–42.
Dolgin, I. M. (1970). Subarctic meteorology. *Ecol. Subarct. Reg. Proc.Helsinki Symp.* pp. 41–62.
Dunbar, M. J. (1976). Climatic change and northern development. *Arctic* **29**, 183–193.
Elliott, D. L. (1979). The current regenerative capacity of the northern Canadian trees, Keewatin, N.W.T., Canada: Some preliminary observations. *Arct. Alp. Res.* **11**, 243–251.

Elsner, R. W., and Pruitt, Jr., W. O. (1959). Some structural and thermal characteristics of snow shelters. *Arctic* **12**, 20-27.
Evans, L. T., ed. (1963). "Environmental Control of Plant Growth." Academic Press, New York.
Federer, C. A., and Tanner, C. B. (1966). Spectral distribution of light in the forest. *Ecology* **47**, 555-560.
Fraser, J. K. (1959). Freeze-thaw frequencies and mechanical weathering in Canada. *Arctic* **12**, 40-53.
Fritts, H. C., Blasing, T. V., and Kutzbach, J. E. (1974). Multivariate techniques for specifying tree-growth and climate relationships and for reconstructing anomalies in paleoclimate. *J. Appl. Meteorol.* **10**, 845-864.
Gates, D. M. (1969). Climate and stability. *In* "Diversity and Stability in Ecological Systems" (G. M. Woodwell and H. H. Smith, eds.), pp. 115-127. Brookhaven Natl. Lab., Brookhaven, New York.
Haag, R. W., and Bliss, L. C. (1974). Functional effects of vegetation on the radiant energy budget of boreal forest. *Can. Geotech. J.* **11**, 374-379.
Hare, F. K. (1950). Climate and zonal divisions of the boreal forest formation in eastern Canada. *Geogr. Rev.* **40**, 615-635.
Hare, F. K. (1954). The boreal conifer zone. *Geogr. Studies* **1**, 4-18.
Hare, F. K. (1959). A photo-reconnaissance survey of Labrador-Ungava. *Can. Dept. Mines Tech. Surv. Geogr. Branch Mem.* 6, 1-64.
Hare, F. K. (1968). The Arctic. *Q. J. R. Meteorol. Soc.* **94**, 439-459.
Hare, F. K. (1970). The tundra climate. *Trans. R. Soc. Can. Sect. 4* **7**, 32-38.
Hare, F. K., and Hay, J. E. (1971). Anomalies in the large-scale annual water balance over northern North America. *Can. Geogr.* **15**, 79-94.
Hare, F. K., and Hay, J. E. (1971). The climate of Canada and Alaska. *In* "World Survey of Climatology, Vol. 11, Climates of North America" (R. A. Bryson and F. K. Hare, eds.), pp. 49-192. Elsevier, Amsterdam.
Hare, F. K., and Ritchie, J. C. (1972). The boreal bioclimates. *Geogr. Rev.* **62**, 334-365.
Hare, F. K., and Taylor, R. G. (1956). The position of certain forest boundaries in southern Labrador-Ungava. *Geogr. Bull.* **8**, 51-73.
Hicklenton, P. R., and Oechel, W. C. (1976). Physiological aspects of the ecology of *Dicranum fuscescens* in the Subarctic. I. Acclimation and acclimation potential of CO_2 exchange in relation to habitat, light, and temperature. *Can. J. Bot.* **54**, 1104-1119.
Hicklenton, P. R., and Oechel, W. C. (1977). Physiological aspects of the ecology of *Dicranum fuscescens* in the Subarctic. II. Seasonal patterns of organic nutrient content. *Can. J. Bot.* **55**, 2168-2177.
Hopkins, D. M. (1959). Some characteristics of the climate in forest and tundra regions in Alaska. *Arctic* **12**, 215-220.
Jarvis, P. G., James, G. B., and Landsberg, J. J. (1976). Coniferous forest. *In* "Vegetation and Atmosphere" (J. L. Monteith, ed.), Vol. II, pp. 171-238. Academic Press, New York.
Johnson, E. A., and Rowe, J. S. (1974). Fire in the subarctic wintering ground of the Beverley caribou herd. *Am. Midl. Nat.* **94**, 1-14.
Kendrew, W. G., and Currie, B. W. (1955). "The Climate of Central Canada." Queen's Printer, Ottawa.
Kershaw, K. A., and Rouse, W. R. (1971). Studies on lichen-dominated systems. II. The growth pattern of *Cladonia alpestris* and *Cladonia rangiferina*. *Can. J. Bot.* **49**, 1401-1410.
Kornás, J. (1972). Corresponding taxa and their ecological background in the forests of temperate Eurasia and North America. *In* "Taxonomy, Phytogeography, and Evolution." (D. H. Valentine, ed.), pp. 37-59. Academic Press, New York.

Krebs, J. S., and Barry, R. G. (1970). The Arctic front and the tundra-taiga boundary in Eurasia. *Geogr. Rev.* **60**, 548-554.
Larcher, W. (1975). "Physiological Plant Ecology." Springer-Verlag, Berlin and New York.
Larsen, J. A. (1965). The vegetation of the Ennadai Lake area, N.W.T.: Studies in subarctic and arctic bioclimatology. *Ecol. Monogr.* **35**, 37-59.
Larsen, J. A. (1971a). Vegetational relationships with air mass frequencies: Boreal forest and tundra. *Arctic* **24**, 177-94.
Larsen, J. A. (1971b). Vegetation of Fort Reliance, Northwest Territories. *Can. Field Nat.* **85**, 147-78.
Larsen, J. A. (1973). Plant communities north of the forest border, Keewatin, Northwest Territories. *Can. Field Nat.* **87**, 241-248.
Larsen, J. A. (1974a). Ecology of the northern continental forest border. In "Arctic and Alpine Environments" (J. D. Ives and R. G. Barry, eds.), pp. 341-369. Methuen, London.
Larsen, J. A., and Barry, R. G. (1974b). Paleoclimatology. In "Arctic and Alpine Environments" (J. D. Ives and R. G. Barry, eds.), pp. 254-276. Methuen, London.
Larson, D. W., and Kershaw, K. A. (1975). Acclimation in arctic lichens. *Nature (London)* **254**, 421-423.
Lechowicz, M. J. (1978). Carbon dioxide exchange in *Cladina* lichens from subarctic and temperate habitats. *Oecologia* **32**, 225-237.
Lee, H. A. (1959). Surficial geology of southern district of Keewatin and the Keewatin Ice Divide, Northwest Territories. *Geol. Surv. Can. Bull.* **51**, 1-42.
Lee, R. (1978). "Forest Microclimatology." Columbia Univ. Press, New York.
Lettau, H. (1969). Evapotranspiration climatonomy: A new approach to numerical predictions of monthly evapotranspiration, runoff and soil moisture storage. *Mon. Weather Rev.* **97**, 691-699.
Lettau, H., and Lettau, K. (1975). Regional climatonomy of tundra and boreal forest in Canada. In "Climate of the Arctic" (G. Weller and S. Bowling, eds.), pp. 210-221. Univ. of Alaska, Fairbanks.
Longley, R. (1972). "The Climate of the Prairie Provinces." Climatolog. Studies No. 13, Atmos. Environ. Serv. Dep. Environ., Canada, Ottawa.
Lydolph, P. E. (1977). "Climates of the Soviet Union: World Survey of Climatology," Vol. 7. Elsevier, Amsterdam.
McFadden, J. D. (1965). The interrelationships of lake ice and climate in central Canada. Tech. Rep. 20, ONR Contract 1202(07), Dept. Meteorol., Univ. of Wisconsin, Madison.
McFadden, J. D., and Ragotzkie, R. A. (1967). Climatological significance of albedo in central Canada. *J. Geophys. Res.* **72**, 1135-1143.
MacFarlane, J. D., and Kershaw, K. A. (1978). Thermal sensitivity in lichens. *Science* **201**, 739-741.
MacHattie, L. B., and McCormack, R. J. (1961). Forest microclimate: A topographic study in Ontario. *J. Ecol.* **49**, 301-323.
Miller, W. S., and Auclair, A. N. (1974). Factor analytic models of bioclimate for Canadian forest regions. *Can. J. For. Res.* **4**, 536-548.
Mooney, H. A., and Billings, W. D. (1961). Comparative physiological ecology of arctic and alpine populations of *Oxyria digyna*. *Ecol. Monogr.* **31**, 1-29.
Munn, R. E. (1970). "Biometeorological Methods." Academic Press, New York.
Newnham, R. M. (1968). A classification of climate by principal component analysis and its relationships to tree species distribution. *For. Sci.* **14**, 254-264.
Nicholson, J., and Bryant, D. G. (1972). "Climate zones of Insular Newfoundland: A Principal Component Analysis." *Can. For. Serv. Publ.* **1299**, 1, 13.

Pearson, Arthur M., and Nagy, J. (1976). The summer climate at Sam Lake, Yukon Territory. *Arctic* **29**, 159–164.
Péwé, T. L. (1966). "Permafrost and Its Effects on Life in the North." Oregon State Univ. Press, Corvallis.
Pruitt, W. O., Jr. (1957). Observations on the bioclimate of some taiga mammals. *Arctic* **10**, 131–138.
Pruitt, W. O., Jr. (1959). Snow as a factor in the winter ecology of the barren ground caribou (*Rangifer arcticus*). *Arctic* **12**, 159–179.
Pruitt, W. O., Jr. (1970a). Some aspects of the interrelationships of permafrost and tundra biotic communities. *In* "Productivity and Conservation in Northern Circumpolar Lands," (W.A. Fuller and P. G. Kevan, eds.) pp. 33–41. Int. Union for Conservation of Nature and Natural Resources, Morges, Switzerland.
Pruitt, W. O., Jr. (1970b). Some ecological aspects of snow. *Ecol. Subarct. Reg. Proc. Helsinki Symp.* pp. 83–100.
Pruitt, W. O., Jr. (1978). "Boreal Ecology." Arnold, London.
Rauner, Yu, L. (1977). "Heat Balance of the Plant Cover." Amerind, New Delhi, India.
Reed, R. J. (1960). Principal frontal zones of the Northern Hemisphere in winter and summer. *Bull. Am. Meteorol. Soc.* **41**, 591–598.
Reed, R. J., and Kunkel, B. A. (1960). The arctic circulation in summer. *J. Meteorol.* **17**, 489–506.
Ritchie, J. C., and Hare, F. K. (1971). Late-Quaternary vegetation and climate near the arctic treeline of northwestern North America. *Quat. Res.* **1**, 331–342.
Rouse, W. R. (1976). Microclimatic changes accompanying burning in subarctic lichen woodland. *Arct. Alp. Res.* **8**, 357–376.
Rouse, W. R., and Kershaw, K. A. (1971). The effects of burning on the heat and water regimes of lichen-dominated subarctic surfaces. *Arct. Alp. Res.* 3, 291–304.
Rowe, J. S. (1972). "Forest regions of Canada." *Can. For. Ser. Publ.* **1300**.
Savile, D. B. O. (1956). Known dispersal rates and migration potentials as clues to the origin of the North American biota. *Am. Midl. Nat.* **56**, 434–453.
Savile, D. B. O. (1963). Factors limiting the advance of spruce at Great Whale River, Quebec, Canada. *Can. Field Nat.* **77**, 95–97.
Savile, D. B. O. (1964). North Atlantic biota and their history. *Arctic* **17**, 138–141.
Savile, D. B. O. (1972). "Arctic adaptations in plants." *Can. Dep. Agric. Res. Br., Monogr.* **6**, 1–81.
Schomaker, C. E. (1968). Solar radiation measurements under a spruce and birch canopy during May and June. *For. Sci.* **14**, 31–38.
Sjörs, H. (1963). Amphi-Atlantic zonation, nemoral to arctic. *In* "North Atlantic Biota and Their History"(A. Löve and D. Löve, eds.), pp. 109–125. Oxford Univ. Press, London and New York.
Stolfelt, M. G. (1963). On the distribution of the precipitation in a spruce stand. *In* "The Water Relations of Plants " (A. J. Rutter and F. H. Whitehead, eds.), pp. 115–126. Wiley, New York.
Stebbins, G. L. (1974). "Flowering Plants: Evolution above the Species Level." Harvard Univ. Press, Cambridge, Massachusetts.
Streten, N. A. (1969). Aspects of winter temperatures in interior Alaska. *Arctic* **22**, 403–412.
Streten, N. A. (1974). Some features of the summer climate of interior Alaska. *Arctic* **27**, 273–286.
Stupart, R. F. (1928). The influence of Arctic meteorology on the climate of Canada especially. *In* "Problems of Polar Research." Am. Geogr. Soc. Special Publ. No. 7. Am. Geogr. Soc., New York.

Suslov, S. P. (1961). "Physical Geography of Asiatic Russia." (N. D. Gershevsky, trans.; J. E. Williams, ed.). Freeman, San Francisco, California.
Szeicz, G., Petzold, D. E., and Wilson, R. G. (1979). Wind in the Subarctic forest. *J. Appl. Meteor.* **18,** 1268–1274.
Tajchman, S. J. (1972). The radiation and energy balances of coniferous and deciduous forests. *J. Appl. Ecol.* **9,** 359–375.
Taylor, K. (1971). Biological flora of the British Isles. *Rubus chamaemorus* L. *J. Ecol.* **59,** 293–306.
Tedrow, J. C. F. (1977). "Soils of the Polar Landscapes." Rutgers Univ. Press, New Brunswick, New Jersey.
Thomas, M. K. (1953). Climatological atlas of Canada. Natl. Res. Counc. Canada, Div. Build. Res., NRC 3151, Ottawa.
Tibbals, E. C., Carr, E. K., Gates, D. M., and Kreith, F. (1964). Radiation and convection in conifers. *Am. J. Bot.* **51,** 529–538.
Tikhomirov, B. A. (1970). Forest limits as the most important biogeographical boundary in the north. *Ecol. Subarct. Reg. Proc. Helsinki Symp.* pp. 35–40.
Treshow, M. (1970). "Environment and Plant Response" McGraw-Hill, New York.
Turnock, W. J. (1955). A comparison of air temperature extremes in two tamarack sites. *Ecology* **36,** 509–511.
Tyrrell, J. B. (1897). Report on the Doobaunt, Kazan and Ferguson rivers and the northwest coast of Hudson Bay. *Geol. Surv. Can. Annu. Rep., New Ser.* **9,** Rept. "F," pp. 5–218.
Vezina, P. E., and Pech, G. (1964). Solar radiation beneath conifer canopies in relation to crown closure. *For. Sci.* **10,** 443–450.
von Humboldt, A. (1807). "Ideen zu einer Geographie der Pflanzen nebst einem, Naturgemalde der Tropenlander." Tubingen.
Vowinckel, T., Oechel, W., and Boll, W. (1975). The effect of climate on the photosynthesis of *Picea mariana* at the subarctic tree line. *Can. J. Bot.* **53,** 604–620.
Walker, J. C. G. (1977). "Evolution of the Atmosphere." Macmillan, New York.
Warren Wilson, J. (1967). Ecological data on dry-matter production by plants and plant communities. *In* "The Collection and Processing of Field Data" (E. F. Bradley and O. T. Denmead, eds.), pp. 77–127. Wiley (Interscience), New York.
Wilson, R. G., and Petzold, D. E. (1973). A solar radiation model for sub-arctic woodlands. *J. Appl. Meteorol.* **12,** 1259–1266.

4   *The Boreal Soils Subsystem*

Soil and air are two distinctly different aspects of the environment. The biosphere exists at the interface between the two—as a layer of vegetation, roots in soil and leaves in air, with accompanying microbial and animal life—sandwiched between the underlying substrate and overlying atmosphere. At this soil–atmosphere boundary layer, organisms are born, grow, reproduce, absorb and later release soilborne nutrients, fix nitrogen, capture solar energy, and otherwise perform life functions as members of species subject to evolutionary changes during the succession of generations. The principal macroenvironmental edaphic factors, namely, soil composition, texture, and structure, are at least partially dependent upon climate—soil formation is influenced by rainfall, temperature, and other climatic factors—and only the geological characteristics of the parent material can be said to be entirely independent of other environmental factors. Atmosphere and soil are, thus, interrelated aspects of the environment of boreal plants, but it is clearly evident that, of the two, soil is by far the more complex, involving a myriad of biochemical as well as geochemical processes. The atmospheric chemical elements and compounds important in plant metabolism are largely limited to oxygen, nitrogen, water, and carbon dioxide. In soil are found many mineral nutrient elements, as well as compounds containing carbon and nitrogen fixed from the atmosphere by the biological processes of photosynthesis and nitrogen fixation. In summary in the atmosphere are found: gaseous elements and compounds, namely, oxygen, nitrogen, carbon dioxide, and water. In the soil are found: macronutrient mineral elements such as potassium, calcium, magnesium, phosphorus, and sulfur, elements fixed from atmosphere, i.e., nitrogen, carbon, and water, micronutrient mineral elements, for example, magnesium, copper, zinc, manganese, iron, and boron, and nonessential mineral elements, namely, aluminum and barium. The natural environment of most higher plants is, thus, essentially a two-phase system, with an exchange of energy and materials and maintenance of an equilibrium between the two phases. Roots and aboveground organs of plants are in close physiological contact. Soil and atmosphere are in chemical and physical equilibrium. The soil is the site of many complex chemical transformations, a large proportion biochemical in nature, for example, *nitrification, denitrification,* and *nitrogen fixation,* as well as *decomposition,* principally of cellulose, lignin, and the many other organic compounds that ultimately decay in or on the soil.

The objective of any study of soil as a subsystem of the boreal ecosystem is to systematize information available on mineral nutrients in soil and to describe the important aspects of what is known about cycling of nutrients in the boreal forest system. To do so, however, it is necessary first to understand and describe the physical and chemical characteristics of boreal soils (see Table I.1, Appendix I). Simple geochemical data on the occurrence of elements in soils, however, are of limited value because they do not indicate the amount of each element existing in chemical form available to plants. Such data, nevertheless, possess significance because they indicate the relative abundance of elements in soil materials and reveal marked excesses or deficiencies in the elements important to plants. It is the cycling of elements in forms available for plant nutrition, however, that ultimately is of most significance to an understanding of plant growth. Plants participate in biogeochemical cycling of elements, including uptake of elements by plant roots, translocation from roots to leaves and other organs, incorporation into the organic compounds of plants, and eventual return of elements to the soil. The physical and chemical structure of boreal soils can initially be described using classical concepts and traditional edaphic terms, and then, imposed on this framework, the chemical and biochemical processes of nutrient cycling, continually in progress in all soils, can be described.

Soils have length, breadth, and depth. In describing a soil it is common practice to refer to its *profile,* the exposed face of soil in a pit, in which different strata can be recognized by differences in color. Features developed during soil formation are described in terms of horizons designated by the letters O, A, B, and C. The O horizon is composed principally of decaying organic matter and large numbers of microorganisms. The A horizon is the uppermost mineral horizon. It is subject to intense weathering and leaching. The B horizon, immediately below the A horizon, contains most of the material leached from the A horizon. The C horizon represents the underlying geological material. Soils vary greatly in the character of these principal horizons, and soil descriptions are essentially summary statements of similarities and differences in terms of structure and appearance of the horizons.

On subsequent pages, a description is undertaken of the methods of soil classification followed in this chapter—the Canadian System of Soil

Classification (Canada Soil Survey Committee, 1978)—and readers unfamiliar with this system are advised to consult the publication. The system is quite different from that used in the United States (Soil Survey Staff, 1967) and in many ways is more tractable for use with boreal soils; it is, moreover, the system employed by the North American country in which boreal soils are found in greatest abundance, and it would be reasonable to follow this system even if it were not superior to the Seventh Approximation of the U.S. Soil Survey.

The soil type widespread in boreal regions is, thus, podzol, a soil that develops characteristically on sand or materials derived from coarse-grained rock, of which granite is perhaps the most widespread example. During podzol formation, soluble materials such as sodium, potassium, and calcium are washed out of the soil by water movement. Iron and aluminum are removed from the A horizon and deposited at depths of a foot or so in the B horizon. There is an accumulation of litter and raw humus on the soil surface. Collecting in the top stratum of the B horizon, termed the Bf, are deposits of decayed organic material. The main zone of iron and aluminum accumulation is also in the Bf horizon. The surface organic material is referred to as the O horizon, a distinction being made between the Of litter layer and the Oh black humus layer. The leaching of iron is not due simply to movement of water but also to substances in plant material capable of reducing iron to the ferrous state. The litter from some plant species leads to more rapid formation of podzols than litter from other species. This evidently is a consequence of the fact that substances known as chelating compounds are present in the litter accumulating on the soil surface; these are derived from the leaves of a number of, if not all, northern coniferous species and probably many shrub species as well, and they have the ability to take up iron and aluminum in what might be termed a relatively loose grip—releasing the metals after the complex has been washed into the upper B horizon where chemical conditions are different than the conditions on the soil surface. The process is understood at least in outline—but not very well in detail—and is probably one of the more important factors in the rapid development of distinct podzol soil horizons beneath forests of coniferous trees.

GENESIS OF PODZOL SOILS

The soils of the boreal forest are so closely interrelated with the other dominant factors of the boreal ecosystem—climate and vegetation—that, when considering the genesis of podzol soils, all three must be considered together as a multifactorial complex. Each aspect of the soil complex can, in some measure, be considered to possess the attributes of both dependent and independent variables, each exerting an influence upon the others and, in turn, being influenced by them. Climate, vegetation, soil parent material, and topography establish the environmental influences that in combination produce typical boreal podzols and associated podzolic soils.

Podzol is the most characteristic soil of the boreal forest ecosystem, although it is a soil type (the result of the complex process of *podzolization*) not confined exclusively to boreal regions. Climate, vegetation, chemical composition of the substrate material, and topography are the environmental influences that produce typical boreal podzol and associated podzolic soils. In outline, podzolization is the movement of iron, aluminum, and organic materials, including organic chelating agents, from an upper A soil horizon to a lower B horizon, and the maintenance then of these characteristic zones in a steady-state equilibrium (Fig. 6).

Intense podzolization occurs throughout the regions of subarctic coniferous forest, in climates characterized by fairly high levels of precipitation and generally low rates of evaporation. Podzolization is particularly intense on acid sandy glacial deposits and weathered sandstone materials and leads typically to the formation of a bleached gray Ae horizon overlying a brown or reddish-brown B horizon (see Tables I.1, I.2, I.3, Appendix I). In areas where processes of podzolization are less intense, where parent materials are higher in clay content or are calcareous in origin, soils more typically belong to the brunisolic soil order or the gray-brown or gray-wooded luvisols. In these soils, a brown or light-brown upper A horizon overlies a brown B horizon enriched in clay as well as organic materials and iron and aluminum oxides (see Table I.2, Appendix I).

To envision soil formation in the sub-Arctic, it is useful to begin by following the process of soil genesis from the time barren rock undergoes the first steps in decomposition until the final stages in the long process of soil formation have been accomplished. Podzolization, we have already seen, is the movement of iron and aluminum compounds, as well as of organic materials, from the A to the B horizon, but before movement of these materials can be accomplished they must—in the case of inorganic materials—be released from parent rocks, and—in the case of organic materials—be manufactured by the activity of living organisms. Weathering is the term used for the gradual breakdown of rock into the fine-grained clay minerals that are the ultimate product; weathering is accomplished by physical action, by chemical processes, and by biochemical activities of organisms. Freezing and thawing of rocks with an appreciable moisture content results in cracking and splintering of larger into smaller pieces. The process exposes surfaces to the atmosphere, with the result that mineral compounds are oxidized, eventually producing the colloidal clays known as kaolinite and montmorillonite—oxides of silicon and aluminum, with an admixture of

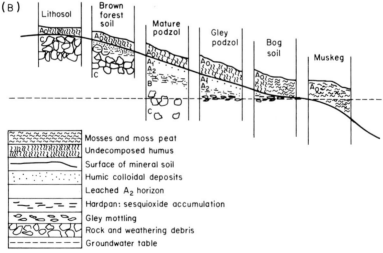

Fig. 7. Peat can accumulate to great depth in northern boreal regions. This peat deposit, exposed by erosion, cracking, and caving, is located along the eastern shore of Colville Lake in Mackenzie Territory, Northwest Territories. Dense spruce forest, not visible here, grows on the surface of the peat deposit and comes to within a few yards of the edge of the peat bank shown above. The photograph was taken from the sandy shoreline of the lake.

Fig. 6A. Morphology of basic boreal soils. Shown is a mature boreal podzol soil typical of the development found on well-drained uplands. The upper horizon is predominantly organic material in various stages of decomposition. The light-colored area just below the upper horizon is the leached layer; it is composed largely of mineral soil from which the iron and aluminum oxides, as well as clay minerals, have been removed. The darker area below the leached horizon is the area of accumulation of the oxides and clay. Below this is the parent material. From Weetman and Webber (1972); photo courtesy G. F. Weetman. Reproduced by permission of the National Research Council of Canada.

Fig. 6B. A diagram of the morphology of the basic boreal soils. At the left is a lithosol, dry and weathered only to a minimal degree, and at the right is the muskeg composed entirely of organic material. The mature podzol, center, has an upper horizon predominantly of organic material in various stages of decomposition. The light-colored area just below the upper horizon is the leached layer, mineral soil from which the iron and aluminum oxides, as well as clay minerals, have been removed. The darker area below the leached horizon is the area of accumulation of oxides and clay. Below this is the parent material.

a wide range of other compounds, including oxides of iron, calcium, magnesium, manganese, and other elements. Decay of organisms that colonize the weathering rock surfaces—bacteria, algae, fungi, lichens, mosses—adds organic matter to the surface of the forming soil. Carbon dioxide given off in respiration of the organisms forms carbonic acid with water, and this speeds the dissolution of minerals of parent rock. Bacterial action markedly hastens weathering processes. Seeds of higher plants eventually find the sites suitable for germination, and roots take hold in upper organic layers, growing into the loose material formed by the weathering action. The soil horizons become more and more distinctly visible, with colors pronounced as time goes by. Eventually, in subarctic regions, the soil deepens, the surface organic layer thickens (Fig. 7), the Ae horizon becomes leached and white in appearance, the B horizon becomes reddish in appearance, and the sand grains become cemented together to form a hardpan. A typical podzol has formed.

THE CHEMICAL PROCESSES

Substantial information has accumulated on the chemical mechanisms involved in podzolization. Alterations of the parent material occur as a result of continual contact with oxygen, carbon dioxide, water, and other substances. Inorganic reactions involve oxidation, reduction, hydration, hydrolysis, carbonation, and a wide variety of exchange reactions. Microorganisms strongly influence the chemical and physical weathering processes. The end products of hydration, hydrolysis, and exchange reactions usually include soluble salts and hydrated aluminum silicates referred to as clay minerals. Soluble materials, clays, organic colloids, and iron and aluminum oxides are leached from the A horizon to the B horizon. Soluble materials such as calcium and magnesium are usually removed entirely from both these horizons by percolating water, whereas clay, colloids, and certain organic compounds are precipitated or flocculated in the B horizon and accumulate there. Since silica and alumina usually make up more than 85% of the parent material, their fate, as well as that of the associated iron and other metals, becomes of critical importance. The basic reaction involved in the formation of hydrated aluminum silicates from parent material is

$$RAl_2Si_2 + H_2O \rightarrow (Al_2Si_2O_7\ldots OH)^-\cdot nR^{2+} + H^+$$

The R^{2+} then forms a salt with ions available in percolating water. The hydrated silicates in colloidal form are carried to the B horizon, flocculate, and attach to the surface of larger particles. The iron silicate compounds are similarly deposited:

$$Fe_2Si_2O_4 + H_2O \rightarrow (SiO \ldots OH)^- \cdot nFe^{2+} + H^+$$

Migration of ferrous ions can take place under reducing conditions, and movement of iron also occurs in the form of ferrous hydroxides. Migration of aluminum can take place in both forms, although it probably occurs largely in hydroxy compounds.

In podzol soils, migration also occurs by means of another mechanism, the ligation of iron and aluminum in the bonds of chelating compounds found abundantly in decomposing litter on the soil surface. Water-soluble materials from conifer needles form chelate complexes with iron and aluminum, and these are carried to the B horizon in percolating water. One of the important chelates involved in this process is evidently a compound known as fulvic acid.

The stability of acid-chelating substances is low, and the iron-dissolving power decreases with an increase in pH. As the materials wash down into less acid soil horizons, the chelate breaks apart and the iron is deposited as $Fe(OH)_3$. The colloids accumulate in the B horizon and coat the silt and sand particles, cementing them into a hardpan. The final result is a surface layer of unincorporated humus, a leached, impoverished Ae horizon, and a dark B horizon.

THE GEOGRAPHY OF PODZOLS

Processes leading to formation of the different soil types are not mutually exclusive; there is some clay in podzol horizons, and some migration of organic material and iron from the Ae of gray-brown and gray-wooded soils. Dominance of one process over the other, however, does result in soils possessing markedly different characteristics. The most intense podzolization is encountered in the more northern boreal areas. Soils of the gray-brown and gray-wooded groups are more characteristic of the southern boreal zone.

In highly podzolized soils the upper mineral Ah and Ae horizons are often nearly depleted of all but organic and siliceous materials. In soils with lesser degrees of podzolization, or on substrates other than the sandy types, a wide variation in the characteristics of the upper horizons is found. The period of time over which weathering processes have been at work is a factor of some consequence. Slightly weathered basalt may yield a loam soil; highly weathered basalt yields a clay. Slightly weathered limestone yields a stony loam; highly weathered limestone yields a red clay. Granite produces a stony, sandy loam in the early stages of weathering and a sandy clay in the ultimate stages. Time, texture, and the proportion of soluble material in the geological parent material are important factors in soil genesis. Microorganisms often play a central role in weathering, with chemical and physical processes mainly modifying microbial activities.

The changes that take place in the formation of classic sandy podzol soils occur largely on the surface of the sand grains. In the Ae and B horizons, for example, the bulk of a sandy soil consists of quartz grains, with a dark reddish-brown coating over the grains in the B horizon. An absence of such coatings on the grains of the Ae horizon gives this horizon its ash-gray appearance. These coatings, or colloidal skins as they are sometimes called, permit podzol soils both to form quickly and to fade relatively rapidly. This situation is quite different from that in gray-brown podzolic and gray-wooded soils, in which the clay-enriched B horizons are relatively stable and permanent.

Basically, two broad principles have been derived from studies on podzol soils. First, the longer the period of elapsed time since the raw parent material became exposed to influences of atmosphere and vegetation, the less apparent the contribution of the parent material to the soil horizons, since the degree of modification is in direct proportion to the elapsed time. Second, under extreme climatic conditions the influence of the parent material upon the soil has a tendency to be minimal. The upper mineral layers of mature podzols are composed of nearly pure silica regardless of the nature of the parent material.

The name *podzol* is derived from a Russian word meaning "under-ash," and initially the term referred not to the entire soil profile but more specifically to the light-colored zone now designated the Ae horizon. Dokuchaiev (1879) suggested that the definition be restricted to the soil type "formed mainly in forests with a significant participation of bog and forest vegetation." Early soil workers confined their attention to the Ae horizon, but a description by Georgievsky (1888) reveals that certain fundamental aspects of podzolization already had been recognized:

> The influence of climate is shown by the geographical distribution of the podzol. Judging by the available literature on the *white earth*, in contrast to the chernozem which is characteristic of the open steppe, it is encountered (at least in its typical form), exclusively in the north and northwest of Russia, i.e., in places with relatively high rainfall, and with an abundance of both forests and bogs. It should not be forgotten that just such an abundance of moisture is one of the essential conditions for podzol formation.

The existence of geographical variation in podzolic soils was first recognized by Glinka (1908), who inferred that podzolization would be carried to varying degrees of completion under differing environmental conditions. He stated that the bleached horizon would be more-or-less strongly developed depending upon prevailing temperatures, duration of growing period, and amount of precipitation, all of which influence the rate of decomposition of organic material accumulating on the soil surface. Clarifying a discussion current in his day concerning the process by which the upper zone becomes bleached, Glinka (1924) expressed the opinion that "in general the podzolic process is basically none other than the leaching from the upper horizons of fine mineral suspensions and the deposition of these suspensions in the B horizons together with a small amount of humus."

Over the enormous extent of the sub-Arctic, across North America and Eurasia, there exists a wide range of variation in the environmental factors that are significant in the formation of soils from one area to another, although the podzolization process and the podzolic characteristics of soils are unifying attributes. The terms "boreal" and "podzol" are, thus, by no means synonymous. Within the boreal zone are found not only soils that are more accurately described as regosols, gleys, peats, brown earth soils, and gray- and brown-wooded soils, but also essentially azonal soils such as lithosols. These differences are often the result of variations in topography and parent materials, which have markedly influenced the podzolization processes.

There are no sharp, easily distinguished changes in soils or major soil-forming processes at work across either the ecotone marking the southern edge of the northern coniferous forest or the northern transition zone where the forest grades into northern tundra. The differences, rather, are subtle and involve such factors as the differing influence upon soil formation of deciduous leaves and of the litter formed by northern conifers. The slower rate of organic decomposition northward is also a factor differentiating northern soils from soils found elsewhere. Indeed, it has been affirmed that the single most distinguishing characteristic of subarctic soils is the amount of organic matter present over the mineral substrate. Well-drained uplands in the boreal forest usually have more organic matter present on the surface than is found either in forests to the south or on tundra to the north (Fig. 8). A generalized map of the dominant characteristics of soils found throughout the circumpolar range of the sub-Arctic is presented in the useful publications by Tedrow (1970, 1977).

The vegetation of the extreme southern limit of the boreal forest can be described as a transition to deciduous forest, with species associated with deciduous forest communities becoming progressively more dominant southward (see Chapter 2). Along with the transition in vegetation

Fig. 8. This soil was found on the slope of a sandy hill at the northern edge of the forest–tundra ecotone and possesses (albeit in somewhat miniature form) all the typical features of a boreal forest podzol; a dark, heavily organic upper surface layer, an ash-gray leached horizon just below the surface layer, and a darker zone where iron and aluminum oxides have accumulated. Podzols occur in a wide variety of forms, from this well-drained type to those usually found in more moist sites where they develop a deep surface horizon of undecomposed humus and accumulated peat.

type is a transition in soils, with podzols giving way to soils characteristic of temperate deciduous forests. Along with the morphological changes that occur over the transition are changes in the intensity of various processes associated with soil genesis, resulting in differences in soil structure and appearance. Decomposition of litter is more rapid. Leaves of temperate deciduous trees are richer than the needles of northern conifers in calcium and other bases, and these materials are rapidly recycled in leaf fall, decay, reabsorption by roots, and movement to leaves again. In the deciduous forest, upper soil layers are enriched in rather than leached of nutrients as in the podzols, and there is a more general dispersal of partly decomposed organic matter. The resulting gray-wooded and gray-brown podzolic soils found in the transition zone, as well as the more southerly brown forest soils of the temperate deciduous forest zone proper, are distinctly different from the northern podzols.

Throughout most of the wide extent of boreal forest in northern Canada, and of Eurasia as well, studies indicate that there generally are large amounts of undecomposed organic material on the surface of soils in both the spruce–feather moss forests and the open spruce–lichen woodlands, as well as in associated lichen–heath tundra communities at the northern edge of the forest–tundra transition zone. Moore (1974), for example, made studies of soils at Cambrian Lake in northern Quebec (lat. 56°30′N, long. 69°15′W), an area where nine vegetational associations representative of both boreal and arctic environments could be easily recognized: These included lichen–heath and shrubby tundra communities on uplands, *Picea*-lichen woodlands on lakeshores and sides of valleys, *Picea*-feather moss forests, and small stands of deciduous trees on wetter valley floors.

In four types of podzolized soils found in the area, Moore found that, while the A and B horizons involved (Ae, Bf) were thin, podzolization processes had been effective in translocating iron, aluminum, and carbon. Soils with podzol features were observed in the area under a wide range of vegetational communities: lichen–heath tundra, tree line shrub, deciduous tree stands, *Picea*-moss forests, and *Picea*-lichen woodlands. The intensity of podzol development, however, was primarily dependent upon parent material, being greatest in quartzite-derived material and least on gneiss-derived deposits with a higher content of ferromagnesian minerals. In general the parent material in the Cambrian Lake area was favorable for podzol development: coarse, freely drained, with ample water for leaching, and with a fairly thick vegetational cover. In wet areas where gley soils might be expected to develop, it appeared that the occurrence of permafrost at shallow depths inhibited the development of reducing conditions necessary for the formation of typical gley soil morphology.

In another study, carried out about 200 km southeast of Cambrian Lake in an area where iron was an abundant constituent of the parent material, Nicholson and Moore (1977) found that podzolization processes were active but weaker than in the Cambrian Lake area and that gleying was more pronounced on favorable waterlogged sites. This area, on the southern edge of the forest–tundra transition zone, is occupied by *Picea*-lichen woodlands on well-drained sites, closed *Picea*-lichen woodlands, and *Picea*–feather moss forests in depressions and valleys. *Picea mariana, Picea glauca,* and *Larix laricina* are dominant trees, with occasional *Abies balsamea.* Lichens and mosses are heavily represented in the ground-layer vegetation, and *Betula glandulosa* and *Ledum groenlandicum* dominate the shrub layer. Trees are stunted and sparse on ridge crests, and here shrubs such as *Vaccinium uliginosum* and *Arctostaphylos alpina* are dominant. No trees are found on the more exposed ridge tops. Muskeg and fen are found in the waterlogged depressions. Nicholson and Moore observed that soils developed from the iron-rich substrate presented difficulties in recognition because of the strong red color found throughout and the consequently indistinct visual differentiation of the horizons. Chemical analyses of the soils, however, showed that iron oxides had moved downward but that relatively high amounts of iron remained in the Ae horizon. Clay minerals had been leached rather thoroughly away from the A and deposited in the Bf horizon. Chemical analyses appeared essential to identify the soils as podzols, since morphological appearances were deceptive if employed without corroborative chemical information. The five soils studied were identified according to the Canadian classification system as an orthic humoferric podzol, a mini humoferric podzol, a degraded dystric brunisol, a gleyed humoferric podzol, and a fera gleysol.

SOIL CLASSIFICATION: BOREAL REGIONS

With the above use of soil names employed in the Canadian soil classification system, it is appropriate to make further mention of the existence of the Canadian System of Soil Classification—but the reader is well-advised to consult the original publication (Canada Soil Survey Committee, 1974), since the system is both comprehensive and complex and cannot be described here in adequate detail. The literature on northern soils is too voluminous to review in a few pages, but any consideration of the soil-forming factors at work would reveal that horizon differentiation is brought about by a combination of the weathering of minerals, accumulation and decay of organic matter, and development of a recognizable soil structure. The generally accepted taxonomic system places soils in categories at different levels of generalization on the basis of horizon characteristics. Six categories are used:

Category 1: The soil type: A subdivision based on the texture of the surface soil.
Category 2: The soil series: the basic unit of classification.

Category 3: The soil family: a grouping of series on the basis of important characteristics of parent material.
Category 4: The subgroup: based on types of profiles long recognized—podzol, chernozem, etc.
Category 5: The great group: a grouping of the subgroups.
Category 6: The order: an organization of the great groups into orders on the basis of major profile similarities.

A brief description here of six orders, on the basis of the soil system adopted in Canada (Ehrlich, 1974), will show the principles on which the classification of boreal forest soils is established.

Luvisolic Order

Developed from basic parent materials under forest, these soils have an impoverished gray layer near or at the surface of the mineral soil, underlain by a dark subsurface horizon enriched in clay. Two great groups in this order are the gray-brown luvisol (gray-brown podzolic) and the gray-wooded soils.

Podzolic Order

These soils are the most common in forest areas of the Canadian shield. They have a gray layer near or at the surface of the mineral soil. The B horizon has an accumulation of organic matter and iron and aluminum oxides. The order includes three great groups: humic podzol, ferrohumic podzol, and humoferric podzol.

Brunisolic Order

Some well-drained soils in the forest region lack the leached gray horizon or the enriched B horizon. These soils are brown and are referred to as brunisolic soils. Four great groups are recognized: melanic brunisol (brown forest) and eutric brunisol (brown-wooded) on calcareous parent material with a high-base status, and sombric brunisol (acid brown forest) and dystric brunisol (acid brown-wooded) on noncalcareous parent material with a low-base status.

Regosolic Order

Well-drained soils lacking noticeable horizon development and with only slightly modified mineral parent material.

Gleysolic Order

Poorly drained soils in which water and lack of oxygen have created reducing conditions. Gleysolic soils include three great groups: humic gleysols (mineral-organic surface horizons), eluviated gleysols (podzolic features), and gleysols (moist or wet regosols).

Organic Order

Widespread in forest and tundra regions, these soils develop under wet conditions and possess an organic layer of considerable depth. There are three great groups: fibrisol, mesisol, and humisol, based on degree of decomposition of organic material.

While soils of the boreal forest do not possess characteristics unique to the region, if a single character were to be considered most representative of northern subarctic soil, it would be the formation of a thick organic mat on the surface. Secondarily, the development of Ae and Bf horizons is of great significance. Over large areas of Canada and elsewhere it is also of pedogenic importance that soils have developed where Precambrian rocks underlie a thin mantle of unconsolidated sand and gravel, and in many areas these rocks often outcrop on ridges or hilltops, furnishing sites on which soils are still in early stages of development if they have progressed much beyond the regosol form. In areas occupied by glacial drift, particularly well-drained uplands and gentle slopes, typical podzol soils develop. There may be some tendency for gray-wooded soils to develop more extensively on fine-textured soils of higher base content than is characteristic of sands derived from Precambrian bedrock, but even in such areas as the so-called clay belt of Ontario the development of podzol soil is of widespread occurrence and appears to be the result of abundant precipitation, optimum drainage conditions, cold temperatures, and predominantly coniferous forest. The type of soil profile that develops depends, in the view of some who have worked in this area (Hills, 1960), not only on the parent material but also on the types of vegetation that have occupied the area and such other changes as are brought about by fires, insect attack, flooding, and other common disturbances involving forest and soil during the natural course of events. On some dry sites, intensively leached podzols are found under hardwood forest, and melanized soils are not unknown beneath conifers. The general rules seem not to apply in such cases, but it is likely that unusual circumstances of one kind or another account for the disparity.

There is currently a rapidly growing volume of information on the more commonly occurring boreal soils, even those in more remote reaches of boreal forest regions, and it is generally apparent that the podzolization process is intense throughout this northern environment. In studies on the more common Newfoundland forest soils, for example, Page (1971) shows that, despite wide climatic, geological, topographic, and forest development differences, two areas—the Avalon peninsula and western Newfoundland—possess soils with large amounts of surface organic material, relatively large amounts of organic matter incorporated into the lower parts of the profile, and generally strong acidic conditions except in areas of western Newfoundland where soils are somewhat less acidic because the parent material is calcareous or shallow noncalcareous material over limestone bedrock.

The soils, in general, can be considered to be of nine common types:

Soil	Extent and vegetational associates
Orthic or sombric brunisol	Infrequent: *Abies balsamea* and hardwoods
Dystric brunisol	Fairly common: *Abies balsamea*
Orthic podzol	Common: *Abies balsamea* and *Picea mariana*
Orthic humic podzol	Common: *Abies balsamea* and *Picea mariana*
Peaty podzol	Common: *Abies balsamea* and *Picea mariana*
Gleyed podzol	Common: *Abies balsamea* and *Picea mariana*
Orthic (humic) gleysol	Common: *Abies balsamea* and *Picea mariana*
Fen peat	Fairly common: *Abies balsamea* and rarely *Picea mariana*
Acid peat	Fairly common: *Picea mariana*

Gleysols and gleyed podzols are common soils on the Avalon peninsula, where tills are fine-textured, compacted, and usually very stony, while orthic and peaty podzols are common soils in western Newfoundland, where tills are less compacted, less stony, and coarser textured. Most soils throughout both areas are relatively mature, the result of long periods without major disturbance. In both sampled areas, growth of *Abies balsamea* and *Picea mariana* is best on brunisols and poorest on gleysols and deep peats, but only a very small portion of the total nutrient supply in any of the soils is in available form. This latter subject is one that will be considered in greater detail in Chapter 8.

By way of comparison, studies at the western extreme of the North American continent show that soils possess essentially the same characteristics as those in the eastern region represented by the studies in Newfoundland. In the upper Mackenzie River area, located generally around the west end of Great Slave Lake, Day (1968) has shown that the soils are closely related to those of Newfoundland. The forest vegetation is principally *Picea mariana* in admixture with *Picea glauca, Abies balsamea, Pinus banksiana, Populus tremuloides, Betula papyrifera,* and, on wetter sites, *Larix laricina*. The glacial till is a stony, gravelly clay. Soils identified by Day in the area range from brunisols to gleysols, the former occupying a somewhat larger proportion of the land surface in this western area than in Newfoundland, at least in part because of the difficulty in classifying some soils technically as podzols though they are largely podzolic in character; upper horizons are podzolic, but lower horizons do not meet certain minimum requirements of a podzol and so are classified as brunisols. The difficulty in interpretation in this instance illustrates the ambiguous nature of soil classification and the recurrent problems encountered by soil scientists in assigning soils to the most appropriate category—and then reaching some general agreement that the classification is a correct one. The science of soils is at present an imperfect one, as most edaphologists would agree, and a good amount of additional research is required before a satisfactory degree of accuracy in soil nomenclature will be attained. This, perhaps, is the principal lesson to be gained from any perusal of current opinion on the general subject of soil classification, particularly when considering the soils of remote regions where only a little intensive research has been conducted.

GENETIC VARIATIONS AND PLANT COMMUNITIES

In regions of cool temperatures and sufficient annual precipitation, podzolization processes are intense. Modifications of the classic picture of podzol formation occur where topography and parent materials create conditions that in one way or another inhibit maximum leaching and translocation. Most commonly the modifications are the result of topography that elevates the water table into the B or the A horizon, so that this zone of permanently wet soil becomes marked by gley mottling. Such a soil is termed a bog soil or gley podzol. At the other extreme, immature podzols are found often on dry upland areas where humus accumulation is thin and leaching is not intense. Here the development of A and B horizons is usually retarded.

Weak development of podzol profiles on dry uplands may also be the result of recent changes in forest composition; fire or clear cutting of coniferous trees evidently permits rapid fading of podzol. Disturbance of the soil profile can be the result of biological activity. This is particularly apparent in the southern boreal zone where ants, rodents, and earthworms are more abundant than in northern areas. There is also an increase in the proportion of base-rich deciduous leaves accumulating in the litter on the forest floor in southern areas.

Within any given area of the boreal region, soils develop into podzols where topography permits, and on rough or rolling terrain the soil catena typically extends throughout the entire range from wet bog soils to dry lithosols, the latter being found on high uplands where soils grade into barren gravel or rock. Although very long periods of time may be involved, the soil catena can probably be related to the successional development of soils, with soils at both ends tending to develop, however slowly, into types more characteristic of central portions of the catena.

Since soil genesis is at least in part a consequence of the kind of vegetation growing on the ground surface, successional development in soils is, by inference, associated with corresponding successional development of the vegetation. Thus, it seems intuitively correct to say that, as soil profiles develop and mature along a catena, coincident changes will occur in the vegetational community. Indeed, the vegetational community and the soil might be said to change together, since in mutual interaction one influences the other. Whether this is, indeed, the case is probably still uncertain, and the concept of a plant community maturing toward ultimate development of a so-called climax is subject to much questioning and debate. The concept, however, is an influential one, and much past effort has been directed toward relating the local distribution of plant communities in forest regions to the soils. It is not difficult, for example, to demonstrate that bog vegetation grows on bog soils. Such obvious examples, however, cannot be used to support views that much more subtle differences in soils are associated with equally subtle differences in vegetational communities. If, however, subtle differences in plant communities could be discerned accurately enough, then differences in soils could be inferred for land management purposes—such as forestry most specifically. There has, as a consequence of this inferred relationship, been a rather extensive effort put forth, most specifically in northern hardwood forests, to relate ground cover vegetation to soils in the hope or expectation that plant communities could be used to differentiate good forest growth sites from poor ones and thus designate areas that could most productively be devoted to forest management. The information, it was hoped, would be useful for large areas where forests had been destroyed by fire. Here it could be employed to delineate habitats most suited to the reforestation of pine, hardwoods, hemlock, and so on, avoiding attempts to grow tree species unsuited to particular local site conditions. Hardwood forest sites were delineated on the basis of plant communities, for example, by

Heimburger (1941) in an area northeast of Montreal. In the northern hardwood forest region of Wisconsin, Wilde et al. (1949) noted distinct plant groupings related to soil type in virgin forests. Other early studies showed that, while the texture of parent material seemed not to have an influence upon the kind of plant communities present, mineral composition, particularly the presence of calcium, was related to species composition of the plant communities growing on the soil (Lemieux, 1963). Beri and Anliot (1965) used classification methods to delineate communities in hemlock forests of West Virginia, and Coffman and Willis (1977) have used indicator species to classify sugar maple and eastern hemlock forests in the upper peninsula of Michigan.

These efforts are based on the assumption that, once the communities growing on different soils have been identified, land managers can employ the information to delineate the habitats most suitable for growth of the different tree species. It is also assumed that at least some species indicative of the different habitats will consistently reestablish on burned or otherwise highly disturbed soil, permitting identification of habitat type. The ultimate test of the usefulness of any such classification system is its accuracy in delineating habitat types following forest disturbance, and it is asserted that the system developed for sugar maple and eastern hemlock forests has demonstrable utility in even highly disturbed stands (Coffman and Willis, 1977).

Site classification systems for spruce forests in Canada have been developed to some extent (Hills, 1960), and work on classification schemes have also progressed elsewhere (Silker, 1965; Layser, 1974; Coffman and Willis, 1977). The studies are of interest, since they relate directly to the question of whether site indicator species are valid indicators of boreal habitat conditions, particularly in view of the fact that the indicator species are expected to be useful following a major forest disturbance that would most likely open the canopy, permitting greater illumination at ground level as well as bringing about changes in nutrient, moisture, and drainage conditions, all of which would encourage light-tolerant pioneer species rather than shade-tolerant species more closely associated with a mature forest environment possessing a closed canopy and stabilized nutrient and moisture cycles.

One of the more interesting early studies in *Picea* forest was that of Wilde et. al. (1954), conducted in the Algoma District of Ontario. They recognized the following broad soil-vegetational categories in the landscape:

Skeletal soils: Rock outcrops and boulder pavements, with reindeer moss (*Cladonia rangiferina*) and a sparse growth of *Ledum groenlandicum* and *Vaccinium myrtilloides* or *V. angustifolium*.

Sandy podzols: On level and pitted outwash supporting pure or mixed stands of *Pinus banksiana* and *Picea mariana*, with some *Betula papyrifera* and *Populus tremuloides*, *Abies balsamea*, *Corylus*, *Lonicera canadensis*, *Vaccinium*, *Cornus canadensis*, *Linnaea borealis*, *Maianthemum canadense*, *Ledum groenlandicum*, and *Epigaea repens*.

Podzolized calcareous loams and clays: The gray-wooded or brunisolic soils, of fluvioglacial and morainic origin, supporting rapidly growing mixed stands of *Picea glauca*, *Abies balsamea*, *Picea mariana*, *Populus tremuloides*, *Populus balsamifera*, and *Betula papyrifera*, with *Lonicera canadensis*, *Clintonia borealis*, *Cornus canadensis*, *Maianthemum canadense*, *Lycopodium*, *Linnaea borealis*, *Aster macrophyllus*, *Streptopus roseus* and, less frequently, *Chiogenes hispidula*, *Galium*, and *Ledum groenlandicum*.

Melanized calcareous loams and clays: Glacial and lacustrian in origin, supporting very rapidly growing stands of *Picea glauca*, *Abies balsamea*, *Populus balsamifera*, and *Populus tremuloides*, with *Viburnum*, *Rhamnus alnifolia*, *Acer spicatum*, *Corylus*, *Lonicera*, *Aralia nudicaulis*, *Streptopus roseus*, *Maianthemum canadense*, *Clintonia borealis*, *Trientalis borealis*, *Galium*, *Viola pubescens*, *Aster macrophyllus*, *Fragaria virginiana*, and *Rubus*.

Muck soils: In drainage lines, supporting alder and willows, and with sedges (*Carex*) and mosses such as *Sphagnum*, *Calliergon*, and *Hypnum*, as well as relatively low densities of *Aralia nudicaulis*, *Coptis trifolia*, *Rubus*, *Viola*, *Galium*, *Vaccinium*, and *Ribes*.

Brown moss (*Hypnum*) peat: Supporting rapidly growing stands of *Picea mariana* with some *Picea glauca*, *Abies balsamea*, *Populus tremuloides*, and *Populus balsamifera*, and with *Hypnum*, *Calliergon*, *Hylocomium*, and *Dicranum* as well as *Sphagnum* mosses, *Cornus canadensis*, *Vaccinium myrtilloides*, *V. angustifolium*, *Maianthemum*, *Rubus*, *Aster macrophyllus*, *Viola*, *Galium*, *Trientalis*, *Coptis*, *Oxycoccus microcarpus*, *Linnaea borealis*, *Ribes*, and *Carex*.

Green moss (*Sphagnum*) peat: Supporting slowly growing stands of *Picea mariana* with occasional *Betula pumila*, *Alnus*, and *Salix*, and with *Chamaedaphne calyculata*, *Kalmia polifolia*, *Andromeda glaucophylla*, *Oxycoccus microcarpus*, *Ledum groenlandicum* and, more rarely, *Carex*, *Eriophorum*, *Rubus chamaemorus*, *Menyanthes trifolia*, and *Sarracenia purpurea*.

It is apparent from these categories that rather broad zonations have been selected for the soil and vegetational delineation, and that rather marked differences could be expected in both soil genesis and plant communities in such widely disparate sites in which topographic position plays as important a role as parent material and other soil-forming factors. It is not apparent, for example, whether soil–vegetation relationships could be useful as indicators within any of the above classifications. The overlap of plant species from one soil category to another is considerable, and it appears questionable that differentiation in all instances even between widely different habitat types would always be possible on the basis of plant species alone. The study is of interest, however, since it represents a pioneer effort in the spruce forest region and, while it would be faulted today on the basis that it was purely observational rather than quantitative, it is nevertheless revealing of the principles upon which vegetational site classification is based. Later workers have shown that within at least some of these essentially topographically differentiated categories there are, indeed, a variety of fertility and moisture regimes, each of which supports a plant community which appears in species composition to be a result of the controlling influence of nutrient and water supplies. There were, for example, a number of exploratory studies of peatland vegetation indicating that moisture and fertility regimes were typically the two main factors controlling species distributions and vegetational patterns (Sjörs, 1948, 1950; Linteau, 1955; Clausen, 1957; Malmer, 1962; Persson, 1962; Damman, 1964; Mueller-Dombois, 1964; Spence, 1964), but until the work of Jeglum (1971, 1972, 1973) the relationships between gradients of these main factors and species behavior were not well documented. Jeglum demonstrated variations in peatland vegetation that were coincident with variations in pH and water level, showing that variations in the environmental factors were related to variations in floristic composition and species densities in the peatland communities. The study demonstrates that vegetational changes are not only associated with moisture gradients but also that there are changes in nutrient status to which plants are responding. There is more than a single factor involved, and the plant response is most likely to the entire environmental complex. Studies indicate that similar gradients occur in many other, if not all, boreal forest communities, and that, indeed, the entire environmental complex can be considered a continuum, with gradients in both plant composition and soils occurring in what is essentially a multidimensional manner.

The plant response to environment is probably more complex than is generally realized, and this is illustrated in studies on Wisconsin plant communities by Curtis (1959) and his colleagues, who show that many herbs and shrubs growing in northern forests show markedly bimodal distributions in response to light and moisture, demonstrating peaks in presence at two points along a wet-to-dry moisture scale—attaining high densities on both moist and dry sites but low densities under mesic conditions. There were species showing an optimum response at every locus along the environmental gradient, and every possible variation was present in species showing bimodal distributions: *Mitchella repens* and *Clintonia borealis*, for example, had generally the highest presence in more mesic forests but prominent declines in presence at the very center of the scale; *Gaultheria procumbens* had marked peaks at both wet and dry ends of the scale; *Maianthemum canadense*, in contrast, showed a broad amplitude of tolerance, occurring at high levels all along the moisture scale. It is obvious, as Curtis points out, that plants of the last-mentioned type have no indicator value with respect to either plant communities or to environmental conditions.

These environmental relationships are dealt with in greater detail in Chapters 6 and 7, and it must suffice here to indicate that whereas the correlations between soils and vegetational communities are always interesting, the variation between anything resembling an accurate one-to-one correspondence between soils and vegetation necessarily leads unbiased observers to question the usefulness of either one as an indicator of the other. Only additional accumulation of relevant data by means of intensive field studies throughout boreal regions will resolve the question of whether use of indicators is universally practical in delineating boreal forest soil types. The studies should identify species most subject to intractable variations, as well as species with patterns of sufficient regularity to be useful as indicators of sites of known potential productivity for the forest tree species. The environmental principles upon which these indicator systems are based have broad ecological implications, and it is not yet certain whether the use of indicator species can be justified in boreal regions on theoretical as well as empirical grounds; these are discussed in greater detail in Chapter 7, but it should be noted here that continuum theory bears importantly on the topic of whether indicator species can be employed in boreal forest classification. Data accumulating as a consequence of field studies of plant communities, particularly of relationships between vegetation and environmental gradients, eventually will permit a realistic assessment of the usefulness of indicator techniques. Many studies employing multivariate analysis, particularly those conducted in the eastern deciduous forest, show a continuous gradation in species from one community to another along environmental gradients, evidently precluding the usefulness of any classification method in delineating communities. Some authors, on the other hand, suggest that the gradations are more apparent than real and result from a pattern of random disturbance that creates an intergrading complex of successional stages which mask the original discrete communities (see Chapter 7). The question will probably not be resolved to the complete satisfaction of those who advocate one or the other view until ample data are available to decide the issue. This is simply an example of a scientific controversy created by the absence of sufficient data to settle the question one way or another.

It is possible that eventually it will be found that the most useful indicator species are those that pioneer areas of disturbance, and not those found most often in mature forests. Some site classification schemes are based on the presence on soils of disturbed areas of species adapted to mature closed-canopy forests—the same species present in the mature forests before the site was disturbed. It must be indicated that species adapted to mature forests may not be indicative of any particular set of environmental conditions outside of a mature forest, and the fact that these species are present on a given plot of land may mean nothing more than that chance circumstances permitted the germination of seeds. Occasionally plants occupy sites that are in no way representative of those most often occupied by the species in mature forests; their presence is misleading if taken to indicate the existence of the environment in a mature forest where the species is ordinarily dominant—or to infer from its presence that a disturbed site is capable of supporting such a mature forest on the basis of an inferred existence of soil types suitable for development of such a forest. Competition, or lack of it, and a host of other factors are involved in germination and growth. Many compensate for one another; a great abundance of one will alleviate the effects of shortages of another. Thus, given conditions difficult to discern, the fact that a species is present or absent on a given site is not an infallible guide to environmental conditions existing there or, in forest management terms, to the quality of the site in terms of potential for tree growth. It is particularly misleading to infer any qualities of a disturbed site from the fact that species are present there that are most often found in one or another mature forest community; it is more likely that the species have invaded the area as a result of circumstances unrelated to the particular qualities of the site—they are best adapted to mature forest environments and must be considered adventive elsewhere.

On the other hand, the species may be the remaining occupants of the area following fire that destroyed the crowns of the trees but not the ground-layer vegetation; other species better adapted to open conditions can be expected to replace them rather quickly. It is conceivable, however, that site quality would be indicated by the early successional, pioneer, or even weedy species that come in following a disturbance, and research along these lines might be productive. There is also the possibility, of course, that indicator techniques will work in some communities and not others, for some kinds of sites and not others, and in some regions and not others. They may, in fact, be the basis of much of the controversy that exists regarding the efficacy of indicator methods.

MOISTURE AND PERMAFROST

On a global basis water is the substance most often in short supply for living plants, although it less frequently limits growth in the relatively moist boreal regions than elsewhere. In boreal regions, evaporation potential is rarely in excess of precipitation, and even sandy upland soils are rarely dry for long periods of time. In the southern boreal forest, vegetational communities found along moisture gradients range from xerophytic *Pinus banksiana* forest on the driest sites to hydrophytic forests of *Picea mariana* on wet sites. Most species have a preferred moisture regime, reflecting this perference in greater vigor and abundance on favored sites. Variations in the soil moisture regime are always accompanied by variations in the vegetation. There is, however, sufficient irregularity in the pattern, at least in boreal *Pinus* forests on coarse soils, to classify sites on the basis of vegetation less useful for forestry purposes than directly on the basis of soil factors (Ritchie, 1961).

The moisture regime of climate and soil is, thus, of significance in the growth of individual plants and in the composition of communities found on the various sites in a given region. Soil moisture is also important in a hydrological sense, since forests affect moisture storage and runoff in streams and rivers; forested areas have a number of well-known and widely recognized attributes in water management and flood control. Snow cover melts slowly in shaded forests, and heavy rain is intercepted by the canopy of the trees, thus preventing torrential surface runoff and erosion.

The water storage capacity of a forest floor is influenced by a number of factors, including proportion of surface covered with mosses, depth and consistency of decaying organic matter, granular structure and noncapillary pore space of mineral horizons, the degree to which hardpan inhibits percolation of water into lower horizons, and the level of the water table. Total elimination of tree cover by fire or other means will inevitably affect a forest soil, but partial disturbance of tree cover that does not scorch or otherwise cause major disruption of the soil evidently will have only a minor effect upon water storage capacity. Thus, Golding and Stanton (1972), for example, showed that there was no significant difference between uncut and partially cut spruce-fir forest in water storage capacity. Both cut and uncut forest stands averaged greater storage capacity than young pine forest regenerating in an area burned 30 years prior to the study; among the differences between the spruce and pine forests was the moss cover, which in the spruce forest was much more extensive than in the pine forest. While spruce stands normally have a much higher proportion of mosses in the ground cover, the great discrepancy between the spruce and pine stands in this study indicated both the degree to which intense fire destroys the soil cover and the influence of mosses in water retention. Partial cutting, according to this evidence, does not greatly disrupt the soil, at least until some as yet unrecognized or unforeseen soil changes occur after an extended period of time as a consequence of the disturbance.

The relationship between water storage capacity of a soil and the density of mosses in the ground cover is reciprocal in the sense that water supply is a major factor controlling growth and habitat preference in mosses. Busby and Whitfield (1978) suggest that studies of response to desiccation would reveal tolerance limits in many moss species. Hence, the habitat in which each species possesses the greatest competitive vigor and tends to attain the greatest density could be identified. Thus, *Hylocomium splendens*, *Pleurozium schreberi*, and *Ptilium crista-castrensis* are circumboreal "feather mosses" which occupy a large proportion of well-drained shady forests in boreal regions. Another moss, *Tomenthypnum nitens*, by way of contrast, forms dense carpets in fens where trees and shrubs are far apart and the water table is close to the surface. In the species studied, 80–90% of the water in saturated moss was held externally. Once the external water evaporated, the mosses responded to continued drying in much the same way as higher plants—by decreasing rates of photosynthesis, a consequence of changing water potential in the tissues. Since there appeared to be little difference among these species in the effect of drying upon net assimilation rate, Busby and Whitfield suggest that relative tolerance to desiccation distinguishes the species in terms of habitat preference.

The significance of the mosses in terms of their unique water storage capability is equaled by their importance as contributors to the organic mat on the surface of boreal podzol soils—establishing not only conditions necessary for podzol formation but also, at least in more northern regions, conditions under which one of the more interesting characteristics of Far Northern soils—permafrost—is established. The permafrost zone is not sharply delimited by distinguishing edaphic conditions, but rather soils become progressively colder northward and, at some point, subsoil temperatures remain below freezing throughout the summer.

Growth of trees and understory vegetation is not, thus, sharply affected at the southern edge of the permafrost zone. The active layer of soil in summer here is sufficiently deep so that temperatures in the rooting zone are not greatly, if at all, reduced as compared to those at similar depths in many areas where no permafrost exists. Progressively northward, however, the thawed active layer becomes more and more shallow until, finally, at the northern edge of the forest–tundra transition zone it averages a depth in the neighborhood of 4 or 5 dm in moss-covered forest soil. In the southern permafrost zone, the permafrost is discontinuous—it is not, in other words, found continuously throughout the landscape but only on well-insulated soils deeply covered with a carpet of living mosses and dead organic matter in various stages of decay. Here, the insulating qualities of the forest and the soil combine with the climatic conditions prevailing to create temperature conditions severely inhibitory to physiological activity in roots of both trees and understory plant species.

Permanently frozen ground underlies a fifth of the world's land surface, and it is a distinctive feature of mountains and polar tundra regions as well as much of the subarctic boreal region. The depth of permafrost ranges to many hundreds of feet below the surface of the earth, and only the annual spring and summer thawing of a shallow surface active layer makes possible the growth of vegetation in regions where permafrost is found. In northern areas largely or entirely occupied by tundra, the permafrost is continuous and found everywhere to great depths. This zone grades into a zone of discontinuous permafrost, largely occupied by boreal forest vegetation, in which permafrost is found at favorable sites, usually spots poorly drained and possessing a thick surface layer of insulating peat. Although permafrost impedes drainage and influences soils in a variety of other ways, it does not prevent growth of forest conifers if summers are warm and long enough to permit thawing of a deep active layer. In many parts of the permafrost zone trees grow to a height of 20 m. The shallow root systems of *Picea mariana* and *Picea glauca* permit growth of these species on sites with an active layer so shallow that it excludes species possessing tap roots, such as *Pinus banksiana*, and most if not all deciduous trees. Annual refreezing of the active layer begins at the surface and progresses downward, trapping a horizontal band of unfrozen ground between the upper and lower frozen strata; hydrostatic pressure in this zone is sometimes released by the forcing of unfrozen material upward through cracks or by a heaving of the frozen surface. The result is a frost boil or a contorted surface on which the trees stand at angles, giving the forest a strangely disjointed appearance. The folding, contraction, fracturing, and flowing of the surface material under these conditions is known as cryoturbation; the result is a continual disruption of both surface and rooting zones. Such permafrost-induced microrelief is very common in many Far Northern forest regions, and the existence of what is termed hummocky permafrost soil is widespread on moderately fine and fine-textured materials, except on steep slopes or frequently flooded alluvial flats. Zoltai and Pettapiece (1974) point out that internal morphology may be quite variable, but in general the hummocks are about 1–2 m in diameter and 20–80 cm high, are circular or somewhat oval in outline, and have convex or flattened tops. Variations in shape, size, and distribution are in response to local differences in texture, drainage, and perhaps microclimate. In forested areas, vegetation completely covers the hummocks, and in the area studied by Zoltai and Pettapiece (northwestern Mackenzie and northern Yukon), there were rather marked differences in the density of tree cover between the tops of hummocks and the sides and troughs. On the tops of hummocks, *Picea mariana* contributes about 5% of the cover; on the sides and troughs it contributes more than 20%. There are corresponding differences between the two sites in proportion of cover contributed by other plant species as well (see tabulation below).

It is also noted by Zoltai and Pettapiece that the heaving soil causes many trees to be tilted, and the proportion of tilted trees appears about equal on the tops and sides and in the troughs between hummocks. Establishment of trees on tops of hummocks evidently is restricted

Species	Percentage cover	
	Hummock tops	Hummock sides and troughs
Cladonia amaurocraea	20	—[a]
Cetraria cucullata	13	—
Vaccinium vitis-idaea	10	—
Ledum decumbens	10	5
Spiraea beauverdiana	10	5
Hylocomium splendens	5	5
Equisetum pratense	5	5
Picea mariana	5	22
Ledum groenlandicum	—	15
Carex cf. *bigelowii*	—	10
Sphagnum fuscum	—	15

[a] Less than 1% or absent.

partly by soil movement and partly by dry conditions resulting from the microrelief. Soil profiles indicate there is a mixing of organic matter into lower horizons by folding, and there is, thus, periodic disruption of normal development in mineral soil horizons by frost action. After a series of especially wet years, or following a severe fire, conditions apparently are most conducive to marked cryoturbation, and soil movement is most intense at such times. The heaving movements shift the trees and cause them to tilt, and this usually—but not always—is in the direction away from the center of the hummock. Annual and perennial herbs, lichens, and mosses can survive on both the tops and sides of hummocks, as well as in the troughs between, but there is a difference in the species most prevalent on each site, as the table shows, indicating that distinct habitats exist in the different locations. Detailed descriptions of these soils, the relationships of each to associated plant communities, and the apparent role of vegetation, frost, and other factors in soil genesis, are discussed by a number of authors who have studied these soils in detail (Benninghoff, 1952; Brown, 1970; Rapp, 1970; Tedrow, 1970; Salmi, 1970; Péwé, 1970; Larsen, 1972; Zoltai and Pettapiece, 1974; Thie, 1974; Pettapiece, 1974, 1975; Brewer and Pawluk, 1975; Crampton, 1974).

SOIL ORGANISMS

In the litter that continually accumulates on the forest floor are found large numbers of organisms including bacteria, actinomycetes, algae, fungi, and a variety of invertebrates. Even estimates of total populations are difficult to obtain because direct microscopic counts fail to distinguish between living and dead organisms or between, for example, bacteria and the spores of actinomycetes. Successful laboratory cultures of organisms for accurate plate counts of the organisms present in the soil are also almost impossible to obtain; every nutrient medium selects for organisms possessing certain physiological characteristics and nutrient requirements. Each medium, thus, favors organisms that grow on it well; other organisms, however, are not entirely excluded. Thus, plate-counting methods, alone or in combination, are limited in usefulness, and the best that can be achieved is a comparison of relative abundance. Counts of invertebrates are likewise hampered by the lack of simple and consistently reliable census techniques.

Because of the problems in obtaining good estimates of numbers of organisms involved, it is not possible to trace in detail the path of organic compounds in litter as it decomposes as a result of the action of bacteria and invertebrate organisms. In many instances, however, there exists a rather good understanding of the processes at work and the kinds of organisms involved in the breakdown of litter into simple organic end products of decay.

Leaves and wood in twigs and branches falling to the forest floor are attacked by many common species of Diptera and Coleoptera possessing symbiotic gut flora that digest cellulose and lignin. Wood is attacked by beetles and fly larvae and is invaded by bacteria in large numbers. Millipedes and isopods proliferate in cracks and on the underside of decaying material.

These events on the ground surface are accompanied by corresponding but concealed events beneath the surface. Roots are subject to the same cycle of life and death as aboveground portions of plants. The total biomass of roots probably nearly approaches the biomass of aboveground plant organs, and thus it is apparent that decay processes deep in the soil are as important as on the surface. They are, however, little understood, and so we must for the present merely make the assumption that events beneath the surface correspond to those known to occur on the surface. There is evidence from temperate zone forests that litter decomposes at a rate directly related to the number of invertebrate animals present; whether this is true also for the boreal forest is not yet known. In studies on soils in northwestern Canada, Brewer and Pawluk (1975) found that surface-derived organic material was distributed to some depth in many soils. Some of this material was evidently the result of packing of fecal pellets of soil fauna, and it was possible to distinguish the pellets of organic-ingesting fauna and soil-ingesting fauna on the basis of the type of material present. The upper horizons of these soils had been thoroughly worked by the soil fauna; the greatest activity was in the surface layer and decreased in intensity progressively to about a 7–10 cm depth. The larger fauna were nearer the surface, and the small forms occurred at lower depths, with some activity found in all the horizons although it was more sporadic at greater depths. In general, they state, accumulation of organic matter and faunal activity decreases sharply with depth, and at lower horizons the soil structure is influenced more by freezing and thawing than by other genetic processes. Soils generally show an increase in organic matter in the lowest horizon overlying the permafrost; this is the result of accumulation of plant fragments, and the evidence seems to indicate that movement of the soil caused by both physical processes and faunal activity results in this accumulation of material at the upper level of the permafrost. A similar zone of high organic content just above the permafrost has been noted in a number of other widely separated areas in the northern regions (Tedrow, 1970, 1977; Zoltai and Pettapiece, 1974; Pettapiece, 1974, 1975).

SOIL MICROORGANISMS

The presence and growth of soil organisms depend primarily on the availability of organic substrates they can utilize. The organic substrates are colonized by microorganisms that fall into a few major well-defined categories: (1) organisms able to use simple soluble carbohydrates, (2) cellulose decomposers, (3) lignin decomposers, and (4) humus fungi.

The primary colonizers are *zymogenous* organisms with the ability to "flare up" rapidly on a newly available organic substrate—dead plants or animals. They have rapidly germinating spores and a rapid growth rate, characteristics giving them a competitive advantage in early colonization of substrates, and they are widespread in soils.

The zymogenous forms can be grouped in the following general categories: (1) decomposers converting the nitrogen in organic compounds into nitrates; (2) fungi growing when organic matter is present; (3) actinomycetes in the vegetative stage; (4) pseudomonads; (5) bacillus forms developing when organic matter is available from decaying plants, animals, or other primary decomposers such as fungi and other bacteria; and (6) soil protozoa, many of which feed on living bacteria.

Distinct from these forms are mycorrhizal associates and nodule-forming microorganisms found in the root tissues of some plant species. Some, like legume nodule bacteria, are capable of fixing nitrogen and are beneficial. The role of others, like endotrophic mycorrhizas, cannot be easily evaluated. A mycorrhizal association is a union between a fungus and the root of a plant; there is no evidence of pathological symptoms, and it is assumed that the association is mutually beneficial.

Substrates for *autochthonous* organisms include humus material, but decomposition occurs without a sudden burst of activity. It is characteristic of autochthonous microorganisms that they exist in stable populations with low levels of activity. There is a lack of detailed information on autochthonous forms, however, and it may be true that no microorganism is consistently autochthonous.

The various horizons of podzol and podzolic soils are often quite distinct in the species of organisms and number of individuals present. Podzol soils with a well-marked B horizon enriched with humus show a markedly greater number of organisms in the A than the B layer. This is shown in some of the profiles examined by pedologists and, in cases where there is little evidence that the B horizon has appreciably different total numbers in comparison with the A or C horizon, there are often differences in the species present. The difficulties involved in such counts, however, are well illustrated by studies on the microbiology of a muskeg in Alberta, carried out by Christensen and Cook (1970). Their tests on ammonification rates in soils showed that up to four times the number of bacteria indicated by plate counts must be present; their conclusion was that three-quarters of the total soil population of ammonification organisms was anaerobic or could not survive on the medium used for the counts. It thus seems likely that anaerobic organisms play an important role in northern soils, and these authors advocate more detailed studies of the anerobic populations in soils of the boreal regions.

Microbiological studies carried out on soils in the Inuvik area of northwestern Mackenzie by Boyd and Boyd (1971) demonstrated the presence not only of bacteria and fungi but also of yeasts, *Streptomyces* species, and protozoa (members of all classes except Sporozoa), as well as rotifers and tardigrades. Higher soil fauna were represented by a variety of bivalves, land snails, and insects. There was a wide variety of both autotrophic and heterotrophic microorganisms present, including *Azotobacter* and *Nitrosomonas,* the former capable of nitrogen fixation and the latter of nitrate reduction. It was of especial interest that relatively large numbers of thermophilic bacteria were encountered and, since little is known of the microhabitats of the soil in which they were found, the authors indicate that additional research will most likely reveal many interesting aspects of the role played by these and other microorganisms in the ecology of the northern regions.

The role of microorganisms in the biochemical processes that take place in soil is a topic intensively studied, and an extensive literature is available in which to pursue the subject. It must suffice here to indicate that microorganisms are able to carry out a wide range of complex chemical activities in soils and that they are actively involved in podzol formation. Their inability to carry out certain processes under boreal conditions is also of great importance—notably their failure to decompose lignin and cellulose rapidly under cold and anaerobic conditions, the consequence of which is the accumulation of deep mats of peat in muskegs and elsewhere. Despite growing interest in northern podzol soils, however, the amount of research carried out is small by comparison with studies on other types of soils, notably those in agricultural or southern forested areas, and much of what is assumed concerning northern podzols is by inference from studies on more southern regions. The northern soils are a field ripe for pioneering studies of almost all kinds, particularly those that will look into the role of microorganisms in the production of individual organic compounds that participate in podzol formation and cycling of nutrients; the latter topic is discussed in additional detail in Chapter 8.

SOIL ANIMALS

Two activities, feeding habits and locomotion have been employed to classify soil animals. In terms of feeding activity, one classification is as follows (Wallwork, 1970):

1. *Carnivores*
   a. Predators, e.g., carabid, pselaphid, scydmanid, and some staphylinid beetles, many mesostigmatid and prostigmatid mites, spiders, harvestmen, pseudoscorpions, scorpions, sun spiders, centipedes, and some nematodes and mollusks.
   b. Animal parasites, e.g., ichneumonids, some staphylinid beetles, parasitic Diptera, and some nematodes.
2. *Phytophages*
   a. Feeding on green plant material above ground, e.g., mollusks and lepidopteran larvae.
   b. Feeding on root systems, e.g., nematode plant parasites, symphylids, larvae of some Diptera, scarabaeid Coleoptera and Lepidoptera, mollusks, and burrowing Orthoptera.
   c. Feeding on wood material, e.g., some termites, beetle larvae, and phthiracaroid mites.
3. *Saprophages*
   Which feed on dead and decaying organic material: for example, lumbricids, enchytraeids, isopods, millipedes, and some hemiedaphic mites, Collembola, and insects. Some of these forms are also probably fecal feeders (coprophages), wood feeders (xylophages), and carrion feeders (necrophages), and are often variously referred to as scavengers, debris feeders, or detritivores.
4. *Microphytic feeders*
   Which feed on fungal hyphae and spores, algae, lichens, and bacteria. Many saprophagous mites and Collembola may also be included here, together with fungus-feeding insects, such as ants, termites, dipteran Mycetophilidae and coleopteran Nitidulidae, nematodes, and certain mollusks and protozoans.

The contribution made by soil fauna to the processes of organic decay depends on the amount of protoplasm, biomass, and rate of metabolism of the soil animals present. Estimates of biomass and metabolism for the fauna of a number of different forest types including spruce stands have been presented by Wallwork (1970). The metabolic contribution of each group varies with the biomass and, since biomass varies with locality, the relative importance of each group in litter breakdown and energy utilization will also vary locally. Some groups compensate for low metabolic rates with a high biomass, and the low biomass of Collembola, Nematoda, and Enchytraeidae for example, may be offset by high metabolic rates. Often the principal effect of soil animals is to increase the activity of the bacteria and fungi, which dominate all other organisms combined in terms of both biomass and metabolic rate.

SUMMARY: CLIMATE AND SOILS

From the foregoing discussion it can be seen that there are close links between the climate and the soils of boreal regions. It is, in fact, the distinct character of the climate that, by and large, determines the character of the soil. The podzolization process is intense in boreal regions as a consequence of cool temperatures and a predominance of precipitation over evaporation. Other characteristics of boreal soils are the slow rate of bacterial decomposition of organic matter in litter, high acidity, low content of strong bases such as calcium, and movement of aluminum and iron compounds from the A to B soil horizon.

These characteristics of climate and of soils have definite influences upon plants and ultimately the animal life of boreal regions. There is, of course, the direct effect of low temperatures upon physiological processes, but equally important are the subtle effects of edaphic characteristics upon energy and nutrient cycling in the system. Often most of the nitrogen and nutrient minerals in the soil are tied up in undecayed organic forms unavailable to plant growth. Even the permanent loss to plant life of nitrogen and mineral nutrients in peat accumulations is common. The rate of nutrient cycling in the boreal ecosystem is critical; unlike the conditions often found in temperate deciduous forests, as time passes and the forest grows to maturity, there is a decreasing supply of nutrients available for annual recycling, since increasing amounts are tied up in undecomposed soil organic matter.

These events have important consequences for the growth of plants in the boreal regions that will be discussed more fully in Chapters 7 and 8. First, however, in the next two chapters we will have a look at the distinctive associations of plant species that make up the vegetational communities of the boreal regions in Canada and the influence of environmental factors upon the composition and structure of these communities. This is a necessary preliminary to the later discussions of community dynamics and nutrient cycling in boreal ecosystems.

REFERENCES

Alexander, M. (1977). "Introduction to Soil Microbiology," 2nd edition. Wiley, New York.
Beals, E. W. (1969). Vegetational change along altitudinal gradients. *Science* **165**, 981–985.
Bel'chikova, N. P. (1970). Data on humus in podzolic and sod-podzolic, natural and cultivated soils in the European part of the USSR. *In* "Microorganisms and Organic Matter of Soils" (M. M. Kononova, ed.), pp. 272–305. U.S. Dep. Agric., Washington, D.C.
Benninghoff, W. S. (1952). Interaction of vegetation and soil frost phenomena. *Arctic* **5**, 34–44.
Beri, R., and Anliot, S. F. (1965). The structure and floristic composition of a virgin hemlock forest in West Virginia. *Castanea* **30**, 205–226.
Boyd, W. L., and Boyd, J. W. (1971). Studies of soil microorganisms, Inuvik, Northwest Territories. *Arctic* **24**, 162–176.
Brewer, R., and Pawluk, S. (1975). Investigations of some soils developed in hummocks of the Canadian subarctic and southern arctic regions. I. Morphology and micromorphology. *Can. J. Soil Sci.* **55**, 301–319.
Britton, M. E. (1966). Vegetation of the arctic tundra. *In* "Arctic Biology" (H. P. Hansen, ed.), 2nd ed., pp. 67–130. Oregon State Univ. Press, Corvallis.
Brown, B. D. K. (1957). Studies on seasonal changes in the temperature gradient of the active layer of soil at Fort Churchill, Manitoba. *Arctic* **10**, 151–183.
Brown, R. J. E. (1970). Permafrost as an ecological factor in the Subarctic. *Ecol. Subarct. Reg., Proc. Helsinki Symp.* pp. 129–139.
Burges, A. (1958). "Micro-Organisms in the Soil." Hutchinson Univ. Library, London.
Busby, J. R., and Whitfield, W. A. (1978). Water potential, water content, and net assimilation of some boreal forest mosses. *Can. J. Bot.* **56**, 1551–1558.
Canada Soil Survey Committee. (1974). "The System of Soil Classification for Canada," rev. ed. Can. Dep. Agric. Publ. 1455. Ottawa.
Canada Soil Survey Committee. (1978). "The Canadian system of soil classification." Can. Dep. Agric. Res. Br. Publ. 1646.
Chang, P. C., and Knowles, R. (1965). Non-symbiont nitrogen fixation in some Quebec soils. *Can. J. Microbiol.* **11**, 29–38.
Christensen, P. J., and Cook, F. D. (1970). The microbiology of Alberta muskeg. *Can. J. Soil Sci.* **50**, 171–178.
Clausen, J. J. (1957). A phytosociological ordination of the conifer swamps of Wisconsin. *Ecology* **38**, 638–646.
Coffman, M. S., and Willis, G. L. (1977). The use of indicator species to classify climax sugar maple and eastern hemlock forests in upper Michigan. *For. Ecol. Manage.* **1**, 149–168.
Cole, L. C. (1960). Competitive exclusion. *Science* **134**, 348–349.
Crampton, C. B. (1974). Linear-patterned slopes in the discontinuous permafrost zone of the central Mackenzie River valley. *Arctic* **27**, 265–272.
Crampton, C. B. (1977). A study of the dynamics of hummocky microrelief in the Canadian North. *Can. J. Earth Sci.* **14**, 639–649.
Curtis, J. T. (1959). "The Vegetation of Wisconsin." Univ. of Wisconsin Press, Madison.
Damman, A. W. H. (1964). Some forest types of central Newfoundland and their relation to environmental factors. *For. Sci. Monogr.* **8**, 1–62.
Damman, A. W. H. (1971). Effect of vegetation changes on the fertility of a Newfoundland forest site. *Ecol. Monogr.* **41**, 253–270.

Day, J. H. (1968). "Soils of the upper Mackenzie River area, Northwest Territories." Can. Dep. Agric., Res. Br., Ottawa.
Day, J. H., and Rice, H. M. (1964). The characteristics of some permafrost soils in the Mackenzie Valley, N.W.T. *Arctic* **17**, 223–236.
Dokuchaiev, V. V. (1879). Cartography of Russian soils. *In* "Collected Works," Vol. 2, p. 226. U.S.S.R. Acad. Sci., Moscow.
Dyrness, C. T., and Grigal, D. F. (1979). Vegetation-soil relationships along a spruce forest transect in interior Alaska. *Can. J. Bot.* **57**, 2644–2656.
Ehrlich, W. A. (1974). Soil classification. *In* "The System of Soil Classification for Canada," pp. 3–6. Can. Dep. Agric. Publ. 1455, Ottawa.
Gagnon, J. D. (1965). Nitrogen deficiency in the York River burn, Gaspe, Quebec. *Plant Soil* **23**, 49–59.
Georgievsky, A. 1888. *Mater. Izuch. Russ. Pochv.*
Gerloff, G. C., Moore, D. D., and Curtis, J. T. (1964). Mineral content of native plants of Wisconsin. Univ. of Wisconsin College Agric. Exp. Sta. Res. Rep. 14.
Glinka, K. D. (1908, 1914). "Soil Science." (In Russian; German translation in 1914) "Die Typen der Bodenbildung, ihre klassifikation under geographische Verbreitung," Verlagsbuch-handlung Gebrüder Borntraeger, Berlin.
Glinka, K. D. (1924). "Disperse Systems in Soils." Kult. Pros. Tr. Tov. Obraz, Leningrad.
Golding, D. L., and Stanton, C. R. (1972). Water storage in the forest floor of subalpine forests of Alberta. *Can. J. For. Res.* **2**, 1–6.
Heilman, P. E. (1966). Change in distribution and availability of nitrogen with forest succession on north slopes in interior Alaska. *Ecology* **47**, 825–831.
Heimburger, C. C. (1934). "Forest-type studies in the Adirondack region." Mem. Cornell Agric. Exp. Sta. 165, pp. 1–122, Cornell Univ., Ithaca, New York.
Heimburger, C. C. (1941). "Forest-Site Classification and Soil Investigation on Lake Edward Forest Experimental Area." Can. Dep. Mines Resources Silvicul. Res. Note 66, Ottawa.
Hills, G. A. (1960a). Regional site research. *For. Chron.* **36**, 401–423.
Hills, G. A. (1960b). The Soils of the Canadian shield. *Agric. Inst. Rev.*, **15**, 41–50.
Hopkins, D. M., and Sigafoos, R. S. (1950). Frost action and vegetation patterns on Seward Peninsula, Alaska, *U.S. Geol. Sur. Bull.* **974-C**.
Hume, G. S. (1954). "The lower Mackenzie River area, Northwest Territories and Yukon." *Geol. Surv. Can. Mem.* **273**.
Ingestad, T. (1962). Macro element nutrition of pine, spruce, and birch seedlings in nutrient solutions. *Madd. Fran.* **51**, 150 pp.
Jeglum, J. K. (1971). Plant indicators of pH and water level in peatlands at Candle Lake, Saskatchewan. *Can. J. Bot.* **49**, 1661–1676.
Jeglum, J. K. (1972). Wetlands near Candle Lake, central Saskatchewan. I. Vegetation. *Musk-Ox* **11**, 41–58.
Jeglum, J. K. (1973). Boreal forest wetlands near Candle Lake, central Saskatchewan. II. Relationships of vegetational variation to major environmental gradients. *Musk-Ox* **12**, 32–48.
Jenny, Hans. (1941). "Factors of Soil Formation." McGraw-Hill, New York.
Johnson, G. H., and Brown, R. J. E. (1964). Some observations on permafrost distribution at a lake in the Mackenzie Delta, N.W.T., Canada. *Arctic* **17**, 163–175.
Johnson, G. H., and Brown, R. J. E. (1965). Stratigraphy of the Mackenzie River delta, Northwest Territories, Canada. *Bull. Geol. Soc. Am.* **76**, 103–112.
Johnson, E. A., and Rowe, J. S. (1974). Fire in the subarctic wintering ground of the Beverley caribou herd. *Am. Midl. Nat.* **94**, 1–14.

References

Larsen, J. A. (1972). Observations of well-developed podzols on tundra and of patterned ground within forested boreal regions. *Arctic* **25**, 153–4.
Layser, E. F. (1974). Vegetative classification: Its application to forestry in the northern Rocky Mountains. *J. For.* **72**, 354–357.
Leahey, A. (1961). The soils of Canada from a pedological viewpoint. *In* "Soils in Canada: Geological, Pedological and Engineering Studies" (R. F. Legget, ed.). R. Soc. Can. Spec. Publ. No. 3, Univ. Toronto Press, Toronto.
Legget, R. F., ed. (1968). "Soils in Canada: Geological, pedological and engineering studies." R. Soc. Can. Spec. Publ. 3, Univ. Toronto Press, Toronto.
Lemieux, G. J. (1963). "Soil-vegetation relationships in the Northern hardwoods of Quebec." Dep. For., For. Res. Br. Contrib. 563, pp. 163–175.
Linteau, A. (1955). Forest site classification of the northeastern section, boreal forest region, Quebec, Canada. Dep. North. Affairs Nat. Res. For. Br. Bull. 118, pp. 1–85.
Mackay, J. R. (1962). Progress of break-up and freeze-up along the Mackenzie River. *Geogr. Bull.* **19**, 103–116.
Mackay, J. R. (1963). The Mackenzie Delta area, N.W.T. Dep. Mines Tech. Surv., Geogr. Br., Mem. 8, pp. 1–202.
Mackay, J. R. (1966). Mackenzie River and Delta ice survey, 1965. *Geogr. Bull.* **8**, 270–278.
Mackay, J. R. (1967). Permafrost depths, lower Mackenzie Valley, Northwest Territories. *Arctic* **20**, 21–26.
Malmer, N. (1962). Studies on mire vegetation in the archean area of southwestern Götaland (south Sweden). II. Distribution and seasonal variation in elementary constituents on some mire sites. *Opera Bot.* **7**, 1–67.
Maxwell, J. A., Dawson, K. R., Tomilson, M. E., Pocock, M. E., and Tetreault, D. (1965). Chemical analysis of Canadian rocks, minerals and ores. *Geol. Surv. Can. Bull.* **115**.
Maycock, P. F., and Curtis, J. T. (1960). The phytosociology of the boreal conifer–hardwood forest of the Great Lakes region. *Ecol. Monogr.* **30**, 1–35.
Moore, T. R. (1974). Pedogenesis in a subarctic environment: Cambrian Lake, Quebec. *Arct. Alp. Res.* **6**, 281–291.
Mueller-Dombois, D. (1964). The forest habitat types in southeastern Manitoba and their application to forest management. *Can. J. Bot.* **42**, 1417–1427.
Nicholson, H. M. and Moore, T. R. (1977). Pedogenesis in a subarctic iron-rich environment: Schefferville, Quebec. *Can. J. Soil Sci.* **57**, 35–45.
Page, G. (1971). Properties of some common Newfoundland forest soils and their relation to forest growth. *Can. J. For. Res.* **1**, 174–192.
Page, G. (1974). Effects of forest cover on the properties of some Newfoundland forest soils. *Can. For. Ser. Publ.* **1332**, 1–32.
Patten, B. C. (1961). Competitive exclusion. *Science* **134**, 1599–1601.
Payette, S., and Morriset, P. (1974). The soils of Sleeper Islands, Hudson Bay, N.W.T. *Can. Soil Sci.* **117**, 352–368.
Persson, A. (1962). Mire and spring vegetation in an area north of Lake Tornetrask, Torne Lappmark, Sweden. II. Habitat conditions. *Opera Bot.* **6**, 1–100.
Pettapiece, W. W. (1974). A hummocky permafrost soil from the Subarctic of northwestern Canada and some influences of fire. *Can. J. Soil Sci.*, **54**, 343–355.
Pettapiece, W. W. (1975). Soils of the subarctic in the lower Mackenzie Basin, N.W.T. Canada. *Arctic* **28**, 35–53.
Péwé, T. L. (1969). The periglacial environment. *In* "The Periglacial Environment" (T. L. Péwé, ed.), pp. 1–10. McGill-Queen's Univ. Press, Montreal, Canada.
Péwé, T. L. (1970). Permafrost and vegetation on flood plains of subarctic rivers (Alaska): A summary. *Ecol. Subarct. Reg. Proc. Helsinki Symp.* pp. 141–142.

Pierce, R. S. (1953). Oxidation-reduction potential and specific conductance of ground water: Their influence on natural forest distribution. *Soil Sci. Proc.* **17,** 61-65.
Pruitt, W. O., Jr. (1978). "Boreal Ecology." Arnold, London.
Putman, D. F. (1951). The pedogeography of Canada. *Geogr. Bull.* **1,** 57-91.
P'yavchenko, N. I. (1967). Some results of station research on the interrelation of forest and bog in western Siberia. *In* "Interrelation of Forest and Bog" (N. I. P'yavchenko, ed.), pp. 1-39. Amerind, New Delhi, India.
Radforth, N. W., and Brawner, C. O. (1977). "Muskeg and the Northern Environment in Canada." Univ. of Toronto Press, Toronto.
Rapp, A. (1970). Some geomorphological processes in cold climates. *Ecol. Subarct. Reg. Proc. Helsinki Symp.* pp. 105-114.
Ray, R. G. (1956). Site-types, growth and yield. *Can. Dep. For., For. Res. Div. Tech. Note* 27.
Ritchie, J. C. (1961). Soil and minor vegetation of pine forests in southeastern Manitoba. *Can. Dep. For., For. Res. Div. Tech. Note* 96.
St. Arnaud, R. J., and Whiteside, E. P. (1963). Physical breakdown and soil development. *J. Soil Sci.* **14,** 267-281.
Salmi, M. (1970). Investigations on palsas in Finnish Lapland. *Ecol. Subarct. Reg. Proc. Helsinki Symp.* pp. 143-153.
Sigafoos, R. S. (1958). Vegetation of northwestern North America as an aid in interpretation of geologic data. *U.S. Geol. Surv. Bull.* **1061-E,** 165-185.
Silker, T. H. (1965). Plant indicators convey species range of accommodations and site-silvicultural-management relations. *Proc. Soc. Am. For. Meet.* 1965 pp. 50-54.
Sjörs, H. (1948). Mire vegetation in Bergslagen, Sweden. *Acta Phytogeogr. Suec.* **21,** 1-299.
Sjörs, H. (1950). Regional studies in north Swedish mire vegetation. *Bot. Notes* 1950 pp. 175-222.
Small, E. (1972). Ecological significance of four critical elements in plants of raised sphagnum peat bogs. *Ecology* **53,** 498-503.
Soil Survey Staff. (1967). "Supplement to soil Classification—A comprehensive system, 7th Approximation. U.S. Dep. Agric. Soil Conserv. Serv. Soil Surv., Washington, D.C.
Spence, D. H. N. (1964). The macrophytic vegetation of freshwater lochs, swamps, and associated fens. *In* "The Vegetation of Scotland" (J. H. Burnett, ed.), Oliver & Boyd Edinburgh.
Tavernier, R., and Smith, G. D. (1957). The concept of braunerde (brown forest soil) in Europe and the United States. *Adv. Agron.* **9,** 217-289.
Tedrow, J. C. F. (1970). Soils of the subarctic regions. *Ecol. Subarct. Reg. Proc. Helsinki Symp.* pp. 189-206.
Tedrow, J. C. F. (1977). "Soils of the Polar Landscapes." Rutgers Univ. Press, New Brunswick, New Jersey.
Thie, J. (1974). Distribution and thawing of permafrost in the southern part of the discontinuous permafrost zone in Manitoba. *Arctic* **27,** 189-200.
Treshow, M. (1970). "Environment and Plant Response." McGraw-Hill, New York.
Van Cleve, K., Viereck, L. A., and Schlentner, R. L. (1970). Accumulation of nitrogen in alder ecosystems developed on the Tanana River flood plain near Fairbanks, Alaska. *Arct. Alp. Res.* **3,** 101-114.
Van Cleve, K., Viereck, L. A., and Schlentner, R. L. (1972). Distribution of selected chemical elements in even-aged alder (*Alnus*) ecosystems near Fairbanks, Alaska. *Arct. Alp. Res.* **4,** 239-255.
Wallwork, John A. (1970). "Ecology of Soil Animals." McGraw-Hill, New York.

Watt, R. F. (1965). Foliar nitrogen and phosphorus level related to site quality in a northern Minnesota spruce bog. *Ecology* **46,** 357–361.

Weetman, G. F., and Webber, B. (1972). The influence of wood harvesting on the nutrient status of two spruce stands. *Can. J. For. Res.* **2,** 351–369.

Whittaker, R. H. (1954). Plant populations and the basis of plant indication. *Angew. Pflanzensoziol.* **1,** 183–206.

Wilde, Sergius, (1946). "Forest Soils and Forest Growth." Chronica Botanica, Waltham, Massachusetts.

Wilde, S. A., and Krause, H. H. (1960). Soil-forest types in the Yukon and Tanana valleys in subarctic Alaska. *J. Soil Sci.* **6,** 22–38.

Wilde, S. A., and Leaf, A. L. (1955). The relationship between the degree of soil podzolization and the composition of ground cover vegetation. *Ecology* **36,** 19–22.

Wilde, S. A., Wilson, F. B., and White, D. P. (1949). Soils of Wisconsin in relation to silviculture. *Wis. Conserv. Bull.* **525-49,** 1–171.

Wilde, S. A., Voigt, G. K., and Pierce, R. S. (1954). The relationship of soils and forest growth in the Algoma District of Ontario, Canada. *J. Soil Sci.* **5,** 22–39.

Wright, J. R., Leahey, A., and Rice, H. M. (1959). Chemical, morphological, and mineralogical characteristics of a chronosequence of soils on alluvial deposits in the Northwest Territories. *Can. J. Soil Sci.* **39,** 32–43.

Zoltai, S. C. (1972). Palsas and peat plateaus in central Manitoba and Saskatchewan. *Can. J. For. Res.* **2,** 291–302.

Zoltai, S. C., and Pettapiece, W. W. (1974). Tree distribution on perennially frozen earth hummocks. *Arct. Alp. Res.* **6,** 403–411.

Zoltai, S. C., and Tarnocai, C. (1971). Properties of a wooded palsa in northern Manitoba. *Arct. Alp. Res.* **3,** 115–129.

5 Boreal Communities and Ecosystems: The Broad View

The goal of ecosystems modeling is to construct computer programs that, to some useful extent, duplicate, represent, or conceptualize the forces and processes at work in natural ecosystems. A model performs as a dynamic mathematical representation of the ecosystem under conditions in which elapsed time is one variable and many or most of the functions are nonlinear. The accuracy with which the model represents the real world determines its predictive usefulness.

This definition is incomplete; it ignores the single most important functional aspect of a natural ecosystem—the fact that it is composed of organisms, populations of individuals represented by from one to many species. Any reasonably good understanding of an ecosystem presupposes a thorough knowledge of the behavior of the individuals of which it is composed, on the assumption that, if individual behavior is understood, then the behavioral characteristics of the population can be inferred. At the present time, the knowledge required to ascertain whether even this assumption is justified is not available. As a consequence, computer modeling of the boreal ecosystem is in the rudimentary stages of development. Moreover, techniques of computer modeling are still undergoing development; until they are perfected, it cannot be hoped that models will function satisfactorily. It does seem reasonable to expect, however, that good models of certain facets of the system will be developed and, perhaps most significant, they will create an awareness of the questions that model development has made it necessary to ask. There is expanding recognition that a wide range of information must be accumulated before modeling can be a success. It is evident that a comprehensive knowledge of boreal community dynamics must be accumulated before modeling can proceed to a stage at which the results are not wildly unrealistic.

It is axiomatic that the identity and behavioral characteristics of organisms—each representative of a species—be known and that the role of individuals as well as species populations in any community under investigation be understood. The species identified will include at least the vascular plants and the higher animals, as well as the more abundant invertebrates, common lichens, fungi, and any other organisms playing significant roles in the ecosystem. In terms of interactions among individuals, the available knowledge is still in outline at best, and practically nonexistent in the case of population phenomena involving such factors as competition and individual or species survival probabilities under any given set of circumstances. Yet these bear importantly on community composition and structure. Not to incorporate into a model the important events that occur in nature is obviously to invite disaster in terms of descriptive or predictive capabilities of the model.

The changes that occur in community structure over periods of time as a consequence of succession, fire, insect infestation, disease, changes in nutrient status of soils, and so on, must also be taken into account in modeling, and all are exceedingly complex from the point of view of individual or species survival in a community. To date it has been possible only to treat the plant biomass as a collective entity, with the biomass of animals described in some trophic relationship to the plants—grazers, browsers, decomposers, predators, and so on. Moreover, descriptive modeling has not yet progressed to a point where the input–output relationships of boreal forest communities are known, so it is evident that understanding of the internal dynamics of the system is elementary at best. In a sense, we are on the outside of the system looking in.

Living organisms regulate, within limits, the course that events will take. They exercise control, within limits again, over the conditions that prevail in the environment. Living things are not passive agents at the mercy of brute-force environmental influences; with the exception of extreme environments (deserts or polar regions, for example) living organisms modify environmental conditions in ways not duplicated by inanimate systems. Living systems are in at least this respect unique, and their complexity, as a consequence, is more pronounced than that of known inanimate systems. There are differences among species in response to levels of nutrients and to light conditions, moisture, and temperature; there are differences among populations of individuals of the same species (ecotypes) in response to these factors. During succession, as conditions change, some species drop out of the community and others make an appearance. A species occurring in abundance at one stage of succession may disappear from the community at another stage.

These examples are illustrative of the kind of information that must be incorporated into a community model—information that, in large measure, remains to be obtained for many boreal communities.

PLANTS AND ENVIRONMENT

There is an extensive literature available on the effects of temperature, light, and moisture upon the growth and developmental patterns of plants. It is thus simplistic to assert that they respond to environmental conditions in ways that are readily demonstrable and consistent throughout entire populations of genetically similar plants. Through selective pressures, each species has developed an individual pattern of response to the various environmental factors. A species maintains its population by close adjustment to physical factors, as well as to such biotic factors as competing plant species, arthropods, insects, ungulate grazers, browsers, and a host of microorganisms. On exploring these generalizations in detail, however, it is easy to become involved in complexities that defy description. The natural environment varies greatly from place to place, even between sites only centimeters apart. Strict genetic uniformity in organisms is a rarity, and physiological and morphological variation within a given species is often very great. All these factors have a bearing upon the success with which a plant of a given species can seize and hold a small portion of the land surface. In boreal regions plant densities are sufficiently high to indicate that at least some degree of competition usually prevails. In communities where competition exists, small differences in response determine the performance of a species in terms of density or frequency relative to that of individuals of other species present. It is environmental response, for example, that gives *Vaccinium vitis-idaea* an advantage over *Vaccinium uliginosum* on the tussock summits of a muskeg, while the latter has the advantage on adjacent upland slopes. The different responses of individual species to environmental conditions can be demonstrated both locally, as in the example above, and regionally, since the composition of plant communities gradually undergoes changes over distances measured in kilometers, changes that can be correlated with regional climatic gradients and other environmental factors. For example, the manner in which an environmental gradient is coincident with the abundance of species in transition zones has been demonstrated at the southern edge of the boreal forest in central North America, where the transition from boreal forest to deciduous forest southward occurs through a northern conifer–hardwood zone around Lake Superior (Maycock and Curtis,

1960). In site preference, species in the forest communities of this region can be arrayed along a moisture gradient, each attaining its highest abundance typically at some point along the gradient and declining in abundance toward wetter or drier conditions. This can be taken to indicate that each species behaves independently of the others in respect not only to moisture conditions but to the total environmental complex. When plotted on a graph with abundance on the vertical axis and moisture conditions on the horizontal, the data for most species typically follow a normal curve. A study of 110 stands in the region showed that no two species had identical tolerance amplitudes, each behaved independently though interrelated with the others, no species occupied a position to the exclusion of the others, and no group of species was distinct in its ecological tolerances. The gradually changing total environmental complex northward was revealed in the gradual change in floristic composition toward greater representation by boreal species (Maycock and Curtis, 1960).

It is a common observation that we often can see a transition in the kind of vegetation found along a line from a meadow to the summit of an adjacent hill. This transition is usually gradual, without abrupt zones of demarcation between, for example, meadow, muskeg, and forest, unless abrupt changes in topography or a history of fire or other disturbance are also involved. The differences between the plant communities on the extreme sites, however, will be striking, and density or frequency tabulations of the species will demonstrate that the communities differ greatly in terms of species present or the abundance in which they occur. When we observe such a continuum, we realize that we are dealing not only with regional climatic adaptations but with microclimatic and edaphic adaptations as well, expressed in competitive advantages for certain species over a distance of a few feet up- or downslope. We can safely infer that the ecological complexity in such situations is very great indeed. When we consider that most species—many of which possess roughly identical geographical distributions—are found with hundreds of other species (at one place or another and at one time or another), then we appreciate that the subject matter of ecology not only deals with physiological adaptations but also with such factors as migration, mutation, gene exchange, the effects of natural disturbance, climatic variation, fire, frost action, grazing, browsing, and so on. Only when we understand the autecology of all species will we understand the synecology of the vegetation. Initial approaches have been through synecology. Species that appear to have similar patterns of physiology and behavior are grouped into smaller, and conceptually more manageable, units. For most of the species comprising the boreal vegetation, this work has only begun, but a fund of knowledge on the more prevalent plant associations has become available. Fortunately the vegetation of vast tracts of land in northern Canadian boreal regions is still in a relatively natural condition, so that it is possible to study relationships between species in native communities relatively undisturbed by human land use.

The author has conducted a fairly extensive program of sampling of major communities in the boreal regions of central Canada, with data also from western and eastern regions, but a discussion of the broad geographical trends in the composition of these sampled vegetational communities will be presented in Chapter 6. In the discussion that follows here, much other material has been taken from early papers on the botany of the regions, in which the emphasis is on description. It is the purpose, in this chapter, to furnish an overall description of the boreal communities, leaving quantitative comparisons until Chapter 6.

THE CIRCUMPOLAR BOREAL FOREST

The circumpolar boreal forest of North America, Europe, and Asia can here be accorded little more than a broadly descriptive summary, which, it is hoped, will convey some general concept of the similarities and differences in the composition of vegetation between one region and another. There is now a growing literature on the botany of boreal regions, but much is published in obscure journals or, in the case of Russian literature, is inaccessible or untranslated. The references given in the text are those the author found most useful for the purposes at hand; they are not always the most recent, some are, in fact, many years old, but they are unusually lucid or they afford, by reason of good bibliographies, an entry into the literature. There is no effort here to compile a comprehensive bibliography, or even a good working one, but the references cited and the additional ones given in the references at the end of the chapter will provide the interested reader with a place to begin the search for more, using the modern reference library facilities now available at most universities.

The region to be considered is a vast one, and as a consequence there can be little attention devoted to minor regional characteristics and peculiarities; yet it is precisely these regional individualities that must be considered in any practical assessment of the boreal forest other than a broad geographical survey. The circumpolar boreal forest is not in itself a community; it is made up of an unknown number of individual communities, regional and local, and each can be considered individually and, in fact, for accuracy must be. In this context, a statement by Rowe

(1961)* is pertinent, and while in subsequent chapters we will refer to his views in greater detail, it should be noted at this point that the concept of community as applied to vegetation is, at best, a loosely defined one, and one laced with problems of definition and context. Rowe writes:

> The status of the community as an object of study has been argued for many years. It is generally agreed that communities are not organisms, and that their internal integration and organization is loose and variable. In this there is the hint that we are dealing with a nonsystem, and the suspicion is reinforced when an attempt is made to apply the scientific viewpoints of physiology and ecology to any selected community. It is apparent that for an individual plant or animal the viewpoints have meaning, likewise for any geographical ecosystem as a volume of earthly space with its aggregate of organisms. But a forest stand (community of trees) has neither physiology nor ecology, except as it is (1) broken down into its individual trees, or (2) placed within the physical ecosystem of which it is a part.... Furthermore, a skeptical eye can be cast on the frequently expressed (though contradictory) notions that, for the derivation of basic principles, vegetation ought to be studied in its most complex forms (in the tropics) or in its simplest forms (in deserts and tundra). Study of the vegetation in one region need not be productive of useful generalizations in other regions. In fact it is doubtful whether there can be a systematic science of ecosystems, let alone of vegetational communities, with formal laws and general principles, for this is not in the nature of geographic phenomenon. Our world is unique, and the remark is appropriate that geography, like history, does not repeat itself (although geographers, and historians, frequently repeat one another). True, geographic phenomena—mountains, lakes, forest-land—are repetitive in their general outlines and can to a point be studied systematically, but geographic uniqueness limits the scope of generalizations about them all. Only the broadest and least useful principles transcend the necessary confines of a geographic framework. Working principles, formed at the local or regional levels, cannot be expected to have world-wide significance.... On theoretical grounds it does not seem likely that vegetational concepts from the context of mid-Europe will fit the forests of the Canadian boreal region.... Universal rules ought to be suspect in forest science; the need is for regional guides derived inductively.... Too great a reverence for "Nature," coupled with intuitive discovery of her "laws of development," may impose an excessive conservatism on silviculture and on soil science, just as the parallel belief in fictional laws of historical necessity has paralyzed political action.

Regional studies are, thus, a necessity for establishing working principles upon which to base ecologically sound forest management and conservation practices, and to demonstrate, on the basis of limited experience in certain areas, how systems analysis can be applied to such studies. It is the purpose in the next few pages to sketch in outline the broad regional differences existing in the boreal forest in terms of species dominant in the various areas selected, ranging from Alaska, across Canada, hence across Eurasia. This will afford at least an introductory hint of the nature of regional vegetational characteristics; one must remember, however, that there are no abrupt regional discontinuities; as

---

*Reproduced by permission of the National Research Council of Canada.

the work of La Roi (1967) and Larsen (1965, 1974), referred to earlier, has shown, there is a gradient of floristic change in the boreal forest communities in North America, and a corresponding gradient must also exist in the vast expanse of territory from Sweden to eastern Siberia. Evidence, moreover, shows that dominant forest tree species change across the stretch of territory from western to eastern Eurasia, and it is assumed that changes occur also in shrubs and herbaceous species. The composition of the vegetational communities is but one aspect of what Rowe (1961) designates the "geographic ecosystem," but it is an aspect for which there are data available and, thus, we have made a start at characterizing at least the biotic material of which the ecosystem is composed.

THE NORTH AMERICAN BOREAL CONTINUUM

Ranging in an irregular crescentric band across North America, east and west through a number of geological and physiographic provinces, north and south across as much as 10° of latitude, and southward great distances along mountain ranges, the boreal forest would be expected to possess a bewildering variety of diverse communities. The concept of complexity is relative, however, and while boreal communities can, indeed, be described as diversified, they are not as intractably so as tropical forests, deciduous forests, and prairies. In appearance the boreal forest is remarkably uniform, a consequence of the fact that the dominant trees are conifers; there are regional variations, but there is also a remarkable absence of the kind of diversity found in, for example, rain forests. Over thousands of square kilometers, the boreal forest is dominated by trees of only four genera—*Picea, Abies, Larix,* and *Pinus,* all coniferous; *Betula* and *Populus* are also present in many areas—both are broad-leaved, and both are represented by two species. In the case of many genera, there are remarkably similar species that have intermediate forms and a variety of subspecies, varieties, and ecotypes. The same is true of many shrub and herbaceous genera. A large number of shrubs and herbaceous species are circumpolar in range; some shrubs, such as willows (*Salix*), for example, embrace wide-ranging species once considered to be two or more distinct entities that are now incorporated into a single species, the consequence of more complete collections that gave taxonomists intermediate specimens that were seen to grade morphologically into those formerly considered representative of two distinct species.

The broad general characteristics of the boreal forest in North America have been admirably summarized by Rowe (1972):

The boreal forest region comprises the greater part of the forested area of Canada, forming a continuous belt from Newfoundland and the Labrador coast westward to the Rocky Mountains and northwestward to Alaska. The white and black spruces are characteristic species; other conifers are tamarack, which is absent only in the far northwest, balsam fir and jack pine, prominent in the eastern and central portions, and alpine fir and lodgepole pine in the extreme western and northwestern parts. Although the forests are primarily coniferous, there is a general admixture of broad-leaved trees such as white birch and its varieties, trembling aspen and balsam poplar; the latter two species playing an important part in the central and south-central portions, particularly in the zone of transition to the prairie. In turn, the proportion of black spruce and tamarack rises northward, and with increasingly rigorous climatic and soil conditions the closed forest gives way to the subarctic open lichen-woodland which finally merges into tundra. In the east a considerable intermixture of species from the Great Lakes-St. Lawrence forest such as eastern white and red pines, yellow birch, sugar maple, black ash, and eastern white cedar occurs.

In his most interesting and revealing study of the regional variation in composition of the boreal forest communities in Canada, La Roi (1967) made the observation that in stands dominated by white spruce, balsam fir, or subalpine fir there was no balsam fir associated with white spruce in Alaska and western Yukon and that there was no white spruce associated with balsam fir in southeastern Newfoundland. Balsam fir occurred in every stand studied by La Roi as far west as Saskatchewan and then dropped out. Paper birch was recorded in every stand southeast of Great Slave Lake, was mostly absent from mountain and foothill stands in northwestern Canada, and appeared in stands in the Alaskan interior. Westward from James Bay, aspen and balsam poplar were present in about two-thirds of the stands, including older ones. Black spruce ranked second after paper birch as an associate of white spruce and fir. Jack pine was present in a few stands in central Canada. Subalpine fir and lodgepole pine occurred with white spruce in the boreal–subalpine taiga transition of the Canadian Rockies. Red spruce and yellow birch occurred with the balsam fir–white spruce in boreal–subalpine taiga transition areas of the Atlantic provinces.

In the stands dominated by black spruce, La Roi found balsam fir present in all stands east of Lake Winnipeg and absent from all stands to the west. White spruce was rare in the stands of the Laurentian uplands and Atlantic provinces but became the most frequently associated species west of Lake Winnipeg. Paper birch was rare in the east and was a common associate in the west.

Discussing the trends in composition of the white spruce and black spruce stands studied in this trans-Canadian transect, La Roi states that

> There are definite geographic trends in the composition of the tree stratum in both series, with the ranges of individual species overlapping to a greater or lesser degree, presumably in accordance with ecological similarities and postglacial migration patterns.... The main differences between the ranges of tree species in the two series

are as follows. Balsam fir occurred farther west in association with white spruce than with black spruce. White spruce was more frequently associated with balsam fir than with black spruce in the East. Balsam poplar was much more restricted in range of occurrence in black spruce than in white spruce stands. By contrast, jack pine and larch were more widely distributed in black spruce stands.

Herb and shrub species show similar trends, as do bryophytes (La Roi and Stringer, 1976), with species prominent in the east often not found in the western stands of white spruce and black spruce in the series. On the other hand, a few species are found essentially throughout the range of the sampled stands. In white spruce stands, these species were *Epilobium angustifolium, Pyrola secunda, Moneses uniflora, Linnaea borealis, Mitella nuda, Cornus canadensis, Goodyera repens,* and *Lycopodium annotinum.* In black spruce stands, these were *Geocaulon lividum, Goodyera repens, Epilobium angustifolium, Pyrola secunda, Cornus canadensis,* and *Linnaea borealis.* Other species did not occur throughout the series, but had very high presence values in one region or another. They are too numerous to list here, but La Roi summarizes by pointing out that more than 70% of the species in both ecosystem types occurred in less than 20% of the sampled stands; 32 species exceeded 40% presence in white spruce–fir stands and 27 in black spruce stands. Only a small fraction of the total sampled flora was consistently associated with either one or another of the ecosystem types, although a substantial number of species were significantly higher in one type than the other. There were 16 vascular plant species that occurred in the understory of more than 60% of the sampled white spruce stands, including (trees) *Picea glauca, Betula papyrifera, Picea mariana;* (shrubs) *Viburnum edule, Rosa acicularis, Ribes triste;* and (herbs and dwarf shrubs) *Linnaea borealis, Cornus canadensis, Pyrola secunda, Mitella nuda, Maianthemum canadense, Rubus pubescens,* and *Moneses uniflora.* In black spruce stands, 12 species occurred in the understory of more than 60% of the sampled stands: (trees) *Picea mariana, Betula papyrifera;* (shrubs) *Ledum groenlandicum, Rosa acicularis;* and (herbs and dwarf shrubs) *Cornus canadensis, Linnaea borealis, Maianthemum canadense, Pyrola secunda, Gaultheria hispidula, Coptis trifolia, Geocaulon lividum,* and *Vaccinium myrtilloides.*

Each species with a high presence value showed, for the most part, a definite range in which it was a member of the given ecosystem type; some species show a wider range in one community than in the other. Moreover, as La Roi points out, there are five reasonably well-defined floristic discontinuities, zones where several species cease to occur (or begin to occur, depending on the direction of travel). It can thus be seen that there are definite patterns of association in the composition of boreal forest communities across the entire expanse of Canada, and that it is possible to describe and compare the interrelationships of the species commonly in the white spruce and the black spruce communities. It is, thus, possible to discern community differences that make it necessary to understand regional community composition and incorporate the unique regional structure in ecosystem models designed to simulate the boreal forest ecosystem. Systems models are necessarily regional and are constrained in descriptive and predictive capabilities to the region on which they are based. It may be possible to adapt a model for one region to the conditions of another, but as yet no model has been sufficiently perfected to determine whether this can be done. Species differences, in short, may or may not affect the system sufficiently to require major changes in a model if it is to be employed in an area other than the one for which it was designed.

From La Roi's detailed study of the regional variation in Canadian boreal white spruce and black spruce ecosystem types, we turn to somewhat more general descriptions of the vegetation found in the various regions of the North American boreal forest zone.

ALASKA

The main northern forest type in Canada, Alaska, and Eurasia is known as the taiga, implying a coniferous northern forest with no admixture of nonconiferous species except birch and aspen. In the interior of Alaska, the principal tree species found on upland areas long undisturbed by fire is the white spruce. When destroyed by fire, the same species may reproduce but, more commonly, paper birch and quaking aspen are the pioneers over newly opened areas. As succession progresses, white spruce enters the birch and aspen stands and eventually, if undisturbed, again attains dominance (Lutz, 1956). Balsam poplar is found along streams and on floodplains. On poorly drained sites, black spruce is invariably dominant and, as Lutz points out, may locally dominate uplands, forming a pure stand in the wake of fire as a consequence of the availability of seed supplies from the serotinous cones of black spruce. When available, seeds from the serotinous cones will reseed areas formerly occupied by white spruce which lack the serotinous characteristic, hence are less readily capable of seeding areas following a fire. Black spruce trees usually retain considerable amounts of viable seed at any given time, protected in closed cones which open after being subjected to the heat of a fire. The seed supply, stored in the tops of the trees, is seldom completely consumed even in the most severe fire; cones may be charred, but the seed remains alive. It is this characteristic that accounts for the abundant regeneration of black spruce sometimes seen after fire over areas where black spruce trees had been present only in relatively small numbers. Black spruce regeneration is most successful, and seedling growth most rapid, on mineral soil receiving full sunlight; thus, fire that consumes the upper layer of soil litter and humus provides an optimum seedbed for reestablishing black spruce. Under forest conditions, black spruce also reproduces by layering, in which new individuals arise at points where lower branches touch the ground. The most extensive stands of black spruce, however, are found on poorly drained habitats where competition from other species is slight. Both paper birch and aspen can regenerate by sprouting following fire, and this ability probably accounts in large measure for the pure birch or aspen stands that dot the landscape throughout the Alaskan interior, forming light-green deciduous islands throughout the darker coniferous forests.

Evidence that upland black spruce forest in at least parts of Alaska tends to become converted to a wet sphagnum bog forest has been obtained by Heilman (1968, 1969). The progression is accompanied by replacement of feather mosses by species of *Sphagnum* and the change is influential in reducing available soil nitrogen and phosphorus, with the result that tree growth is slowed and the spruce stand degenerates, ultimately to become a sphagnum bog with a few stunted black spruce. Fire usually sooner or later destroys the community and the cycle is repeated. The process is discussed in greater detail under the subject of succession in boreal forest stands in Chapter 7.

In sampling mature undisturbed closed-crowned stands of spruce in southern central Alaska, La Roi (1967) found that there were usually a few black spruce in the stands dominated by white spruce, and a few white spruce in the black spruce stands, with *Betula papyrifera* and *Populus balsamifera* also represented in low numbers. The species found in abundance in white spruce stands include *Alnus crispa, Rosa acicularis, Ribes triste, Viburnum edule, Empetrum nigrum, Shepherdia canadensis, Vaccinium vitis-idaea, Pyrola asarifolia, Mertensia paniculata, Equisetum arvense, E. sylvaticum, Galium septentrionale, Epilobium angustifolium, Pyrola secunda, Linnaea borealis, Cornus canadensis,* and *Lycopodium annotinum.* The species found in abundance in the black spruce stands included *Salix arbusculoides, S. glauca, S. bebbiana, S. myrtillifolia, Ledum groenlandicum, Empetrum nigrum, Arctostaphylos rubra, Vaccinium uliginosum, Pyrola asarifolia, Equisetum scirpoides, Mertensia paniculata, Vaccinium vitis-idaea, Geocaulon lividum, Cornus canadensis,* and *Linnaea borealis.*

In the transition region of northern Alaska–Yukon, where the boreal conifer forest grades into tundra, studies on the upper Firth River valley show that here, too, white spruce and black spruce are the dominant species, with tamarack common and poplar, willow, and birch in shrubby forms occurring along streams often far into the tundra. Open woodlands on alluvium are dominated by white spruce, and the understory cover includes many species with arctic affinities: *Vaccinium uliginosum, Rhododendron lapponicum, Arctostaphylos rubra, Salix reticulata, Dryas integrifolia, Silene acaulis, Senecio lugens, Polygonum viviparum, Pedicularis capitata, Hedysarum* species, *Salix alaxensis,* and *S. richardsonii,* as well as *Ledum decumbens, Empetrum nigrum,* and *Vaccinium vitis-idaea.* A moss mat dominated by *Hylocomium alaskanum* may be up to several inches deep. Lichens include *Cetraria richardsonii* and *Thamnolia vermicularis.*

THE CORDILLERA

Eastward from Alaska are the northern ranges of the Rocky Mountain system occupying the Yukon and western Mackenzie, known as the Mackenzie and the Richardson mountains. The mesophytic upland forests on all but the north slopes are dominated by white spruce, ranging in more xerophytic forest phases upward from this altitude to the timberline. Associated with the white spruce are white birch and a number of shrubs and herbs including *Alnus crispa, Ribes triste, Rosa acicularis, Shepherdia canadensis, Viburnum edule,* and *Vaccinium vitis-idaea.* Forests on the north-facing slopes in the region are dominated by black spruce, with common ground species including *Salix glauca, Alnus crispa, Ledum groenlandicum, Vaccinium uliginosum, Arctostaphylos rubra,* and *Andromeda polifolia.* Southward along the Cordillera, into the subalpine zone of Alberta and of the western United States, the forest dominants are *Picea engelmannii, Abies lasiocarpa, Larix lyallii,* and *Populus balsamifera.*

Southern extensions of the subarctic coniferous forest of North America stretch far into the United States, as far as Arizona in the West and North Carolina in the East along the mountain chains traversing the continent from north to south. In the northern Canadian Rockies, as outlined above, the principal species include white and Engelmann spruce, black spruce, subalpine fir, and larch, as well as jack pine and lodgepole pine. Farther to the south in the Rocky Mountains, especially in southern Canada, northern Idaho, Montana, and Wyoming, the subalpine forest dominants include Engelmann spruce, lodgepole pine, limber pine, whitebark pine, lowland white fir, Douglas fir, subalpine larch, mountain hemlock, and western larch, as well as aspen and a few other species. The altitudinal ranges of the various species change with latitude, but in the region of the Medicine Bow Mountains of Wyoming, for example, the subalpine spruce–fir forest reaches its best development between 9500 and 11,000 ft elevation. Both spruce and fir may extend downward in moist canyons to elevations as low as 8200 ft, but the best stands are found between 9800 and 10,600 ft elevation (Oosting and Reed, 1944). The next zone below the spruce–fir forest is predominantly one of lodgepole pine with aspen a frequent associate. Occasional stands of ponderosa pine and Douglas fir are found on the lower margin of the belt of lodgepole pine; below this is shrubby foothill vegetation.

NORTHWESTERN MACKENZIE–YUKON REGION

In terms of zonal distribution of the boreal vegetational communities in Canada, Ritchie (1962) points out that the zonal sequence (north to south) of vegetation in Canada is tundra, forest–tundra, open coniferous forest, and closed coniferous forest; this sequence is followed in Labrador–Ungava and from the western shores of Hudson Bay to the northwest corner of the district of Mackenzie. He adds that the pattern is not maintained in Alaska and the Yukon, where it is obscured by mountainous topography, and that the pattern is also interrupted in northern Ontario where the marine deposits of the Hudson Bay lowlands and the lacustrine deposits of the clay belt region bear extensive areas of fen, bog, and black spruce forest. In Labrador–Ungava, the open forests have both a more continuous and a more luxuriant lichen mat, presumably related to the greater summer precipitation of the eastern region, which restricts forest fires and stimulates lichen growth.

The vascular plant flora of the boreal forest includes *circumpolar species*, with areas more or less continuous in the Northern Hemisphere, including species represented in parts of the range by subspecies; and *American species*, with areas confined to North America and Greenland; and *amphi-Atlantic species*.

In relation to latitude, there are four main categories, with several combinations: *Arctic species* confined to the arctic region; *subarctic species* centered about the northern limit of forested vegetation; *boreal species* centered about the coniferous forest belt; and *temperate species* confined in their main range to regions of grassland and broad-leaved forest.

Ritchie has analyzed the floristic composition of the boreal forest in detail and points out in summary that the circumpolar species are predominantly (68.7%) arctic or subarctic plants while the American species are largely boreal or temperate (74.0%). Almost one-half of the total of both categories combined is composed of arctic-subarctic species.

In northwestern Mackenzie the spruce forest reaches the northernmost extent of its range in North America, occupying the islands and marginal slopes of the Mackenzie River delta almost to the Arctic Ocean (Fig. 9). Environmental conditions along the river and on the delta apparently are somewhat ameliorated over the surrounding uplands; for northward along the river from approximately Inuvik the uplands be-

Fig. 9. The northernmost forest on the North American continent is found where the Mackenzie River enters the Arctic Ocean. Spruce forest on the delta extends a number of miles north of the edge of the forest on the surrounding uplands. The photograph was taken looking northwest over the delta from the east side of the Mackenzie River. Unbroken tundra extends eastward, out of the photograph to the right. The ground cover is composed of willows, dwarf birch, ericoid shrubs, lichens, and an assortment of arctic herbaceous plant species.

come increasingly devoid of trees, while the islands of the delta remain covered with forest. A relatively abrupt line between forest and tundra on the delta islands occurs between Reindeer Depot and Richards Island. To the west of the delta region, spruce occupies the slopes of the foothills of the Richardson Mountains only a considerable distance south of its northern limit on the delta itself. From the delta the forest border trends east and then south toward the Anderson River. Writing on the general aspect of the vegetation of the delta area, Cody (1965) describes the region north of Inuvik:

> The general progression from south to north is from continuous forest to open forest, to trees and tundra with willows and birch scrub, to open tundra, although there are extensive tracts of tundra lying south of tree line and many outliers of trees on the tundra.

Cody points out that higher well-drained coastland supports a dry tundra in which the species represented include *Dryas, Cassiope, Poa* species, and *Trisetum*. Lower areas contain *Carex lugens* and *Eriophorum*. He adds also that the steep west-facing slopes of the Caribou Hills adjacent to the east branch of the Mackenzie support many species with a very restricted distribution in this area, including *Shepherdia canadensis, Selaginella sibirica, Myosotis alpestris* ssp. *asiatica, Lathyrus japonicus,* and *Silene repens* ssp. *purpurata*. On limestone hills near Inuvik, other local species include *Linum lewisii, Woodsia glabella, Cystopteris fragilis,* and *Galium boreale*.

Inuvik is located a few miles within the northern edge of the forest border. The summits of the hills around Inuvik are forested with both black and white spruce, the latter being most conspicuous along the shorelines and the islands of the delta. South of the forest border, black spruce is abundant inland from the river, occupying upland as well as lowland sites. In low muskeg, the distance between black spruce attaining tree size (12 in. basal area at breast height or larger) often exceeds 50 ft, although individuals of smaller size may average about 15 ft apart. In such areas, typical of extreme northern black spruce communities, tamarack becomes an important tree by definition, since it more frequently than black spruce attains a size greater than 12 in. basal area at breast height, and many tamaracks are larger than the largest spruce. Southward from the tundra, the vegetational sequence is, in general, from tundra to tundra with willow and ground birch, to willow and ground birch, to open woodland, and finally to continuous woodland.

It has been pointed out that the dominant environmental feature of this entire region is permafrost, since it underlies all soils at a depth depending upon texture, topographic position, soil moisture, and vegetation. Permafrost at Arctic Red River is about 350 ft thick. Near Fort McPherson it is about 400 ft thick. In the Mackenzie Delta, however, its depth is highly variable, depending upon heat sources provided by lakes and channels, but depths of 350 ft also occur. The mean annual ground surface temperature at a site near Inuvik has been estimated −4.6°C. At Arctic Red River and Fort McPherson the temperature on June 23, 1966, at depths of 2 and 2.5 ft was between −2.7°C and 3.0°C, with the permafrost table at a shallower level (Mackay, 1967). A thawed zone is always found under lakes and streams that have a mean bottom temperature greater than 0°C. The moderating influence of a river has been shown to extend as deep as 100 ft beneath the river and 300 ft back from the shoreline. Beneath a shallow lake (4–5 ft maximum depth) in the Mackenzie Delta, sediments below the lake were unfrozen to a depth of 230 ft (at which bedrock was encountered in the drill hole). At places around the lake, permafrost extended at least 115 ft below the ground surface. In September, the active layer was at a depth of 8.6 ft at the edge of the lake, dipping at a sharp angle to more than 22 ft below the surface at a point 13 ft out from the shore. The active layer was between 4 and 5 ft deep at a point 10 ft inland from the shore, decreasing to about 2–3 ft at a distance 40–50 ft from the lake (Johnston and Brown, 1965).

The forest of the Inuvik area is relatively rich in species. The stands of white spruce, most frequent on uplands or south-facing slopes, include scattered tamarack and aspen. A high proportion of the ground is covered by *Arctostaphylos uva-ursi, A. rubra, Carex* species, *Dryas integrifolia, Empetrum nigrum, Ledum groenlandicum, Potentilla fruticosa, Juniperus communis, Rubus chamaemorus, Salix* species, *Vaccinium uliginosum, V. vitis-idaea,* and *Hedysarum alpinum.* Lichen species include *Cladonia alpestris, C. rangiferina, Cetraria nivalis,* and *C. islandica,* and common mosses are *Rhytidium rugosum, Dicranum fuscescens,* and *Ptilidium ciliare.*

Black spruce on uplands on north-facing slopes have an understory community consisting most abundantly of *Andromeda polifolia, Arctostaphylos rubra, Betula glandulosa, Carex* species, *Cassiope tetragona, Dryas integrifolia, D. octopetala, Equisetum scirpoides, Empetrum nigrum, Ledum groenlandicum, Pedicularis labradorica, Potentilla fruticosa, Pyrola grandiflora, Rhododendron lapponicum, Salix richardsonii, Tofieldia pusilla, Vaccinium uliginosum,* and *V. vitis-idaea.* Lichens include *Cladonia alpestris, C. rangiferina, Cetraria cucullata,* and *Cornicularia aculeata.* The more moist black spruce stands, often in depressions of the landscape, include the above species in some abundance and, in addition, *Oxycoccus microcarpus, Pinguicula villosa, Pedicularis lapponica,* and *Drosera rotundifolia. Cladonia alpestris* is abundant, and *Sphagnum* mosses dominate although the mosses mentioned above are also present along with *Aulacomnium tur-*

gidum. Ritchie (1977) has determined the percentages of cover of various communities in an area near Inuvik, and his approximations show that 45% of the area is occupied by trees or tall shrubs, with the following proportions of each: *Picea*, 35%; *Larix*, 1%; *Betula*, 2%; *Alnus*, 4%; *Salix*, 2%; and *Shepherdia canadensis*, less than 1%. The ground cover community is occupied by the following proportions of species: ericads, 20%; dwarf birch, 7%; *Dryas*, 5%; *Salix*, 4%; *Juniperus*, 2%; Leguminosae, 3%; lichens, 32%; mosses, 4%; and other herbaceous species, 16%.

In general, the species composition of black spruce stands in the area is relatively homogeneous, and Black and Bliss (1978) found that stands varying in age from about 15 to 300 years after fire were not radically different in composition, although the species showed changes in frequency or density values as a result evidently of increased canopy closure with stand age.

Areas south and east of Great Bear Lake have forest communities not markedly different from those in the area around Inuvik, although there is a reduction in species with distinct arctic affinities. The major communities occupying the largest proportion of the undisturbed land surface at Colville Lake, for example, are white spruce on rolling, well-drained uplands and black spruce in the lowland areas. Disturbance by fire has been extensive. Common species include *Andromeda polifolia, Arctostaphylos alpina, Betula glandulosa, Empetrum nigrum, Equisetum scirpoides, Ledum decumbens, L. groenlandicum, Petasites frigidus, Pyrola secunda, Rubus chamaemorus, Salix glauca, Salix planifolia, Vaccinium uliginosum,* and *V. vitis-idaea*. The Keller Lake area in the Cartridge Mountains, south of Great Bear Lake, is forested with relatively dense black spruce which covers both the uplands and lowlands in the gently rolling glacial drift. A relatively dense white spruce lichen woodland is found more rarely, almost invariably on level areas of sandy outwash. The stands of black spruce are often dense, grading perceptibly into wet lowland muskegs with widely scattered, dwarfed black spruce. Jack pine is present but is rare and found on occasional high, sandy uplands. The forested areas are principally black spruce upland with white spruce lichen woodland confined to flat sandy outwash near the shorelines of lakes. The lower areas are sparsely treed open muskeg with a high density of ground lichens, principally *Cladonia* species, and mosses. Species present include those listed above, with the addition of some that are more characteristic of southern areas: *Galium septentrionale, Linnaea borealis, Pyrola secunda, Rosa acicularis, Zigadenus elegans, Cornus canadensis, Petasites palmatus, Rubus acaulis,* and *Salix planifolia*.

To the west of the Mackenzie River in the vicinity of Norman Wells the very gently rolling alluvial plain is about 20 mi wide, and beyond it rise rather sharply the east slopes of the Mackenzie Mountains. The botany of the region was surveyed extensively by Porsild, whose publications (1945, 1951) provide good general descriptions of the area and the history of its exploration, as well as lists of the plants collected.

Plant community studies conducted in this region by the author reveal that at Florence Lake the floor of the valley is forested with stands often including relatively large white spruce. The lower slopes have scattered stunted black and white spruce, and a community with an admixture of boreal and arctic species prevails over much of the surface. In contrast, at Carcajou Lake, which is both deeper in the mountain system and at a higher elevation, a richer arctic component in the vegetation is apparent, and the forested areas in the Carcajou valley tend to be dominated by black spruce. At Carcajou Lake the complex topography and irregular patterns of atmospheric factors—principal among which probably are wind, moisture distribution, radiation, and evapotranspiration—render the pattern of communities over the landscape too complex for easy interpretation. A few general relationships, however, can be readily discerned. The most dense spruce stands on well-drained uplands, for example, are on east- and southeast-facing slopes; this is apparently a response to moisture, since such areas have lower evaporation rates in late afternoon when temperatures are highest, thus creating a relatively more mesic environment. Snow accumulates in winter on eastern slopes, providing greater moisture supplies in spring and adding to the protection of the trees from desiccating effects of winter winds. On low and relatively moist north-facing slopes, spruce is present only as scattered, dwarfed individuals, if at all. The driest slopes are those facing either south or west (Fig. 10).

Wide-ranging arctic and boreal species are found in close juxtaposition to one another throughout the area. To the east, around Ennadai or Dubawnt lakes, for example, *Geocaulon* and *Cassiope* possess range limits that are far removed from one another, but at Florence Lake these two species are found in the same community along with such other arctic representatives as *Dryas, Silene, Rhododendron, Polygonum,* and *Lupinus,* all of which, to the east, are found ordinarily only far to the north of extensive stands of spruce. At Fort Reliance, *Dryas* and *Rhododendron* occupy the same communities, but here this unusual combination is found only on Fairchild Point, for reasons that remain obscure (Larsen, 1973).

In the Carcajou Lake area, transects have been run along a continuous cline from the bottom of a long slope to the summit, over a total distance of perhaps 4000–5000 ft and encompassing an elevation range of about 1500 ft. This series of transects reveals that a number of species are

Fig. 10. At about 5000–6000 ft elevation in the eastern slopes of the Mackenzie Mountains, to the west of Norman Wells, Northwest Territories, the spruce is confined to protected declivities in the terrain. Presumably because of the moisture conditions, the trees grow most vigorously on the east-facing slopes (right center) where they are exposed to direct sunlight only during the morning hours when evaporation rates are low. They grow less vigorously on the colder north-facing slopes and on the drier west-facing slopes. Tundra occupies the ridges and summit areas.

ubiquitous in the tundra tussock community, and that others demonstrate a clear preference for either lower or higher ranges. Thus, species that elsewhere are wide-ranging and abundant are here present throughout the transects with high frequencies: *Saussurea angustifolia, Ledum decumbens, Salix glauca, Betula glandulosa, Dryas integrifolia, Empetrum nigrum, Pedicularis labradorica, Vaccinium uliginosum, Vaccinium vitis-idaea, Arctostaphylos alpina*, a few *Carex* species, and *Eriophorum*.

Species confined to lower slopes where moisture supplies are probably consistently higher, at least during periods with little precipitation, are *Picea mariana* seedlings (so defined because of size rather than mode of reproduction), *Astragalus alpinus,* and *Cassiope tetragona.* On higher slopes there is greater representation of species with arctic affinities, and included in this group are *Saxifraga* species, *Tofieldia coccinea, Carex atrofusca, Carex misandra, Rumex arcticus, Salix arctica, Pedicularis* species, and *Potentilla* species. The highest transect, run on the summit of a high, windswept ridge, reveals a relatively depauperate community closely resembling those found on high rock fields in the Arctic.

High north- and northeast-facing slopes in the Carcajou Lake area often possess a rich vegetation of shrubs, herbaceous species, mosses, and lichens, most of which are either wide-ranging or have arctic affinities. On such slopes, particularly on the lower portions, decumbent or markedly dwarfed spruce may be present. Despite the presence of spruce, however, these communities commonly lack many of the species often associated with spruce forests elsewhere and possess, instead, an aggregation of species found most often in the tundra communities of the summit or the near-summit slopes.

In comparing communities of three areas—Colville, Florence, and Carcajou lakes—striking differences exist in species composition between the forest–tundra transition communities of the interior plains (Colville) and the forest and tundra of the east-facing slope of the Mackenzie Mountains. On the high, exposed tundra slopes around Carcajou Lake there is an aggregation of arctic species occurring with high frequency, which are absent entirely from Colville and Florence lakes. This rather exclusive group of species includes *Armeria maritima, Carex atrofusca, Hierochloe alpina, Luzula confusa, Luzula nivalis, Kobresia simpliciuscula, Lychnis apetala, Potentilla nivea, Polygonum bistorta, Salix arctica, Juncus albescens,* and other species distinctly arctic and alpine in their geographical range.

Another group, arctic in affinities but extending commonly into communities at or near the continental forest border, includes species found both at high elevations around Carcajou Lake and at lower elevations around Florence Lake: *Cassiope tetragona, Lupinus arcticus, Hedysarum alpinum, Anemone parviflora, Pedicularis capitata, Pedicularis labradorica, Pedicularis lanata, Polygonum viviparum, Pyrola grandiflora,* and *Astragalus alpinus.*

Finally, of interest are species, wide-ranging throughout all areas, found in the communities of the interior plains as well as in the communities of the highest elevations around Carcajou Lake. These species include *Andromeda polifolia, Arctostaphylos alpina, Carex membranaceae, Dryas integrifolia, Empetrum nigrum, Eriophorum vaginatum, Ledum decumbens, Pyrola secunda, Rhododendron lapponicum, Salix reticulata, Salix glauca, Vaccinium uliginosum,* and *Vaccinium vitis-idaea.*

Only a few species are noteworthy because they occur in the spruce forests at Colville Lake and appear not to extend into the communities at Florence and Carcajou lakes with sufficient density to be recorded in quadrats. This group, which must be considered not to extend into the slopes in sufficient numbers, at least, to be considered important, includes *Larix laricina, Salix planifolia,* and *Oxycoccus microcarpus.* Most species in the Colville Lake communities are wide-ranging northern boreal species found throughout the forest and northward to the southern edge of the arctic tundra. There is a greater total number of species in the black spruce community at Florence Lake as compared with black spruce stands at Colville, Ennadai, Dubawnt, or Artillery Lake to the east. At the higher altitudes at Carcajou Lake, there is again a marked increase in the number of species in the black spruce community; in this latter area, the number of species in the community is 50–100% greater than in the forest–tundra ecotone of the interior plains region. These relationships can be seen in the following tabulation of the number of species in black spruce communities at the locations indicated. White spruce community data are included when available.

It is apparent that black spruce stands located at the northern edge of the forest–tundra ecotone (Ennadai, Artillery) possess fewer species than stands located a distance of 20 mi or more southward and within the forest (Fort Reliance, Inuvik). In areas such as Dubawnt Lake, located far north of the present-day forest border but nevertheless retaining a few relict stands of what was once more extensive forest, the same paucity of species is apparent. Colville Lake is within the forest but apparently sufficiently near the forest border to demonstrate the depauperate effect.

Area	Number of species in black spruce stands	Number of species in white spruce stands
Ennadai Lake	29 (6)[a]	24 (2)
Dubawnt Lake	22 (1)	
Artillery Lake	22 (3)	39 (2)
Artillery Lake[b]	29 (5)	20 (2)
Ft. Reliance	46 (6)	50 (7)
Inuvik	35 (2)	19[c] (1)
Colville Lake	19 (3)	31 (3)
Florence Lake	34 (2)	59 (4)
Carcajou Lake	44 (3)	

[a] Number of stands included in the sample.
[b] "Pike's Portage" area at the south end of Artillery Lake.
[c] Mackenzie River delta forest.

The species that account for the increase in total numbers in the Carcajou Lake stands are arctic species; at the higher altitudes few species are lost from the stands, and arctic species are gained, for an increase in total number. There exists a continuum in which the relatively rich spruce understory community at lower elevations is replaced gradually at increasing elevation by an equally rich arctic tundra community. There is no zone between the two occupied only by species wide-ranging and ubiquitous throughout both forest–tundra and low-arctic zones.

If it is the behavior of regional air masses and the configuration of the frontal zones in summer that determine the structure and position of the forest–tundra ecotone in northern central Canada (Larsen, 1972, 1973, and see Chapter 2), then we should expect to find no depauperate zone on mountain slopes. The altitudinal climatic gradient shows no zonal air mass or frontal alignments that might be said to correspond to those found in the continental interior. There is, instead, along the altitudinal gradient, a rather uniform change in climatic parameters, and only such an event as persistent cloudiness at a given level would tend to create a zonal effect. At the relatively low elevations of the east slope of the Mackenzie Mountains, clouds often persist at heights over the entire range, but they are usually well above the level capable of inducing cloud–forest effects. The frequency of mist is probably uniform over most of the moutain area, and fog is a local phenomenon of lake basins and stream valleys.

It is thus apparent that rather fundamental differences exist between the altitudinal and latitudinal zonations in vegetation, the result evidently of differences not only in the synoptic air mass climatology of the two gradients but also probably of more subtle differences in vapor pressure and perhaps in the quality of incoming radiation or other factors. There are, however, distinct and interesting parallels in vegetational changes along the two gradients, and it seems that further study of these differences and similarities and their correspondence with the differing physical characteristics of the two environmental gradients will yield interesting information on the ecology of northern vegetation in general.

SOUTHWESTERN MACKENZIE AND NORTHERN ALBERTA

The interior plain of central Canada stretches from the foothills of the Cordillera eastward to the western edge of the Canadian shield (Bostock, 1970), encompassing large areas of western Mackenzie and northern Alberta. The most common of the forest tree species in the region are black spruce, white spruce, jack pine, lodgepole pine (*Pinus contorta* var. *latifolia*), aspen, and balsam poplar, and less common are tamarack and balsam fir. White spruce is typically found in the greatest abundance on uplands and along rivers, with black spruce more commonly predominating at lower elevations over the landscape. Black spruce occupies a greater proportion of the total land surface than is indicated in the reports and maps of the original explorers of the region (Fig. 11), and

Fig. 11. Looking into a closed-canopy black spruce stand from the edge of a cutting, the late afternoon sunlight can be seen to penetrate what otherwise would be a forest interior too dark to photograph well even in full sunlight. Opening of the canopy will eventually permit heavier growth of shrubs and herbaceous species, now almost entirely absent.

one infers that the inaccuracies are due to the fact that travel was restricted largely to the waterways where white spruce is more abundant. Inland a mile or more from the water, black spruce often is the dominant tree in both upland and lowland areas, sharing the land surface with jack pine and white spruce. The influence of fire is seen everywhere, and the better adaptation of black spruce for regeneration following fire may help account for black spruce prevalence in areas more distant from the moist shorelines of rivers and lakes. Muskegs and wet lowlands have black spruce predominating, with tamarack a frequent associate, along with *Ledum groenlandicum,* willows, and an assemblage of other frequent species (Raup, 1946; Moss, 1953b; Thieret, 1964).

Taxonomic studies have shown that a number of the members of important northern genera are highly polymorphic as a consequence of hybridization and introgression among closely related species. As a consequence, the difficulties encountered in field identification of species during ecological work are considerable. Thus, *Betula glandulosa* and *B. glandulifera* hybridize to form *B.* × *sargentii,* and the latter can then hybridize with *B. papyrifera* to form *B.* × *arbuscula* (Dugle, 1966). These produce a hybrid swarm of plants with bewildering morphology. They are, moreover, often important members of the plant communities in northern and northwestern Canada, and have long been a source of confusion because of the intermediate characteristics possessed by individual plants. The willow *Salix glauca* is another representative highly variable species that Argus (1965, 1973) has studied extensively, coming to the conclusion that it is advisable to treat the species in a broad sense. Argus recognizes four geographical variants, even though, as he points out, specimens resembling each of the variants occur within the range of the others. The genus *Picea* is another in which taxonomic problems confront the field ecologist. In the spruces it is apparent that both environmental stress and introgressive hybridization are at work to produce clinal variation in morphology, with the resulting uncertainty in identification of individual trees in the field (Roche, 1969; La Roi and Dugle, 1968). This difficulty appears to be least severe in the case of *Picea mariana* and *P. glauca,* although forms with confusing intermediate characteristics have been reported in central Canada (Dugle, 1971), in forest border areas of the forest–tundra ecotone in northern Canada (Larsen, 1965), and elsewhere. Hybridization between *P. mariana* and *P. glauca* was long considered to be virtually absent in nature although the species are sympatric across most of North America. Reports of apparent hybridization have been questioned on the basis of subsequent detailed morphological analyses (Gordon, 1976; Parker and McLachlan, 1978). Taxonomic studies should be undertaken, however, from northern forest border areas where, at the edge of the range of the species, the possibility of hybrids finding suitable habitats for survival may be higher than is the case in central parts of the range of the two species (Roche, 1967).

There are other genera in which a similar situation exists, including *Oxycoccus* (*Vaccinium oxycoccus*), *Rosa, Carex,* and perhaps others. For the purposes of the studies undertaken by the author and described in Chapter 6, field identifications of plants in the groups presenting difficult taxonomic problems have necessarily been provisional. For the purposes of statistical analysis, however, the specimens have, at times quite arbitrarily, been assigned to one species or another on the basis of subjective judgment. It should be understood that the assignment of a specimen as, for example, *Betula glandulosa* does not preclude its possessing morphological characteristics intermediate with *B. glandulifera* or *B. papyrifera* or both. The same applies to *Oxycoccus microcarpus, Rosa,* and the difficult *Carex* groups. In the listing of species presented in Appendix VI, the word "complex" is used in conjunction with species that are members of difficult groups, and it should be understood that the problems encountered with these species in the ecological studies were insurmountable and the results should be interpreted accordingly. Thus, *Betula glandulosa* specimens from the southern parts of the boreal forest region probably possess *B. glandulosa* characteristics, or may indeed actually be *B. glandulosa,* and in the far northwest or north may be represented by forms intermediate with *B. nana*. Hultén (1968) shows *B. glandulosa* and *B. nana* ssp. *exilis* with overlapping range in northern Canada, and he indicates that *B. nana* ssp. *exilis* forms complete introgression with *B. glandulosa* where the ranges overlap. He adds that hybrids with the tree birches are not rare. The result in terms of taxonomic accuracy in ecological field studies leaves much to be desired, but it also reveals the clinal nature of the regional variations in these difficult species groups, and this, in itself, may be of some scientific value. The groups in which this occurs with greatest frequency are those that include the species *Betula glandolosa, Oxycoccus* species (*Vaccinium oxycoccus, V. microcarpus*), *Salix glauca, Rosa acicularis, Carex rotundata,* and perhaps *Picea* and *Pinus,* particularly *Picea glauca,* as well as *Pinus banksiana,* in the West (where intergradation with *P. engelmannii* and *P. contorta,* respectively, is frequent). The question of *Picea glauca* and *P. mariana* at the northern forest border in the central regions remains open. The designation *Picea glauca* here refers to any and all varieties and intermediates. In northern Alberta and southwestern Mackenzie, white spruce communities range in character from those in which the spruce trees are extremely crowded, with the associated vegetation sparse and the ground cover composed chiefly of spruce needles, to open stands in which the trees are some distance apart and the ground vegetation is characterized by a prominent shrub stratum in which *Shepherdia canadensis* locally northward and *Viburnum, Rosa,* and *Ribes* species are common. The rich herb component includes *Linnaea borealis, Rubus pubescens, Mertensia paniculata, Fragaria* species, *Pyrola* species, *Epilobium angustifolium, Calamagrostis canadensis, Cornus canadensis, Mitella nuda, Viola renifolia,* and *Petasites palmatus.* In the more open stands there is usually a mixture of aspen and poplar as well as willow, often *Salix bebbiana,* and paper birch. The feather moss forest stands are intermediate between these two extremes in tree density; they have a ground cover consisting of extensive mats of mosses with *Hylocomium splendens* and *Calliergonella schreberi* predominating and such herbaceous species as *Galium septentrionale, Maianthemum canadense, Vaccinium vitisidaea, Equisetum* species, *Geocaulon lividum,* and *Aralia nudicaulis* prevalent. The shrub *Alnus crispa* is also present in many of the stands (Moss, 1953b). The closed canopy admits enough light for feather mosses but excludes most shrubs and herbaceous species. As Moss (1953b) points out, it is doubtful that the feather moss forest is self-perpetuating. The mosses replace shrubs and herbaceous species present in younger and more open associations and, eventually, a feather moss forest is almost invariably subject to destruction by fire and its successional history is then repeated.

In the feather moss–white spruce stands of more southern areas, the mat of mosses is often 6 in. deep in places, and the ground cover is composed of a rather long list of secondary species, including *Equisetum sylvaticum, Equisetum scirpoides, Lycopodium annotinum, Goodyera repens, Habenaria obtusata, Orchis rotundifolia, Corallorrhiza trifida, Calypso bulbosa, Ribes lacustre, Ribes triste, Arctostaphylos rubra, Moneses uniflora,* and *Peltigera aphthosa.*

A white spruce–feather moss stand along the Kakisa River in southwestern Mackenzie as described by Thieret (1964) possesses trees of more than 28 in. diameter at breast height, 130 ft tall, and more than 180 years old:

> They cast a dense even shade. The floor is deeply carpeted with *Hylocomium splendens* into which the walker sinks three or four inches at every step. The moss and peat layer is thick enough so that . . . I struck frozen peat (about 16 inches down in mid-July) before I was able to reach mineral soil. Reproduction of the spruce is good. *Cladonia rangiferina* occurs in small scattered patches, and *Peltigera aphthosa* is common. *Alnus crispa,* to about eight feet tall, forms a more or less definite understory. Parasitic on the roots of the alder, and very rare, is *Boschniakia rossica,* the only Orobanchaceae known in the region. Other plants in the forest are much scattered and include the shrubs *Juniperus communis* var. *depressa, Rosa acicularis, Shepherdia*

canadensis, Arctostaphylos rubra, Ledum groenlandicum, Vaccinium vitis-idaea var. *minus, Linnaea borealis* var. *americana,* and *Viburnum edule;* and the herbs *Cystopteris montana, Carex concinna, Corallorhiza trifida, Cypripedium guttatum, Geocaulon lividum, Actaea rubra, Hedysarum mackenzii, Mitella nuda, Moneses uniflora, Pyrola asarifolia,* and *P. grandiflora.*

The majority of predominating species range widely throughout both northern and southern portions of the interior plains occupied by spruce forest but, as Raup (1946) points out, many species of secondary abundance drop out northward; the following are occasional to common around Lake Athabasca but were not observed, at least in numbers, at the eastern arm of Great Slave Lake: *Lycopodium tristachyum, Lycopodium obscurum* var. *dendroideum, Schizachne purpurascens, Oryzopsis pungens, Cypripedium acaule, Goodyera repens* var. *ophioides, Mitella nuda, Amelanchier florida, Sorbus scopulina, Prunus pensylvanica, Cornus canadensis, Aralia nudicaulis, Chimaphila umbellata* var. *occidentalis, Arctostaphylos uva-ursi* var. *adenotriche, Vaccinium canadense, Trientalis borealis, Lonicera glaucescens, Campanula rotundifolia, Tanacetum huronense* var. *floccosum.*

Jack pine is common throughout the region and extends northward almost to the forest border in some areas. Fairly extensive dense stands exist along Pike's Portage route between Great Slave and Artillery lakes, just south of the forest border. It is perhaps noteworthy that farther east, south of Ennadai Lake, jack pine ceases to be a component of the forest many miles farther south of the northern tree line than is the case in Mackenzie. It is, however, uncommon in the northeastern Great Slave Lake area and cannot be considered a major component of the forest, although it becomes more abundant southward. As for the floristic component of the spruce forest in the northern parts of the region, a number of species of northern affinity found in the open spruce woods of Great Slave Lake do not appear in abundance in the forests around Lake Athabasca, although some occur in muskegs (Raup, 1946): *Tofieldia palustris, Cypripedium passerinum, Dryas integrifolia, Dryas drummondii, Hedysarum alpinum* var. *americanum, Oxytropis viscida, Ledum decumbens, Pedicularis labradorica, Solidago multiradiata, Erigeron compositus* var. *trifidus.* Raup also notes that more than one-half the northern species not found near Lake Athabaska inhabit the more xerophytic phases of the spruce woods, while nearly all southern species that fail to extend north to Great Slave Lake represent the more mesophytic phases of the Lake Athabasca spruce.

Northward the more common species found in white spruce forest, as for example in the stands on Fairchild Point in the eastern arm of Great Slave Lake, include *Empetrum nigrum, Arctostaphylos rubra, Vaccinium vitis-idaea, V. uliginosum, Ledum groenlandicum, L. decumbens, Alnus crispa,* and *Betula occidentalis*, and the species found in lesser abundance include *Equisetum scirpoides*, *Equisetum pratense*, *Tofieldia palustris*, *Cypripedium passerinum*, *Calypso bulbosa*, *Arenaria lateriflora*, *Aquilegia brevistyla*, *Rosa acicularis*, *Pyrola secunda*, *Pyrola virens*, *Pyrola asarifolia* var. *incarnata*, *Gentiana amarella*, *Linnaea borealis* var. *americana*, and *Senecio cymbalaroides* var. *borealis*, as well as the following tha occur more abundantly on ridges: *Calamagrostis purpurascens*, *Geocaulon lividum*, *Dryas integrifolia*, *Rosa acicularis*, *Epilobium angustifolium*, *Arnica lonchophylla*, and *Senecio cymbalaroides* var. *borealis*.

Distinct from the upland stands of white spruce in the northern parts of the interior plains are those stands occupying the ancient lake beaches and sand plains that border existing lakes. The parklike spruce forest (Fig. 12) characteristic of these areas in the northern forest-tundra trans-

Fig. 12. Ancient beach at the eastern extremity of Great Slave Lake, several hundred feet above the present level of the lake water; the sandy former beach, parts of which are devoid of trees, is covered with a thin forest, and in the foreground the sand is covered with lichens, principally *Stereocaulon*.

tion zone has trees widely spaced and branched to the ground. Along with black and white spruce are found paper birch and the following shrub and herb species: *Empetrum nigrum, Arctostaphylos uva-ursi, Juniperus communis* var. *montana, Calamagrostis purpurascens, Salix bebbiana, Betula occidentalis, Pulsatilla ludoviciana, Saxifraga tricuspidata, Geocaulon lividum, Empetrum nigrum, Epilobium angustifolium, Arctostaphylos uva-ursi, Vaccinium uliginosum, Vaccinium vitis-idaea* var. *minus, Pedicularis labradorica,* and *Solidago multiradiata.* The *Cladonia* lichens, especially *C. rangiferina, C. alpestris,* and *C. islandica,* are conspicuous components of the ground cover.

White spruce forest in the upper Mackenzie River area may occur in pure stands or may be infiltrated with varying densities of jack pine, aspen, and poplar, with tamarack and black spruce present in numbers in some stands. White spruce can also occur in stands dominated by one or more of these other species. Under such complex circumstances it is difficult to delineate communities clearly even in an arbitrary manner, for the purpose of identification and comparison. Black spruce forest tends to occupy the wettest of the forested areas, ranging from deep peat-filled (or peat- and marl-filled) depressions to shallower depressions, gentle slopes, and level land wherever the moisture relations favor black spruce, but field observation often reveals black spruce dominant in a stand close to a stand dominated by white spruce, often on adjacent and apparently identical sites. The ground community under each may be distinctive, in other cases it may not. Instances such as one described by Thieret are frequent—in which an understory community "typical" of a black spruce stand is found, for example, beneath a stand of jack pine. Obviously, influences of succession, or chance invasion following disturbance by fire or other causes, govern these community relationships, and they are not sufficiently understood at the present time to permit realistic description.

Black spruce forest, however, clearly dominates lowland areas, with the pattern of lowland forest and open muskeg following the microtopography of peat accumulation; open muskeg is characteristically occupied by *Sphagnum* mosses, ericaceous shrubs such as *Chamaedaphne calyculata, Ledum groenlandicum, Ledum decumbens,* and such other common forest species as *Rubus chamaemorus.* The abundance of *Ledum decumbens,* as well as *Rhododendron lapponicum* and *Vaccinium uliginosum,* more arctic than boreal in geographical affinities, raises interesting questions. Here these shrub species attain an unusually large size, in comparison to dwarf stature in the Arctic. Whether this is a genotypic or phenotypic response is apparently not yet known. They are here at the utmost southern extent of their range, as shown in the maps presented by Porsild (1957). That they should locally attain such high stature and density is perplexing.

In the black spruce–feather moss community, the trees grow closely together and frequently include a scattering of white spruce, aspen, poplar, and various willows; the mosses covering the forest floor usually include *Hylocomium splendens, Calliergonella schreberi,* and *Aulacomnium palustre,* with such other species as *Hypnum crista-castrensis, Camptothecium nitens,* and *Sphagnum* species occurring commonly but in somewhat lesser abundance. *Cladonia* lichen species are usually prevalent. Ground cover vegetation also usually includes *Ledum groenlandicum, Vaccinium vitis-idaea, Cornus canadensis, Petasites palmatus, Linnaea borealis, Mitella nuda, Rubus pubescens,* and *Carex* species. The lower branches of the trees have a covering of arboreal lichens, including *Usnea hirta* and *Parmelia saxatilis*. Other species usually common in the black spruce–*Hylocomium* forest include *Peltigera aphthosa, Juniperus communis, Zigadenus elegans, Mitella nuda, Rosa acicularis, Rubus pubescens, Hedysarum* species, *Moneses uniflora,* and *Linnaea borealis*.

In the more southern localities of the interior plains black spruce–feather moss forests generally develop on level upland terrain, and black spruce–peat moss communities are more frequent on depressions in the landscape. Any natural successional trends are generally obscured by the frequent fires that bring about renewal of early stages, although it often appears that the vegetation of recently burned areas is characterized simply by early replacement of species prevalent in the forest at the stage of development when burning occurred. Total destruction of upper peat soil horizons, however, will result in initial pioneering by a herbaceous species such as *Epilobium angustifolium*. As Moss (1953b) points out, whether there is a natural succession of black spruce–feather moss forest to white spruce dominance is problematical, although it has been postulated.

The black spruce–peat moss community is also called a bog forest or a muskeg forest. The trees are usually less crowded than in the black spruce–feather moss association, and tamarack is a common associate as are a number of species of willows. *Sphagnum* mosses dominate the forest floor, although the mosses common in the feather moss forest occur in low densities. The more abundant shrubs and herbaceous species as listed by Moss (1953b) include *Ledum groenlandicum, Vaccinium vitis-idaea, Rubus chamaemorus, Smilacina trifolia, Betula glandulosa, Vaccinium oxycoccos, Equisetum sylvaticum, Eriophorum spissum,* and *Carex* species.

As indicated by Raup (1946), the black spruce–peat moss forest probably covers more land surface in the Precambrian portions of the

Athabasca–Great Slave lake region than it does farther westward, since development of this community is dependent upon undrained depressions in the land surface that are more common in Laurentian country. In addition to the species listed above as common in the community, a wide range of secondary species is found and these, according to Raup (1946) include: *Equisetum sylvaticum, Equisetum scirpoides, Carex gynocrates, Carex disperma, Carex media, Carex capillaris, Eriophorum opacum, Maianthemum canadense, Habenaria hyperborea, Habenaria obtusata, Orchis rotundifolia, Listera borealis, Corallorhiza trifida, Spiranthes romanzoffiana, Salix pyrifolia, Salix glauca, Salix myrtillifolia, Betula glandulosa, Geocaulon lividum, Ranunculus lapponicus, Ranunculus gmelini, Drosera rotundifolia, Parnassia palustris, Ribes hudsonianum, Ribes triste, Mitella nuda, Rubus acaulis, Rubus chamaemorus, Cornus canadensis, Moneses uniflora, Pyrola secunda, Pyrola asarifolia, Pyrola virens, Chamaedaphne calyculata, Arctostaphylos rubra, Vaccinium oxycoccus, Vaccinium vitis-idaea, Vaccinium uliginosum, Linnaea borealis.*

In many areas, such as the vicinity of Yellowknife, Precambrian outcropping rock surfaces are a dominant feature of the landscape and, as Thieret (1964) points out, they constitute a distinctive habitat for forest vegetation. The bases of the outcrops are overlain with mineral soil, and here are found stands of white and black spruce, aspen, paper birch, and willows. *Alnus crispa, Rosa acicularis, Shepherdia canadensis,* and *Viburnum edule* are found surrounding the outcropping bare rocks. On the rocks are a number of species of such genera as *Cladonia, Cetraria, Parmelia,* and the rock tripes *Actinogyra* and *Lasallia.* Common fruticose lichens include *Cladonia alpicola, C. cornuta, C. degenerans, C. metacorallifera, C. mitis, C. pyxidata, C. rangiferina, C. uncialis, C. verticillata, Actinogyra muhlenbergii, Lasallia pensylvanica, Cetraria nivalis, Parmelia centrifuga, P. stenophylla,* and *P. sulcata.*

At the other topographic extreme, the wettest sites are characterized by *Sphagnum* species with *Cetraria* and *Cladonia* lichens and with *Ledum groenlandicum,* mixed northward with *L. decumbens.* Other species in such habitats include *Equisetum palustre, Selaginella selaginoides, Smilacina trifolia, Tofieldia glutinosa, Ranunculus lapponicus, Drosera rotundifolia, Parnassia multiseta, Rubus chamaemorus, Oxycoccus microcarpus, Vaccinium vitis-idaea, Pinguicula vulgaris,* and *Senecio lugens.*

A number of other forest communities occupying relatively large areas in parts of the western and central Canadian southern boreal forest have been described, including tamarack bogs, pine forests, and stands of balsam fir, trembling aspen, and poplar. Moss points out that, although dominated by tamarack, the former are floristically similar to the black spruce-peat moss community. They develop in areas subject to periodic flooding, and *Sphagnum* mosses dominate the understory. Tamarack is intolerant of shade, and thus there may be a tendency for succession to progress toward a black spruce-peat moss forest, but this would require a long period of time during which the stand would have to be unaffected by fire—probably an unlikely situation even considering the usually moist state of the substrate.

Pine forests are abundant in the southern portions of the boreal zone in Alberta and range northward to the area around Yellowknife on sandy hills, ridges, and granitic outcrops with coarse soil in the depressions. As Moss (1953b) describes the community, jack pine predominates, although paper birch and spruce, alder, and various other species of shrubs are common. In these pine woods, the primary species in the ground layers of both types include *Arctostaphylos uva-ursi, Cetraria nivalis,* and *Cladonia rangiferina. Picea glauca* is sometimes abundant, and in the lower strata *Artemisia frigida, Amelanchier florida, Saxifraga tricuspidata, Picea mariana, Vaccinium canadense,* and *V. vitis-idaea* commonly occur.

There appears to be some differentiation in the species making up the jack pine communities on sandy and on rocky substrate types but, according to Raup (1946) and Thieret (1964), species commonly found on both sandy and rock substrates include the following: *Lycopodium annotinum, Festuca saximontana, Poa glauca, Agrostis scabra, Calamagrostis purpurascens, Elymus innovatus, Maianthemum canadense, Goodyera repens, Salix bebbiana, Geocaulon lividum, Rosa acicularis, Shepherdia canadensis, Epilobium angustifolium, Aralia nudicaulis, Galium boreale, Linnaea borealis, Campanula rotundifolia, Equisetum scirpoides, Juniperus communis, J. horizontalis, Bromus pumpellianus, Oryzopsis asperifolia, O. pungens, Carex aenea, C. foenea, Zygadenus elegans, Calypso bulbosa, Corallorhiza trifida, Populus balsamifera, P. tremuloides, Salix glauca, Alnus crispa, Betula glandulosa, B. papyrifera, Arenaria capillaris, Anemone multifida, A. patens, A. parviflora, Aquilegia brevistyla, Ribes lacustre, Amelanchier alnifolia, Fragaria virginiana, Potentilla fruticosa, Astragalus americanus, Hedysarum alpinum, H. mackenzii, Lathyrus ochroleucus, Oxytropis splendens, Empetrum nigrum, Hudsonia tomentosa, Cornus canadensis, Pyrola asarifolia, P. secunda, P. virens, Arctostaphylos uva-ursi, Ledum groenlandicum, Vaccinium vitis-idaea, Apocynum androsaemifolium, Pedicularis labradorica, Galium septentrionale, Lonicera dioica, Viburnum edule, Arnica lonchophylla, Aster ciliolatus, A. sibiricus, Erigeron glabellus, Hieracium umbellatum, Senecio tridenticulatus,* and *Solidago spathulata.*

Raup (1946) adds that the jack pine forest is the simplest, floristically, of any of the woodlands of the region. The long list above is a composite formed from studies in several different regions in Wood Buffalo Park and around Lake Athabasca. When a single site is considered by itself, it is commonly found to have a very small number of species. A northern Cordilleran floristic influence is distinctly noticeable in the Athabasca Lake country. Raup notes that the Caribou Mountain plateau north of the lower Peace River has stands of *Picea mariana* and *Pinus contorta* var. *latifolia* with a thick mat of woodland mosses. Lodgepole pine woods are also found on the uplands near Lesser Slave Lake. As Thieret points out, jack pine forest on ridges and xerophytic flatlands may be regarded as a pyric or edaphic climax, maintained by xerophytism of the habitat but also by recurrent fires. It is common to see a mature pine forest on one part of a sandy ridge and a burned-over area with dead trees and a dense growth of small young pines on an adjacent part of the ridge.

The most mesophytic forests of the Athabasca–Great Slave lake region are in the extreme southern part, as Raup states, and are dominated by a mixture of white spruce and fir. According to Moss the community is rare for the region, although it is fairly prevalent near Lesser Slave Lake. He adds that it is doubtful that all the balsam fir of the region can be assigned to *Abies balsamea*; there is the possibility that many areas have *Abies lasiocarpa* or intermediate forms as well. In these stands, *Alnus*, *Cornus*, and *Viburnum* are prevalent shrubs, and the following herbaceous species are commonly associated, according to Raup (1946): *Equisetum pratense, Maianthemum canadense, Cinna latifolia, Orchis rotundifolia, Habenaria obtusata, Habenaria hyperborea, Calypso bulbosa, Listera borealis, Corallorhiza trifida, Caltha palustris, Thalictrum venulosum, Mitella nuda, Ribes triste, Ribes hudsonianum, Rubus pubescens, Rosa acicularis, Rhamnus alnifolia, Viola palustris, Viola renifolia, Epilobium angustifolium, Aralia nudicaulis, Cornus stolonifera, Pyrola asarifolia, Pyrola virens, Pyrola secunda, Trientalis borealis, Viburnum edule, Linnaea borealis*.

Trembling aspen forms pure stands on areas of more mature forest that have burned, and Moss states that the balsam poplar (*P. balsamifera*) also occurs in nearly pure stands, especially on moist sites such as river flats. In contrast to balsam poplar, aspen thrives over a wide range of edaphic conditions and is often found as the dominant species on areas that appear ideally suited to poplar. Species found in association with aspen and poplar include *Salix bebbiana, Cornus stolonifera, Alnus rugosa, Rosa* species, *Rubus idaeus, Lonicera involucrata, Calamagrostis canadensis, Aster ciliolatus, Rubus pubescens, Mertensia paniculata, Fragaria* species, *Equisetum* species, *Epilobium angustifolium, Thalictrum venulosum, Galium septentrionale, Vicia americana,* and *Pyrola asarifolia*.

THE CANADIAN SHIELD REGION

Considering the large area involved, the Canadian shield region west and south of Hudson Bay possesses a relatively uniform vegetational cover. Topographic and climatic variations, nevertheless, have a profound influence upon plant community structure in terms of differences between one place and another in abundance of the species in communities, particularly along climatic gradients north and south. Forest tree species are few in number and the same as those found in the interior plains to the west of the edge of the shield—black spruce, white spruce, aspen, poplar, jack pine, balsam fir, white birch, and tamarack. The Hudson Bay lowlands should perhaps not be included in this discussion of the shield region, but they are in a sense part of the shield since geological origins are so closely linked; the region is, moreover, surrounded by the shield on three sides and by Hudson Bay on the fourth. It is an immense area of flat coastal plain with black spruce and tamarack woodlands in areas of muskeg, and with white spruce, balsam fir, aspen, poplar, and birch along river levees. For these reasons, as well as chorographic convenience, the Hudson Bay lowlands will be discussed here in the same context as the shield.

Ritchie (1962) establishes a zonal division of the Precambrian shield of four broad types of vegetational communities, each grading into the other along a south-to-north cline: closed coniferous forest, open coniferous forest, forest–tundra, and tundra. Lowland vegetation is predominantly bog or fen. He adds that it would be of interest to determine the nature of the stable vegetation developed on equivalent sites throughout the region, delineating zones within which the communities develop as an expression of climatic regimes. He suggests that the two main environmental factors governing the distribution of species are climate and topography. Zonal categories are aligned roughly in parallel with such climatic factors as length of growing season and the mean daily temperature of the warmest month. Within each zone the variation in vegetation is apparently the result of local topography and succession following fire. In the Hudson Bay lowlands, the topography or the drainage pattern governs the nature of the vegetation, while climate, disturbance, and historical development are important as local secondary factors. Ritchie adds*:

> However, in many areas it seems that the influence of one factor has been masked by that of others. It is not clear at present whether the open coniferous (spruce) forest

*Reproduced by permission of the Arctic Institute.

zone is primarily a climatic type or whether its preponderance in an area is related to an abundance of well-drained, coarse mineral substrata. Elsewhere in northwestern Canada, Raup (1946) has suggested the latter correlation and Sochava (1956) makes a similar comment regarding the open lichen forests of Finno-Karelian S.S.R. and the Russian Plain. Also, there are parts of the Caribou River section where it is evident that tundra vegetation is secondary, the result of fire. In the Churchill section it is likely that the regional climate is favourable enough for fair tree growth, but the generally poorly drained substrata of the area seldom bear tree growth of more than stunted stature. Also, the prevalence of treeless vegetation on the peat substrata of the Broad River section might well be related to the presence of continuous permafrost. . . . However, these are speculative comments, and it is clear that ecological (*sensu stricto*) investigations are needed.

There are, thus, a number of regions that can be distinguished in central Canada on the basis of topographic and climatically related vegetational characteristics. The central plateau is rugged and rocky, largely upland, covered with jack pine and black spruce forest. The more southern Superior region is topographically varied, with balsam fir–white spruce forests often luxuriant in valleys, mixed with greater amounts of jack pine on uplands and black spruce in lowlands. Elsewhere in northern Ontario, Manitoba, and Saskatchewan, the forest vegetation on flat, poorly drained land is dominated by black spruce and tamarack. On the better drained sites are found white spruce, aspen, and poplar, with abundant balsam fir and white birch. The ridges and sandy hills in the more southern areas are often covered with jack pine or trembling aspen. Northward, toward the northern forest border, black spruce becomes increasingly dominant on the uplands as well as being present in the lowland areas, the latter increasingly occupied northward by sparsely treed or open and treeless muskeg. At given latitudes in these areas, aspen, poplar, and balsam fir reach northern range limits and drop out of the communities. Throughout the region, essentially the same forest sequence occurs from south to north: In the south are found richer forests in which balsam fir, white spruce, aspen, poplar, birch, and black spruce occur in abundance; black spruce and jack pine become increasingly important northward where terrain is rocky or sandy and severe climatic conditions prevail (Fig. 13). Near the northern forest border, black spruce is the dominant species, with white spruce found more rarely in special habitats favoring its existence.

At the southern edge of the boreal forest, where it exists as outliers in small local favorable habitats in northern Minnesota, Wisconsin, and Michigan, the forest community is transitional to the northern conifer–hardwood forest, and, as Curtis (1959) points out, practically no remaining examples of the original boreal forest exist, although extensive secondary stands, some in relatively late stages of regeneration, are found

Fig. 13. Jack pine stands occupy large areas in northcentral Manitoba; this sandy hillside with well-drained soil is found in the area to the east of Reindeer Lake. Note the low density of shrubs and herbaceous species and the fairly heavy growth of lichens (lighter areas on the surface of the ground).

on heavy, wet soils of flat or low topography. Data on the composition and structure of boreal forest stands in the region (Maycock, 1956, 1957, 1961; Curtis, 1959; Maycock and Curtis, 1960) show that in most stands balsam fir is the most important species, far more so than white spruce. White pine, white cedar, and white birch all possess importance greater than that of white spruce in the stands studied by Maycock and Curtis. Three species of aspen and three of maple are other important members of the community. Black spruce and tamarack dominate the lowland stands. The structural outlines of the boreal stands are summarized by Curtis (1959) as follows*:

> The boreal forest in Wisconsin is present in three partially distinct types—the first, the old stands of relatively pure conifers with balsam fir and white spruce as the

*Reproduced by permission of the Regents of the University of Wisconsin.

major dominants, associated with large quantities of white pine, red pine, or white cedar as found especially along Lake Superior; the second, the mixed conifer-hardwood stands, particularly on inland mesic sites, with the shade-tolerant hardwoods gradually replacing the spruce and fir; and the third, the young stands of dense balsam fir and white spruce under an aging and decadent canopy of trembling aspen or white birch, as found throughout the range. The majority of the hardwood species are found in the second type. The hardwoods associated with the first type are white birch, mountain-ash (*Sorbus americana*), red maple (*Acer rubrum*), and mountain maple (*A. spicatum*), while the third type usually has only white birch or one of the aspens or poplars, with balsam poplar (*Populus balsamifera*) occasionally reaching significant levels of importance. Obviously, the last two types do not represent distinct and stable entities but reflect successional recovery from recent disturbances.

It is of interest that many of the species found in the forests along the north shore of Lake Superior are found also in Europe. Agassiz (1850) was the first to note that the following species are found in both regions: *Drosera rotundifolia, Fragaria vesca, Circaea alpina, Epilobium angustifolium, Saxifraga aizoön, Linnaea borealis, Vaccinium oxycoccus, V. vitis-idaea, V. uliginosum, Arctostaphylos uva-ursi, Loiseleuria procumbens, Andromeda polifolia, Ledum groenlandicum, Pyrola rotundifolia, P. secunda, Chimaphila umbellata, Trientalis borealis, Empetrum nigrum, Polygonum viviparum, Juniperus communis*, and a number of grasses and sedges as well as mosses and lichens. In some instances, more recent studies have given the plants in the two regions specific or subspecific designations, but many remain the same. It is also the case, of course, that there are species pairs in the two regions in which similarities are so great that taxonomists suspect that they were a single species at one time in the past and that divergent evolution has since then rendered them distinct. In the coniferous forests of North America and Europe, some species are held in common. As Curtis (1959) points out, Oberdorfer (1954) listed 109 species in the spruce–fir community of south Germany, of which 66 were in common with the coniferous forests of northern Wisconsin. These include *Athyrium filix-femina, Dryopteris phegopteris, Goodyera repens, Monotropa hypopithys, Oxalis acetosella*, and *Pyrola secunda*. More than one-half of the 101 genera listed were present in both areas, among them such typical groups as *Abies, Picea, Acer, Pinus*, and *Sorbus* among the trees; *Lonicera, Rubus, Ribes, Sambucus*, and *Vaccinium* in the shrubs; and *Anemone, Aquilegia, Corallorhiza, Listera, Maianthemum, Melampyrum, Petasites*, and *Viola* in the herbs.

It is evident that the boreal forest grades into the eastern deciduous forests of the United States, and the resulting intermingling of species adapted primarily to one or the other leads to a complex ecological picture not fully understood at the present time. The bimodal behavior of some of the tree, shrub, and herb species in the boreal forest of

Wisconsin and the Great Lakes region (Maycock and Curtis, 1960; Curtis, 1959) has been noted in Chapter 4; this behavior serves to indicate the unexpected complications that arise in detailed studies of species behavior in these forests. As Curtis points out*:

> The exact nature of the compensating factors present which allow this bimodal behavior are unknown, but they probably include light, acidity, and nitrogen supply. In any case, the phenomenon is an expression of a basic ecological principle which states that plants become more sensitive to differences in soil moisture as temperatures increase. Within Wisconsin, for example, we find a number of species occurring throughout the entire moisture gradient here in the boreal forest stands, fewer such species (although largely the same ones) with that behavior in the northern pine-hardwoods, and only very few that range from willow-cottonwood to bur oak-black oak in the southern counties. Farther south, in the Mississippi valley, virtually no species are found in common between the cypress-tupelo (*Taxodium-Nyssa*) swamps and the adjacent hilltop forests of blackjack oak (*Q. marilandica*) and post oak (*Q. stellata*) as in Missouri and Arkansas. In the tropics, there may be several sets of species present between the two extreme conditions of moisture. The same principle can be seen at work on an altitudinal basis in mountainous regions. Thus, in Whittaker's (1956) nomograms of species behavior in the Great Smoky Mountains, a considerable shift in species composition occurred with decreasing moisture at low elevations, while at high elevations, the spruce-fir forest was found on nearly the full range of the moisture gradient.

It seems likely, according to Curtis, that a conifer swamp in northern Wisconsin may ultimately give rise to a terminal forest of white spruce and balsam fir; it is possible, he says, that conifer swamps of Wisconsin should be considered wet stages of the boreal forest and that they are in a position to develop into a spruce–fir forest by successional processes. Curtis concludes, however, that: "Such a view is useful in certain respects, but should not be held too rigorously, because the supporting evidence available is inadequate for firm judgements." It is evident that one of the more interesting problems, and perhaps one of the more difficult to obtain a satisfactory solution to, is that of describing the community dynamics of boreal species in an environment capable of supporting broadleaf forest over much of the land surface, allowing boreal species to survive in only such marginal habitats as fens and bogs.

SHIELD COMMUNITIES; SPECIES COMPOSITION

On the Precambrian shield both east and west of Hudson Bay, the general north–south zonation of vegetation—tundra, open coniferous forest, and closed coniferous forest—extends across southern central Mackenzie, northern Saskatchewan, northern Manitoba, Ontario,

*Reproduced by permission of the Regents of the University of Wisconsin.

Quebec, Ungava, and Labrador in a sequence similar in all general aspects to that found in the western portions of northern Canada. The varied topography and surficial geology ranges from bare outcropping rock to the familiar features of heavily glaciated terrain such as drumlins, moraines, eskers, and large areas of relatively flat, sandy glacial drift and outwash. The result is a mosaic of vegetational communities over the region somewhat more varied in structural and dominance patterns than is found on the plains to the west, although the floristic composition of the boreal vegetation is, with some exceptions, quite similar in both the shield region and the interior plains.

It seems reasonable to affirm Ritchie's conclusion, based on vegetational distribution, landforms, and climatic data, that the two basic environmental factors involved in geographic distribution of the vegetation are climate and topography. Zonal categories in the region west of Hudson Bay can be aligned roughly along climatic lines trending southeast–northwest with length of the growing season, mean daily temperature of the warmest month, and other climatic parameters usually strikingly coincident with the vegetational zonation (Ritchie, 1956).

Variation in vegetation can often be seen to be the result of the influence of local topography and of disturbance by fire. In the Hudson Bay lowlands and the Ontario clay belt, the distinctive nature of the vegetation appears to be the consequence of level topography, poor drainage, and, consequently, wet or waterlogged soils. Ritchie adds, however, that in many flat, sandy areas of better drainage it is not clearly apparent whether the open spruce forest is primarily the result of climatic influences or the abundance of coarse, well-drained mineral substrates. In the Hudson Bay lowlands and clay belt, however, the climate appears sufficiently favorable for forest growth, put poorly drained substrates furnish no sites capable of producing trees of more than stunted stature. Ecological studies are usually needed to determine clearly which of the factors—climate, soil, or topography—is of greatest importance in a given area. All certainly are involved to some degree, and it is difficult to identify those with primary influence in a given area.

In areas to the south of the forest–tundra transition and the open coniferous forest (known also as lichen woodland), closed spruce forests, so-called because the spacing of trees allows little light to penetrate to the forest floor, form the main cover type. Black spruce is the more important tree occupying upland sites. In poorly drained areas where peat accumulates, spruce forest grows on *Sphagnum* moss muskeg. Over most of the region, stands of old, mature closed spruce forest on moderately drained soils are rare, the consequence of frequent fire, and in most areas the trees are no larger than 4–9 in. (10–23 cm) in trunk diameter. White birch and trembling aspen are also found in some numbers, and white spruce, balsam fir, and balsam poplar are common in wetter areas. Jack pine is often dominant on ridges and sand plains. It is of some significance that the occurrence of white spruce requires habitats with particularly favorable soil and microclimate, evidently often associated with alluvium and river shorelines. Where there is no evidence of recent fire, trees in black spruce forests attain heights of 14-20 m, with trunks 20-30 cm in diameter 1 m from the ground. In stands of this type studied by Ritchie (1956), other trees or tall shrubs are present only in breaks in the canopy. The soil possesses an upper horizon of partly decomposed moss 6 cm in depth below a carpet of mosses 3-5 cm in depth. Permafrost is encountered in areas of northern central Manitoba in July at a depth of 1.5 m, and depths are, of course, shallower northward and greater and more discontinuous to the south.

It is clear, Ritchie adds, that, for this region, the closed spruce forest is the most advanced stage of forest succession on moderately drained mineral substrates, and the range of upland communities seen over the landscape represent phases in succession after fire. He cites as evidence the presence of large white birch stumps in pure spruce forests, abrupt transitions between open pine and closed spruce forest on uniform physiographic sites, and large numbers of young spruce trees in mixed forest communities. Degree of destruction by fire influences the structure and composition of the recolonizing vegetation. It appears that jack pine, white birch, and several shrub species, by virtue of a more rapid rate of initial growth, become dominants following fire. As a closed forest develops, conditions become favorable for black spruce, which replaces the pioneer shade-intolerant dominants. Decreased illumination results in the elimination of herbs in the ground layer, which are replaced by a dense carpet of mosses. This mature forest is known also as a "conifer–feather moss forest" (Hustich, 1949) and a "black spruce–feather moss association" (Moss, 1953a,b). As Hustich writes: "If there are such things as climax forests in nature, the spruce feather moss forest is certainly one of them." Over much of the region, the southern spruce forest zone is characterized by black spruce on upland sites rather than white spruce; the latter are rare and local on sites with favorable soil and exposure. In comparable areas in Quebec, Hustich (1949) indicates that here, also, black spruce is more common than white spruce, the latter being restricted to particularly favorable localities. In more southern regions, trembling aspen comes in after fire in areas with good soils and high insolation. Rocky areas of low insolation favor birch. Both aspen and white spruce become less common northward, and the two species show a coincident pattern of abundance to a considerable extent.

Throughout the area of Ritchie's study, *Picea glauca* was of sporadic occurrence, never forming more than local stands in a forest of primarily black spruce. Small white spruce communities studied in northern Manitoba by Ritchie had widely spaced trees, 15–20 m tall, and ground vegetation associated with the spruce on well-developed dry, loamy soil included *Ribes oxyacanthoides, Amelanchier alnifolia, Prunus pensylvanica, Aralia nudicaulis, Shepherdia canadensis, Viburnum edule, Epilobium angustifolium, Linnaea borealis, Salix bebbiana, Alnus crispa, Cornus stolonifera, Cornus canadensis, Mertensia paniculata, Pyrola asarifolia, Mitella nuda, Fragaria vesca, Rosa acicularis, Petasites palmatus,* and *Calamagrostis canadensis.*

Occupying most of the landscape in the study area is a closed-canopy black spruce forest in which jack pine and paper birch occur only occasionally but are, nevertheless, constant features of the community; *Salix bebbiana* is also common, and other frequent species include *Alnus crispa, Ledum groenlandicum, Rosa acicularis, Ribes glandulosum,* and *Vaccinium vitis-idaea* (Fig. 14). Less common but nevertheless regular associates in the community are *Mertensia paniculata, Epilobium angustifolium, Geocaulon lividum, Equisetum sylvaticum, Petasites palmatus, Achillea millefolium, Cornus canadensis, Arctostaphylos uva-ursi, Linnaea borealis, Lycopodium complanatum, L. annotinum,* and *Mitella nuda.* An assemblage of mosses includes *Pleurozium schreberi, Hylocomium splendens, Dicranum rugosum,* and *D. bergeri.* Wetter areas have *Sphagnum* species and *Aulacomnium palustre.*

The open forest dominated by jack pine usually has paper birch as a frequent associate, and there is a shrub stratum of *Salix* species, *Alnus crispa,* and, often, young black spruce. These stands represent an early phase of the regeneration community characteristic of mesic sites after fire. *Ledum groenlandicum* and *Vaccinium vitis-idaea* are common in the ground-layer vegetation. High outcrop ridges are a common feature of the Canadian shield, and the summits of these, with weathered rock and thin drift, often possess a pine forest in which the trees grow in crevices and on shallow mineral soil. Black spruce often grows along the edges of outcrops where sand and drift materials form sloping ecotonal zones. The ground vegetation often includes *Vaccinium myrtilloides, Empetrum nigrum, Corydalis sempervirens, Carex deflexa,* and *C. brunnescens.* Other distinct communities in the region include stands of mature white birch forest on organic terraces and muskegs which form in flat depressions between ridges. Open peat bogs are often seen in the center of muskegs, forming a wet central area with *Menyanthes trifoliata, Epilobium* species, and *Carex* species. This is surrounded by muskeg in which *Chamaedaphne calyculata* and *Betula glandulosa* are common, and other frequent species include *Kalmia polifolia, Rubus chamaemorus, Smilacina*

Fig. 14. Dense growth of *Ledum groenlandicum* is characteristic of relatively undisturbed, open, mesic black spruce stands throughout the southern boreal forest of central Canada. The stand photographed here is a few miles inland from the north shore of Lake Superior.

trifolia, Oxycoccus microcarpus, Salix species, *Pinguicula villosa, Drosera rotundifolia, Carex limosa, Myrica gale,* and *Andromeda polifolia.*

Descriptions of the Hudson Bay lowlands and extensive portions of the clay belt do not differ markedly from those of muskeg and peat bog; the former are essentially vast areas of bog, fen, and lowland black spruce. The pattern is widespread in northern Ontario where the marine deposits of the Hudson Bay lowlands and the lacustrine deposits of the clay belt region support extensive areas of fen, bog, and lowland black spruce forest (Hustich, 1957; Baldwin, 1958; Sjörs, 1958; Rowe, 1972; Ritchie, 1956, 1959, 1960a,b,c, 1962). As Ritchie points out, the vegetation of the Hudson Bay lowlands to the south of the Nelson River appears to be very similar to that of the lowlands in Ontario, but north of the Nelson River a northern subarctic regime is apparent from the increase in treeless communities on deep peat. There are similarities of community structure and composition between the lowlands of northern Ontario and lowlands throughout northern Canada, as well as the lowland areas in Alaska described by Drury (1956).

Carlton and Maycock (1978) describe the clay belt region as poor in relief, except where intrusions of rock project through the clay or where eskers wind over the landscape. Glacial outwash and sand plains are numerous. As they describe the forests of the region, many of the communities are composed of mixtures of tree species, including white spruce, balsam poplar, black spruce, balsam fir, trembling aspen, paper birch, and jack pine. White spruce and balsam poplar are found most often in the river valleys. Black spruce occupy a wide range of habitats from wet lowland bogs to very dry rock outcrops. Jack pine forms even-aged stands over large areas on outwash plains. Well-developed balsam fir forest is relatively infrequent, found most often on good, moist upland soils. Balsam fir, however, is abundant in stands dominated by other species, often in association with paper birch and trembling aspen.

A conspicuous feature of the boreal upland conifer forest of the shield region is the relative paucity of the flora. This is emphasized by a comparison between the studies of Swan and Dix (1966) carried out in northern Saskatchewan and the studies of Maycock and Curtis (1960) in the Great Lakes region. There were 42 tree species in the boreal stands studied in northern Wisconsin and Michigan, compared to 8 at Candle Lake, Saskatchewan. As Swan and Dix point out, the rich forest flora in the Great Lakes forest may be accounted for by the mingling of floral elements from two large floristic regions, the boreal and the deciduous forests. The forest of northern Saskatchewan contains few of the species representative of the deciduous forest, and even many species found in abundance in the Great Lakes forest are absent from the Candle Lake forests.

Although the area studied by Swan and Dix is essentially beyond the edge of the Precambrian shield, with bedrock of an upper Cretaceous shale, the area is covered with glacial drift and, as such, is representative of much of the topography of the shield region. They were able to establish no relationships between vegetation structure and three environmental factors considered to be a measure of available soil moisture—drainage regime estimates, moisture measurements, and water-retaining capacity. It was apparent that soil moisture rarely was a limiting factor in the development of upland forest at Candle Lake. Of the factors that influenced forest development, shade cast by the forest canopy seemed to be particularly important. Swan and Dix write*:

> Black spruce, and to a lesser extent balsam fir, cast the densest shade and support the smallest numbers of understorey vascular species and the highest per cent moss cover. The pH is also conspicuously low in black spruce stands and its role in excluding vascular species from these stands is not known. However, the comparatively low number of vascular species in balsam fir stands where pH is considerably higher than that in black spruce suggests that shade density is the controlling factor. Conversely, aspen, white birch, jack pine and, to a lesser extent, white spruce, have relatively thin canopies and support a well-developed understorey of herbs and shrubs and a scant moss cover. Thus, the plants themselves appear to exert a significant influence in determining the sub-canopy species structure of a site.

At Candle Lake, aspen, black spruce, and jack pine occupy a large percentage of the total area, often in stands in which one of the species contributes two-thirds or more to the canopy. Upland black spruce stands are usually relatively small; the canopy permits little light penetration, and a depauperate herb and shrub ground community is the result. Mosses form a rich, green carpet, often a foot deep. Jack pine occurs most commonly on sand plains, often over areas of many hundreds of square miles. The thin jack pine canopy allows a comparatively rich ground community to develop. White spruce and white birch are common over the landscape, but rarely do they dominate a stand; both are closely associated with aspen. Balsam fir is rarely a dominant, even though it is evidently the only species capable of reproducing under its own shade.

Analysis of the data obtained at Candle Lake led Swan and Dix to the conclusion that the structure of the plant communities could not be described by a linear gradient; a multidimensional scheme was necessary to display the relationships. They write*:

> The structural complexity of this forest is not unique in the boreal forest; similar complexity being found in areas remote from our location.... In general terms, canopy and sub-canopy structures is interpreted as a continuously varying pattern in which the distributions of many species overlap, each species having, however, its optimum performance in a limited part of the ordination.

Swan and Dix point out further that the boreal forest of the Athabasca–Great Slave lake region (Raup 1946), Alberta (Moss, 1953a,b, 1955), and northwestern Manitoba (Ritchie, 1956), appears similar in structure and composition to the upland forest of Candle Lake. Parklike

---

*Reproduced by permission of the British Ecological Society.

white spruce–lichen stands and jack pine stands with a sparse understory of herbs and shrubs were not found at Candle Lake. These are found only at latitudes far north of Candle Lake, and their absence may be explained by climatic differences.

The shield region is no exception to the rule that fire is a major influence in forest composition; few if any areas fail to show such characteristic features as even-aged stands of pioneer species or sharp boundaries between new young and old mature stands, the boundary marking the edge of a burn. Observations by Swan and Dix in the Candle Lake area suggest that white spruce usually develops beneath a canopy of aspen. Balsam fir becomes established only through succession, usually following white spruce. In conclusion, Swan and Dix cite Rowe (1961) as presenting an accurate summary of the successional relationships of the tree species throughout the boreal forest:

> In short, there are no species in the Canadian western boreal forest possessing in full the silvical characteristics appropriate to participation in a self-perpetuating climax. The boreal forest is a disturbance forest, usually maintained in youth and health by frequent fires to which all species, with the probable exception of fir, are nicely adapted.

An exception to this general rule is perhaps to be found in the boreal forest stands of the Great Lakes region where, in northern Minnesota, Wisconsin, and Michigan, remnants of original boreal forest are found as outliers in topographically favorable habitats. Many of the wide-ranging boreal species are found in abundance in northern conifer–hardwood forests as well as in areas north of Lake Superior where original coniferous forest in places exists in a relatively undisturbed state. Curtis (1959)* points out that:

> Little observational evidence exists concerning the stability of the nearly pure, mature conifer stands in Wisconsin. In the major range of the boreal forest in Canada, it is apparent that catastrophe is of frequent occurrence, caused either by fire, wind, or the spruce budworm. As a result, there is a tendency for entire stands to be destroyed and replaced simultaneously, with resultant even-age forests over wide areas. In Ontario, there may be a direct replacement of spruce-fir forest by spruce-fir forest, with no intervening stage. More frequently, white birch and trembling aspen form an intermediate and short-lived cover crop, particularly after fire. On both the dry sand soils and the wet or poorly drained flatlands, there may be an initial stage of black spruce and jack pine. Both of these species can readily replace themselves after fire, so the type tends to be semi-permanent in areas subject to frequent fire. Examples of spruce-fir invasion of old jack pine stands can also be found in Wisconsin. . . .

In his studies of forest stands in the Great Lakes region, Maycock (1961) found that, in the upper peninsula of Michigan, balsam fir, white spruce, sugar maple, aspen, paper birch, white pine, and eastern white

*Reproduced by permission of the Regents of the University of Wisconsin.

cedar were dominant members of the tree canopy, with the first three exhibiting the greatest ecological importance. Maycock has found that the total list of 18 tree species could be roughly separated into those that are important members of the boreal forest northward and those that are members of the deciduous forest southward. Of the boreal tree species, only balsam fir has a higher representation in the reproduction layer than in the tree canopy. Maycock points out that heavy reproduction in the sapling sizes does not necessarily mean increased representation in future generations. Such is the case with balsam fir. It does indicate, however, that the species has at least the capability of maintaining its place in the forest under favorable conditions. This is not so with *Picea glauca*, he adds. Representation in the reproduction layer is reduced by at least 25% and by as much as 83%, and it is doubtful that *Picea glauca* will increase in importance in the future regardless of how suitable growing conditions may become. *Populus tremuloides* and *Betula papyrifera* are completely absent in the reproduction layer or are significantly reduced. Similarly, *Pinus strobus* is almost entirely without reproduction in the stands Maycock studied. In the study areas, 25 species of shrubs and herbs were present, only 6 with frequency values of 5 or greater in all stands. *Rubus parviflorus* and *Lonicera canadensis* were the most important shrubs; only the first appeared in all stands. Herbaceous species in all stands were *Aster macrophyllus, Maianthemum canadense, Mitella nuda, Carex pedunculata*, and *Aralia nudicaulis*. Understory species found in the stands but which reached greatest abundance in deciduous forests to the south included *Cornus rugosa, Geum canadense, Aralia racemosa, Solidago flexicaulis, Prenanthes alba, Caulophyllum thalictroides, Heuchera richardsonii, Anemone canadensis,* and others.

The list of ground-layer species found in these forests is too long to include here, but the prevalent species are listed by Curtis (1959) and include a rather large number of species that in Wisconsin attain their highest presence values in the boreal forest community: *Actaea rubra, Aster macrophyllus, Cornus canadensis, Clintonia borealis, Corylus cornuta, Diervilla lonicera, Galium triflorum, Lonicera canadensis, Maianthemum canadense, Pyrola secunda, Rubus parviflorus, R. strigosus, Streptopus roseus, Trientalis borealis, Vaccinium angustifolium, V. myrtilloides,* and *Viola conspersa*.

From Wisconsin northward to the southern portions of the Keweenaw peninsula, deciduous forest dominated by *Acer saccharum, Betula lutea,* and *Tilia americana* forms the dominant vegetation (Maycock, 1961). This community persists where soils are of considerable depth and mesic conditions prevail. Farther toward the north, forest composed almost entirely of boreal coniferous species (*Abies balsamea, Picea glauca*) are found where conditions permit competition with the deciduous tree species. Throughout the peninsula, mixtures of the dominants of these two forest communities, often with *Pinus strobus* included with the other conifers, occur. In low-lying, wet areas *Larix laricina, Thuja occidentalis, Abies balsamea,* and *Picea mariana* are the prevalent species.

Farther northward, across the expanse of Lake Superior, the forest along the north shore is more thoroughly dominated by boreal coniferous species. Recalling his journey with Louis Agassiz along the northern shoreline of Lake Superior, Cabot (Agassiz, 1850) describes the relatively undisturbed forest in these words:

> In geographical position the lake would naturally seem to lie within the zone of civilization. But on the north shore we find we have already got into the Northern Regions. The trees and shrubs are the same as are found on Hudson's Bay; spruces, birches and poplars; the Vaccinia and Labrador tea. Still more characteristic are the deep beds of moss and lichen, and the alternation of the dense growth along the water, with the dry, barren, lichenous plains of the interior. Here we are already in the Fur Countries; the land of voyageurs and trappers; not from any accident, but from the character of the soil and climate. Unless the mines should attract and support a population, one sees not how this region should ever be inhabited.
>
> This stern and northern character is shown in nothing more clearly than in the scarcity of animals. The woods are silent, and as if deserted; one may walk for hours without hearing an animal sound, and when he does, it is of a wild and lonely character; the cry of a loon, or the Canada jay, the startling rattle of the arctic woodpecker, or the sweet, solemn note of the white-throated sparrow. Occasionally you come upon a silent, solitary pigeon sitting upon a dead bough; or a little troop of gold-crests and chickadees, with their cousins of Hudson's Bay, comes drifting through the tree-tops. It is like being transported to the early ages of the earth, when the mosses and pines had just begun to cover the primeval rock, and the animals as yet ventured timidly forth into the new world.

He adds that the spruce woods were very dense and in places "encumbered with fallen birch trunks, as if the spruces had usurped the place of a birch forest."

Quadrat studies by the author in forests of the north shore dominated by balsam fir and white and black spruce indicate that the more prevalent ground-layer species include *Clintonia borealis, Cornus canadensis, Equisetum sylvaticum, Gaultheria hispidula, Ledum groenlandicum, Linnaea borealis, Maianthemum canadense, Mitella nuda, Petasites palmatus, Rubus pubescens, Trientalis borealis, Vaccinium angustifolium, V. myrtilloides,* and seedlings of balsam fir, white birch, and yellow birch. In stands dominated by black spruce, fewer numbers of species of both trees and ground-layer plants were noted, and dominant among the latter were *Ledum groenlandicum, Vaccinium myrtilloides, Gaultheria hispidula, Cornus canadensis,* and black spruce seedlings. Among the species noted in the region by Agassiz were a number occurring also in Europe, and his listing includes *Drosera rotundifolia, Fragaria vesca, Circaea alpina,*

Epilobium angustifolium, Saxifraga aizoön, Linnaea borealis, Vaccinium vitis-idaea, V. uliginosum, Arctostaphylos uva-ursi, Loiseluria procumbens, Andromeda polifolia, Pyrola rotundifolia, P. seconda, Chimaphila umbellata, Empetrum nigrum, Polygonum viviparum, and *Juniperus communis,* as well as *Ledum groenlandicum, Trientalis borealis,* and a number of species of sedges, grasses, mosses, and lichens. In most of the forest stands, mosses are of importance in the understory, more so than lichens, in contrast to the communities of the northern boreal zone, but in southern areas arboreal lichens attain great importance, often covering the lower branches of the trees with a dense growth (Fig. 15). Species of vascular plants with northern boreal affinities include *Shepherdia canadensis, Ribes triste, Linnaea borealis, Cornus canadensis, Mitella nuda, Pyrola secunda, Lycopodium clavatum, L. annotinum, Trientalis borealis, Petasites palmatus,* and so on.

Sampling by the author (Larsen, unpublished data) in black spruce

Fig. 15. Arboreal lichens hang in festoons from the branches of spruce in the southern boreal forest of central Canada. The lichen species shown here are principally of the genus *Usnea.*

stands along the north shore of Lake Superior showed that the following species were of greatest importance in the ground layer (values are frequency in 20 quadrats, two stands averaged): *Ledum groenlandicum,* 92; *Gaultheria hispidula,* 77; *Vaccinium myrtilloides,* 92; *Cornus canadensis,* 32; *Picea mariana* (seedlings), 27; *Carex trisperma,* 22; *Kalmia polifolia,* 15; *Equisetum sylvaticum,* 12; *Vaccinium oxycoccos,* 12; and *Rubus chamaemorus,* 10. Lichens were found with a 30% frequency, and mosses were ubiquitous. Present were *Betula papyrifera* (seedlings), *Dryopteris disjuncta, Larix laricina* (seedlings), *Linnaea borealis, Listera cordata, Lycopodium annotinum, Melampyrun lineare, Monotropa uniflora,* and *Smilacina trifolia.*

In mixed wood stands with balsam fir, white spruce, paper birch, and aspen commonly dominating the arboreal stratum, there was a much wider range of important species. In five stands, frequency values in 20 quadrats (averaged for the five stands) were as follows: *Cornus canadensis,* 86; *Clintonia borealis,* 54; *Maianthemum canadense,* 46; *Linnaea borealis,* 44; *Equisetum sylvaticum,* 30; *Vaccinium angustifolium,* 27; *Gaultheria hispidula,* 27; *Abies balsamea* (seedlings), 26; *Trientalis borealis,* 25; *Vaccinium myrtilloides,* 22; *Mitella nuda,* 21; *Rubus pubescens,* 22; *Petasites palmatus,* 18; *Ledum groenlandicum,* 17; *Lycopodium annotinum,* 15; *Aralia nudicaulis,* 14; *Diervilla lonicera,* 12; *Smilacina trifolia,* 10; *Oryzopsis asperifolia,* 9; *Streptopus roseus,* 7; *Galium borealis,* 7; *Mertensia paniculata,* 7; *Sorbus decora,* 6; *Monotropa uniflora,* 6; *Lycopodium complanatum,* 6; *Apocynum androsaemifolium,* 5; *Amelanchier alnifolia,* 4; *Aster ciliolatus,* 4; *Coptis trifolia,* 4; *Gaultheria procumbens,* 4; and *Podophyllum peltatum,* 4. Mosses were abundant, but lichens were rare in the stands. The following species were present but not in abundance: *Alnus crispa, Alnus rugosa, Aster* sp., *Betula lutea* (seedlings), *Betula papyrifera* (seedlings), *Carex trisperma, Epigaea repens, Fragaria virginiana, Geocaulon lividum, Goodyera repens, Grass* sp./spp., *Habenaria* sp., *Impatiens biflora, Listera cordata, Lycopodium obscurum, Melampyrum lineare, Mentha arvensis, Moneses uniflora, Picea mariana* (seedlings), *Pteridium aquilinum, Ribes triste, Rosa acicularis, Rubus idaeus, Solidago* sp., *Viburnum edule, Viola* sp., and lichens.

BOREAL FORESTS OF EASTERN CANADA

Eastern Canada, including the area from the Maritimes and the Gaspé Peninsula northward to the extreme tip of Ungava, occupies in excess of ½ million square miles, equivalent to the central area occupied by Manitoba, Saskatchewan, and southern Keewatin. As would be expected, the variation in vegetation throughout such a vast region is considerable, and we are fortunate a large and generally excellent botanical and ecological literature is available for the area. The most extensive general survey of the vegetation and physiography of the Labrador–Ungava region is that of Hare (1959), who employed aerial photographic interpretation and ground studies to carry out what he termed a reconnaissance survey of the vegetational cover of the region. He recognized five major vegetational community types and constructed cover maps on the basis of these—boreal forest, lichen woodland, forest–tundra ecotone, sedge tundra, and rock tundra—in order of their general occurrence from south to north across the boreal forest and into the northern tundra.

In discussing the coincidence between vegetational cover types and physiography, Hare points out that "air-photo interpretation in wilderness areas can never be regarded as *either* cover-type or physiographic interpretation, in mutual isolation. What is actually interpreted is an amalgam of both; the characteristic associations between cover-type and land form are used as an aid in interpretation." The purpose of his work was to reveal the structure of a large, representative section of the boreal forest formation and to present a generalized regional reconnaissance of the region in terms of the major physiognomic, vegetational (and physiographic) types, without reference to the detailed composition of the plant communities, although the dominant trees are named at least to genus. Within the five major zonal subdivisions are about 20 cover types, a moisture-based series from rock–desert to bog and fen. The vegetation types found on the various sites in any given area are in general a response to topography, hence the moisture regime. The major zonal subdivisions, ranging from tundra in the north to relatively dense coniferous or mixed forest in the south, is evidently a response to climate. Boreal forest throughout the world is, in general, located within a zone characterized by potential evaporation values of 35–52 cm, with the forest–tundra transition occurring in the range 31–35 cm. Values of potential evaporation for the Labrador–Ungava region coincide closely with vegetational zonation.

For the other more southern regions of eastern Canada there exist no general reconnaissance surveys of the type undertaken by Hare in Labrador–Ungava. Much of the area has suffered disturbance, either from agriculture or extensive logging, hence natural undisturbed vegetation exists only in rare, sheltered or inaccessible spots. For many of the southern areas, however, recent descriptions of the existing vegetation are now available. Some of the regional floras are of especial value, particularly those providing a summary of the historical, physiographic, and climatic factors that bear upon the present-day distribution of the species and the composition of the forest communities. It is not possible here to undertake a review of all these publications, since those of value are numerous (Hustich, 1949, 1950, 1951a,b, 1954, 1962, 1966; Scoggan, 1950; Linteau, 1955; Rowe, 1972; Lemieux, 1963; Grandtner, 1963, 1967; Maycock, 1963; Harper, 1964; Wilton, 1964; Damman, 1964, 1965; Jurdant, 1964; Maycock and Matthews, 1966; Davis, 1966; Douglas and Drummond, 1955; Ives, 1960; Allington, 1961; Hughes, 1964; Fraser, 1956).

GASPÉ–MARITIME REGION

From the various forest classifications of the Gaspé Peninsula and Maritime provinces it is apparent that a rather wide variety of forest types is to be found in the region despite its limited geographical extent. Rowe (1972), for example, classifies the boreal coniferous region of New Brunswick alone into six major divisions, Nova Scotia into seven, and Newfoundland into five. The Great Lakes–St. Lawrence region of mixed deciduous species, hemlock, and pine extends eastward, moreover, into the Gaspé Peninsula and New Brunswick, further extending the variety of forest species that can be expected to occur with varying degrees of frequency in one or more of the forest community types. Boreal conifers are predominant in the Mt. Albert and Mt. Jacques Cartier areas of the Gaspé Peninsula, the highlands section of the northern peninsula of Cape Breton Island, and, less extensively, the southern portion of the northern peninsular section of Newfoundland. Surficial geology of the sampled areas is relatively uniform, consisting in the case of both the Gaspé and Newfoundland study areas of coarse, unassorted granitic drift and in Cape Breton of a somewhat more sandy drift. Bedrock in the mountainous areas of the Gaspé and of Newfoundland is most abundantly granitic, and in the Cape Breton highlands drift overlies Paleozoic sedimentary and metamorphic bedrock with granitic intrusions.

The forests in all these areas have strong similarities. In each, the major forest cover type is coniferous and balsam fir is present in highest densities throughout most of the moderately sloping, well-drained upland surfaces. In Newfoundland, black spruce is somewhat more abundant in places, but balsam fir is the major cover type over a large proportion of the land surface. Aspen and balsam poplar ordinarily are not of great importance, and the species associated with balsam fir most abundantly are black spruce, white spruce, and paper birch. Pure black spruce stands are comparatively rare, in sharp contrast to the forests found farther to the north in northern Quebec, Labrador, and Ungava.

The discussion here will be concerned with the balsam fir–mixed wood community of the Gaspé and Cape Breton area, the black spruce communities of the Cape Breton highlands, the balsam fir and spruce communities of Newfoundland, and the spruce forests of Ungava and Labrador. The data given in Tables 8–14 were obtained by the author in a reconnaissance of the regional vegetation and are included here since they are not incorporated into the regional community comparisons presented in Chapter 6.

TABLE 8

Mixed Wood (Balsam Fir)[a] Gaspé Peninsula and Cape Breton

Species[b]	Average frequency
Abies balsamea (seedlings)	45
Aralia nudicaulis	18
Aster sp.	35
Betula papyrifera (seedlings)	8
Clintonia borealis	78
Coptis groenlandica	29
Cornus canadensis	79
Dryopteris austriaca	12
Dryopteris robertianum (*Gymnocarpium r.*)	12
Epigaea repens	6
Gaultheria hispidula	16
Kalmia angustifolia	9
Linnaea borealis	48
Maianthemum canadense	44
Mitella nuda	14
Oxalis montana	65
Pteridium aquilinum	16
Pyrola secunda	6
Rubus pubescens	9
Solidago macrophylla	5
Trientalis borealis	33
Vaccinium myrtilloides	9
Vaccinium vitis-idaea	10
Mosses	100

[a] Number of stands averaged: 4.
[b] Present: *Acer rubrum* (seedlings), *Actaea rubra*, *Amelanchier* sp., *Andromeda polifolia*, *Carex interior*, *Carex pseudo-cyperus*, *Epilobium angustifolium*, *Galium triflorum*, *Goodyera repens*, *Ledum groenlandicum*, *Listera cordata*, *Lycopodium* sp., *Ribes glandulosum*, *Ribes lacustre*, *Thalictrum* sp., *Thelypteris palustris*, *Thuja occidentalis* (seedlings), *Vaccinium angustifolium*, *Viburnum trilobum*, *Viola* sp., lichens.

THE FORESTS OF BALSAM FIR

In the eastern balsam fir or mixed wood forests, of greatest interest are many species, including *Cornus canadensis, Maianthemum canadense, Aralia nudicaulis, Rubus pubescens, Clintonia borealis, Trientalis borealis, Coptis trifolia,* and *Mitella nuda,* which are also of great importance in central Canada (see Table 8). Two species, *Rubus pubescens* and *Galium triflorum,* increase significantly in their average frequency westward. A few species more frequent in the east, notably *Kalmia angustifolia, Chiogenes hispidula, Vaccinium angustifolium,* and species of the genera *Oxalis* and *Solidago,* are of consistently high importance only in eastern communities.

Black spruce tends to increase in importance in balsam fir stands at higher elevations on the slopes of Mt. Albert and Mt. Jacques Cartier, grading into a nearly pure black spruce zone at about 2500 ft. Here the individual trees are of stunted growth form, although not actually dwarfed; maximum height is within the range of 20–30 ft. At least one species, *Epigaea repens,* appears largely confined in great abundance to this upper zone, and *Kalmia* and *Vaccinium* became noticeably more frequent.

CAPE BRETON AND NEWFOUNDLAND

The boreal forest of the Gaspé–Maritimes region is increasingly dominated by black spruce northward, grading into the predominantly black spruce forest that covers the forested areas of Labrador–Ungava. The greater part of the Maritime provinces, exclusive of Newfoundland, possesses forests closely related to those of the Great Lakes–St. Lawrence region. Labrador–Ungava communities are more properly part of the broad continental belt of boreal spruce forest; these relationships are particularly apparent in the understory (see Tables 9–11). Species characteristic of the northern forest increase in frequency northward, and those more characteristic of the southern mixed wood or deciduous forests increase in frequency southward. The complexity of the Gaspé and Maritime provinces is augmented by the complexity of the topography and the variety of soils. Disturbance by logging and fire is a complicating aspect of the more densely populated regions, and it is conceivable that the opportunity to study relationships between climate and the original natural vegetation is fast disappearing if not already gone. Northward in Labrador–Ungava, however, vast tracts of largely undisturbed forest still remain.

TABLE 9

Black Spruce[a] Cape Breton Island–Newfoundland[b]

Species[c]	Average frequency
Abies balsamea (seedlings)	18
Alnus crispa	15
Carex trisperma	12
Carex sp./spp.	32
Chamaedaphne calyculata	18
Clintonia borealis	32
Coptis groenlandica	45
Cornus canadensis	67
Gaultheria hispidula	15
Kalmia angustifolia	55
Kalmia polifolia	23
Ledum groenlandicum	28
Linnaea borealis	22
Picea mariana (seedlings)	48
Rhododendron canadense	35
Trientalis borealis	10
Scirpus caespitosus	8
Smilacina trifolia	13
Vaccinium angustifolium	35
Vaccinium microcarpus	22
Vaccinium myrtilloides	29
Lichens	50
Mosses	65

[a] Number of stands averaged: 3.
[b] Cape Breton stands are in the northeastern portion of the island; the Newfoundland stand is in the western central region.
[c] Present: *Acer rubrum, Carex paupercula, Drosera rotundifolia, Epigaea repens, Geocaulon lividum, Equisetum sylvaticum, Larix laricina* (seedlings), *Maianthemum canadense, Pyrola secunda, Rosa* sp., *Rubus chamaemorus, Sorbus decora, Vaccinium vitis-idaea, Viburnum* sp.

Forests of the northern part of central Newfoundland have been described by Damman (1964) as predominantly balsam fir on well-drained, nutrient-rich sites, with black spruce restricted to nutrient-poor soils. According to Damman's observations, fire may result in the occupation of former balsam fir sites by black spruce, although the richest balsam fir sites may be given over to white birch following destruction of the balsam fir stand. In eastern Newfoundland there is an interesting distribution of forest on ribbed moraine; as described by Delaney and Cahill (1978), a stunted and poorly developed balsam fir and black spruce forest community develops on south slopes, white birch and balsam fir develops on the summit of north slopes, and good balsam fir forest develops on lower north slopes.

LABRADOR–UNGAVA

One of the early—and still one of the most useful—descriptions of the vegetation of the Labrador peninsula is that of A. P. Low, who, along with others such as Tyrrell, must be accorded membership in that small band of literate adventurers who explored the interior of the various regions of Canada. The following description of the presettlement vegetation of the Labrador peninsula is derived from the journals of Low, published in the annual reports of the Canadian Geological Survey (Low, 1896, 1897).

> The forest is continuous over the southern part of the peninsula to between latitudes 52° and 54°, the only exceptions being the summits of rocky hills and the outer islands of the Atlantic coast. To the northward of latitude 53°, the higher hills are treeless and the size and number of the barren areas rapidly increase. In latitude 55°, more than half the surface of the country is treeless, woods being found only about the margins of small lakes and in the valleys of rivers. Trees also decrease in size until, on the shores of Ungava Bay, they disappear altogether. The Leaf River which empties into the bay a few miles north of the Koksoak River, is the northern limit of forest trees on the west side of Ungava Bay... a line drawn a little south of west, from the mouth of the Leaf River to the mouth of the Nastapoka River on Hudson Bay, would give a close approximation to the northern tree limit of western Labrador.
>
> The tree-line skirts the southern shore of Ungava Bay and comes close to the mouth of the George River, from which it turns south-south-east, skirting the western foot-hills of the Atlantic coast range, which is quite treeless, southward to the neighborhood of Hebron, in latitude 58°, where trees are again found in protected valleys at the heads of the inner bays of the coast. At Davis Inlet, in latitude 56°, trees grow on the coast and high up on the hills, the barren grounds being confined to the islands and the headlands, which remain treeless to the southward of the mouth of Hamilton Inlet.

Low points out that the most abundant species of trees are black and white spruce, balsam fir, tamarack, white birch, trembling aspen, balsam poplar, jack pine, and white cedar. Only the first four are found in any abundance northward in the peninsula, although balsam poplar and birch are found down the Kaniapiskau River to the Koksoak and down the George River to its outlet on Ungava Bay. The distribution of balsam fir and white birch is similar to that of poplar along the Kaniapiskau, but neither reaches as far north along the George River. Aspen, cedar, and jack pine are confined to favorable areas in the southern half of the peninsula. In commenting upon the distribution of black spruce, Low states that:

Black spruce is the most abundant tree of Labrador and probably constitutes over ninety percent of the forest. It grows freely on the sandy soils which cover the great Archaen areas, and thrives as well on the dry hills as in the wet swampy country between the ridges. On the southern watershed the growth is very thick everywhere, so much so that the trees rarely reach a large size. To the northward, about the edges of the semi-barrens, the growth on the uplands is less rank, the trees being in open glades, where they spread out with large branches resembling the white spruce. The northern limit of the black spruce is that of the forest belt; it and larch being the last trees met with before entering the barrens.

Low's description of the forest–tundra ecotone is noteworthy in that he is particularly conscious of the dense thickets, undoubtedly the consequence of painful experience; it was often necessary to travel through such thickets when his party portaged the difficult rapids. His description of the forest–tundra border:

> Throughout the forest belt, the lowlands fringing the streams and lakes are covered with thickets of willows and alders. As the semi-barrens are approached, the areas covered by these shrubs become more extensive, and they not only form wide margins along the rivers and shores of the lakes, but with dwarf birches occupy much of the open glades. The willows and birches grow on the sides of the hills, above the tree line, where they form low thickets exceedingly difficult to pass through. Beyond the limits of the true forest, similar thickets of Arctic willows and birches are found on the low grounds, but on the more elevated lands they only grow a few inches above the surface. In the southern region, the undergrowth in the wooded areas is chiefly Labrador tea (*Ledum latifolium*) and "laurel" (*Kalmia glauca*), which grow in tangled masses, from two to four feet high, and are very difficult to travel through. In the semi-barrens this undergrowth dies out, and travel across country is much easier in consequence. In the southern regions the ground is usually covered to a considerable depth with sphagnum, which northward of 51° is gradually replaced by the white lichens or reindeer mosses (*Cladonia*), which grow freely everywhere throughout the semi-barren and barren regions.

The *Ledum* to which Low refers, of course, now bears the name *Ledum groenlandicum,* and the laurel is *Kalmia angustifolia.* Low also mentions the more abundant species found in the forested area, including discussions of the habitat preferences of species which might normally draw the attention of a portaging voyageur such as *Rubus* and *Vaccinium* species. During his explorations in 1892, A. H. D. Ross accompanied Low for the purpose of making a large collection of plants. These were identified by J. M. Macoun, and the list is included in Low's report to the Canadian Geological Survey.

Subsequent to Low's explorations, other workers penetrated the Labrador wilderness, and their collections form the basis of the published floras that include the plant species of Labrador and Quebec. Descriptions of the northern-central forests form the basis of the work of Hustich (1949, 1951a,b, 1953, 1954, 1962) who also provides a detailed review of previous work and presents maps of the range limits of the important tree species.

NORTHERN CENTRAL QUEBEC

In discussing the Schefferville area (which at that time retained its former name, Knob Lake) Hustich (1954) found that analysis of the forest vegetation of the area on the basis of forest or site type was difficult, not only in view of the information he was able to obtain during a fairly limited time in the field but also because of the nature of the vegetation:

> The area is situated between taiga proper and forest-tundra, between mountains and lowlands of varying topography. In such regions the forests tend to be less homogeneous, with many types represented in a comparatively small area. However, certain forest-types are common: spruce lichen forests, spruce lichen dwarf shrub forests, and tamarack swamps. Fairly common also are rich white spruce forests (i.e., forests with good timber-sized white spruce, rich in herbs), particularly in the lowlands on sedimentary bedrock.

The first two forest types can be included in a more general category, lichen woodlands. With the black spruce muskegs, there are five principal vegetational types in the Schefferville area: lichen woodlands, rich spruce forests, tamarack and black spruce bogs, black spruce muskegs, and hilltop tundra areas. The rich spruce forests grade into white spruce lichen woodlands (Fig. 16). The black spruce stands grade at their wetter extremes into forests dominated increasingly by tamarack and, on their drier edges, into lichen woodland. On slopes, the spruce declines in stature and density toward the summit. The forests around Schefferville are in areas of sedimentary rocks or on till strongly influenced by the sedimentaries. Beyond the Labrador trough, in areas of granitic bedrock and drift, there are few or no white spruce stands and, indeed, white spruce individuals appear to occur only very rarely. This is taken to indicate, as Hustich (1954) infers, that white spruce is confined to rather special habitats. He states that in the Schefferville area:

> The white spruce forests rich in herbs are common in the lowlands and are typical of more or less calcareous areas with good drainage... forests with timber-sized white spruce and herbs such as *Mitella nuda, Cornus canadensis, Coptis groenlandica, Trientalis borealis, Pyrola uniflora* and other species, which in the northern forests represent "rich elements" in the vegetation. The number of species is greater in these forests than in others with the occasional exception of open tamarack swamps on calcareous soil. Reproduction is poor because of the dense undergrowth of mosses and herbs.

Fig. 16. Widely spaced spruce trees with lichens and shrubs as a ground cover form the characteristic structure of the lichen woodland. Area shown here is near Schefferville in northern central Quebec. Note the young spruce in the foreground.

Hustich also states that, in general, the number of species declines to the north and that northward a more typically northern aggregation of species appears in communities:

> The forest vegetation tends to be poorer in vascular species towards the north, which is natural. Another common and typical feature is that forest plants, which farther south are restricted to certain habitats, in the subarctic forest are more widespread.... Note the wide range of forest types and habitats of Labrador tea (*Ledum groenlandicum*), crowberry (*Empetrum hermaphroditum**), dwarf birch (*Betula glandulosa*) and mountain cranberry (*Vaccinium vitis-idaea*), for instance, in the Knob Lake area. These plants occur in the southern part of the taiga on bogs, but also in the north on dry lichen heaths and in moist forests. This variability of the site-requirements of certain common plants is also well-known from northern Europe.

In the interior of Labrador, climatic influences, oceanic on the one side and continental (modified by Hudson Bay) on the other, result in dif-

*Empetrum nigrum.

ferences between east and west, of which the most readily apparent is the proportionally greater extent of lichen woodland to the east of Hudson Bay. In a discussion of the position of the arctic front over Canada, Barry (1967) points out:

> It is noteworthy that in this region (Labrador-Ungava) where the location of the climatological front is, according to the present results, uncertain, the forest boundary is replaced by an extensive forest-tundra ecotone and the treeline for individual species extends almost to Ungava Bay. . . .

In the Schefferville area, the dominant tree in the open lichen woodland is white spruce, but it can scarcely be considered the dominant species since it is so widely spaced and, additionally, since the individual trees rarely exceed a maximum height of 25–30 ft or attain a basal area at breast height of more than 35–40 in. The dominant in terms of the ground cover is *Cladonia alpestris,* virtually continuous and broken only where a rock protrudes or where a rare spot of frost action has bared the substrate. *Ledum groenlandicum* is also a dominant ground species. Other species include *Betula glandulosa, Vaccinium uliginosum, V. angustifolium,* and *Cornus canadensis,* and white spruce seedlings are found more rarely (see Table 10).

As described by Wilton (1964), over most of the forested terrain of Labrador the tree cover is open-growing black spruce with a few stunted larch. White spruce is found on well-drained soils and in some localities is the dominant species. The common shrubs are *Ledum groenlandicum, Betula glandulosa,* and *Vaccinium angustifolium.* Spruce–feather moss forests are also common, being found southward on middle slopes. As trees grow to maturity in such forests, mosses become increasingly important in the understory and, if undisturbed for 200 years or more, build the raw humus soil up to great depths.

In a study on biomass production in the Schefferville area, Rencz and Auclair (1978) describe the lichen woodlands as developing on well-drained, nutrient-poor soils, with the more fertile sites dominated by a somewhat more closed forest and having a somewhat thicker moss understory. *Cladonia alpestris* is the more abundant dominant in the open lichen woodland, with *C. gracilis* increasing in importance in forests with greater shade. *Dicranum fuscescens* is a common moss species. Common shrubs are *Betula glandulosa* and *Ledum groenlandicum.* The lichen woodland described by Rencz and Auclair is characterized by a strong dominance by black spruce, with white spruce restricted to relatively deep soils.

After a day or two of clear weather, the *Cladonia* mat is dry and crisp, although the soil beneath remains moist for long periods even though

TABLE 10

Labrador–Quebec Spruce (Lichen Woodland)[a] Schefferville Area

Species[b]	Average frequency
Betula glandulosa	46
Cornus canadensis	73
Empetrum nigrum	55
Geocaulon lividum	5
Ledum groenlandicum	80
Lycopodium selago	11
Picea glauca (seedlings)	6
Vaccinium angustifolium	61
Vaccinium uliginosum	59
Vaccinium vitis-idaea	43
Lichens	99
Mosses	38

[a] Number of stands averaged: 4.
[b] Present: *Aster* sp., *Carex* sp./spp., *Coptis groenlandica*, *Epilobium angustifolium*, *Equisetum sylvaticum*, grass sp./spp., *Kalmia polifolia*, *Listera cordata*, *Lycopodium alpinum*, *Petasites palmatus*, *Picea mariana* (seedlings), *Ribes glandulosum*, *Sanguisorba canadensis*, *Solidago macrophylla*.

no rain falls. This accounts, no doubt, for the survival capabilities of *Ledum* and *Betula* on these dry (well-drained upland) sites. The *Cladonia* mat is sufficiently deep so that the impression of a footprint made when the lichen is dry will remain visible for long periods of time. The mat is between 3 and 4 in. deep over a large portion of the surface.

The terrain around Schefferville is gently rolling, possessing large areas of upland with slope angles ranging from 0° to 3°–4°. These are almost uniformly occupied by open lichen woodland, perhaps better developed here over large areas than elsewhere in Canada. The deep carpet of lichens must be, at least in part, the consequence of a lack of heavy grazing by caribou. Such extensive areas of open woodland with a deep carpet of lichens are a most impressive spectacle. Lowland areas are given over to more dense black spruce forests with a relatively more varied understory of herbs and shrubs.

Fire unquestionably plays a role in the ecology of the lichen woodland, since burns are found in scattered places throughout the forest, but it does not seem to be a necessary factor in the perpetuation of the open forest structure. In open lichen woodlands, spruce seedlings and saplings are widely spaced but appear to exist in densities ample to replace old trees lost to age and windfall. They are rarely if ever sufficiently abundant to appear capable eventually of creating a forest community as dense as the spruce communities found in more lowland areas. In the Schefferville area, open lichen–white spruce woodland appears to be the most abundant cover type, although areas in which black spruce is the dominant tree are common and mixtures of the two can be found without difficulty. Indeed, the number of community types described by the various authors of publications on the Ungava vegetation is somewhat in excess of what might be considered useful, since all grade into one another and it is often difficult to make a decision as to the category a given stand should be assigned.

BLACK SPRUCE COMMUNITIES

The lowland forest occurs widely throughout Labrador and Ungava, becoming increasingly frequent northward; it is found on flat or gently sloping and poorly drained topography, and, as Wilton points out, sometimes stretches for endless miles although the pattern is continuously broken by open boglands and low ridges. Shrub species that predominate throughout most of the region include *Ledum groenlandicum* and *Chamaedaphne calyculata*. Southward, *Kalmia angustifolia* is a common associate in the understory. In the Schefferville area, however, this species is not found frequently, and the visible dominants usually include *Ledum, Empetrum, Equisetum sylvaticum, Rubus chamaemorus, Vaccinium uliginosum, Smilacina trifolia,* and *Salix* species. The latter appear with somewhat higher frequency in wet depressions where *Carex* species are more abundant. On dry moss hummocks, lichens often form light-colored patches, and vascular species such as *Rubus, Chiogenes, Kalmia, Vaccinium,* and *Empetrum* are found in high frequency.

In most stands the trees average 8–12 ft apart and attain a maximum of no greater than 50–60 in. in basal area at breast height. Height averages not much greater than 30 ft.

Black spruce communities 45 mi northwest of Schefferville, on a topography of gently rolling unassorted gravelly drift, were by far the most dominant in the area. No white spruce was seen. The forest was undisturbed over large areas, and black spruce ranged to maximum sizes of 40–60 in. basal area and 40 ft in height. In comparison with the communities at Schefferville, there appear to be fewer species in the lowland black spruce communities but more in the open lichen woodlands. This may be the consequence of a long succession without disturbance into what Hustich terms the spruce dwarf shrub lichen woodland, a rich community which he believes eventually develops from open lichen woodland (see Tables 11 and 12).

Fraser (1956) confirms observations that black spruce increases in frequency in all forest communities toward the northern forest border. At one study area, Marymac Lake (lat. 57°N, long. 68°33′W), he notes that only occasional white spruce trees were encountered and that they appeared to be near their northern limit. The forest stands were composed almost entirely of *Picea mariana* which attained a height of 35–40 ft. It is notable that here, too, the substrate is glacial drift rather than the sedimentary rocks of the Labrador trough. Although extending this far northward, the trough lies to the west of Fraser's study area at Marymac Lake. Fraser further states:

> Because this station was located so near the northern limit of the boreal woodland, it was expected that differences in the composition and structure of the association would exist. For example, it was thought that possibly the density of the trees would be less. However, both the structure and density were approximately the same as at the other stations.

TABLE 11

Labrador–Quebec Black Spruce (Lichen Woodland)[a] Rainy Camp

Species[b]	Average frequency
Betula glandulosa	17
Carex sp./spp.	10
Empetrum nigrum	45
Kalmia polifolia	32
Ledum groenlandicum	85
Lycopodium selago	18
Petasites palmatus	8
Picea mariana (seedlings)	22
Rubus chamaemorus	12
Vaccinium angustifolium	75
Vaccinium microcarpus	8
Vaccinium uliginosum	43
Vaccinium vitis-idaea	40
Lichens	65
Mosses	100

[a] Number of stands averaged: 3.

[b] Present: *Aster* sp., *Coptis groenlandica*, *Cornus canadensis*, *Equisetum sylvaticum*, *Gautheria hispidula*, grass sp./spp., *Listera cordata*, *Lycopodium alpinum*, *Ribes glandulosum*, *Sanguisorba canadensis*, *Smilacina trifolia*, *Solidago macrophylla*.

TABLE 12

Labrador–Quebec Black Spruce[a] Rainy Camp

Species[b]	Average frequency
Betula glandulosa	40
Carex microglochin	15
Carex paupercula	40
Carex trisperma	22
Carex vaginata	12
Carex sp./spp.	45
Chamaedaphne calyculata	40
Coptis groenlandica	47
Cornus canadensis	18
Empetrum nigrum	87
Equisetum sylvaticum	62
Gaultheria hispidula	25
Geocaulon lividum	32
Kalmia polifolia	72
Ledum groenlandicum	30
Linnaea borealis	22
Petasites palmatus	42
Picea mariana (seedlings)	32
Rubus acaulis	15
Rubus chamaemorus	80
Salix arctica	17
Salix sp.	30
Scirpus caespitosus	20
Smilacina trifolia	40
Solidago macrophylla	12
Vaccinium microcarpus	55
Vaccinium uliginosum	100
Lichens	37
Mosses	100

[a] Number of stands averaged: 2.

[b] Present: *Agrostis* sp., *Bartsia alpina, Carex pauciflora, Carex saxatilis, Carex* sp., *Equisetum palustre, Erigeron* sp., *Eriophorum angustifolium, Eriophorum spissum, Geum canadense,* grass sp./spp., *Habenaria dilatata, Juncus castaneus, Listera cordata, Lycopodium selago, Parnassia palustris, Polygonatum biflorum, Potentilla tridentata, Ranunculus lapponicus, Solidago multiradiata, Tofieldia pusilla, Trientalis borealis, Vaccinium angustifolium, Vaccinium vitis-idaea, Viola adunca* var. *minor.*

It is of interest that virtually the same conditions exist at the forest border west of Hudson Bay, where lowland black spruce densities continue to be relatively uniform northward to the edge of the forest border. This is true in the Ennadai Lake area, at Artillery Lake, and at Colville Lake— all sites along, or just within, the continental (latitudinal) forest border between Hudson Bay and its western limit near the mouth of the Mackenzie River. Thus, it appears that, in terms of structure and composition, the northward variation in at least the arborescent stratum of the boreal forest community is remarkably uniform over long distances. Differences that do exist are found in the species composition and species frequency values in the understory. What appear to be anomalous distributions in the case of white spruce can at least usually be attributed to the marked preference of this species for alluvium or slightly basic soils. Black spruce, on the other hand, appears rather uniformly distributed in terms of latitudinal variation, with areas of similar climate usually bearing black spruce forests of similar general appearance, although understory species and species frequencies may differ. Likewise, along the entire range of the forest across the continent, a latitudinal displacement of equal distance from the northern or southern border will result in what is roughly a predictable and equal change in the overall appearance of the forest. Moreover, while white spruce may approach the northern forest border as an occasional individual, it is black spruce that is the dominant at the forest border itself, comprising in at least most instances the dominant tree species in the forest border communities.

Southward, white spruce increases in frequency and dominance, particularly on those especially favorable sites already mentioned, and southward, also, the other species associated more commonly with the southern boreal forest increase in importance, notably jack pine, balsam fir, aspen, balsam poplar, cedar, and other species often present with greater frequency in hardwood forests. It is of interest, too, that the spruce–feather moss forest community is increasingly rare northward. In the central part of the Labrador peninsula it reaches its limit of occurrence somewhere between Wabush and Schefferville (see Table 13). Hustich (1949) describes the very impressive feather moss forests in the following words:

> If there are any such things as climax forests in nature, the spruce feather moss forest is certainly one of them. In the southern height-of-land areas this forest type seems to be fairly common; the important subtype is black spruce-dominated. Here the feather mosses, *Hylocomium splendens, Pleurozium schreberi,* and *Ptilium crista-castrensis* form a dominating moss cover, with scattered, mostly sterile individuals of various herbs and dwarf shrubs. The forest is usually homogeneous, old (150–200 years) and there is hardly any regeneration at all, the moss cover being too thick and too dry. I have seldom seen white spruce forests of this kind, but balsam fir and white birch often intermingle with black spruce. Old jack pine forests, where black spruce intrudes and gradually becomes dominant, also often belong to this forest type.... It is usually a good pulpwood and timber forest, because of its dense even-aged stands; it appears locally all over the taiga, but is probably most common in the southern height-of-land areas.

TABLE 13

Labrador–Quebec Black Spruce–Feather Moss Woodland[a] Wabush

Species[b]	Average frequency
Gaultheria hispidula	25
Ledum groenlandicum	80
Picea mariana (seedlings)	30
Vaccinium angustifolium	80
Vaccinium vitis-idaea	95
Lichens	50
Mosses	100

[a] Number of stands: 1.
[b] Present: *Abies balsamea, Cladonia alpestris, Polytrichum* sp.; feather mosses.

That this forest is representative of what is termed climax is now open to question, and Hustich implies as much when he states that reproduction is poor or entirely lacking because of the dense moss cover that prevents regeneration of spruce. In another paper, Hustich (1954) also indicates that, in heavy moss, spruce seeds may germinate but the roots fail to reach mineral soil through the thick moss and humus, with the result that they are depressed and eventually die. The consequence is that the forest is not self-perpetuating, although its ultimate fate in the region is uncertain since stands are almost invariably destroyed by fire before succession can proceed to any other stage.

ALTITUDINAL TREE LINE

The tree line in the area around and north of Shefferville is especially interesting because it occurs rather uniformly at about 760 m above sea level on all the hills in the area that reach or exceed this altitude. Harper (1964) has compared the tree line on Sunny Mountain at Schefferville and points north and south (see tabulation on next page).

As Harper points out, the low temperatures of the water of Ungava Bay apparently work a cooling influence on the lands bordering the bay and for some distance to the south, evidently as far as Lac Aulneau. The increase in altitude toward the south of the tree line is clearly apparent, and as one continues southward this line rises until, in the mountains of Gaspé Peninsula and New England, it is found only a little below 4000 ft. The alpine tundra community on the summit of Mt. Jacques Cartier on the Gaspé Peninsula, at an altitude of 4160 ft, is represented by very

Location	Latitude	Distance from next point north (mi)	Altitude (ft)	R[a]
Fort Chimo	58°05'	—	100	—
Lac Aulneau	57°01'	73	900	11.0
Sunny Mountain	55°03'	140	1900	7.1
Lorraine Mountain	53°06'	141	2550	4.6

[a] R is the approximate increase in altitude of the tree line (in feet) per mile of latitude in comparison to the next locality to the north.

few species of high frequency, some of which, found just below the summit proper on the south slopes, are more representative of boreal forest than of alpine tundra. Farther to the north, however, on the hills around Schefferville, many more species of the Canadian eastern Arctic are represented in the communities of summits and south-facing slopes. Harper points out that 25 species are found on the summit of Sunny Mountain that are not found elsewhere in central Ungava, including *Salix herbacea, Luzula confusa, Sibbaldia procumbens, Oxyria digyna, Carex scirpoidea, Cassiope hypnoides, Arenaria humifusa, Stellaria longipes, S. crasifolia, Juncus trifidus,* and others, all of arctic affinities. In the Schefferville area, the largest proportion of the land surface is covered with forest, but the proximity of the Arctic is apparent in the composition of the communities on the hill summits.

Both Sunny and Irony mountains at Schefferville possess associations of species representing both arctic and boreal affinities. A high ridge on Irony Mountain, for example, possesses a rich association of both arctic species and such boreal representatives as *Salix vestita, Chiogenes hispidula, Mitella nuda, Betula glandulosa, Solidago,* and *Viola.* A single clump of spruce in a protected swale on the summit is surrounded by a mat largely composed of species with definite arctic affinities, including *Arctostaphylos alpina, Salix uva-ursi, Dryas integrifolia, Carex bigelowii, C. scirpoidea, Loiseleuria,* and *Scirpus caespitosus,* as well as *Empetrum* and *Vaccinium uliginosum.* The summit rock fields include, in addition to these species, others with distinct arctic affinities such as *Diapensia* and *Luzula,* as well as the arctic moss *Rhacomitrium* (see Table 14).

Thus, it can be seen that there exists a rather definite distribution of species along a vertical gradient up the slopes of the hills in the Schefferville area, and that this corresponds at least roughly to the zonation occurring over many degrees of latitude along a north-south line in the same region. Species with the highest frequencies in the most southern

TABLE 14

Labrador–Quebec Rock Field[a] Schefferville (Mounts Sunny, Gerin, and Irony)

Species[b]	Average frequency
Arctostaphylos alpina	26
Bartsia alpina	8
Betula glandulosa	24
Carex bigelowii	36
Carex capillaris	6
Carex scirpoidea	11
Carex sp./spp.	67
Coptis groenlandica	6
Cornus canadensis	3
Dryas integrifolia	18
Empetrum nigrum	44
Grass sp./spp.	4
Linnaea borealis	3
Loiseleuria procumbens	11
Lycopodium selago	21
Phyllodoce caerulea	11
Polygonum viviparum	7
Pyrola grandiflora	6
Rhacomitrium lanuginosum	4
Rhododendron lapponicum	14
Salix herbacea	11
Salix uva-ursi	20
Salix vestita	12
Saxifraga aizoön	5
Scirpus caespitosus	8
Solidago macrophylla	4
Solidago sp.	4
Trientalis borealis	3
Vaccinium angustifolium	3
Vaccinium vitis-idaea	56
Vaccinium uliginosum	65
Lichens	100
Mosses	71

[a] Number of stands averaged: 7.
[b] Present: *Arenaria groenlandica, Arnica alpina, Carex glacialis, Carex* sp., *Carex vaginata, Cassiope hypnoides, Cerastium* sp., *Diapensia lapponica, Hierochloe alpina, Juncus trifidus, Ledum groenlandicum, Linnaea borealis, Luzula confusa, Luzula parviflora, Oxyria digyna, Mitella nuda, Rubus chamaemorus, Polygonatum biflorum, Senecio pauciflorus, Sibbaldia procumbens, Viola adunca* var. *minor*.

parts of the boreal forest, such as *Trientalis, Linnaea, Cornus,* and *Coptis,* are found at the lower levels of the slope; *Betula* and northern representatives of *Solidago* and *Viola* are intermediate; and on the upper levels are such pronounced arctic species as *Arctostaphylos, Dryas, Carex bigelowii, Loiseleuria,* and *Luzula confusa.*

It is interesting to speculate on the climatic factors responsible for this marked zonal effect on the vegetation of Sunny, Irony, and Gerin mountains in the Schefferville area, but speculation cannot substitute for the information that could be gained by climatic data obtained on the slopes of these three hills throughout a single summer season. If it were possible to establish weather-recording stations at similar topographic sites and at comparable elevations from the southern limit of the Labrador peninsula to the coast in the north, it is almost certain that a most significant mass of data would be obtained of great interest to bioclimatologists and of real importance. It is astonishing how much speculative effort can be avoided by obtaining a few pertinent facts.

THE NORTHERN FOREST BORDER

In the vast northern interior of Canada, the transition between forest and tundra, called variously the continental arctic tree line, the northern forest border, and the northern forest–tundra ecotone, lies roughly along a WNW–ESE axis extending from the Mackenzie Delta in the northwest to the western shore of Hudson Bay at the mouth of the Churchill River. Along this entire length, the forest border swings northward at some points and southward at others, forming interdigitations of forest into tundra along river valleys and tundra into forest along expanses of uplands; and between these extremes there are the ubiquitous lichen woodlands—forests of widely spaced spruces with a ground carpet of lichens, scattered northern shrubs, and individuals of a number of common herbaceous species (Fig. 17).

Vegetational studies on the forest–tundra ecotone in Keewatin and Mackenzie, Northwest Territories, central Canada (Larsen, 1965, 1971a,b, 1972a,b), reveal that tundra communities here are characterized by fewer plant species than are present in the communities either north or south of this transition zone. Moreover, species making up the transition zone communities are geographically wide-ranging, many occupying habitats throughout both the northern edge of the forest and the low-arctic tundra. Although these species, in aggregation, may be said to characterize the tundra communities of the regions, many are so ubiquitous throughout both forest and tundra (Larsen, 1971b) that their

Fig. 17. Northern edge of the forest–tundra ecotone in the Ennadai Lake area of southern Keewatin, Northwest Territories. Spruce and tamarack grow in favorable habitats where muskeg grades into upland and in the lee of eskers (rock pavement of wind-eroded esker summit in foreground).

ecological relationships, particularly those involving climatic conditions, pose some interesting and perplexing problems.

There are rather definite relationships between forest border and regional climatic characteristics (Larsen, 1965, 1974). The zone of floristically depauperate tundra communities extends for some distance north of the forest border, and it appears that this distinct vegetational zone coincides with a correspondingly distinct climatic zone. Frontal conditions occur with decreasing frequency northward and southward. Plant species adapted to survival in the region dominated by arctic masses occur with increasing presence, frequency, or both, in vegetational communities along a traverse from the forest northward into the region occupied nearly continuously by arctic air. Apparently it can be inferred from the remarkable width of the transition zone characterized by depauperate vegetational communities that the climatic frontal conditions in Keewatin occupy a broader north–south band than they do farther to the west.

Climatic variation across the boundary between the two regions is sufficient to account, at least in large measure, for the vegetational differences, since geological and topographical characteristics are generally uniform throughout (Lee, 1959). On the other hand, a self-perpetuating system that augments the influence of the climatic parameters may be at work within the forest or the forest–environment complex. A comparison of the radiation characteristics and energy budgets of forest and tundra has revealed differences of sufficient magnitude to warrant further study of the possibility that energy relationships on a gross forest–atmosphere scale are involved.

Energy transfer relationships within the forest itself effectively contribute to maintenance of the forest where it now exists. The forest canopy undoubtedly serves as both a windbreak and energy trap, resulting in higher late summer soil- and ground-level air temperatures in the forest than will occur in tundra, even when air temperatures a short distance above forest and tundra vegetation are identical. Hence, the closed forest canopy tends to create a ground-level environment more ameliorated in terms of growing conditions for mesophytic plants than that to which tundra vegetation is exposed, even on adjacent and similar topographic sites. Species comprising the spruce understory community are obviously capable of survival and reproduction in the forest environment, but many are equally obviously unable to advance into areas occupied by tundra. Even very small environmental differences must be critical, and these must account for the abrupt nature of the forest border (Fig. 18).

It remains to be considered whether the floristically depauperate communities of the ecotone region at and near the forest border may not be, at least partially, the consequence of movement of the forest border within recent geological time. Evidence indicates that the forest retreated from a former position, at or near Dubawnt Lake, somewhat more than 3500 years ago. It appears to have fluctuated north and south of its present position at the south end of Ennadai Lake since that time, having produced fossil soil profiles observed at the south end of the lake. These periods of soil formation were separated by a period in which aeolian sand blew over the profiles from nearby eskers, indicating an interval during which the area was drier, and probably colder, than it is at the present time. Forest occupied the Dimma Lake area, about 50 mi north of Ennadai Lake, as well as the north end of Ennadai Lake, about 900 years ago. Evidence from pollen deposited in a peat bank at the edge

Fig. 18. The tree line may often be a visible entity on higher hillsides at the northern edge of the forest–tundra ecotone. The forest shown here, along the south end of Ennadai Lake (territory of Keewatin, Northwest Territories) is here quite dense, and the trees, although not more than 7–10 m in height, form a closed-canopy forest along the margins of the lake. On the upper hill slope, the forest reaches its environmental limit about 100–150 m above lake level. Dwarfed spruce clones are seen in the foreground. Alder, willow, dwarf birch, ericoid shrubs, and lichens dominate the ground vegetation.

of Ennadai Lake indicates that *Picea* has not moved north of Ennadai Lake within the past 630 years (see Nichols references in Chapter 2). Since many arctic species probably are incapable of persisting beneath a spruce canopy, even though (improbably) climatic conditions might otherwise be favorable, it is likely that these species would be absent from an area recently occupied by spruce forest until sufficient time had elapsed for migration and recolonization. Evidence (Savile, 1956, 1964, and personal communication) indicates, however, that arctic species are capable of very rapid migration into environmentally suitable areas, since many opportunities exist for dispersal by animals and by physical events in arctic and subarctic regions.

In the light of this evidence, it appears that the absence of species of more arctic affinities at Ennadai, for example, cannot be attributed to insufficient time for migration. It is of interest in this regard, also, that *Diapensia lapponica,* an arctic species, was found atop an unusually high hill (400–500 ft above lake level) in the Ennadai area. Thus, at least one arctic species occupies a rare site which, microclimatically, must resemble arctic areas farther north. Significantly, it has failed to become a generally frequent component of the rock field communities at Ennadai, an ecological role it plays in the vegetation farther northward. One must assume that this, presumably, is because *Diapensia* is ill-adapted to the even very slightly more subarctic environment to be found on lower hills in the Ennadai area.

There exist one series of climatological observations that bear on the nature of the frontal zone in the region under consideration. Ragotzkie (1962) and McFadden (1965) report on observations of dates of freeze-up and breakup of lakes in central Canada, providing maps showing (for various periods in spring and fall) the position of the zone in which some lakes are frozen and some are not (i.e., the zone between the line north of which all lakes are frozen and the line south of which all lakes are open). It is of interest that this zone is notably wider in the Keewatin area in which the above vegetational data were obtained than in areas farther west. The floristically depauperate zone is correspondingly wider in the Keewatin area than it is in the area, for example, around the eastern arm of Great Slave and Artillery lakes. In the Keewatin area, jack pine is absent from the forest for a considerable distance (perhaps 100 mi) south of the forest–tundra ecotone at Ennadai Lake, while it is found in abundance at one point along the portage between the east arm of Great Slave and Artillery lakes. Additionally, *Rhododendron lapponicum* and *Dryas* species are found at Fort Reliance within the spruce forest, but to the east along the Kazan River they are not found at least in any abundance until one travels many miles north of the forest–tundra ecotone. The same is true of other species of arctic affinities. It is thus apparent that in the west where the frontal zone characteristically occupies a narrower belt than in the east, boreal and arctic species overlap ranges, while in the east, where the frontal zone is wide, there exists a wide gap between species typically arctic or boreal in affinities. Here there exists a wide belt in which only the more ubiquitous species are found in sufficient abundance to appear regularly with high frequencies in transects. It will be of interest, and perhaps of considerable ecological significance, to explore these relationships further and attempt to determine the physiological characteristics that account for the distinctive response to the climatic conditions that prevail on the part of the species involved. The presence of this zonation can be interpreted as a consequence of a corresponding zonation in climate, and in view of the general monotony of the terrain throughout this vast region (Fig. 19) this is, indeed, the simplest and most generally satisfactory explanation. Farther north, Cody and Chillcott (1955) noted that the breakup at Musk-ox Lake (slightly north of the north arm of Aylmer Lake) was 2 weeks behind that at Matthews Lake about 95 mi to the southwest, demonstrating a rather sharp climatic zonation in the region.

Spruce stands found north of the forest border presumably represent relicts surviving from a period when the border lay farther north than it does at the present time (Larsen, 1965; Nichols, 1967; Elliott, 1979a,b).

Fig. 19. At the northern edge of the forest–tundra ecotone the spruce forest (middle distance) is restricted to favorable sites, which often are found at the transition zone between open muskeg (foreground) and the gentle slopes of more upland areas (barely visible beyond the spruce stand in the distance). The species of the muskeg are principally cotton grass, sedges, and dwarf birch. The area shown is at the northern end of Ennadai Lake, Northwest Territories (territory of Keewatin).

Fig. 20. These spruce trees show the effects of harsh growing conditions in a Far Northern area. Note the trees displaced at angles by soil–frost action, the undecayed dead and fallen stems and branches, and the trees with branches missing from the side exposed directly to the prevailing winds in winter. Reproduction is primarily by layering.

Small clumps of black spruce, probably clones since reproduction appears to be primarily by layering, thus constitute a potential source of propagules for afforestation of the entire area, and the question arises as to the characteristics of the environment, of spruce, or of both, that prevent the spread of spruce forest over most of the available land surface (Fig. 20).

It is of interest that trees identified in the field as white spruce (*Picea glauca*) occur abundantly on some sites, both on eskers and along small streams running through draws between hills, both at Ennadai (Larsen, 1965; Elliott, 1979a,b), and in other areas at or near the forest border in central Canada (Ritchie, 1959) and eastern Canada (Hustich, 1966), but not in the area studied by Argus some distance southwest of the forest border. Dwarfed white spruce, identified by cone size and shape, have been photographed by the author on the shore of Dubawnt Lake, and rather tall specimens of what were identified in the field as white spruce, additionally, have been found on eskers 25-30 mi north of Ennadai Lake along the Kazan River, as well as at a point just west of Yathkyed Lake (lat. 62°44′N, long. 98°38′W). Similarly, white spruce also is found north of the forest border along the shores of Artillery Lake. These outliers of a species, with apparently more restricted environmental tolerance than black spruce, pose interesting ecological, and perhaps paleoclimatological, questions, as well as genetic ones, since what appears to be evidence of introgression between black and white spruce also occurs here. Parker and McLachlan (1978), however, show that there are reasons to question the evidence supporting the existence of hybridization between *Picea glauca* and *P. mariana* in northwestern Ontario, and other reports of natural hybridization between white and black spruce (Larsen, 1965; Roche, 1969; Dugle and Bols, 1971) in other areas warrant reexamination. The morphological characteristics that make speciation difficult may be the result of the harsh environment at the northern forest border rather than gene exchange. On the other hand, it seems reasonable to expect some degree of introgression at the edge of the range of two similar species (Anderson, 1949), and the question at present remains open whether such has occurred between black and white spruce in northern central Canada.

The persistence of permafrost at shallow depths also appears to be a factor rendering the environment a marginal one for spruce, and permafrost levels may account for the failure of spruce clumps to regenerate after disturbance in Far Northern areas. The traditional concepts of forest succession must be inapplicable; once forest has been eliminated, there appear to be no species, or aggregations of species, capable of ultimately creating conditions which will again permit ecesis by spruce and the spruce community unless a change in climate has ameliorated the total environmental complex (see Larsen, 1965). The high heat-exchange capacity of water results in deeper active layers along shorelines, and this may account for the persistence of spruce northward along waterways and rivulets in regions where upland and inland areas are devoid of even dwarfed trees, except perhaps the rare relicts seen to persist at times on what appears to be most unlikely upland sites. A number of other factors also known to affect the growth of trees at the forest border are presented by Savile (1963).

There are areas of forest border, however, in which spruce reproduction by seed apparently does occur. Elliott and Short (1970), for example, found that northernmost trees occurred at Napaktok Bay on the Labrador coast and, although the forest of white spruce appeared very old, with many fallen dead trees, reproduction was taking place, with a wide variety of age classes of trees present. The forest, hence, appears to be in equilibrium with the present climate, in contrast to the northern trees of Keewatin. On the higher uplands, the stunted white spruce were, however, producing no cones. No seedlings were seen in these upper areas, although there were saplings present estimated to be about 100 years old. Black spruce and balsam fir have been reported in the Napaktok Bay forest, but these were not seen by Elliott and Short, who indicate they may be found with more extensive exploration although they were not apparent during aerial reconnaissance.

In the Thelon River basin, at Warden's Grove, there appears to be some sexual reproduction occuring in both white and black spruce. Large numbers of seedlings have been reported on the periphery of many stands and in the better-drained meadows. Reproduction in clumps of dwarf spruce in the upper Hanbury drainage and in exposed locations around Warden's Grove appears to be by vegetative layering (C. Norment, personal communication).

In the Artillery Lake area, Clarke (1940) found small clumps of spruce far beyond the forest proper and concluded that this zone (where spruce persists as small, isolated clumps) was broader than the transition zone at the forest border. He found that tree sparrows followed the small clumps of spruce far into the tundra, and that snow buntings apparently found the southern limit of their breeding range at the northern limit of the range of spruce as a species. Despite the broad extent of this zone, however, he concluded that

> because of the few species involved it is perhaps not worthy of recognition as a valid life zone. There are, nonetheless, so many tiny clumps of spruce that were any climatic change to occur making it possible for trees to occupy exposed situations in that region the occupation would be rapid.

The open forests of the Far North in Central Canada have a continuous lichen mat, but one not as luxuriant as that found to the east in Ungava, presumably related to the greater summer rainfall in the eastern region. Grazing on lichens by barren-ground caribou in northern Manitoba and Saskatchewan may also be involved in this difference. The main feature of the open coniferous forest in northern central Canada (Ritchie, 1960a,b,c) is the prevalence on upland sites of stands of well-spaced conifers, with a ground cover consisting of a lichen mat closely associated with *Vaccinium vitis-idaea, Empetrum nigrum, Betula glandulosa, Salix myrtillifolia,* and other willow species (Fig. 21). Black spruce is the common tree on drift ridges and hills. On eskers and floodplains, white spruce is common. The tundra upland vegetation is floristically similar to that of the tundra zone, according to Ritchie, the chief species being *Vaccinium uliginosum, V. vitis-idaea, Loiseleuria procumbens, Empetrum nigrum, Cladonia mitis, C. rangiferina,* and *C. nivalis.* In lower areas there are found scrubby stands of black spruce with *Betula glandulosa.* Areas with slightly impeded drainage have small black spruce trees with an understory of *Ledum groenlandicum, Vaccinium vitis-idaea, Rubus chamaemorus,* and *Sphagnum* species; the wet areas are treeless and have *Eriophorum spissum, Carex stans, C. capitata, C. rariflora,* and *Scirpus caespitosus.*

Fig. 21. Sandy esker ridges in the foreground, and gently rolling hills of sandy glacial till in the background, are forested with spruce in the northern boreal forest of southern Keewatin, Northwest Territories. The south end of Kasba Lake is shown in the distance.

A lichen woodland is found at places along the portage route between Great Slave and Artillery lakes, with widely spaced spruce and an understory dominated by *Stereocaulon* and *Cladonia* and with denser aggregations of *Vaccinium vitis-idaea* and *Empetrum nigrum* beneath the trees. The substrate is primarily well-drained sand and gravel. The greater proportion of black spruce stands across the portage, however, occupy areas where till and weathering products have accumulated in the declivities between outcropping hills. The scattered dwarf spruce seldom exceed 20 ft in height or 12 in. basal area at breast height. There are crustose lichens on the rocks, and *Rhacomitrium* frequently surrounds the bases of rock outcrops. An occasional white birch is present. In areas of outcropping rocks, plants are confined to rock fissures and small deposits of weathered material. In such areas, it is difficult to discern any aggregations of plants that might be considered a community. On hill summits where a larger deposit of gravel exists, however, plant cover is nearly continuous, and an association of largely decumbent, xerophytic species can often be found.

Farther westward, in the Winter Lake area, north of Yellowknife, the landscape is characterized by rolling hills of deep glacial till and outcrops occur rarely. The summits of the highest hills rise about 400 ft above the surface of the lake, and the relief in the area, greater than that prevalent in the surrounding country, may account for the extension of spruce groves into tundra within the valley of the Snare River system northeast of the general regional position of the edge of the spruce forest. Black spruce forest, composed of dwarfed spruce attaining a maximum height of perhaps 10–12 ft, is found over favorable flat, lowland areas and protected hollows. White spruce groves are found along streams, on eskers, and on higher hills in protected declivities. The generally rough terrain in the valley of the Snare River and the chain of lakes including Winter and Roundrock lakes very likely accounts, additionally, for the existence of areas of deep snow accumulation, affording winter protection for spruce in the area. The terrain is similar in this respect to the Ennadai Lake area. In the immediate vicinity of Winter Lake, spruce groves of more than an acre in size and often larger are found virtually all the way to the eastern extremity of the lake. Beyond this, to the northeast, however, tundra dominates the landscape exclusively.

South of Great Bear Lake, in the Cartridge Mountains, the area is largely forested, with relatively dense black spruce covering both uplands and lowlands of the gently rolling glacial drift. A relatively dense white spruce–lichen woodland is found more rarely, almost invariably in level areas of sandy outwash. The lowland stands of black spruce are often dense, grading perceptibly into the wet lowland muskegs with

TABLE 15
Frequency in 20-Quadrat Transects of Species Common in Forest Border Black Spruce Communities of the Areas Named[a,b]

Species	Mackenzie Mts. (east slope)	Colville Lake	Inuvik	East Keller (treed muskeg)	East Keller (lichen woodland)	Winter Lake	Lynn Lake-Kasmere-Kasba-So. Ennadai	Dubaunt Lake	Yathkyed (presence in area indicated by x)
Alnus crispa	33		3		2	40	6		
Andromeda polifolia	14	12	50	38		5	2		
Anemone parviflora	61		3						
Arctostaphylos alpina					20				
Arctostaphylos uva-ursi	12	22	68	28	17		17	5	x
Betula glandulosa	49	47	48	28		48	20	40	x
Calamagrostis canadensis			3					60	x
Carex sp./spp.	68	42	50	50	2	50	18	20	x
Carex scirpoidea	22		20	13					
Carex stans				5					
Carex vaginata	4		10					18	
Dryas integrifolia	76		40					18	
Empetrum nigrum	28	32	35	18	12	20	53	43	x
Equisetum arvense				5				18	x
Equisetum scirpoides	14	12	93	10	12		11	3	
Eriophorum spissum	19					10			
Geocaulon lividum	8			30	65		2		
Grass sp./spp.	7	2	3	13	47		11	38	

Species	1	2	3	4	5	6	7	
Juniperus communis	6					5	5	x x
Larix laricina (seedlings)			5	13		5	10	
Ledum decumbens	17	40	15	75	80	10	25	
Ledum groenlandicum	32	66	78	80	75	88	40	
Linnaea borealis			3					x
Pedicularis labradorica	16			8	2	4		
Pedicularis sp.	13				24			
Picea mariana (seedlings)[c]		20	18				38	x
Pinguicula vulgaris	4		63	10				
Potentilla fruticosa	29		38	18	2			
Pyrola grandiflora	2		30		3		28	x x
Pyrola secunda		8	5	3	13		8	
Rosa sp./spp.				10	75			
Rubus acaulis				8		3		x
Rubus chamaemorus		75	3	33	73	47	75	
Salix glauca	3	18	3		8	1	10	
Salix planifolia		7	3	10				x
Saussurea angustifolia	6							
Shepherdia canadensis	4	2	3		3			
Solidago multiradiata	2				2			
Tofieldia pusilla	5			15			3	
Vaccinium uliginosum	66	68	23	45	40	63	40	x
Vaccinium oxycoccos		7	90	48	28	23	10	x
Vaccinium vitis-idaea	37	93	78	80	100	88	43	x
Lichens	99	75	100	98	75	47	NK[a]	NK
Mosses	NK	100	100	100	100	58	NK	NK

[a] From Larsen (unpublished data).
[b] Frequency is expressed in percentage of quadrats in which species occurs.
[c] Source from seedlings or layering not noted.
[d] NK, not known.

widely scattered and dwarfed black spruce. Jack pine is present but is rare and found on occasional high, sandy uplands.

A comparison of the relative frequency of dominant species in the black spruce communities of a number of areas between Hudson Bay and the western Cordillera is presented in Table 15. The data show that many of the dominant species retain importance in the understory of black spruce stands of the northern forest border across the continent (see also Tables 17–20 for northern Quebec) and that the forest border ecotone is remarkably uniform in appearance and species composition over several thousand miles of land surface in North America.

LICHENS AND MOSSES OF THE FOREST FLOOR

Lichen species are frequently the most conspicuous component of the ground-layer vegetation in northern boreal regions, especially in areas near the northern forest border where they virtually cover the ground surface between trees. A special term, *lichen woodland*, has been accorded the forests in which lichens occur with such abundance. Quantitative measurements of lichen importance in the vegetation of northern boreal regions indicate that they possess the same order of importance in terms of biomass (weight) as trees and other higher plants, and they play a crucial role in water regimes of northern forest soils (Scotter, 1964; Kershaw and Rouse, 1971; Rouse and Kershaw, 1971; Rencz and Auclair, 1978; Ahti, 1977). Physiological adaptations of lichens give them the capability to colonize sites where competition is of low or moderate intensity because of unfavorable water or temperature regimes for other plants (Smith, 1962; Ahmadjian, 1967; Hale, 1967; Kershaw, 1972; Lechowicz et. al., 1974; Ahti, 1977). In terms of biomass, lichens growing on the boles and branches of trees are also of considerable importance in many boreal communities (Scotter, 1964; Jesberger and Sheard, 1973; Kershaw, 1977; Holmen and Scotter, 1971; Culberson, 1955; Lambert and Maycock, 1968). It seems likely that lichens must also be of significance among factors that influence the germination of seeds and the growth of seedlings.

In his discussions of lichens in North America, Thomson (1967, 1972, 1979) points out that there is a paucity of data on the relative abundance with which the various lichen species occur in boreal communities. Virtually no studies have been conducted to discern whether correlations exist between variations in ground lichen species frequency and variations in environmental factors, especially regional climates. A large proportion of the species are wide-ranging or circumboreal, and it can be inferred that they possess a wide range of environmental tolerance, but the degree to which they are tolerant of specific environmental factors has yet to be revealed. The work of Lechowicz and associates (see References) indicates that photosynthetic temperature adaptations, as well as relationships to moisture supply, underlie the biogeography of at least the species of *Cladonia* studied. Jesberger and Sheard (1973) found that lichen species growing on the bark of trees showed preferences for certain tree host species and that they also had marked preference for moisture regimes of the geographical regions in which they were found.

The lichens of the greatest importance in boreal forests are those of the *Cladonia* subgroup called *Cladinae*, both in terms of the proportion of ground surface occupied and the total biomass involved. The *Cladinae* are the so-called reindeer lichens because of their significance as a staple winter food for the Eurasian reindeer and the North American caribou. In terms of the number of species of the genus represented in forest stands *Cladonia* species are also of the greatest significance. This is demonstrated in Appendix IV, Tables IV.1 and IV.2, in which the ground lichens present in stands of black spruce and of jack pine in northern Saskatchewan are listed; the species of the genus *Cladonia* outnumber all other species combined. In Appendix IV, Tables IV.3 and IV.4, the lichen species found growing on the boles and twigs of black spruce in stands located in Manitoba are also listed and their frequency of occurrence given.*

In his volume, "The Lichen Genus *Cladonia* in North America," Thomson (1967) points out that distributional patterns of *Cladonia* lichens appear to be comparable with those of higher plants and quite probably are determined by the kind of substrate on which the lichen grows as well as by the same soil temperature and moisture relationships that affect the distribution of higher plants. It appears also that atmospheric moisture is very important in affecting the distribution of lichens. The greatest number of species of *Cladonia* in North America (north of Mexico) have circumpolar distribution patterns, according to Thomson. The most northerly of the circumpolar species include *Cladonia acuminata, C. stellaris (alpestris), C. amaurocraea, C. bacilliformis, C. bellidiflora, C. carneola, C. coccifera, C. cyanipes, C. decorticata, C. ecmocyna, C. lepidota, C. macrophylla, C. norrlinii, C. phyllophora, C. pleurota, C. pseudorangiformis, C. subfurcata,* and *C. sylvatica.* These species are found in the arctic, the subarctic, and the boreal coniferous forests of North

---

*All lichen samples were obtained by the author in mature stands of spruce and pine, and appreciation is hereby extended to Professor John W. Thomson of the University of Wisconsin for the lichen identifications.

America and may extend southward in the Rocky Mountains in the west, in the Adirondacks and the Presidential Range in the east and, in some cases, in cooler microclimatic spots in the vicinity of Lake Superior. Some circumpolar species occupy exceedingly broad distributional bands, being found not only in the arctic and in the boreal forest but also in the temperate broadleaved forests, Thomson points out. Such broadly ranging species include *Cladonia cariosa, C. chlorophaea, C. crispata, C. cornuta, C. gonecha, C. gracilis, C. grayi, C. merochlorophaea* (probably), *C. mitis, C. pyxidata, C. rangiferina, C. scabriuscula, C. squamosa, C. subulata, C. turgida, C. uncialis,* and *C. verticillata*. A very large proportion of the circumpolar species do not appear to range quite as far north as the preceding but may be found in the boreal coniferous forests and southward into the temperate broadleaf forests, their principal abundance being in the United States and their ranges in Canada mainly limited to the southern portion of the boreal coniferous forests. Such species include *Cladonia bacillaris, C. botrytes, C. cenotea, C. coniocraea, C. conista, C. cryptochlorophaea* (probably), *C. deformis, C. digitata, C. fimbriata, C. floerkeana, C. furcata, C. glauca, C. macilenta, C. nemoxyna, C. ochrochlora, C. parasitica, C. pityrea,* and *C. polycarpoides*.

Reindeer lichens normally grow on organic and inorganic soils rather than on bare rock surfaces, bark, or decayed wood. *Cladinae* do not require the proximity of rock, but one preferred habitat is rock outcrop covered by a thin veneer of humus and minerals—the result, to a great extent, of the absence of competition from vascular plants and mosses in such habitats. On wet soils, mosses and vascular plants force the lichens out. In bogs, particularly on hummocks, they thrive on moist peat. Oligotrophic, sandy, gravelly, stony, or rocky soils, however, offer the best habitats for *Cladinae*. In his discussion of *Cladonia* ecology, Ahti (1961) points out that light is one of the most significant growth-limiting factors; few lichens are able to live in habitats as shaded as those occupied by common boreal forest mosses. The habitats of *Cladina* are treeless or thinly wooded; in dense forests they occur more rarely. The main source of water for *Cladinae* is atmospheric humidity. In general, lichens are considered pioneer plants in successional relationships, and reindeer lichens are the most successful in competition with vascular plants and mosses. Growth rates allow successful competition with common boreal forest mosses, but on some sites strong competition results in a mosaic of lichens and mosses. In oceanic climates mosses are prevalent, and in continental regions lichens dominate mosses. If the growth rate of reindeer lichens is continuously greater than decomposition, a white "lichen peat" gradually accumulates; the site is easily colonized by mosses which cause paludification, with the result that the mosses take over the site from the lichens. This succession is probably confined to areas associated with permafrost; elsewhere the accumulation of lichen peat does not take place, or it occurs too slowly to cause eventual paludification. Reindeer lichens, however, also are found in some stable communities in advanced stages of succession. The driest sites are often covered with lichen woodland or tundra heath dominated by *Cladonia stellaris (alpestris)* and the other common *Cladonia* species. The difficulties of identification of *Cladonia* species in phytosociological work has limited the numbers of botanists willing to make the effort required for detailed studies (Ahti, 1964).

Kershaw (1977) points out that no comprehensive efforts have been made to relate the ecology of the lichens in boreal regions to the growth and development of the tree species. Studies by Kershaw, however, have shown that marked changes in microclimate take place during the development of *Stereocaulon paschale* and *Cladonia stellaris* woodlands during restoration following fire.

All early accounts of mature northern boreal woodland describe a ground cover of almost pure *Cladonia stellaris (alpestris)*, with *C. mitis* and *C. rangiferina* present in scattered clumps. The widespread nature of *Cladonia stellaris* woodland is well established, but Kershaw demonstrates the importance of *Stereocaulon* woodlands as a dominant feature on upland sites in northern central Canada. He points out the probability that there is a more-or-less continuous zone of *Stereocaulon* woodland north and south of lat. 60°N, running from west of Churchill to Great Slave Lake. The predominantly *Cladonia stellaris* woodlands in Ontario and Quebec presumably reflect the major climatic differences between this region and that of the Northwest Territories. Kershaw adds that the northern Scandinavian woodlands have an extensive cover of *Stereocaulon paschale;* it has been suggested that here it is a pioneer species and the first colonizer after fire or heavy grazing by reindeer. This is in contrast to *Cladonia stellaris*, which is considered a typical climax species.

Collections of lichens from various communities in the area north of the east arm of Great Slave Lake were obtained by the author during the course of community sampling, and in the black spruce communities the following lichens were found to be common associates: *Alectoria nitidula, A. ochroleuca, Cetraria cucullata, C. islandica, C. nigricascens, C. nivalis, Cladonia stellaris (alpestris), C. amaurocraea, C. coccifera, C. mitis, C. rangiferina, C. uncialis, Cornicularia aculeata, C. divergens, Nephroma arcticum, Parmelia physodes, Peltigera aphthosa* var. *aphthosa, P. polydactyla, P. pulverulenta, Pertusaria panyrga, Stereocaulon alpinum,* and *S. paschale*.

This area at the east end of Great Slave Lake is within 20 mi of the northern edge of the forest; tundra begins at the south end of Artillery Lake north of a sharp ecotonal region. Most black spruce stands sampled in the area were either in the immediate vicinity of Fort Reliance or along the Pike's Portage route to Artillery Lake. White spruce is also fairly abundant in the Fort Reliance area, and lichens found commonly in the white spruce understory communities are as follows: *Alectoria nitidula, Cetraria cucullata, C. islandica, C. nigracascens, C. nivalis, C. richardsonii* (on Crystal Island in Artillery Lake), *Cladonia stellaris, C. amaurocraea, C. crispata, C. deformis, C. ecmocyna, C. gracilis* var. *chordalis, C. gracilis* var. *dilatata, C. lepidota, C. mitis, C. pyxidata* var. *neglecta, C. rangiferina, C. uncialis, Cornicularia aculeata, C. divergens, Stereocaulon paschale,* and *S. tomentosum.*

MOSSES IN THE COMMUNITIES

Mosses, unlike lichens, have been accorded a rather important place in studies devoted to the ecology and forestry of boreal regions, particularly in more southern portions of the forest where they attain greater importance than in the forests northward. This is reflected in the local terminology by which many of the southern spruce forests are known. Such terms as "spruce–feather moss forests," and "spruce–peat moss forests" refer to the mosses *Hylocomium splendens* and *Pleurozium schreberi,* and to the *Sphagnum* species that constitute a continuous vegetational carpet in these respective forest types.

In such forests, mosses very obviously are an important component of the vegetation, and studies were long ago instituted into such influences as inhibition of germination and growth of forest tree seedlings. In most southern forests, moreover, mosses far outweigh lichens in importance; only occasionally will a small patch of lichens be seen growing on the virtually continuous carpet of mosses. Northward the role of lichens becomes increasingly important. In lichen woodlands of the forest–tundra ecotonal zone, lichens attain and surpass mosses in importance. In far northern forest–tundra ecotonal zones, only on wet areas—open muskeg—unsuited for tree growth, do mosses retain dominance. Here *Sphagnum* species form a substrate upon which the northern *Ericaceae* and *Carex* species grow in profusion.

The spruce–feather moss forest is an association characteristic of the southern boreal forest. It ceases to be of significance northward in the zone where open lichen woodland begins to appear in abundance. Beyond this transition zone, *Sphagnum* peat mosses become increasingly important, especially in areas where moisture is plentiful. Northward—beyond the line of forest—*Sphagnum* persists as a component of plant associations far into the tundra, especially around the edges of low, wet meadows and tundra ponds and lakes. Savile (1961), in describing the vegetation of the Hazen Lake area of northeastern Ellesmere Island, notes that a number of mosses attain importance in the plant communities of this Far Northern area.

There is little doubt that mosses are as ecologically important as lichens in many plant communities but, as in the case of lichens, quantitative studies of their importance are scarce. Moss (1953) described white spruce–feather moss forests in northern Alberta in which the ground cover was a continuous carpet of *Hylocomium splendens, Pleurozium schreberi,* and other mosses. He described the developing stages of the white spruce stands as characterized by an understory in which the following species were prominent: *Viburnum edule, Rosa* species, *Ribes* species, *Linnaea borealis, Rubus pubescens, Mertensia paniculata, Fragaria* species, *Pyrola* species, *Epilobium angustifolium, Calamagrostis canadensis, Cornus canadensis, Mitella nuda, Viola renifolia,* and *Petasites palmatus*. These species declined in importance as the stand matured, to be replaced by mosses, including the species mentioned above and the following in addition: *Ptilium crista-castrensis, Dicranum* species, *Aulacomnium palustre, Orthotrichum elegans, Orthotrichum obtusifolium, Pylaisia polyantha, Polytrichum juniperinum, Mnium* species, *Eurhynchium strigosum, Drepanocladus uncinatus, Brachythecium salebrosum, Thuidium recognitum, Pohlia nutans, Sphagnum capillaceum,* and *Rhytidiadelphus triquetrus*.

There are variations in black spruce forests which, like the white spruce forests described, depend upon terrain, age and density of the spruce trees, and past history of fire. Moss* describes the black spruce–feather moss association:

> This community might well be named a *Picea mariana-Hylocomium splendens* association, to denote the chief moss of the community. The dominant tree, black spruce, generally grows in close formation and is frequently accompanied by white spruce, aspen, poplar, and various willows (*Salix* spp.) The floor is carpeted with "feather" mosses, notably *Hylocomium splendens,* while *Pleurozium schreberi* and *Aulacomnium palustre* are usually present, often abundant. More sporadic in occurrence are the mosses *Ptilium crista-castrensis, Camptothecium nitens,* and *Sphagnum capillaceum*. Accompanying the mosses are various lichens, notably *Peltigera aphthosa* and *Cladonia* spp. Associated species include *Ledum groenlandicum, Vaccinium vitis-idaea* var. *minus, Rosa* spp., *Ribes* spp., *Equisetum arvense, Equisetum scirpoides, Cornus canadensis, Petasites palmatus, Linnaea borealis* var. *americana, Mitella nuda, Rubus pubescens,* and *Carex* spp.

Another black spruce association occupying large areas is described by Moss* as a black spruce–peat moss association:

*Reproduced by permission of the National Research Council of Canada.

In technical language this is a *Picea mariana-Sphagnum* association.... The dominant black spruce is usually less crowded than in the former association. The most common tree associates are tamarack, paper birch, and certain willows. The floor is characterized by bog mosses, especially *Sphagnum capillaceum* and *Sphagnum fuscum* and by species of *Cladonia*. Certain mosses, *Hylocomium splendens*, *Aulacomnium palustre*, and *Polytrichum juniperinum* occur sporadically. The most common flowering plants are *Ledum groenlandicum*, *Vaccinium vitis-idaea* var. *minus*, *Rubus chamaemorus*, and *Smilacina trifolia*. Less constant species include *Betula glandulosa*, *Vaccinium oxycoccus*, *Equisetum sylvaticum*, *Eriophorum spissum*, and *Carex* spp. The most striking structural feature of this community is the uneven floor of *Sphagnum* mounds and the nearly continuous cover of the low shrub, *Ledum groenlandicum*, commonly known as Labrador tea.

In describing the general differences between the two black spruce communities, Moss points out that an initial distinction can be made on the basis of topography:

> The black spruce-feather moss association seems generally to have developed on quite shallow depressions, or fairly level terrain, through sedge-grass-willow stages and without much peat formation. In contrast, the black spruce-peat moss association has arisen in deeper depressions, through acid bog (muskeg) stages and with the production of a considerable thickness of *Sphagnum* peat.... Natural succession is to a black spruce-feather moss community, for there is abundant evidence in many of the drier bog forests of the replacement of *Sphagnum* and *Ledum* by feather mosses.... But this trend is quite generally offset by the pyric factor, which brings about retrogression to earlier stages of the bog forest sere, through destruction not only of the vegetational cover but also of the upper peat layers.

The marsh and bog vegetation of the area is also described by Moss in a subsequent publication (Moss, 1953b).

In other communities of the region, Moss points out that *Larix* stands are quite common, usually developing in bogs dominated by *Drepanocladus* and *Carex* species and, as the trees become larger, eventually developing an understory in which the mosses *Camptothecium nitens* and *Aulacomnium splendens* are usually represented in abundance with *Sphagnum* species. The common vascular plants in this community usually include *Vaccinium oxycoccos*, *Rubus acaulis*, *Galium labradoricum*, *Potentilla palustris*, *Caltha palustris*, *Equisetum palustre*, *Menyanthes trifoliata*, and *Carex* species. The community often merges with the black spruce–*Sphagnum* community described above. Another community fairly common in the region occupies the other habitat extreme—dry, sandy, and gravelly areas. This community, the pine–feather moss faciation, according to Moss, is characterized by *Hylocomium splendens* and *Calliergonella schreberi* in the understory, along with *Linnaea*, *Pyrola*, *Cornus canadensis*, *Polytrichum*, and *Cladonia*. Drier open areas are colonized by *Arctostaphylos uva-ursi*, *Vaccinium vitis-idaea*, *Elymus innovatus*, and

Oryzopsis pungens, with *Alnus crispa*, *Rosa* species, *Salix* species, and *Maianthemum* found generally throughout.

Groenewoud (1965) utilized association analysis to reveal some of the relationships among the understory mosses and herbs, shrubs, and trees in the white spruce communities of Saskatchewan and has shown some rather interesting interactions among the strata. He found, for example, that there was a general decline in the herbaceous flora with increased competition. Under a dense canopy, some species disappear altogether. Others persist in only small numbers, and these include *Linnaea borealis*, *Cornus canadensis*, *Maianthemum canadense*, *Pyrola virens*, *Petasites palmatus*, and *Mertensia paniculata*. *Linnaea borealis* is an example of a species that is drought-resistant, thriving under dry, open conditions, but other species require moist conditions and are better able to withstand shade. Still other species, of which *Equisetum* is an example, do not thrive under intense shade even though they require moist conditions. The mosses *Hylocomium splendens*, *Pleurozium schreberi*, and *Ptilium crista-castrensis* often form a continuous cover where conditions are both moist and shady.

Other interesting aspects of the white spruce–feather moss forest are revealed in Groenewoud's study. Stands with a very high proportion of feather mosses were not significantly different from stands with a lush herbaceous ground cover when the pH of the soil, the height growth, and the nitrogen content of the foliage of the white spruce trees were considered. There is, thus, no discontinuity in habitat conditions between stands with mosses and those without. Groenewoud considers the absence of herbaceous species in the spruce–feather moss stands to be one extreme of a continuously variable spectrum of herbaceous vegetation. He notes that there is a highly significant positive correlation between the moss cover and the basal area of the trees, suggesting that competition plays an important role in the community. He has found, also, that most of the sites with a heavy moss cover have a very high tree root concentration in the uppermost soil horizons, and that the latter, as a consequence, are very dry except immediately after a rain. The high pH is an indication that leaching is negligible. The dry conditions and the low light intensity resulting from the dense canopy depress the herbaceous species and permit heavy growth of mosses. Groenewoud further points out that these relationships tend to discount the successional concepts put forth by Moss (1953a) and others. Although communities intermediate between the feather moss stands and those with a lush herbaceous and shrub cover are present in the area, they are not intermediates in a developmental series but are "stable" communities occupying different habitats; the communities intermediate between the feather moss and the shrub types follow various environmental gradients.

Rowe (1956) shows that a relative abundance of understory species follows the sequence from tall shrubs through medium shrubs, herbs, and finally mosses as the progression in the overstory of tree species goes from poplar to spruce.

Rowe also points out that the series *Corylus cornuta* → *Aralia nudicaulis* → *Cornus canadensis* → *Pleurozium schreberi* (tall shrub → tall herb → low herb → moss) is "clearly one of increasing tolerance to shade." He points out, further, that the influence of moisture is also clearly apparent, with a sequence of arboreal and understory species in central Saskatchewan that can be related to the moisture class of the forests. As the moisture classes "moist," "very moist," and "wet" are approached, the species *Picea mariana, Abies balsamea, Populus balsamifera,* and *Larix* increase in importance, along with a characteristic assemblage of shrubs and herbs as well as lichens and mosses. The well-defined moss-type forest at the wet end of the moisture series is characterized by an almost continuous cover of such mosses as *Hylocomium splendens, Climacium americanum, Aulacomnium palustre,* and *Mnium* species, often quite abundant throughout, and with *Sphagnum* species and *Camptothecium nitens* on the deepest peats beneath an overstory of black spruce.

Northward in the black spruce forests, in areas such as northern Manitoba and Saskatchewan and the forested southern portions of the western parts of the Northwest Territories, mosses occupy an even more important role than they do in the southern regions discussed. In the closed black spruce forest of northern Manitoba, for example, Ritchie (1956) has found high presence values for *Ceratodon purpureus, Dicranum bergeri, Dicranum rugosum, Pohlia nutans, Pleurozium schreberi, Hylocomium splendens, Ptilium crista-castrensis,* and *Ptilidium ciliare.* The dominant moss is *Pleurozium schreberi,* and he indicates that by far the greater area of the ground is covered by a dense carpet of these mosses. *Pohlia, Ceratodon,* and *Ptilidium,* along with some lichens, occur exclusively at the bases of spruce trees, extending up the trunk for short distances. Mineral soil exposed by fallen spruces is, according to Ritchie, invariably colonized by a dense carpet of *Polytrichum piliferum.* Lichens, particularly the more common *Cladonia* species, assume local dominance in areas where the sunlight reaches the forest floor with greater brilliance than generally. Under alders and in other moist places, particularly in depressions where moisture accumulates, *Sphagnum capillaceum* and *Aulacomnium palustre* are also common. On dry ridges and in jack pine forests, *Polytrichum juniperinum* var. *alpestre* is found, and *Polytrichum commune* is observed in hollows where an organic soil has developed. The chief moss species on hummocks in muskeg areas are *Sphagnum* species along with *Aulacomnium palustre* and *Camptothecium nitens*. Eastward, in Quebec, the moss species of the forest understory have been employed extensively in the designation of forest site types, and in particular Linteau (1955) has associated the lesser vegetation—lichens, mosses, herbs, and shrubs—with the various tree species found in forests of the region. From Quebec northward into Labrador and eastward to Newfoundland, mosses retain their importance in forest communities with little change except the addition of one or two species to the lists of those attaining dominance (Brassard and Weber, 1978). La Roi and Stringer (1976) described the bryophyte components of 34 mature forest stands dominated by white spruce (*Picea glauca*) and/or balsam fir (*Abies balsamea*) and of 26 stands dominated by black spruce (*Picea mariana*). The stands were located on undisturbed upland sites from Alaska to Newfoundland. The study reported on the bryophytes found in the stands for which La Roi provides a description of the vascular plants in a publication previously mentioned (La Roi, 1967). The average cover of the terrestrial bryophyte stratum was 42% in the white spruce–fir and 67% in the black spruce stands, while the average combined cover of understory vascular plants was 38% in the white spruce–fir and 21% in the black spruce stands. Bryophyte cover exceeded subordinate vascular cover in 21 white spruce–fir and 24 black spruce stands. Clearly an adequate understanding of the structure and function of the two forest community types requires study of the bryophyte components, as La Roi and Stringer point out.

At one extreme are the black spruce–feather moss forests, where the tree stratum is uniform and moderately dense and the terrestrial bryophyte stratum is almost continuous. At the other extreme are the white spruce–shrub–herb forests, where the tree stratum is irregular and more open, the broadleaf shrub and herb–dwarf shrub strata are dense and species-rich, and the terrestrial bryophyte stratum is very patchy.

In regions with plentiful precipitation and high summer humidity, bryophytes usually cover most of the ground not densely inhabited by vascular plants, and often a substantial part of the lower trunk space as well. In very dry regions, bryophytes are usually scarce on the ground beneath the crowns and on the trunks of mature spruce and fir. There is no apparent geographical trend in the total bryophyte species richness of white spruce–fir stands, according to La Roi and Stringer, nor is there an apparent geographic trend in total bryophyte, liverwort, or moss species richness of black spruce stands.

Thirteen bryophytes have a high presence and/or occur almost throughout the entire continental range of white spruce–fir stands. *Hylocomium splendens* and *Pleurozium schreberi* are the leading members of this group. Twelve bryophytes have a high presence and/or occur almost throughout the range of black spruce stands. *Pleurozium schreberi* and *Hylocomium splendens* are again the characteristic dominants. There are no species that can be regarded as totally restricted to one substrate type, but many have affinities for one or a few types. *Ceratodon purpurpeus* is the only species with >50% relative frequency on mineral soil. Three species have strong affinities for humus, occur frequently on mineral soil, but avoid wood and bark: *Aulacomnium palustre*, *Polytrichum juniperinum*, and *Ptilidium ciliare*. Seven species have strong affinities for needles and detritus, occur much less frequently on wood and bark, and all except the last one listed occur only rarely on mineral soil: *Dicranum polysetum*, *Hylocomium splendens*, *Bazzania trilobata*, *Pleurozium schreberi*, *Ptilium crista-castrensis*, *Rhynchostegium serrulatum*, and *Rhytidiadelphus triquetrus*. Members of the largest group, with 17 species, have strong affinities for wood and/or bark, occur less frequently on humus, and occur still less frequently on mineral soil. Included are *Cephalozia media*, *Tetraphis pellucida*, *Lepidozia reptans*, *Pylaisiella polyantha*, and *Ptilidium pulcherrimum*. Four species with an equal affinity for all substrate categories are *Brachythecium salebrosum*, *Drepanocladus uncinatus*, *Eurhynchium pulchellum*, and *Pohlia nutans*.

La Roi and Stringer indicate that the equilibrium state in forest floor subsystems seems to result from a complex of interactions among vascular plants, bryophytes, and lichens; fresh substrate materials are assimilated by established bryophytes and lichens or colonized by other species. Local variations in light, temperature, moisture, and nutrients modify and regulate activity in forest floor subsystems dominated by mosses and lichens just as they do in systems dominated by vascular species, and an understanding of the former is as important as an understanding of the latter in boreal systems modeling.

Busby *et al.* (1978) point out that, while mosses form a conspicuous feature of the vegetation of boreal regions in North America and Eurasia, little is known about the environmental factors that control growth rates and habitat limits. Water, however, is a fundamental factor controlling growth and distribution of mosses; mosses lack root systems and have no mechanism for water storage. Metabolic activity is, hence, closely linked to the water regime. Busby *et al.* studied the water relationships of the mosses *Tomenthypnum nitens*, *Hylocomium splendens*, *Pleurozium schreberi*, and *Ptilium crista-castrensis*, the last-mentioned three collectively designated "feather mosses." All four species are circumboreal and occupy large areas. *Tomenthypnum* is robust and forms dense communities in fens throughout its range. Feather mosses vary in the proportions of the various species in communities but are found throughout the boreal regions of North America and Eurasia. *Hylocomium* has a wide distribution and occurs in a variety of habitats. *Pleurozium* has a more restricted range in North America and mainly occurs in boreal regions; it predominates in the feather moss communities of the Great Lakes region. *Ptilium* has the most restricted geographical distribution and is rarely prominent.

As with lichens, collections of mosses from the boreal and arctic regions are not numerous or extensive; within recent years, however, the number has increased greatly, and compilations of data from existing collections have begun to appear, among them data on the mosses of Labrador (Brassard and Weber, 1978).

It becomes evident that successful categorization of forest communities in a manner that describes both their ecological relationships and their species composition is a task that will require greater knowledge of the role of moss species in community dynamics than is presently available. This is surely a field in which it appears that the exploration of possible new avenues of description and characterization might be useful.

As pointed out above, most of the field research that constitutes the basis of the discussions in the following chapter did not take mosses specifically into account. Some data were obtained, however, on the more common mosses found in certain areas, and these species are listed below according to the area and community type in which they were found in some abundance. The areas named are located on the map presented in the following chapter. All identifications have been provided by courtesy of Howard Crum of the University of Michigan.

Otter Lake (North-Central Saskatchewan)

Black spruce stands: *Sphagnum* species, *Hylocomium alaskanum*, *Pleurozium schreberi*, *Polytrichum juniperinum* var. *alpestre*, and *Ptilium crista-castrensis*. *Hylocomium* and *Ptilium* were also found in white spruce and aspen stands.

Jack pine stands: *Dicranum bergeri*, *D. scoparium*, and *Pleurozium schreberi*.

Wapata and Black Lakes (Northern Saskatchewan)

Black spruce stands: *Sphagnum* species, *Pleurozium schreberi*, *Polytrichum commune*, *Dicranum polysetum*, *Ptilidium ciliare*, and *Calliergon stramineum*.

White spruce stands: *Hylocomium alaskanum* and *Drepanocladus uncinatus*. *Sphagnum* species were usually not present in the white spruce stands.

Yellowknife (Great Slave Lake)

Black spruce stands: *Sphagnum* species, *Ptilidium ciliare*, *Dicranum bergeri*, *Pleurozium schreberi*, *Drepanocladus uncinatus*, *Polytrichum commune*, *Polytrichum juniperinum* var. *alpestre*, *Aulacomnium palustre*, *Aulacomnium turgidum*, *Hylocomium alaskanum*, and *Ditrichum flexicaule*.

White spruce stands: *Hylocomium alaskanum* and *Ptilidium ciliare*.

Fort Providence (Great Slave Lake)

Black spruce stands: *Sphagnum* species, *Tomenthypnum nitens*, *Aulacomnium palustre*, *Campylium stellatum*, *Ditrichum flexicaule*, and *Catoscopium nigritum*.

White spruce stands: *Tomenthypnum nitens*, *Hylocomium alaskanum*, *Pleurozium schreberi*, and *Aulacomnium palustre*.

South Artillery Lake (Southeastern Mackenzie)

Black spruce stands: *Aulacomnium turgidum*, *Hylocomium alaskanum*, and *Dicranum angustum*. *Sphagnum* species were abundant and dominant.

Aylmer Lake (Tundra) (Eastern Mackenzie)

Low meadow communities: *Dicranum elongatum*, *Aulacomnium turgidum*, *Dicranum angustum*, *Drepanocladus revolvens*, and *Sphenolobus minutus*. *Spagnum* species were abundant.

Rock field communities: *Polytrichum piliferum*, *Aulacomnium turgidum*, *Rhacomitrium lanuginosum*, *Ptilidium ciliare*, *Ceratodon purpureus*, *Dicranum angustum*, *Pohlia nutans*, *Polytrichum juniperinum*, *Sphagnum* species, *Polytrichum commune*, and *Aulacomnium palustre*.

Tussock muskeg communities: *Sphagnum* species, *Polytrichum juniperinum* var. *alpestre*, and *Dicranum elongatum*.

Ennadai Lake (Southern Keewatin)

Black spruce stands: *Sphagnum* species, including *Sphagnum capillaceum* var. *tenellum* and *Sphagnum recurvum*, *Aulacomnium turgidum*, *Aulacomnium palustre*, *Polytrichum juniperinum*, *Polytrichum juniperinum* var.

alpestre, Drepanocladus exannulatus, Drepanocladus uncinatus, Pleurozium schreberi, Dicranum fuscescens, Dicranum bergeri, Hylocomnium splendens, and *Sphenolobus minutus.*

Tussock muskeg communities: *Sphagnum* species, including *Sphagnum fuscum* and *Sphagnum subsecundum, Hylocomium splendens, Bryum* sp. *Dicranum elongatum, Sphenolobus minutus, Aulacomnium turgidum, Ptilidium ciliare, Calliergon sarmentosum,* and *Meesia tristicha.*

Rock field communities: *Polytrichum juniperinum* var. *alpestre, Ptilidium ciliare,* and *Aulacomnium turgidum.*

Esker summits and slopes: *Dicranum elongatum, Polytrichum juniperinum, Aulacomnium turgidum, Polytrichum piliferum, Dicranum fuscescens, Ptilidium ciliare,* and *Rhacomitrium canescens.*

The geographical range of many moss species is very wide. They apparently can tolerate a wide range of conditions within a given area, providing only that there is sufficient moisture available and some light, however small in amount. It is their adaptation to moist, dark habitats that enables mosses to survive in deep forests where few higher plants can survive. In more xeric environments, species so adapted can persist despite periods of drought, taking advantage of limited water supplies when they are available. *Polytrichum* found on esker summits in the Ennadai Lake area, for example, obviously perform the very useful function of consolidating surface layers of sand which otherwise would quite rapidly be eroded by wind. Once consolidated by the mosses, such sites can be colonized by more hardy esker species such as *Empetrum* and grasses which, with *Polytrichum,* are the advance stabilizers of the bare esker sands.

It is an interesting question whether mosses would be good indicators of climatic differences from place to place as a consequence of their general ubiquity, wide range, and apparently fairly close adjustment to environmental conditions. The same may also be true of the ground-inhabiting lichens, and here the wider variety of forms and the greater number of species would afford an even better chance of there being a few of value for this purpose. Looman (1964a,b) has shown that the distribution of ground lichen species correlates with climate elsewhere.

In the case of corticolous lichens—bark- and twig-inhabiting species—there may be species that will be found to have considerable utility for purposes of climatic indication. Here we have, or at least can select, forms that grow on a given tree species and even on trees of a given age, on given site types, with a given moisture content, etc., hence we have almost complete control of such conditions of substrate as might otherwise greatly influence the distribution of the individuals.

Only climate will vary, hence the complexity of the situation under investigation is greatly reduced. It may be shown ultimately that corticolous lichens are even—perhaps by far—better indicators of climatic conditions, and of climatic differences from place to place, than the trees or the higher plants of the understory, shrubs and herbs, and grasses.

Rather than simplifying our problem when we seek ways to utilize the natural forest vegetation, particularly as a climatic tool—as an indicator of climate—mosses actually render the problem quite complex. This is largely because they condition the environment—particularly for the forest understory species—in ways that are quite apparent but are hard to assess in terms of impact and influence upon growth or reproduction. Wagg (1964), for example, has discussed the role of mosses in the forest understory, particularly as it affects white spruce regeneration on the Peace and Slave river lowlands. Utilizing his own work and the work of others in his discussion, Wagg points out that *Hylocomium splendens* and to a lesser degree *Pleurozium schreberi* and *Ptilium crista-castrensis* are responsible for the almost continuous carpet within many of the white spruce stands in this area. The presence of this mat of mosses has a rather wide spectrum of influence throughout the community. First of all, since soil frost is prevalent, the generally low temperatures in the soil may limit germination and development of seedlings. Moreover, the annual growth of the moss continually increases the depth of the raw humus layer and thereby elevates the lower limit of the soil active layer. This has an effect ultimately on mature trees, since eventually lower roots will be encased permanently in frozen ground, requiring either the development of new adventitious roots or, failing this, acceptance of an early demise. Adventitious roots, however, develop from stem and branches of trees when these parts have been covered by mosses that furnish the needed moisture, aeration, darkness, and warmth.

Feather mosses provide at best a poor seedbed, a characteristic commented upon many times by foresters both in North America and Scandinavia and Russia. Cutting of the forest trees does not immediately improve conditions for the growth of seedlings, since feather mosses in such areas wither and die or remain in a semidry condition around the stems of surviving or regenerating shrubs. In openings, the feather mosses finally become desiccated and fragmented, ultimately being incorporated into the peat detritus and humus of the forest floor. Meanwhile, they have tied up nutrients that would otherwise have been released rapidly to the forest floor and made available for reabsorption by seedlings and the roots of the plants making up regenerating vegetation. The presence of the mosses, thus, accounts not only for a characteristic set of conditions prevailing beneath a forest canopy but also for the conditions persisting in an area long after the canopy has been removed. In a natural expanse of boreal landscape, of course, there are examples of every stage of maturity and regeneration, and a wide spectrum of possible avenues of development from freshly opened canopy to mature feather moss forest exists. The opportunities to elucidate these various possible avenues have as yet been limited; little apparently is known about them. Investigations such as the one mentioned above by Wagg, however, ultimately will serve to furnish useful information concerning what remains at present an underdeveloped area of ecology but one that has legitimately been given increasing emphasis in recent years.

The relationships between mosses and other plants in *Sphagnum*–black spruce forests are probably even less well understood than those in feather moss–spruce forests. Here the moss cover is not only equally extensive but in many of the areas exists to a considerable depth, thus constituting the total edaphic environment to which the roots of plants are exposed. In many areas of forested muskeg, *Sphagnum* peat attains depths of 3 ft and more, and carbon dating for palynological purposes has shown that the peat at these depths has persisted in a relatively unmodified state for many centuries. As one might expect, environmental conditions in such an area are quite different from those prevailing on more upland sites where a wide range of conditions is found as a consequence of the varying depth of the organic layer.

Holmen (personal communication to Harvey Nichols, University of Wisconsin) points out that there are both dioecious and autoecious species of *Sphagnum* mosses. In the dioecious group there are many species, often dominating a moss understory carpet but rarely found with sporophytes. Of the autoecious forms, there are fewer species, rarely dominating and found with sporophytes more or less from one year to another, with changes from year to year usually too slight to merit consideration. On the other hand, in some years the dioecious forms produce abundant sporophytes, although in other years it is difficult to find a single sporophyte.

In Denmark, where Holmen has made the larger part of his observations, successful fertilization in dioecious species will depend upon the weather conditions of the fall and winter. A mild winter with little snow and only short periods of temperatures below zero will ensure large quantities of spores the following summer. Development of the sporophyte, however, may be further endangered by inclement weather during June and July of the following summer, with cool temperatures during this period tending to inhibit sporophyte development. In Greenland and Alaska, Holmen notes, sporophytes even of the autoecious species are rare. The growth of sporophytes apparently cannot continue through the low June and July temperatures prevailing in these regions.

These ecological characteristics are of significance in the regeneration of *Sphagnum* mosses in areas where forests have been destroyed by fire or other causes. They are, of course, also of significance in palynological interpretations of past climatic changes by means of shifts in the proportions of *Sphagnum* spores in the peat strata under study. Thus, as Holmen states, high spore counts may signal the appearance of wetter conditions if summer temperatures are also high. Wetter conditions, with lower summer temperatures, however, would result in some intermediate, and, at present, indeterminate degree of spore production. Prolonged moister conditions would most likely change the species composition of the *Sphagnum* components of the plant communities, perhaps to more rarely fruiting forms. In a period of climatic shift that transformed open bog or muskeg to *Sphagnum*-black spruce muskeg forest, the production of spores might initially increase. But over a subsequent extended period, a shift in species composition would probably occur as a result of the higher prevailing temperatures and changes in precipitation regimes. To determine the changes accurately, one would have to know what species were present in the area before the environmental change took place.

It is, thus, apparent that not only does the presence of a moss carpet have considerable influence on the capacity of a spruce forest to persist and regenerate, but it is itself responsive to changes in climate and weather variations from year to year, and presumably other environmental changes as well. These result in shifts in species composition and in changes in the reproductive and, one assumes, regenerative capabilities of mosses in the understory. At present, little is known concerning the potential effect of environmental factors upon mosses or, conversely, the effect upon the environment and surrounding plant community of changes in the structure of the moss component. Sphagnum mosses are difficult to identify, and it may be some time before anything resembling an intensive and comprehensive study of their ecological relationships is forthcoming. It goes without saying that such a study, of both sphagnum mosses and feather mosses, would be a worthwhile undertaking.

THE APPALACHIAN EXTENSION

Coniferous forest extends southward along the mountains of the eastern United States just as it does along the Rocky Mountains in the western United States. Whether either the eastern or the western extension of the forest into the southern mountains can be appropriately considered an extension of the boreal forest is open to discussion (Whittaker, 1956), but species are found in Appalachian spruce–fir forests that range far north into the boreal forests of Canada and for this reason are given brief mention here. While it is possible to distinguish the Appalachian from the northern Canadian boreal communities on the basis of species composition, it is impossible, as Oosting and Billings (1951) point out, to separate the two in southeastern Canada where transition from black and white spruce-dominated forests to red spruce (*Picea rubens*) forest occurs. Red spruce and Fraser fir (*Abies fraseri*) are the dominant trees of the southern Appalachians, along with other such species as *Betula lutea, Sorbus americana,* and *Acer spicatum.* There are shrub and herbaceous understory species that range from Canada southward and are found throughout the Appalachian extension; other species are found only in the northern portion of the Appalachian range (the Catskill Mountains) or in the southern portions (the Great Smoky Mountains) as revealed in studies by McIntosh and Hurley (1964), Oosting and Reed (1944), Crandall (1958), and others.

EXTENT OF THE EURASIAN BOREAL FOREST

It is, of course, absurd to attempt to treat the boreal vegetation of Eurasia in any less detail than has been accorded that of North America; the experience of the author, however, is limited to North America and, in addition, the literature available in translation, hence accessible to most North American readers, is also of limited extent. The available literature, moreover, does not present community data in a form that would be statistically comparable to that obtained by the author in North America, and so no comparisons can be undertaken in the presentations of Chapter 6. The reader interested in greater detail concerning the Eurasian forest should consult the references cited in the following very brief outline of the composition of the Eurasian boreal forest.

As in North America, the boreal forest of Eurasia stretches across the entire extent of the continent, east to west. The Scandinavian boreal forest is a coniferous forest of *Picea abies* and *Pinus silvestris*, with the former tending to occupy the more moist and the latter the drier sites. In the mountains a subalpine forest is typically present, with stands of birch (*Betula tortuosa*) in which are mixed such species as mountain ash (*Sorbus aucuparia*), juniper (*Juniperus communis*), and dwarf birch (*Betula nana*).

In the more heavily forested central boreal areas to the east, the wide belt of coniferous forest is not homogeneous across European Russia and Siberia, since marked regional differences are as apparent as in North America, presumably as a consequence of the different climatic conditions in the various parts of the continent (see Chapter 3). Tseplyaev (1965) divides the broad taiga into the following sections on the basis of the dominant tree species and the general character of the forest: the Karelian taiga, the taiga of European Russia, the West Siberian taiga, the Yenisei taiga, the taiga of the Tunguska, the Angara taiga, and the Yakut taiga. The numerous mountain ranges in the various parts of Eurasia are considered separately.

In the forest–tundra transition zone of East-European Russia, the dominant tree species is Siberian spruce (*Picea obovata*), with some birch. The forest–tundra communities of western Siberia have Siberian larch as the principal tree, with spruce and alder in significant amounts. The central Siberian forest border has Siberian larch as the dominant tree species, and the Far Northeastern regions have forest communities dominanted by Dahurian larch and Japanese stone pine (*Pinus pumila*) mixed with Asiatic poplar and birch, particularly in the valleys. The forest–tundra extends from the Kola peninsula to the eastern edge of the Chukot area, varying in width from 20 to 200 km and occupying in total area perhaps 500,000 km^2.

In European Russia the northern taiga is dominated by rather thin stands of spruce, birch, and pine, including *Picea excelsa* and *Pinus silvestris*; to the northwest *Picea obovata* and *Larix* are present, and in the northeast *Abies sibirica* and *Pinus sibirica* are found. The characteristic forest is termed a "green-moss" community, with few shrubs and herbs and with *Pleurozium schreberi* and *Hylocomium proliferum* as ground cover dominants.

Forested areas of the northern Ural mountains have Siberian larch, Siberian spruce and birch (*Betula tortuosa*) as common tree species, grading along the upper boundaries into thickets of scrub Manchurian alder (*Alnus fruticosa*). The larch grows on a wide variety of different substrates, both in river valleys and at the upper limit of the forest.

In western Siberia, the region to the east of the Ural Mountains, the landscape is largely an undulating plain of glacial till with occasional ridges and morainic hills. Some large areas are bog, poorly drained, and appear to be the beds of ancient glacial lakes. The climate is continental, and water is always abundant because of the surplus of precipitation over evaporation; surface flow is constant. Siberian larch (*Larix sibirica*), Sukachev's larch, and Siberian cedar are among the important tree species, extending in range from the northern forest where, with *Picea obovata*, they are dominant, to the southern forest–steppe transition zone where they occur in forests dominated by other species. East of the Ural

Mountains, as Sukachev (1928) was among the first to point out, it is not easy to distinguish between areas given over to European spruce and Siberian spruce, since over a large area the two species present a series of intermediate forms.

In the central region, Siberian larch attains heights of 150 ft and 6-ft diameters under optimum conditions. Siberian pine (*Pinus sibirica*) becomes increasingly important southward, occurring in pure stands or mixed with larch, spruce, fir, cedar, birch, and aspen. At the southern border of the boreal forest there are transitional belts of birch and aspen, and throughout the central and western Siberian region there are admixtures of Siberian larch and spruce with pine and larch.

Taking in the Yenisei, Tungus, and Angara regions described by Tseplyaev, central and eastern Siberia possess forests rather markedly different from those of western Siberia. The drainage is generally more efficient, the climate is drier and more extremely continental, and the forest is dominated in one region or another by Siberian spruce, cedar, fir, pine, and Dahurian larch (*Larix dahurica*), admixed with a few individuals of the species dominant in the west. Sochava (1956) describes typical Siberian spruce forest–tundra as an open community of trees ranging to 10 m in height. In composition and structure of the ground-layer community, they are very similar to the communities in North America, and often the same species are involved or at least closely related taxa. In the western region of Siberia most of the conifer species have coincident northern range limits; in the east, however, the northern limits of the species vary widely, with fir and Siberian pine absent from the communities of the more northern forests. Dahurian larch resembles the North American *Larix laricina* in that it can withstand severe cold, grows on a wide variety of soils, and is adapted by reason of a shallow root system to permafrost soils.

SUMMARY

In the introductory paragraphs to this chapter, it was indicated that an understanding of an ecosystem presupposes a rather through knowledge of the behavior of individuals of the species comprising the community. Community composition, in terms of the species present, and structure, which is the proportion with which each species is represented, is then a basic element in the knowledge required for any study of the biological system. In this chapter, the fundamental outlines of the vegetation of boreal regions have been described—largely on the basis of the dominant tree species and the most common and widespread of the understory species. This is not quite the kind of information required for model building; it is too general and sweeping in nature but, as in a canvas by Breugel, for comprehension of the detail in a corner, it is necessary to see the whole picture. It is preferable to begin with a view of the whole broad canvas and then to narrow one's attention progressively to the manageable unit—the individual community. In the next chapter, we narrow the field of vision to the plant communities and their composition and structure, in one region—the central Canadian boreal forest.

REFERENCES

Agassiz, L. (1850). "Lake Superior: Its Physical Character, Vegetation, and Animals." Gould, Kendall, and Lincoln, Boston (Reprinted 1974 by Krieger, Huntington, New York).
Ahmadjian, V. (1967). "The Lichen Symbiosis." Ginn (Blaisdell), Boston, Massachusetts.
Ahti, T. (1959). Studies of the caribou lichen stands of Newfoundland. *Ann. Bot. Soc. (Vanamo)* **30**, 1–44.
Ahti, T. (1961). Taxonomic studies on reindeer lichens (*Cladonia*, subgenus *Cladina*). *Ann. Bot. Soc. (Vanamo)* **32**, 1–160.
Ahti, T. (1964). Macrolichens and their zonal distribution in boreal and arctic Ontario, Canada. *Ann. Bot. Fenn.* **1**, 1–35.
Ahti, T. (1977). Lichens of the boreal coniferous zone. *In* "Lichen Ecology" (M. R. D. Seward, ed.), pp. 147–181. Academic Press, New York.
Ahti, T., Scotter, G. W., and Vanska, H. (1973). Lichens of the reindeer preserve, Northwest Territories, Canada. *Bryologist* **76**, 48–76.
Allington, K. R. (1961). The Bogs of central Labrador–Ungava: An examination of their physical characteristics. *Geogr. Analer* **43**, 403–417.
Anderson, E. G. (1949). "Introgressive Hybridization," 2nd ed. Hafner, New York.
Argus, G. W. (1965). The taxonomy of the *Salix glauca* complex in North America. *Contrib. Gray Herb. Harv. Univ.* **196**, 1–142.
Argus, G. W. (1966). Botanical investigations in northeastern Saskatchewan: The subarctic Patterson–Hasbala lakes region. *Can. Field Nat.* **80**, 119–182.
Atlas of Canada. (1957). "Major physiographic regions." Can. Dept. Mines Tech. Surv.
Bakuzis, E. V., and Hansen, H. L. (1965). "Balsam Fir." Univ. of Minnesota Press, Minneapolis.
Baldwin, W. K. W. (1953). Botanical investigations in the Reindeer–Nueltin lakes area, Manitoba. *Annu. Natl. Can. Bull.* **128**. 110–142.
Baldwin, W. K. W. (1958). Plants of the clay belt of northern Ontario and Quebec. *Natl. Mus. Can. Bull.* **156**, 1–324.
Barry, R. G., 1967. Seasonal location of the arctic front over North America. *Geogr. Bull.* **9**, 79–95.
Black, R. A. and Bliss, L. C. (1978). Recovery sequence of *Picea mariana–Vaccinium uliginosum* forests after burning near Inuvik, Northwest Territories, Canada. *Can. J. Bot.* **56**, 2020–2030.
Bliss, L. C. (1979). Vascular plant vegetation of the southern circumpolar region in relation to antarctic, alpine, and arctic vegetation. *Can. J. Bot.* **57**, 2167–2178.

Bostock, H. S. (1970). Physiographic subdivisions of Canada. *In* "Geology and Economic Minerals of Canada," 5th edition. Econ. Geol. Rep. 1, Dept. Energy, Mines, and Resources, Ottawa.

Brassard, G. R., and Weber, D. P. (1978). The mosses of Labrador, Canada. *Can. J. Bot.* **56**, 441-466.

Brown, R. T., and Curtis, J. T. (1952). The upland conifer-hardwood forests of northern Wisconsin. *Ecol. Monogr.* **22**, 217-234.

Buell, M. F., and Niering, W. A. (1957). Fir-spruce-birch forest in northern Minnesota. *Ecology* **38**, 602-610.

Busby, J. R., Bliss, L. C., and Hamilton, C. D. (1978). Microclimate control of growth rates and habitats of the boreal forest mosses, *Tomenthypnum nitens* and *Hylocomium splendens*. *Ecol. Monogr.* **48**, 95-110.

Carleton, T. J., and Maycock, P. F. (1978). Dynamics of boreal forest south of James Bay. *Can. J. Bot.* **56**, 1157-1173.

Clarke, C. H. D. (1940). A biological investigation in the Thelon Game Sanctuary. *Natl. Mus. Can. Bull.* **96**, 135pp.

Clausen, J. J. (1957). A phytosociological ordination of the conifer swamps of Wisconsin. *Ecology* **38**, 638-646.

Cody, W. J. (1965). Plants of the Mackenzie River delta and reindeer grazing preserve. Can. Dept. Agric. Plant Res. Inst. pp. 1-56.

Cody, W. J., and Chillcott, J. G. (1955). Plant collections from Matthews and Muskox lakes, Mackenzie District, N.W.T., Canada. *Can. Field Nat.* **69**, 153-162.

Cottam, G. and Curtis, J. T. (1956). The use of distance measures in phytosociological sampling. *Ecology* **37**, 451-460.

Crandall, D. L. (1958). Ground vegetation patterns of the spruce-fir area of the Great Smoky Mountains National Park. *Ecol. Monogr.* **28**, 337-360.

Crum, H. (1976). "Mosses of the Great Lakes Forest," rev. ed. Univ. of Michigan Herbarium, Ann Arbor.

Culberson, W. L. (1955). The corticolous communities of lichens and bryophytes in the upland forests of northern Wisconsin. *Ecol. Monogr.* **25**, 215-231.

Curtis, J. T. (1959). "The Vegetation of Wisconsin." Univ. of Wisconsin Press, Madison.

Damman, A. W. H. (1964). Some forest types of central Newfoundland and their relation to environmental factors. Can. Dep. For. For. Sci. Monogr. 8, pp. 1-62.

Damman, A. W. H. (1965). The distribution patterns of northern and southern elements in the flora of Newfoundland. *Rhodora* **67**, 363-392.

Daubenmire, R. (1978). "Plant Geography: With Special Reference to North America." Academic Press, New York.

Davis, R. B. (1966). Spruce-fir forests of the coast of Maine. *Ecol. Monogr.* **36**, 79-94.

Delaney, B. B., and Cahill, M. J. (1978). A pattern of forest types on ribbed moraines in eastern Newfoundland. *Can. J. For. Res.* **8**, 116-120.

Dix, R. L., and Swan, J. M. A. (1971). The roles of disturbance and succession in upland forest at Candle Lake, Saskatchewan. *Can. J. Bot.* **49**, 657-676.

Douglas, M. C. V., and Drummond, R. N. (1955). Map of the physiographic regions of Ungava-Labrador. *Can. Geogr.* **5**, 9-16.

Drew, James V., and Shanks, R. E. (1965). Landscape relationships of soils and vegetation in the forest-tundra ecotone, upper Firth River valley, Alaska-Canada. *Ecol. Monogr.* **35**, 285-306.

Drury, W. H., Jr. (1956). "Bog flats and physiographic processes in the Upper Kuskokwim River Region, Alaska." *Contrib. Gray Herb. Harv. Univ.* **196**, 1-130.

Dugle, J. R. (1966). A taxonomic study of western Canadian species in the genus *Betula*. *Can. J. Bot.* **44**, 929-1007.

Dugle, J. R., and Bols, N. (1971). Variation in *Picea glauca* and *P. mariana* in Manitoba and adjacent areas. At. Energy Comm. Can. Publ. AECL3681, pp. 1–63.

Dyrness, C. T., and Grigal, D. F. (1979). Vegetation-soil relationships along the spruce forest transect in interior Alaska. *Can. J. Bot.* **57**, 2644–2656.

Elliott, D. B. (1979a). The stability of the northern Canadian tree limit: current regenerative capacity. Ph.D. thesis, Univ. of Colorado, Boulder.

Elliott, D. B. (1979b). The current regenerative capacity of the northern Canadian trees, Keewatin, N.W.T., Canada: Some preliminary observations. *Arct. Alp. Res.* **11**, 243–251.

Elliott, D. L., and Short, S. (1979). The northern limit of trees in Labrador. *Arctic* **32**, 201–206.

Fraser, E. M. (1956). The lichen woodlands of the Knob Lake area of Quebec–Labrador. McGill Subarct. Res. Pap. 1., McGill University, Montreal.

Gordon, A. G. (1976). The taxonomy of *Picea rubens* and its relationship to *Picea mariana*. *Can. J. Bot.* **54**, 781–813.

Grandtner, M. M. (1963). The southern forests of Scandinavia and Quebec. Laval Univ. For. Res. Found. Contrib. 10, Quebec, Canada.

Grandtner, M. M. (1966). "La Vegetation Forestiere du Quebec Meridional." Presses Univ., Laval, Quebec.

Grandtner, M. M. (1967). Les resources vegetales des Hes-de-la-Madeleine. Laval Univ. For. Res. Found. Bull. 10, Quebec, Canada.

Groenewoud, van, H. (1965). Analysis and classification of white spruce communities in relation to certain habitat features. *Can. J. Bot.* **43**, 1025–1036.

Hale, M. E., Jr. (1967). "The Biology of Lichens." Arnold, London.

Hare, F. K. (1959). "A Photo-Reconnaissance Survey of Labrador–Ungava." Can. Dep. Mines Tech. Surv., Geogr. Br., Mem. 6. 1–83.

Harper, F. (1964). "Plant and Animal Associations in the Interior of the Ungava Peninsula." Mus. Nat. Hist. Misc. Publ. No. 38, University of Kansas, Lawrence.

Heilman, P. E. (1966). Change in distribution and availability of nitrogen with forest succession on north slopes in interior Alaska. *Ecology* **47**, 825–831.

Heimburger, C. C. (1941). Forest site classification and soil investigation on Lake Edward experimental area. Can. For. Br., For. Res. Div. Silvicul. Res. Note 66. Ottawa.

Heinselman, M. L. (1970). Landscape evolution, peatland types, and the environment in the Lake Agassiz Peatlands Natural Area, Minnesota. *Ecol. Monogr.* **40**, 235–261.

Hicklenton, P. R., and Oechel, W. C. (1976). Physiological aspects of the ecology of *Dicranum fuscescens* in the Subarctic. I. Acclimation and acclimation potential of CO_2 exchange in relation to habitat, light, and temperature. *Can. J. Bot.* **54**, 1104–1119.

Hicklenton, P. R., and Oechel, W. C. (1977). Physiological aspects of the ecology of *Dicranum fuscescens* in the Subarctic. II. Seasonal patterns of organic nutrient content. *Can. J. Bot.* **55**, 2168–2177.

Holmen, K., and Scotter, G. W. (1971). Mosses of the reindeer preserve, Northwest Territories, Canada. *Lindbergia* **1**, 34–56.

Horton, K. W., and Lees, J. C. (1961). Black spruce in the foothills of Alberta. Can. Dept. For., For. Res. Div., Tech. Note 110.

Hughes, O. L. (1964). Surficial geology, Nichicun-Kaniapiskau map area, Quebec. Can. Dept. Mines Tech. Surv. Geol. Br. Bull. 106, pp. 1–20.

Hultén, E. (1968). "Flora of Alaska and Neighboring Territories." Stanford Univ. Press, Stanford, California.

Hustich, I. (1970). On the phytogeography of the eastern part of central Quebec–Labrador peninsula, II. *Commentat. Biol. Soc. Sci. Fenn.* **30**, 1–16.

Hustich, I. (1949). On the forest geography of the Labrador peninsula. A preliminary synthesis. *Acta Geogr.* **10**, 1–63.

Hustich, I. (1950). Notes on the east coast of Hudson Bay and James Bay. *Acta Geogr.* **11**, 1–83.

Hustich, I. (1951a). The lichen woodlands in Labrador and their importance as winter pastures for domesticated reindeer. *Acta Geogr.* **12**, 1–48.

Hustich, I. (1951b). Forest-botanical notes from Knob Lake area in the interior of Labrador peninsula. *Natl. Mus. Can. Bull.* **123**, 166–217.

Hustich, I. (1953). The boreal limits of conifers. *Arctic* **6**, 149–162.

Hustich, I. (1954). On forests and tree growth in the Knob Lake area, Quebec–Labrador peninsula. *Acta Geogr.* **13**, 1–60.

Hustich, I. (1955). Forest-botanical notes from the Moose River area, Ontario, Canada. *Acta Geogr.* **13**, 3–50.

Hustich, I. (1957). On the phytogeography of the subarctic Hudson Bay lowland. *Acta Geogr.* **16**, 1–48.

Hustich, I. (1962). A comparison of the floras on subarctic mountains in Labrador and in Finnish Lapland. *Acta Geogr.* **17**, 1–24.

Hustich, I. (1966). On the forest–tundra and the northern tree lines. *Ann. Univ. Turku Ser. A2.* **36**, 7–47.

Ives, J. D. (1960). The deglaciation of Labrador-Ungava—An Outline. *Cah. Geogr. Quebec* **8**, 323–343.

Jesberger, J. A., and Sheard, J. W. (1973). A quantitative study and multivariate analysis of corticolous lichen communities in the southern boreal forest of Saskatchewan. *Can. J. Bot.* **51**, 185–201.

Johnson, P. L., and Vogel, T. C. (1966). "Vegetation of the Yukon Flats region, Alaska." Res. Rep. 209, Cold Regions Research and Engineering Laboratory, U.S. Army Material Command, Hanover, New Hampshire.

Johnston, G. H., and Brown, R. J. E. (1965). Stratigraphy of the Mackenzie River delta, Northwest Territories, Canada. *Geol. Soc. Am. Bull.* **76**, 103–112.

Jurdant, M. (1964). Carte phytosociologique et forestiere de la forest experimentale de Montmorency. Can. Dep. For. Publ. 1046F, pp. 1–73.

Kershaw, K. A. (1972). The relationship between moisture content and net assimilation rate of lichen thallii and its ecological significance. *Can. J. Bot.* **50**, 543–555.

Kershaw, K. A. (1977). Studies on lichen-dominated systems. XX. An examination of some aspects of the northern boreal lichen woodlands in Canada. *Can. J. Bot.* **55**, 393–410.

Kershaw, K. A., and Rouse, W. R. (1971). Studies on lichen-dominated systems. I. The water relations of *Cladonia alpestris* in spruce–lichen woodland in northern Ontario. *Can. J. Bot.* **49**, 1389–1399.

Krylov, G. V. (1973). "Priroda Taiga Zapadnoi Bibiri" ("Nature of the Taiga of Western Siberia"). Novosibibirsk. (In Russian).

Kujala, V. (1952). Vegetation. A general handbook on the geography of Finland. *Fennia* **72**, 209–234.

La Roi, G. H. (1967). Ecological studies in the boreal spruce–fir forests of the North American taiga. *Ecol. Monogr.* **37**, 229–253.

La Roi, G. H., and Dugle, J. R., (1968). A systematic and genecological study of *Picea glauca* and *P. engelmannii*, using paper chromatograms of needle extracts. *Can. J. Bot.* **46**, 649–687.

La Roi, G. H., and Stringer, M. H. L. (1976). Ecological studies in the boreal spruce–fir forests of the North American taiga. II. Analysis of the bryophyte flora. *Can. J. Bot.* **54**, 619–43.

Lacate, D. S., Horton, K. W., and Blyth, A. W. (1965). Forest conditions on the lower Peace River. Can. Dep. For. Publ. 1094, pp. 53. Ottawa.

Lambert, J. D. H., and Maycock, P. F. (1968). The ecology of terricolous lichens of the northern conifer–hardwood forests of central eastern Canada. *Can. J. Bot.* **46**, 1043–1077.

Larsen, J. A. (1965). The vegetation of the Ennadai Lake area, N.W.T.: Studies in arctic and subarctic bioclimatology. *Ecol. Monogr.* **35**, 37–59.

Larsen, J. A. (1966). Relationships of central canadian boreal plant communities: Studies in subarctic and arctic bioclimatology, II. Tech. Rep. 26, Office Naval Res. (NR 387–002, Nonr 1202(07), pp. 1–37. Dept. Meteorol. , Univ. of Wisconsin, Madison.

Larsen, J. A. (1967). Ecotonal plant communities north of the forest border, Keewatin, N.W.T., Central Canada. Tech. Rep. No. 32, Office Naval Res. (NR 387–022, Nonr 1202(07), pp. 1–16. Dep. Meteorol. Univ. of Wisconsin.

Larsen, J. A. (1971a). Vegetational relationships with air mass frequencies: Boreal forest and tundra. *Arctic* **24**, 177–193.

Larsen, J. A. (1971b). Vegetation of the Fort Reliance and Artillery Lake areas, N.W.T. *Can. Field Nat.* **85**, 147–167.

Larsen, J. A. (1972a). Growth of spruce at Dubawnt Lake, Northwest Territories. *Arctic* **25**, 59.

Larsen, J. A. (1972b). The vegetation of northern Keewatin. *Can. Field Nat.* **86**, 45–72.

Larsen, J. A. (1973). Plant communities north of the forest border, Keewatin, Northwest Territories. *Can. Field Nat.* **87**, 241–248.

Larsen, J. A. (1974). Ecology of the northern continental forest border. *In* "Arctic and Alpine Environments" (J. D. Ives and R. G. Barry, eds.), pp. 341–369. Methuen, London.

Larson, D. W., and Kershaw, K. A. (1975a). Acclimation in arctic lichens. *Nature (London)* **254**, 421–423.

Larson, D. W., and Kershaw, K. A. (1975b). Studies on lichen-dominated systems. XI. Lichen–heath and winter snow cover. *Can. J. Bot.* **53**, 621–626.

Lechowicz, M. J., and Adams, M. S. (1974a). Ecology of *Cladonia* lichens. I. Preliminary assessment of the ecology of terricolous lichen–moss communities in Ontario and Wisconsin. *Can. J. Bot.* **2**, 55–64.

Lechowicz, M. J., and Adams, M. S. (1974b). Ecology of *Cladonia* lichens. II. Comparative physiological ecology of *C. mitis, C. rangiferina,* and *C. uncialis. Can. J. Bot.* **52**, 411–422.

Lechowicz, M. J., Jordan, W. P. and Adams, M. S. (1974). Ecology of *Cladonia* lichens. III. Comparison of *C. caroliniana,* endemic to southeastern North America, with the three *Cladonia* species. *Can. J. Bot.* **52**, 565–573.

Lee, H. A. (1959). Surficial geology of southern district of Keewatin and the Keewatin Ice Divide, Northwest Territories. *Geol. Surv. Can. Bull.* **51**, 1–42.

Lemieux, G. J. (1963). Soil–vegetation relationships in the northern hardwoods of Quebec. Can. Dep. For. For. Res. Branch Contrib. 563 pp. 163–175.

Linteau, A. (1955). Forest site classification of the Northeastern coniferous section, boreal forest region, Quebec. Can. Dep. Nat. Res. For. Br. Bull. 118, pp. 1–85.

Looman, J. (1964a). Ecology of lichen and bryophyte communities in Saskatchewan. *Ecology* **45**, 481–491.

Looman, J. (1964b). The distribution of some lichen communities in the prairie provinces and adjacent parts of the Great Plains. *Bryologist* **67**, 209–224.

Low, A. P. (1889). Report on exploration in James Bay and country east of Hudson Bay, drained by the Big, Great Whale and Clearwater Rivers, 1887 and 1888. *Geol. Nat. History Surv. Can. Annu. Rep.* **3,** Rep. J, pp. 6–80.

Low, A. P. (1896). Report on explorations in the Labrador peninsula, along the East Main, Koksoak, Hamilton, Manicuagan and portions of other rivers, in 1892–93–94–95. *Geol. Surv. Can. Annu. Rep.* **8,** Rep. L, pp. 5–387.

Low, A. P. (1897). Report on a traverse of the northern part of the Labrador peninsula from Richmond Gulf to Ungava Bay. *Geol. Surv. Can. Annu. Rep.* **9,** Rep. L, pp. 1–43.

Lutz, H. J. (1956). Ecological effects of forest fires in the interior of Alaska. *U.S. Dept. Agric. Tech. Bull.* **1133**, 1–121.

McFadden, James D. (1965). "The interrelationships of lake ice and climate in central Canada." Tech. Rep. 20. (Nonr 1202(07)), Dept. Meteorol. Univ. of Wisconsin, Madison.

McIntosh, R. P., and Hurley, R. T. (1964). The spruce–fir forests of the Catskill Mountains. *Ecology* **45**, 314–326.

Mackay, J. R. (1967). Permafrost depths, lower Mackenzie Valley, Northwest Territories. *Arctic* **20**, 21–26.

Maini, J. S. (1966). Phytoecological study of sylvotundra at Small Tree Lake, N.W.T. *Arctic* **19**, 220–243.

Maycock, P. F. (1956). Composition of an upland conifer community in Ontario. *Ecology* **37**, 846–848.

Maycock, P. F. (1957). "The Phytosociology of Boreal Conifer–Hardwood Forests of the Great Lakes Region." Ph.D. Thesis, Univ. of Wisconsin, Madison.

Maycock, P. F. (1961). The spruce–fir forest of the Keweenaw peninsula, northern Michigan. *Ecology* **42**, 357–365.

Maycock, P. F. (1963). The phytosociology of the deciduous forests of extreme southern Ontario. *Can. J. Bot.* **41**, 379–438.

Maycock, P. F., and Curtis, J. T. (1960). The phytosociology of boreal conifer–hardwood forests of the Great Lakes region. *Ecol. Monogr.* **30**, 1–35.

Maycock, P. F., and Matthews, B. (1966). An arctic forest in the tundra of northern Quebec. *Arctic* **19**, 114–144.

Moore, P. D., and Bellamy, D. J. (1974). "Peatlands." Springer-Verlag, Berlin and New York.

Moss, E. H. (1953a). Forest communities in northwest Alberta. *Can. J. Bot.* **31**, 212–252.

Moss, E. H. (1953b). Marsh and bog vegetation in northwestern Alberta. *Can. J. Bot.* **31**, 448–70.

Moss, E. H. (1955). The vegetation of Alberta. *Bot. Rev.* **21**, 493–567.

Nichols, H. (1967). Pollen diagrams from subarctic central Canada. *Science* **155**, 1665–1668.

Oberdorfer, E. (1954). "Suddeutsche Pflanzengesellschaften." Fischer, Jena.

Oosting, H. J., and Billings, W. D. (1951). A comparison of virgin spruce–fir forest in the northern and southern Appalachian system. *Ecology* **32**, 84–103.

Oosting, H. J., and Reed, J. F. (1944). Ecological composition of pulpwood forest in northwestern Maine. *Am. Midl. Nat.* **31**, 182–210.

Oosting H. J., and Reed, J. F. (1952). Virgin spruce–fir forest in the Medicine Bow Mountains, Wyoming. *Ecol. Monogr.* **22**, 69–91.

Parker, W. H., and McLacklan, D. G. (1978). Morphological variation in white and black spruce: Investigation of natural hybridization between *Picea glauca* and *P. mariana*. *Can. J. Bot.* **56**, 2512–2520.

Porsild, A. E. (1945). The alpine flora of the east slope of the Mackenzie Mountains. *Natl. Mus. Can. Bull.* **101**, 1–35.

Porsild, A. E. (1951). Botany of southeastern Yukon adjacent to the Canol Road. *Natl. Mus. Can. Bull.* **121**, 1–344.

Porsild, A. E. (1957). Illustrated flora of the Canadian arctic archipelago. *Natl. Mus. Can. Bull.* **146**, 1–209.
Ragotzkie, R. A. (1962). "Operation Freeze-Up: An Aerial Reconnaissance of Climate and Lake Ice in Central Canada." Tech. Rep. No. 10 [Nonr 1202(07)], Dep. Meteorol., Univ. of Wisconsin, Madison, 24 pp.
Raup, H. M. (1947). The botany of southwestern Mackenzie. *Sargentia* **6**, 1–275.
Raup, H. M. (1934). Phytogeographic studies in the Peace and upper Laird river regions, Canada. *Contrib. Arnold Arbor. Harv. Univ.* **6**, 1–230.
Raup, H. M. (1941). Botanical problems in boreal America. *Bot. Rev.* **7**, 147–248.
Raup, H. M. (1946). Phytogeographic studies in the Athabasca–Great Slave lake region. II. *J. Arnold Arbor. Harv. Univ.* **27**, 1–85.
Rencz, A. N., and Auclair, A. N. D. (1978). Biomass distribution in a subarctic *Picea mariana–Cladonia alpestris* woodland. *Can. J. For. Res.* **8**, 168–176.
Ritchie, J. C. (1956). The vegetation of northern Manitoba. I. Studies in the southern spruce forest zone. *Can. J. Bot.* **34**, 523–561.
Ritchie, J. C. (1959). The vegetation of northern Manitoba. III. Studies in the Subarctic. *Arct. Inst. North Am. Tech. Pap.* 3, pp. 1–56.
Ritchie, J. C. (1960a). The vegetation of Northern Manitoba. VI. The lower Hayes River region. *Can. J. Bot.* **38**, 769–788.
Ritchie, J. C. (1960b). The vegetation of northern Manitoba. V. Establishing the major zonation. *Arctic* **13**, 211–229.
Ritchie, J. C. (1960c) The vegetation of northern Manitoba: IV. The Caribou Lake region. *Can. J. Bot.* **38**, 185–199.
Richie, J. C. (1962). A geobotanical survey of northern Manitoba. *Arct. Inst. North Am. Tech. Pap.* 9, pp. 1–47.
Ritchie, J. C. (1977). The modern and late Quarternary vegetation of the Campbell–Dolomite uplands near Inuvik, N.W.T., Canada. *Ecol. Monogr.* **47**, 401–423.
Roche, L. (1969). A genecological study of the genus *Picea* in British Columbia. *New Phytol.* **68**, 505–554.
Rønning, O. I. (1969). Features of the ecology of some arctic Svalbard (Spitsbergen) Plant communities. *Arct. Alp. Res.* **1**, 29–44.
Rouse, W. R., and Kershaw, K. A. (1971). The effects of burning on the heat and water regimes of lichen-dominated subarctic surfaces. *Arct. Alp. Res.* **3**, 291–304.
Rowe, J. S. (1956). Uses of undergrowth plant species in forestry. *Ecology* **37**, 461–473.
Rowe, J. S. (1959). Forest regions of Canada. *Can. Dep. North. Affairs Nat. Res. For. Br. Bull.* **123**, pp. 1–71.
Rowe, J. S. (1961). Critique of some vegetational concepts as applied to forests of northwestern Alberta. *Can. J. Bot.* **39**, 1007–1017.
Rowe, J. S. (1972). Forest regions of Canada. *Can. For. Serv. Publ.* **1300**.
Savile, D. B. O. (1956). Known dispersal rates and migratory potentials as clues to the origin of the North American biota, *Am. Midl. Nat.* **56**, 434–453.
Savile, D. B. O. (1961). The botany of the northwestern Queen Elizabeth Islands. *Can. J. Bot.* **39**, 909–942.
Savile, D. B. O. (1963). Factors limiting the advance of spruce at Great Whale River, Quebec. *Can. Field Nat.* **77**, 95–97.
Savile, D. B. O. (1964). North Atlantic biota and their history, *Arctic* **17**, 138–141.
Savile, D. B. O. (1972). Arctic adaptations in plants. *Can. Dep. Agric. Res. Br. Monogr.* **6**.
Scoggan, H. J. (1950). The flora of Bic and the Gaspé Peninsula. *Natl. Mus. Can. Bull.* **115**.
Scoggan, H. J. (1957). Flora of Manitoba. *Natl. Mus. Can. Bull.* **140**.

Scotter, G. W. (1962). Productivity of boreal lichens and their possible importance to barren-ground Caribou. *Arch. Soc. Fenn. (Vanamo)* **16**, 155–161.
Scotter, G. W. (1964). Effects of forest fires on the winter range of barren-ground caribou in northern Saskatchewan. *Wildl. Manage. Bull.* **1**, 1–111.
Sjörs, H. (1959). Bogs and fens in the Hudson Bay Lowlands. *Arctic* **12**, 3–19.
Sjörs, H. (1961). Forest and peatland at Hawley Lake, northern Ontario. *Natl. Mus. Can. Bull.* **171**, 1–31.
Smith, D. C. (1962). The biology of lichen thalli. *Biol. Rev.* **7**, 89–102.
Sochava, V. B. (1956). Tyemnokhvoynye lesa (The dark coniferous forests). *In* "Rastitel'nyi Pokrov S.S.S.R." ("The Plant Cover of the U.S.S.R."), (E. M. Lavrenko and V. B. Sochava, eds.), pp. 130–216. Moscow.
Sukachev, V. N. (1928). Principles of classification of the spruce communities of European Russia. *J. Ecol.* **16**, 1–18.
Suslov, S. P. (1961). "Physical Geography of Asiatic Russia." (N. D. Gersheoksy, transl.). Freeman, San Francisco, California.
Swan, J. M. A., and Dix, R. L. (1966). The phytosociological structure of upland forest at Candle Lake, Saskatchewan. *J. Ecol.* **54**, 13–40.
Thieret, J. W. (1964). Botanical survey along the Yellowknife Highway, Northwest Territories, Canada. II. Vegetation. *SIDA Contrib. Bot.* **1**, 187–239.
Thomson, J. W. (1953). Lichens of arctic America. I. Lichens from west of Hudson's Bay. *Bryologist* **56**, 8–35.
Thomson, J. W. (1967). "The Lichen Genus *Cladonia* in North America." Univ. of Toronto Press, Toronto.
Thomson, J. W. (1972). Distribution patterns of American arctic lichens. *Can. J. Bot.* **50**, 1135–1156.
Thomson, J. W., Scotter, G. W., and Ahti, T. (1969). Lichens of the Great Slave Lake region, Northwest Territories, Canada. *Bryologist* **72**, 137–177.
Tolmachev, A. I. (1968). "Distribution of the Flora of the U.S.S.R." Israel Prog. Sci. Translations, Jerusalem.
Tseplyaev, V. P. (1965). "The Forests of the U.S.S.R." Israel Prog. Sci. Translations, Jerusalem.
Van Cleve, K., Viereck, L. A., and Schlentner, R. L. (1970). Accumulation of nitrogen in alder ecosystems developed on the Tanana River flood plain near Fairbanks, Alaska. *Arct. Alp. Res.* **3**, 101–114.
Van Cleve, K., Viereck, L. A., and Schlentner, R. L. (1972). Distribution of selected chemical elements in even-aged alder (*Alnus*) ecosystems near Fairbanks, Alaska. *Arct. Alp. Res.* **4**, 239–255.
Wagg, J. W. B. (1964). White spruce regeneration on the Peace and Slave River lowlands. Can. Dep. For. For. Res. Br. Contrib. 617, Pub. No. 1069, pp. 1–35.
Whittaker, R. H. (1956). Vegetation of the Great Smoky Mountains. *Ecol. Monogr.* **26**, 1–80.
Whittaker, R. H. (1966). Forest dimensions and production in the Great Smoky Mountains. *Ecology* **47**, 103–121.
Whittaker, R. H. (1970). "Communities and Ecosystems." Macmillan, New York.
Wilton, W. C. (1964). The forests of Labrador. Can. Dep. For., For. Res. Br. Contrib. 610, pp. 1–72.
Wolfe, J. A. (1972). An interpretation of Alaskan Tertiary floras. *In* "Floristics and Paleofloristics of Asia and Eastern North America." (A. Graham, ed.), pp. 225–233. Elsevier, Amsterdam.
Yurtsev, B. A. (1972). Phytogeography of northeastern Asia and the problem of transberingian floristic interrelations, *In* "Floristics and Paleofloristics of Asia and Eastern North America." (A. Graham, ed.), pp. 19–54. Elsevier, Amsterdam.

Zoltai, S. C. (1972). Palsas and peat plateaus in central Manitoba and Saskatchewan. *Can. J. For. Res.* **2**, 291–302.

Zoltai, S. C., and Pettapiece, W. W. (1974). Tree distribution on perennially frozen earth hummocks. *Arct. Alp. Res.* **6**, 403–411.

Zoltai, S. C., and Tarnocai, C. (1971). Properties of a wooded palsa in northern Manitoba. *Arct. Alp. Res.* **3**, 115–29.

6 Relationships of Canadian Boreal Plant Communities

The boreal forest is one of the principal native vegetational provinces of the North American continent, occupying at least as great a proportion of the land surface as any of the other major vegetation zones. While the boreal forest is a relatively homogeneous vegetational type, readily identifiable throughout its broad expanse on the basis of dominant coniferous trees and a number of wide-ranging shrubs and herbs, there are nevertheless differences in the composition and structure of boreal communities from one area to another, with greater differences appearing as the distances between reference areas increase.

Methods now available for vegetational continuum analysis were used to eludicate similarities and differences among examples of understory associations found in four basic forest types located in various areas of a region within the central North American boreal forest (Fig. 22). Efforts have been made, additionally, to demonstrate that community differences between one area and another within this region appear to represent a response to environmental differences and that they appear strongly related to climatic gradients.

The data employed have been obtained principally from the understory association of black spruce stands, with additional references to aspen, jack pine, and mixed wood (primarily balsam fir and white spruce) stands. This discussion is concerned chiefly with data from the black spruce stands, although material from other types is included for comparison. The main emphasis is upon phytosociological relationships, although references to climatic, geological, and edaphic factors are made to establish some necessary controls and assumptions.

The concept is not a new one that natural vegetational communities are aggregations of native plants that behave essentially independently of one another. The native plants respond to the environment in a manner determined by genetic endowment, and each plant, with its charac-

Fig. 22. Study areas in central Canada for vegetational community analyses described in the text.

teristic range of environmental tolerance, interacts with the individuals of its own and all other species to the extent that its (and their) presence modifies the environmental factors to which each is responsive. Associations differ from one another in two basic measurable ways: species composition (whether a species is present or absent) and community structure expressed in the abundance with which a species is represented by individual plants (frequency). Thus, community composition will be different under differing environmental conditions, and the extent of community differences can be employed as an index of the extent of differences between two areas.

That the distribution of native vegetational communities or biotic provinces is related to regional climate dates as a scientific concept at least to von Humboldt (1807), and a large number of biotic and climatic classifications have been based on rather readily observed gross relationships (Kendeigh, 1954). The study reported here represents an effort (1) to describe variations in plant community structure related to spatial distribution in the study region, (2) to select species that contribute most importantly to these variations, and (3) to demonstrate that community variations over distance can be related to variations in other environmental factors and that some proportion of this vegetational community variation is related to climatic variation, although other contributing factors must also include variations in soils, ecotypic variations in species and, perhaps, such added influences as geological history and postglacial migration patterns of species. Little in the way of control measures can now be devised for these other contributing factors largely because of the relative lack of concrete information concerning their effect upon geographical distribution of species.

A vast body of literature exists on the vegetation of the boreal forest, and no effort is made here to mention all pertinent references. A number of studies have indicated the feasibility of the approach taken herein, the more significant of which have been discussed in other chapters. Rowe (1956), for example, made a study of the relationships between understory community composition and habitat conditions and categorized the understory species into groups based on specific moisture preferences and on the tree species with which they were most often found in association. Maycock (1957) studied the boreal conifer–hardwood forests of the Great Lakes region and found that "the most notable changes are the greater simplicity of floras passing northward and increasing richness passing southward." Maycock and Curtis (1960) expressed the concept in more general terms when they stated: "A gradually changing environmental complex northward results in gradually changing floristic composition toward greater representation by boreal species." Curtis (1959) presents a table showing that differences between stands become greater as distances between them increase, and that differences are at least roughly proportional to distance. This characteristic is also revealed in greater detail by the studies of La Roi (1967). Pertinent also, are the publications of Clausen (1957) and Christensen *et al.* (1959) showing that conifer swamp communities differ along a north–south gradient and that "no two southern stands attain the high degree of similarity found between more northern stands."

The literature that relates plant community differences to differences in environment is a large one and cannot be wholly reviewed here. A statement by Whittaker (1954) might be taken as representative:

> Both factors of environment and properties of communities are isolates from the ecosystem. When a factor and a property have been chosen for study and found

related, we may sometimes interpret the latter as dependent and the former as more nearly independent. The distinction between independent and dependent variables in the ecosystem should not, however, be taken too literally.

In reference to boreal vegetational and climatic relationships, Raup (1941) points out that, while it has been long assumed that climatic controls are dominant in maintaining this vast vegetational province, no quantitative work has appeared that describes the relationship.

The research reported here does not attempt a precise assessment of the degree to which regional vegetational differences are a response to climatic ones but is exploratory of the correlations that can be obtained between plant community differences and known climatic parameters in a region where detailed information on either is largely lacking.

PROCEDURE

Field sampling procedures and data analysis methods were employed to (1) demonstrate a reasonable degree of uniformity and comparability in the characteristics of the arborescent strata of the sampled stands, and (2) to discern between-area similarities and differences in presence and percentage frequency of occurrence of understory community species, in addition to identifying the species contributing the greatest proportion of variation between stands. Subsequent analysis relates these variations in species frequency and presence to geographical, hence climatic, position in the study region and delineates other characteristics of the variation between communities that might be considered related to the environmental variations.

A basic assumption underlying the sampling method employed in this study is that the age and structure of the tree strata constitute indexes to the understory environment and that the environmental tolerance of the dominant tree species establishes the range and limits of variation in the environmental factors to which the understory community is exposed. While the environment is not described or defined, and with presently available knowledge cannot be, it must be assumed to range only within limits established by the range of tolerance of the dominant trees. It is recognized that ecotypic variation exists in species and, were appearances not similar, ecotypic variants might well be recognized as so different in their environmental responses as to constitute different species. However, it must also be recognized that, taking ecotypic variation into account, the range of environmental tolerance of a species is still within definable limits, with extremes at the limits of tolerance of the ecotypes. Since some of these responses are phylogenetically conservative, particularly adaptation to prevailing environmental temperatures, it can be assumed that they effectively establish the range limits of many species. It must also be assumed that plant species growing together in a community exhibit ranges of tolerance that overlap in the environment where they are found growing together.

Another assumption is that the duration of postglacial time has been sufficient for most species to accomplish migration to the edge of their environmental tolerance in terms of geographical distribution. While the opposite assumption has formed the basis for some work in the past, to the writer the evidence seems to indicate that such an assumption can safely be made in regard to at least the greater proportion of the species under consideration (see Chapter 2).

All the areas of study are within the glaciated region, thus providing as comparable a range of geological parent material as can probably be attained over such a large geographical area. Parent materials in study areas are morainic, lacustrine, or unassorted drift. The contribution that soil differences may make to these regional variations in plant community structure is at present unknown. So little information is available on the soil factors limiting distribution of the species, or on the conditions permitting optimal growth or competitive advantage, that the possible influence of soils here receives minimal consideration. An effort has been made to obtain a measure of control over soil factors in the comparisons, however, by obtaining analyses of soils of stands in each area. That the areas are at least grossly similar in soil characteristics can be seen from Table 16 in which data on particle size distribution, pH and, in some instances, nutrient content are presented for each community type in many of the areas.

It is, hence, assumed that under the influence of similar forest vegetation types there have developed at least roughly comparable edaphic conditions, at least in the upper soil horizons, as a result of podzolization. This assumption finds support in the literature on northern soils (see Chapter 4). For example, Wilde (1946) points out that "the soil profiles which develop under unbalanced or extreme climatic conditions are influenced by the parent material only to a limited extent." Wilde (1958) also affirms that the "ecological effects of the parent soil material ... diminish with distance from the temperate climate and with the age of the soil." It must be recognized that the depth of the surface organic horizons varies from stand to stand, and from site to site within stands, and that the influence of this added variable is extremely difficult to measure effectively. In summary, however, it is assumed that the existence of a virtually continuous moss and peat cover, strong leaching potentials, and high acidity of surface horizons tends to create generally

TABLE 16A

Soil Analyses from Representative Stands in Designated Areas[a,b]

	Ennadai	Lynn Lake	Lynn Lake	Ilford	Ilford	Ilford	Rocky lake	Waskesiu	Clear lake	Raven Lake	Raven Lake	Klotz Lake	Klotz Lake	Remi Lake	Trout Lake	Bemidji	West Hawk
Spruce (Dominated by *Picea mariana*)																	
Surface pH	4.8	4.8	3.3	4.4	5.5			4.5	5.6	4.3			6.0		5.5	5.7	
pH, 6–12 in.	4.9		4.2	4.9	7.4			5.3	5.6	4.8			7.4	5.3	5.3	5.6	
Texture, 6–12 in.																	
Sand (%)	80	64	om	om	50	60	67	55	om	34	64	22	38	12	om	19	
Silt (%)	15	29	om	om	40	34	24	36	om	60	26	70	52	48	om	55	
Clay (%)	5	7	om	om	10	6	9	9	om	6	10	8	10	40	om	26	
Jack pine																	
Surface pH			4.1						5.5	4.5					4.1		5.1
pH, 6–12 in.			4.6						5.8	5.2					5.3		5.8
Texture, 6–12 in.																	
Sand (%)			87				20		44	33					84		81
Silt (%)			7				60		20	60					9		10
Clay (%)			6				20		36	7					7		9
Aspen																	
Surface pH								4.8							5.6		5.1
pH, 6–12 in.								5.2							6.8		5.8
Texture, 6–12 in.																	
Sand (%)								57				44	32		71		59
Silt (%)								30				42	42		17		30
Clay (%)								13				14	26		13		11

[a] In soils in which organic matter only is present in the upper 12 in. are designated by "om"; in instances where mineral soils were available in this upper stratum, the textural data are given.
[b] Analysis conducted courtesy of the Wisconsin Soil Survey, Francis Hole.

TABLE 16B

Soil Analyses for Nutrient Minerals in Stands in the Designated Areas[a]

Area	Depth (in.)	Nitrogen	Phosphorus	Potassium
		(calculated on basis of pounds per acre)		
Black Spruce				
Ilford	0–6	75	1	75
	12	100	8	80
	24	125	2	30
	0–2	175	7	120
	6	125	1	50
	18	75	7	35
Waskesiu	0–2	125	63	95
	6	50	30	160
	18	85	210	130
Clear Lake	0–6	225	14	120
	12	200	5	20
	36	225	7	10
Raven Lake	0–2	125	9	130
	6	150	11	50
	18	75	150	20
Trout Lake	0–2	250	16	105
	6	275	16	40
	18	225	35	25
Bemidji	0–2	150	7	90
	9	150	8	60
	36	150	7	50
Jack Pine				
Clear Lake	0–2	275	130	390
	12	150	220	510
	30	150	1	260
Raven Lake	0–2	175	9	120
	6	100	28	75
	18	50	31	50
Trout Lake	0–2	175	22	95
	6	75	75	85
	18	75	75	50
Aspen				
Waskesiu	0–2	175	77	150
	6	75	90	105
	18	50	175	200
West Hawk	0–2	175	12	145
	6	125	13	200
	18	125	120	280
Trout Lake	0–2	125	130	85
	6	125	75	120
	18	75	60	130

[a] Analyses conducted courtesy Wisconsin Soil Survey and Francis Hole.

comparable conditions in the horizons with which the understory herbs and shrubs are most directly in contact.

Some additional evidence tends to substantiate this latter assumption and to indicate that soil differences in the area under consideration are not sufficient in magnitude to be of paramount importance in accounting for vegetational differences. Three areas that appear to be distinct from the others in terms of soil pH—Ilford, Klotz, and Remi—nevertheless conform in their vegetational similarities and differences to the pattern that would be expected were their community structural characteristics determined more by climate than by soil. Also, while both Klotz and Remi possess substrates characteristic of the clay belt region (Baldwin, 1958), their vegetational characteristics are quite distinctly different, with a higher representation of the northern mesic understory species (Curtis, 1959) in Klotz stands than in the stands at Remi. It is pertinent that climatic differences between Klotz and Remi may be of sufficient magnitude to account in a primary fashion for the vegetational differences.

STATISTICAL METHODS

Sampling was conducted in the areas designated by dots and names on the map in Fig. 22. In each of these areas, a search was made (employing aerial photographs for initial guidance) for mature stands of the four major forest community types included in this study, black spruce, jack pine, aspen, and balsam fir–white spruce (mixed wood). In some areas, such as God's Lake, only two sets of data, one for black spruce and another for jack pine, were obtained; in other areas, much more extensive sampling was conducted.

Basic field procedures included use of the point quarter method for determining characteristics of the arborescent stratum of each stand (Cottam and Curtis, 1956), and species frequency tabulations were made for herbs and shrubs in 1-m^2 quadrats. Sampling was conducted at 20 points along a compass line, with 30 paces between points. Each point established the corner of a quadrat, thus tree and quadrat data coincide at each point. Stand selection in northern areas was made on the basis that a stand appeared representative of the forest community over a large proportion of the land surface. In southern areas, where black spruce stands are rare, only mature and relatively undisturbed stands were sampled. In stands where cores were taken to determine tree ages, a core was taken from the tree in the forward left quadrant at each point. The tree data were used to obtain relative density, average basal area, number of trees per acre, and importance values for the various tree species. Also calculated were the density and percentage frequency of saplings and the percentage frequency of seedlings, herbs, and shrubs occurring in the quadrats. Tree data were used to compare the general characteristics of the stands (Table 17). Relative values demonstrating the degree of dominance of the arboreal species were obtained by dividing the values for importance value (IV), density, basal area, etc., of the dominant species by the totals of these values for all tree species in the stand. For example, if the IV for black spruce is 150, for jack pine 75, and for aspen 75, then the relative value of the black spruce (column 9, Table 17) would be 0.50:

Black spruce	150	0.50
Jack pine	75	300/150
Aspen	75	
	300	

This procedure was followed in obtaining all values in Table 17, columns 6 and 9. Data for more than one stand were averaged.

The tabulation of herb and shrub frequencies occurring in each stand was processed to obtain a matrix of coefficient of similarity ($2w/a + b$) values. From these, ordinations were constructed. Factor analyses, R mode and Q mode, were used to analyze herb and shrub data, employing in each instance the standard correlation matrix rather than a similarity index. In the factor analysis, 66 species that occurred in at least two stands, with a frequency value of at least 15 in one, were employed in the computation. A list of these species is given in Table 18.

Detailed descriptions of the methods for constructing the ordination and of computing factor analysis are not given here, since these techniques are described and references provided in Greig-Smith (1964). The ordination is essentially a representation by spatial distribution of similarities between stands, with stands most similar to one another being located in close juxtaposition on the ordination. The R-mode factor analysis aggregates into groups (factors) those species with frequency values that behave similarly throughout the matrix. The Q-mode groups the areas that have a high degree of similarity in regard to the presence and frequency of the species found within the plant communities. Thus, Lynn Lake and Brochet (located close together) are always grouped in the same factors because of the high degree of similarity between the plant communities found in these areas. By pairing the R-mode and Q-mode analyses, it is possible to arrive at an estimation of the species groupings characteristic of the various area groupings; thus, for example, the species in R-mode factor 2 are found to give the Lynn

TABLE 17

Characteristics of Forest Stands in Various Study Areas[a]

	Number of stands		Average age of trees	Number of stands trees aged	Range of ages (Average age in even-aged stands)	Sapling importance	Relative to all species			
	With vegetation data	With tree data					Basal area per acre	Density	Importance value	
Black spruce stands										
Ennadai[b]	8	5	141	3	128–162	0.98	0.87	0.80	0.84	
Brochet	2	3	74	3	69–84	0.80	0.78	0.75	0.75	
Lynn Lake	10	11	109	8	58–155	0.85	0.95	0.95	0.92	
Waskesiu	1	1	61	1	61	0.54	0.52	0.55	0.50	
Rocky Lake	6	7	85	7	46–192	0.77	0.60	0.69	0.75	
Clear Lake	1	1	112	1	112	0.70	0.92	0.86	0.85	
Ilford	5	5	96	4	52–146	0.84	0.98	0.97	0.92	
God's Lake	1	1				0.70	0.96	0.95	0.94	
West Hawk	2	2	60	3	46–70	0.89	0.98	0.98	0.98	
Raven Lake	2	2	76	3	58–100	0.91	0.87	0.92	0.86	

Location									
Klotz Lake	3	4	85	3	67–95	0.79	0.82	0.85	0.81
Remi Lake	1	1	146	1	146	0.72	0.98	0.96	0.94
Bemidji	1	1	84	1	84	0.65	0.90	0.83	0.85
Trout Lake	3	3	33	1	33	0.81	0.62	0.70	0.78
Aspen stands									
Waskesiu	3	3	54	2	42–66	0.65	0.83	0.84	0.80
Rocky Lake	2	2	47	2	31–63	0.20	0.70	0.69	0.64
Clear Lake	2					1.00	1.00	1.00	1.00
West Hawk	2	2	62	2	60–65	0.64	0.59	0.60	0.39
Raven lake	1	1	69	1	69	0.09	0.74	0.58	0.59
Klotz Lake	3	2	82	1	82	0.0	0.27	0.45	0.57
Remi lake	1					0.0	0.77	0.70	0.66
Trout Lake	1					0.10	0.64	0.53	0.55
Jack pine stands									
Brochet	1	1	86	1	86	0.30	0.77	0.68	0.64
Lynn Lake	6	6	50	4	24–86	0.72	0.82	0.90	0.90
Rocky Lake	2	2	104	1	104	0.0	0.61	0.53	0.53
Clear Lake	1	1	66	1	66	0.04	0.60	0.51	0.53
God's Lake	1	1				0.48	0.97	0.97	0.95
West Hawk	3	3	53	3	51–56	0.35	0.77	0.82	0.80
Raven Lake	1	1	61	1	61	0.06	0.73	0.63	0.63
Trout Lake	1	1				0.30	0.96	0.94	0.91
Klotz lake	1	1	79	1	79	0.11	0.55	0.55	0.50
Remi Lake	4	4	78	3	57–97	0.83	0.51	0.62	0.51

[a] Data obtained in stands were used for analyses described in text. Unpublished data from Larsen.
[b] South end of Ennadai Lake only.

TABLE 18

List of Species in Factor Analysis

Abies balsamea	Lathyrus ochroleucus
Alnus crispa	Ledum decumbens
Alnus rugosa	Ledum groenlandicum
Anemone quinquefolia	Linnaea borealis
Aralia nudicaulis	Maianthemum canadense
Arctostaphylos rubra (alpina)	Mertensia paniculata
Aster ciliolatus	Mitella nuda
Aster macrophyllus	Oryzopsis asperifolia
Betula glandulosa	Petasites palmatus
Betula papyrifera	Picea mariana
Carex brunnescens	Potentilla palustris
Carex chordorrhiza	Pyrola asarifolia
Carex paupercula (limosa)	Pyrola secunda
Carex saxatilis	Rosa acicularis
Chamaedaphne calyculata	Rubus acaulis
Clintonia borealis	Rubus chamaemorus
Coptis groenlandica	Rubus pubescens
Cornus canadensis	Salix bebbiana
Diervilla lonicera	Salix discolor
Drosera rotundifolia	Salix herbacea
Empetrum nigrum	Salix myrtillifolia
Epigaea repens	Salix planifolia
Epilobium angustifolium	Smilacina trifolia
Equisetum arvense	Solidago hispida
Equisetum sylvaticum	Streptopus roseus
Eriophorum spissum	Trientalis borealis
Fragaria virginiana	Vaccinium angustifolium
Galium boreale	Vaccinium myrtilloides
Galium triflorum	Vaccinium oxycoccus
Gaultheria procumbens	Vaccinium uliginosum
Geocaulon lividum	Vaccinium vitis-idaea
Kalmia polifolia	Viburnum edule
Larix laricina	Viola renifolia

Lake, Brochet, Ilford Q-mode grouping factor 1 and/or 7 its individuality. These R-mode and Q-mode pairings must be conducted subjectively, but in most instances the relationship is apparent. As is frequently the case, however, in the Q-mode analysis several factors may appear to apply to a given R-mode factor, and it is often difficult to make a decision as to which pairings seem most appropriate. Since the differences between the relevant Q-mode factors are minor ones, for the purposes of this discussion all the pertinent factors have been included.

The S/D index was computed using the average coefficient of similarity between areas (S) and a the distance index (D) between areas (in computation, 0.1 times the actual distance in miles was used). The S/D index was designed to utilize the assumption that the coefficient of similarity values between areas decreases as the distance increases, and that any deviation from this rule indicates the existence of distortions in what would otherwise be a uniform continuum field (in which similarities decrease proportionally to distance in all directions). With the data employed here, the range of coefficient of similarity values is 8.6 (Ennadai–Klotz) to 54.5 (Lynn–Brochet), and the distance index values (0.1 times the actual distance in miles) range from 100 (Ennadai–Trout) to 7 (Lynn–Brochet). An arbitrary range of "expected" S/D values was established on a graph (Fig. 23), and all points falling outside this range were considered as indicating continuum field distortions.

The variability in topographic preference of the dominant species of the arborescent strata along a north–south cline is a factor for which no meaningful control has been devised. In the south, for example, black spruce is found in lowland stands. In the central area, it generally dominates the landscape regardless of topographic position. In the northern portion of the study region, it is found on uplands. Hence, not only are there area-to-area environmental differences associated with day length, climate, and substrate, but also differences in topographic preference of the dominant tree species, and thus in drainage patterns and moisture supplies. It appears probable, however, that the last-mentioned difference may be the result of a temperature response. It has been shown, for example (Curtis, 1959), that average spruce bog environmental temperatures in Wisconsin are lower than those of the surrounding upland forests, and it appears that they are generally comparable to the temperatures of more upland areas in regions farther north, just as more southern areas with northern exposures have temperatures more generally comparable to those of northern areas with southern exposures than to those of adjacent sites. While such relationships seem intuitively reasonable, little or no data exist that effectively demonstrate that this is actually the case, and it must be acknowledged that the relationships are at present speculative even though they appear reasonable. According to this view, however, northern upland sites and southern lowland sites have more temperature-related environmental factor ranges in common than similar topographic sites in the two regions; hence similarities are greater than topographic differences might indicate.

The sampling method employing frequency tabulations of species occurring in 20 quadrats spaced 30 paces apart along a compass line is one that tends to minimize the effects of minor within-stand variations in community structure by essentially averaging the data from all sites

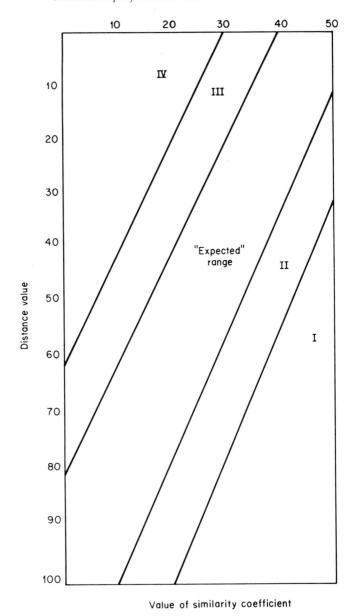

Fig. 23. As distances increase, similarity between study site communities declines. The figure shows the arbitrary "expected" relationship between distance between study sites and the similarity between the plant communities. Values falling outside the expected range indicate that study site communities are more (or less) similar than would be expected for the distances involved.

encountered in a transect. Quadrats of the size employed and transects of this length (somewhat longer than ½ mi) tend to produce data more broadly descriptive of an area than those employed in forest site classification techniques. Between-stand differences within areas are, nevertheless, seen to occur, and it appears that the largest and most apparent differences occur in the more southern areas. In these areas, the largest number of species is present, hence the largest number of combinations of community structure is possible. Additionally, it is here that the larger number of environmental combinations, such as air mass frequency patterns and soil types, occurs. This wider variety in structure of the southern communities is apparent on the black spruce ordination, where the somewhat greater clumping of stand data obtained in the extreme northern area (Ennadai) contrasts with the wider scattering of the stands in southern areas. Despite the variability within areas, however, the regional differences remain overriding in importance, and the stands within areas are clumped sufficiently so that between-area differences are apparent. Perhaps, differentiation in this exploratory study is not as clear as will be obtained eventually with larger samples and with more exacting standards of stand selection.

RESULTS

The extent to which the stands attain comparability in terms of inspection of the data on tree species, age, and dominance characteristics can be derived from Table 17. The importance value, density, and dominance values indicate that black spruce stands, with one exception, are clearly dominated by mature black spruce and, in the exception, Waskesiu, where the dominance is not as great as in the other stands, black spruce is nevertheless equal in dominance to all other species combined. While the aspen and jack pine stands show a somewhat greater range of variation, these species constitute in each instance the single major dominant and, although many stands have a number of associates, none appear to attain positions of significant importance for the purposes here.

The ordination of coefficient of similarity values for the spruce stands (Fig. 24) points out the greater similarities between stands located close together geographically. The ordination of aspen, jack pine, and mixed wood stands (Fig. 25) demonstrates greater variability in the understory communities of the jack pine stands in comparison to those of the aspen and mixed wood types and wider differentiation on what appears to be a geographical basis. The aspen and mixed wood stands are closely

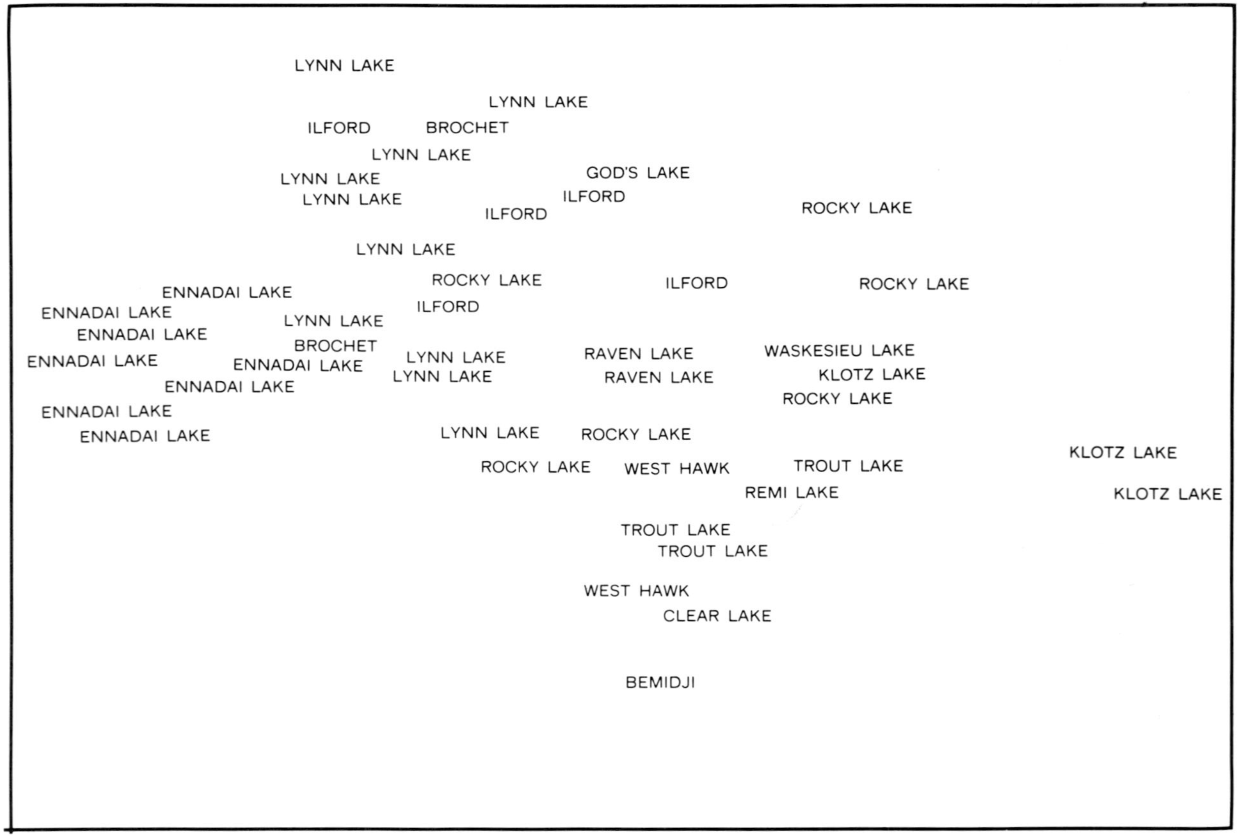

LYNN LAKE
LYNN LAKE
ILFORD
BROCHET
LYNN LAKE
LYNN LAKE
LYNN LAKE
GOD'S LAKE
ILFORD
ILFORD
ROCKY LAKE
LYNN LAKE
ROCKY LAKE
ILFORD
ROCKY LAKE
ENNADAI LAKE
ENNADAI LAKE
ILFORD
LYNN LAKE
ENNADAI LAKE
BROCHET
ENNADAI LAKE
ENNADAI LAKE
LYNN LAKE
LYNN LAKE
RAVEN LAKE
RAVEN LAKE
WASKESIEU LAKE
KLOTZ LAKE
ROCKY LAKE
ENNADAI LAKE
ENNADAI LAKE
ENNADAI LAKE
LYNN LAKE
ROCKY LAKE
ROCKY LAKE
WEST HAWK
TROUT LAKE
REMI LAKE
KLOTZ LAKE
KLOTZ LAKE
TROUT LAKE
TROUT LAKE
WEST HAWK
CLEAR LAKE
BEMIDJI

Fig. 24. Ordination of coefficient of similarity values for black spruce stands. Stands most like one another are closest together on the ordination, hence the stands from the study areas are grouped, and stands from study areas near one another are closest together. The end points on the first and second axes—Klotz Lake and Ennadai Lake on the first axis and Lynn Lake [illegible] are most distant.

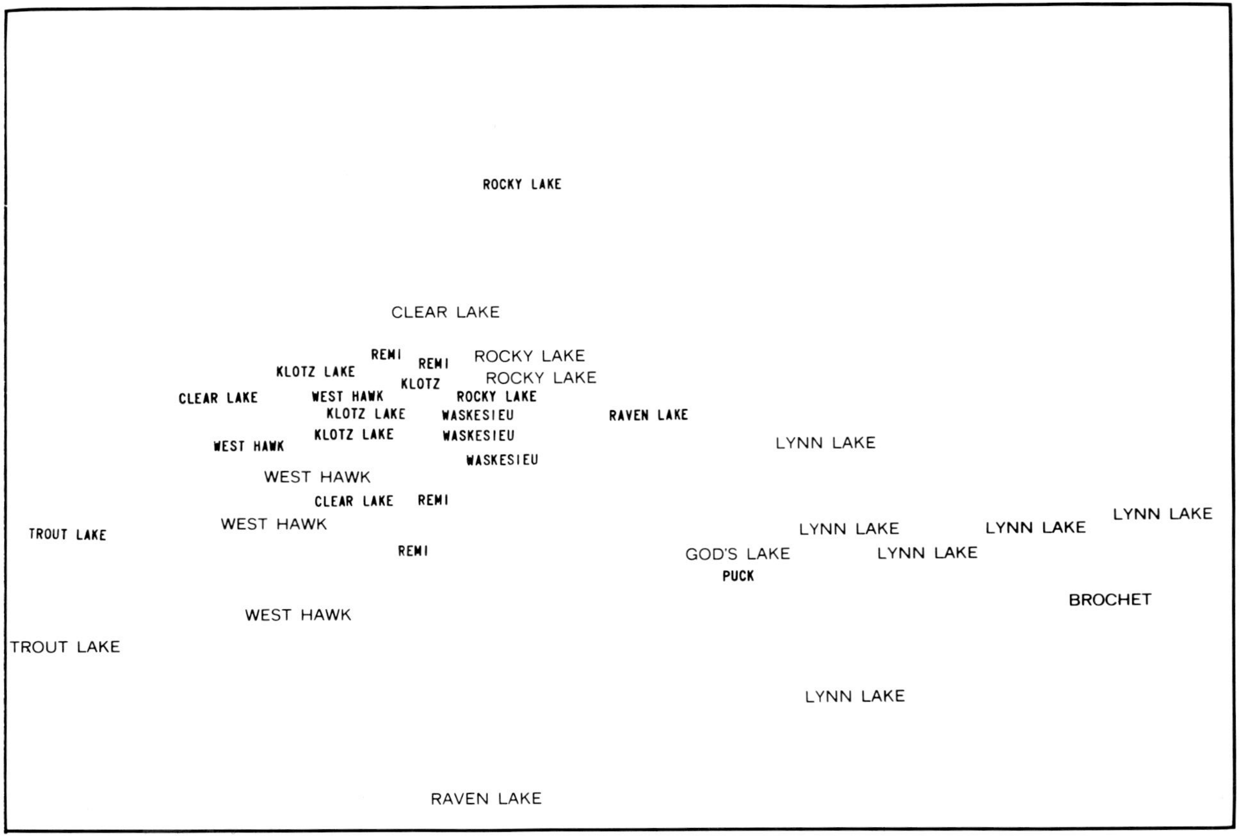

Fig. 25. Ordination of coefficient of similarity values for jack pine and aspen stands; pine stands are in the larger type, and aspen stands in the smaller type. See Fig. 24 for an explanation of placement of stands in relationship to one another.

clumped in the center of the ordination and appear less responsive in structure to geographical position.

With another approach, it is possible to further explore the increased similarity between black spruce communities of northern areas in comparison to the communities of southern areas. When average similarities between areas are correlated with distance between areas, the following list of correlation coefficients is obtained:

Area	r
Ennadai	−0.82
Klotz	−0.73
Ilford	−0.64
God's	−0.62
Brochet	−0.52
Lynn	−0.47
West Hawk	−0.40
Clear	−0.38
Trout	−0.33
Raven	−0.30
Bemidji	−0.24
Remi	−0.05
Waskesiu	−0.05
Rocky	−0.02

The two areas constituting the end points of the first axis on the ordination of black spruce stands, Ennadai and Klotz, are those showing the highest correlation (inverse) between similarities and distance with all other stands. With the exception of the Klotz area, all the areas showing the highest correlations with distance are areas in the northern portion of the region, while those with low correlation values are southern areas. Thus, the degree of similarity between any southern area and the other areas bears a lesser relationship (or none at all) to the distances by which the areas are separated. The reverse is true of the northern areas, which show a generally good correspondence between similarity coefficients and distance.

The position of Klotz on this table is of considerable interest. The communities at Klotz possess a rather distinct individuality in terms of the number of northern mesic forest species present with rather high frequency. These apparently are present with steadily declining presence and frequency values with increasing distance from Klotz. These species achieve for the Klotz area an individuality in the southern portion of the region as distinct as that of Ennadai in the north. Both areas possess strong inverse relationships between similarity and distance. Both apparently function as nodal points in the distribution of the two groups of species contributing significantly to variation between stands, the one group of great significance in the south and the other in the north, and the importance of each declining proportionally with distance from the nodes.

The *S/D* index provides a simple quantitative expression for these relationships. When the concept of a uniform continuum field (in which all between-area distances were inversely proportional to similarities) was employed as a basis for the *S/D* expression, it was possible to determine which area pairs possessed an *S/D* index falling outside the range of values to be expected, hence were indicative of distortions in the hypothetical continuum field. From the chart (Fig. 26) showing the alignment of stand pairs possessing "unexpected" *S/D* index values (those falling outside the "expected" range) it is apparent that stand pairs showing greater-than-expected similarity for the distances in-

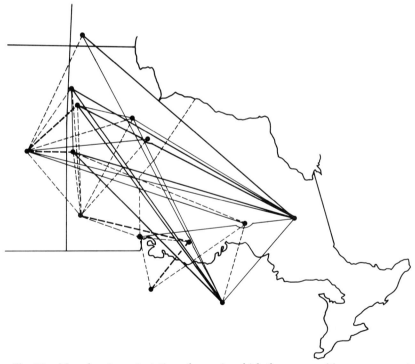

Fig. 26. Map showing orientation of areas in which the communities possess greater or lesser similarity for the distances involved. Heavy, continuous lines indicate greater similarity than expected (area II in Fig. 23), light dashed lines indicate low similarity for distance and heavy dashed lines indicate very low similarity for distance (area IV in Fig. 23). Note that the stands northwest–southeast of one another are of greater similarity than expected; those northeast–southwest of one another are of lesser similarity than expected.

volved are aligned generally on roughly northwest–southeast axes, while those showing generally lower similarity than would be expected for the distances are aligned in directions crossing these axes. There is, moreover, greater variability in the direction of the between-stand axes in the latter group, and the areas from which these axes originate most numerously are those found in the southern and in the southwestern portions of the region. These are relationships that would be anticipated if the areas of greatest similarity were found at positions along the same or similar climatic isopleths, hence generally oriented northwest or southeast of one another.

R-mode factor analysis of black spruce stands (Appendix II, Table II.1), indicates that a rather large number of species groups are required to account for the variance. With an arbitrary cutoff point of 0.05 in added variance explained, the first six factors contribute slightly more than 58% of the variance. Had a cutoff point of 0.025 been adopted, the first 14 factors would have contributed slightly more than 84% of the variance. Despite the low added variance contributed by the latter factors in the analysis, the species grouped therein subjectively appear to reflect with considerable accuracy the regional vegetational differences.

While there exists no major repetition of species in the R-mode analysis, there are pronounced similarities in many of the Q-mode groupings. For example, factors 1, 4, 6, and 12 contain both Rocky and Ilford; Lynn and Brochet are included in two of these factors; West Hawk is present in three. This apparently indicates that at least four different groups of species can be considered characteristic components of the communities in these areas, and each group differs from the others in representation and distribution of species among areas. Since there is no repetition of species in the R-mode analysis, it can be assumed that different species make up these groups.

From the analysis, it can be seen that three rather distinct geographical zonations are present, with various species groups representative of the vegetational communities within each. These are (1) the northern zone, with species groups represented primarily at Ennadai, (2) the central groupings, characteristic of Lynn, Brochet, Ilford, and other central areas, and (3) the southern groupings, composed of various combinations of species in the southern areas.

DISCUSSION

From climatic descriptions (see Chapter 3), it is apparent that isopleths of such traditional climatic parameters as mean temperature, temperature extremes, degree-days, mixing ratios, precipitation, limit of permafrost, dates of first and last frost, and so on, generally follow a northwest–southeast orientation in this region. Thus, areas falling to the northwest and southeast of one another are likely to have more similar climatic regimes than those falling (at equal distances) northeast and southwest of one another. To demonstrate that vegetational relationships in this region reflect climatic parameters, it becomes necessary to show that, given sufficiently similar substrates, the vegetation of areas northwest–southeast of one another in this region possesses greater similarity than that of areas oriented northeast–southwest of one another. That such is the case in the region under study is brought out by the vegetational relationships revealed by the correlations between distance and similarity and the S/D index. It seems reasonable to conclude that a correspondence can be discerned in this region between climate and the species composition of the vegetational communities under consideration in this study.

In general, these principles hold for black spruce, aspen, mixed wood, and jack pine understory communities. Aspen and mixed wood communities show greater clumping on the ordination than either jack pine or black spruce communities, hence appear tolerant of a narrower range of environmental variation. Black spruce and jack pine communities show the greatest variability and are found commonly throughout a greater range of geographical and environmental conditions.

In summary, from this study it appears that climatic factors can be invoked as contributing to the variation in composition and structure of indicated plant communities over the region under study, although it is at present not possible to ascribe the degree to which the various aspects of environmental influence are responsible for vegetational community differences. Adaptations of plant species not only to climate but to substrates and other plants (as they influence moisture, nutrient, and light conditions) are involved in competitive community relationships, and only more complete knowledge of all these factors, along with knowledge of physiological tolerances of individual plant species, will finally yield an understanding of these relationships that can be expressed in terms of a single equation. So far, efforts have been made only to infer similarities in environment from similarities in plant communities. The relationship is multifactorial, and with the present lack of knowledge of the effect of even relatively gross soil differences (and even of the existence of differences in most instances), little can be said concerning the relative importance of the various factors. In the study reported here, however, a measure of control over edaphic and microclimatic factors has been attempted, and the result indicates that some proportion of the remaining variability between communities of different areas within the study region can be ascribed to climatic influences.

It is significant that factor analysis of spruce understory community data reveals the existence of certain species groups, each of which is composed of species found (and occurring with what is essentially predictable frequency) in the communities of one of the three zones in the region under study. It becomes apparent that vegetational community structure can be rather definitely related to climatic parameters, and that the climatic patterns of the region possess discernible correlations with the vegetational communities found there, particularly the understory communities of the spruce forest, which were studied most intensively in the work reported.

BROAD CLIMATIC RELATIONSHIPS

Studies by the author and others (Larsen, 1965, 1971a,b, 1973, 1974; Bryson, 1966) have demonstrated a striking correspondence between climatic parameters and the position of the North American forest border. In such studies, an assumption is made that the isopleth under consideration delineates an ecotonal region where one or more parameters of the environmental complex exceed the tolerance limits of the trees. In all studies of these relationships, the advantage of working with the forest border is obvious. It can be visualized and mapped, and the geographical position determined with a fair degree of accuracy. But the fact that here, in the forest–tundra ecotone, large numbers of shrub and herbaceous species also show changes in abundance along environmental gradients has seldom if ever been taken into account. It was the purpose of the research described here to show that changes in the abundance of some of these other species in the vegetation are as striking as the more visible changes that occur in the arboreal stratum. There are also, as a result of demonstrable correlation between frequency of species and frequency of prevailing air mass type, implications lending support to the continuum concept of vegetation (Curtis, 1959; McIntosh, 1967).

THE STUDY

The study was conducted using data obtained from more than 40 sites mostly in the area to the north of the southern limit of the boreal forest

(Fig. 27). At each of the study sites, plant communities to be sampled were selected on the basis that they fell into one of the following seven general categories: black spruce forest, white spruce and/or balsam fir (mixed wood) forest, jack pine forest, aspen forest, rock field (fell field) tundra, tussock muskeg tundra, and low-meadow tundra.

In each instance, the basis for selection of a stand for sampling was initially visual, employing aerial photographs (coverage available from the Canadian National Air Photo Library) and then inspection on the ground. This initial visual selection had as its basis the criterion that stands appear homogeneous over at least a large enough area to enclose a transect. In all but a few instances, the stands were a great deal larger than this; in most areas, one or more of the communities occupied a relatively large proportion of the total landscape. In the case of the forest communities, the trees and saplings were sampled by the point quarter method (Cottam and Curtis, 1956), and for a stand to be included in the analysis the data obtained had to establish clearly that the indicated dominant species (black spruce, white spruce, and/or balsam fir, etc., see above) possessed an importance value (Curtis, 1959) greater than that of all the other tree species in the sample combined. Tundra sampling sites were selected by the same visual means. Topographic position was the main criterion for selection in the case of tundra communities: Rock fields occupy hill summit areas, tussock muskeg occupies intermediate slopes and has a layer of peat and plant detritus overlying the upper horizon of inorganic soil material, and low meadows are lowlands usually inundated at least in spring and early summer by 1-2 in. of water. Frequency tabulations for the forest understory and tundra communities were obtained using quadrats 1 m^2 in shape and size, arranged equidistant (usually 30 paces apart) along a compass line. The number of different study areas in which the various communities were sampled were as follows: black spruce, 25; white spruce, 12; jack pine, 12; aspen, 10; rock field, 14, tussock muskeg, 9; low meadow, 7. Principal component analysis was used to extract from the data those species that behave similarly with respect to average frequency values from one study area to another. Additionally, frequency of occurrence for the plant species was correlated with air mass frequencies. In the analysis, the frequencies of four air mass types employed (arctic air, Pacific air, Alaska–Yukon air, and southern air) were included in the matrix of vegetational values and accorded the same weight as the plant species frequencies. Rather pronounced relationships were revealed between air mass frequencies and the frequencies of a number of species occurring in the communities (Larsen, 1971a,b).

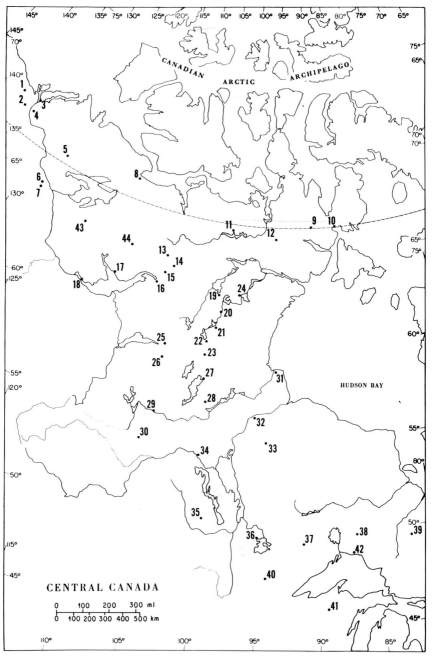

Fig. 27. Broad region of study areas; this network of areas encompasses 44 sites at which sampling of the vegetation was conducted. The areas are as follows: 1, Trout Lake, Northwest Territories; 2, Canoe Lake; 3, Reindeer Station; 4, Inuvik; 5, Colville Lake; 6,

SPECIES BEHAVIOR

It is probably of considerable significance that all species in the black spruce community possessing high positive correlations with arctic air also show similarly high correlations with this air mass type as part of the white spruce community. In the white spruce community, however, a larger group of species is highly correlated with Pacific air. Species in the aspen community show little correlation with either arctic or Pacific air, but a significant group is positively correlated with Alaska–Yukon air and negatively with southern air. There are few species in the jack pine community that are positively correlated with arctic or Alaska–Yukon air, but a group of nine species is correlated positively with southern and Pacific air.

The three tundra communities also possess species groups demonstrating high correlations with air mass types, and in each community the largest group showing this characteristic is the one correlated with the frequency of arctic air masses.

Since the frequencies of the air mass types are inversely correlated with one another, it is to be expected that species positively correlated with one air mass type are negatively correlated with another—or with other—type(s). It can also be noted that, for example, species correlated positively with arctic air are also correlated positively with Alaska–Yukon air and that species correlated with Pacific air are often also correlated with southern air. Thus one is led to the conclusion that the plant species are not responding directly to these specific air mass types but rather to a northern and southern climatic component or, in other words, geographical orientation with respect to air mass source regions. It must be indicated that the climatology of the region is not to be described as simply a combination of the air mass frequencies as presented here, since there are other factors to be considered, among them such influences as the direction of resultant winds, wave structure of frontal zones, and air mass modification during movement from region

---

Florence Lake; 7, Carcajou Lake; 8, Coppermine; 9, Curtis Lake; 10, Repulse Bay; 11, Pelly Lake; 12, Snow Bunting Lake; 13, Aylmer Lake; 14, Clinton-Colden Lake; 15, Artillery Lake; 16, Fort Reliance; 17, Yellowknife; 18, Fort Providence; 19, Dubawnt Lake; 20, Dimma Lake (Kazan River); 21, Ennadai Lake; 22, Kasba Lake; 23, Kasmere Lake; 24, Yathkyed Lake; 25, Black Lake; 26, Wapata Lake; 27, Brochet; 28, Lynn (Zed) Lake; 29, Otter Lake; 30, Waskesiu Lake; 31, Churchill; 32, Ilford; 33, God's Lake; 34, Rocky Lake; 35, Clear Lake; 36, West Hawk Lake; 37, Raven Lake; 38, Klotz Lake; 39, Remi Lake; 40, Bagley (Pike Bay); 41, Trout Lake; 42, Lake Superior; 43, Cartridge Mountains ("East" Keller Lake); 44, Winter Lake.

to region; air mass frequency is but one of the measurable parameters indicative of the prevailing climatic regime in a given area.

Another parameter of interest is the correlation between eigenvector coefficients and distance from the forest border. Since the forest border coincides with a climatic frontal zone (Bryson, 1966), this relationship demonstrates the increasing or decreasing importance of species as this frontal zone is approached from central portions of the air mass regions.

The dominance relationships of the species highly correlated in frequencies with the frequencies of the air mass types are of particular interest. In the black spruce and white spruce communities and in the rock field community especially, the species with the highest correlations with air mass frequencies are consistently among the dominant species in the sampled stands. To illustrate this, the average frequency values in black spruce community stands of species highly correlated with arctic air are given in Table 19, and the dominance relationships of these species in the individual stands sampled are shown in Table 20a and b.

It is apparent that in the black spruce community the species showing the highest relationships to arctic air mass frequencies are those that attain dominance in the community, at least in the more northern areas. Dominant species in the rock field community are those with a high degree of correlation with Pacific air; these are species that attain dominance in the more southern tundra areas. In the tussock muskeg community the relationship becomes less apparent; none of the species showing high correlation with an air mass type are among the first 10 dominants in the community. Dominance and air mass relationships in the low-meadow community are less discernible (Larsen, 1971a,b).

It is apparent that the frequencies of certain species found in each of the seven communities possess striking correlative relationships to the frequencies of air mass types. There are 63 species that possess high correlations of frequency values with the frequency of an air mass type. Some of these 63 species are of importance in only one community, and others are present in two or more communities or in as many as five communities, as in the case of *Vaccinium vitis-idaea*. These latter species are found in both forest and tundra over a wide geographical area.

It is of interest, additionally, that only in the black spruce and white spruce forest communities and in the rock field tundra community do species with high air mass correlations attain a consistently dominant position in the sampled stands. As field ecologists well know, variability in vegetation is often so great as virtually to preclude high correlations between vegetational composition and environmental parameters. It is perhaps of ecological significance that northward there are communities

TABLE 19
Average Frequencies in Stands of Each Study Area of Species Highly Correlated Positively with Arctic Air Mass Frequencies[a] Black Spruce Understory Communities

Area	Betula glandulosa	Empetrum nitrum	Ledum decumbens	Ledum groenlandicum	Salix glauca	Vaccinium uliginosum	Vaccinium vitis-idaea	Average	Number of samples (stands)
Clinton-Colden	90	100	100	60	30	90	100	81	1
Kazan River	85	65	65	95		65	95	67	1
Artillery Lake	53	71	39	73	8	73	98	59	8
Colville Lake	47	32	40	67	18	68	93	52	3
Ennadai Lake	50	49	24	60	5	75	85	50	18
Inuvik	53	35	15	73	3	90	76	49	2
Ft. Reliance	7	41	8	71	19	60	77	41	7
Dubawnt Lake	65	43	25	40	10	40	43	39	2
Florence–Carcajou	40	28	17	32	3	66	37	33	5
Churchill		50		40		40	100	33	1
Yellowknife	2	8	11	83	8	7	96	31	5
Wapata Lake		2		86			96	26	16
Otter Lake	10			81			88	26	5
Black Lake		8		90			71	24	5
Ilford	3	9		80			75	24	9
Lynn Lake	1	1	2	74		1	79	22	16
Ft. Providence	1	25	17	77	3	25		21	5
God's Lake				75			40	16	1
West Hawk Lake				88			45	16	2
Rocky Lake	12			41			33	12	9
Waskesiu Lake				47			37	12	3
Remi Lake				85				12	1
Trout Lake				80				11	3
Raven Lake				62			2	9	5
Klotz Lake				40			3	6	4
Clear Lake				37			7	6	3
Bagley–Pike Bay				43				6	2

[a] The exception is *Ledum groenlandicum* included here to illustrate significance as a widespread and generally abundant species.

TABLE 20A
Dominance Relationships of Species in Black Spruce Communities[a]

Species	Number of areas in which species occurs as a dominant	Number of individual stands according to rank of species			
		1st	2nd	3rd	4th
Ledum groenlandicum	25	44	35	18	8
Vaccinium vitis-idaea	19	63	22	10	
Picea mariana (seedlings)[b]	13	2	4	13	20
Rubus chamaemorus	12	7	8	8	9
Vaccinium oxycoccus (complex)	10	4	4	5	2
Betula glandulosa	11	3	3	6	9
Empetrum nigrum	10	2	8	4	8
Cornus canadensis	9	7	2	5	3
Linnaea borealis	8	2	3	3	6
Vaccinium uliginosum	8	9	8	10	7
Vaccinium myrtilloides	7	1	2	10	3
Geocaulon lividum	7		3	4	6
Chamaedaphne calyculata	7	3	2	3	2
Equisetum sylvaticum	6	4	1	6	5
Equisetum scirpoides	6	2	2	5	1
Equisetum arvense	6	1	1	2	5
Kalmia polifolia	6	1	2	1	5
Maianthemum canadense	5	2	1	2	3

[a] Thirty-nine additional species unlisted are each one of first four dominants in at least one stand.
[b] Reproduction by seedlings or layering not differentiated.

TABLE 20B

Species in Various Communities Showing a High Incidence of Dominance and Also a High Correlation with Air Mass Frequency

Community	Number of stands in which species occurs	Number of stands in which species ranks 1st, 2nd, 3rd, and 4th, in frequency value				Correlation with air mass type
		1st	2nd	3rd	4th	
Black spruce						
Vaccinium vitis-idaea	104	63	22	10		0.65 (arctic)
Empetrum nigrum	56	2	8	4	8	0.69 (arctic)
Vaccinium uliginosum	48	9	8	10	7	0.56 (arctic)
White spruce						
Vaccinium vitis-idaea	23	6	4	3	2	0.65 (arctic)
Empetrum nigrum	18		4	3	3	0.76 (arctic)
Betula glandulosa	12	4	1	1	1	0.67 (arctic)
Ledum groenlandicum	25	1	3	1	1	0.51 (arctic)
Cornus canadensis	23	5	1	3	1	0.48 (Pacific)
Rubus pubescens	15	3	1	2	3	0.67 (southern)
Jack pine						
Vaccinium vitis-idaea	20	11	4	5		0.68 (Alaska-Yukon)
Maianthemum canadense	9	3			3	0.84 (Pacific)
Rock field						
Vaccinium vitis-idaea	65	32	13	6	2	0.86 (Pacific)
Ledum decumbens	45	12	8	9	9	0.68 (Pacific)
Vaccinium uliginosum	50	9	4	9	7	0.58 (Pacific)
Arctostaphylos alpina	44	6	7	8	2	(Pacific)
Empetrum nigrum	48	4	7	7	10	0.83 (Pacific)
Loiseleuria procumbens	32		1	6	2	0.60 (Pacific)

in which at least some species may be found to be correlated with macroclimatic parameters. From the evidence available, it appears that the greatest difficulty in correlating environmental factors with vegetational structure is encountered in the communities of the lower slopes and meadows and, by inference, the shorelines and marshes. Perhaps one could have predicted that this would be the case, since the physical characteristics of abundant water might tend to override other environmental influences. The significance for paleoclimatological work is also apparent.

Perhaps most interesting are species that correlate with opposing air mass types in different communities. Each is correlated with arctic air when it occurs in either the black or white spruce community and with Pacific air (with one exception) when it is a component of the rock field or tussock muskeg community. Because of this regularity in pattern, the explanation is readily apparent. These are ubiquitous species that range widely through forest and tundra. They increase in frequency northward through the forest and then decrease northward beyond the forest border in the tundra communities. In the former case they correlate positively with arctic air, and in the latter positively with Pacific air. The species of this group, including *Empetrum nigrum*, *Ledum decumbens*, *Rubus chamaemorus*, and *Vaccinium uliginosum*, for example, attain their highest frequencies in the communities of the forest–tundra ecotone where they are dominants in a floristically depauperate vegetation (Larsen, 1967, 1971). Whether uniquely adapted or simply widely tolerant of harsh and variable conditions, they provide an interesting group for further study.

REGIONAL ENVIRONMENTAL ANALYSIS

Another study undertaken by the author was based on a regionalization of the Canadian boreal forest and tundra based on climatic, topographic, edaphic, and other physical data, and on a demonstration that species dominant in the vegetation were distributed geographically in a rather striking relationship to the regionalization scheme.

A number of conventional parameters for general measurement of environmental conditions were employed to organize the study areas into environmental regions, thus grouping areas possessing basic similarities. The parameters employed were each distributed as continuities and used to (1) designate areas bounded by sharp gradients of change in several or perhaps many environmental characteristics, (2) designate areas of continuous but gradual transition, or (3) designate an intermediate situation. In the case of the last-mentioned two instances it becomes difficult to distinguish readily regions with definitely delineated boundaries, although places some distance from one another may differ distinctly. Dickinson (1970) has aptly summarized the theoretical aspects of the approach, and more need not be said here concerning its rational basis.

A number of conventional parameters for the general measurement of environmental conditions were employed in a principal component analysis to group the various study areas into environmental description, making possible a rational delineation of areas possessing basic similarities. The conventional environmental parameters employed were as follows:

1. Soil type. A generalized categorization of soils was employed. Soils of the study areas were classed as being of the gray-wooded or other more southern soil types, well-developed podzols, or soils of the tundra.

2. Average elevation of the land surface above sea level. Data obtained from a Canadian topographic survey map series.

3. Maximum relief. Data obtained from a combination of aerial photographic examination and information provided on Canadian topographic survey maps.

4. Total annual solar radiation. Data obtained from published sources.

5. Number of days with minimum temperatures above 32°F. Data from published Canadian meteorological summaries, interpolated for study areas between weather-recording stations.

6. Number of days with mean temperatures above 32°F. Data source same as above.

7. Frequency of arctic air masses. Data obtained from Bryson (1966) and Larsen (1971).

8. Frequency of Pacific air masses. Data source as above.

9. Frequency of Alaska–Yukon air masses. Data source as above.

10. Distance from forest border. Data obtained by straight-line measurement of distance from study site to forest border.

11. The ratio R/Lr. Here R is the total incoming annual solar radiation in langleys, and L is the annual precipitation in centimeters. The value r is the latent heat of water, calculated at 540 cal/ml.

12. Average summer precipitation: average precipitation received during June, July, and August.

The first analysis utilized all 44 study areas, and the 4 study areas with great relief dominated the analysis, with most of the variance (85%) explained in the first factor. The second factor was dominated by elevation above sea level, grouping otherwise widely diverse areas. In the second factor, total variance explained was brought to 97%.

The second analysis employed data from 40 study areas, omitting the 4 areas in the Cordillera. It is considerably more revealing of the environmental relationships throughout the entire region of study.

The grouping of areas according to the principal component analysis is presented in Table 21. Isopleths demonstrating the spatial distribution of the areas grouped together are shown in Fig. 28.

It is apparent that a significant differentiation of regional environmental variation can be readily established, employing a complex of physical environmental parameters in an analysis of this type (Larsen, 1972d)

TABLE 21

Major Study Areas Grouped into Zonal Divisions Designated by Analysis

1. Outlying lowland stands in the northern conifer–hardwood forest
    - Trout Lake (Wisconsin)
    - Bagley (Minnesota)
2. Southern edge of the southern boreal forest
    - Raven Lake
    - West Hawk Lake
    - Clear lake
3. Southern boreal
    - Remi Lake
    - Klotz Lake
    - Rocky Lake
    - Waskesiu
    - Otter Lake
    - Lake Superior
4. Northern boreal
    - God's Lake
    - Ilford
    - Lynn Lake
    - Brochet
    - Black Lake
    - Wapata Lake
    - Kasmere Lake
    - Yellowknife
    - Providence
5. Forest–tundra transition zone
    - Kasba Lake
    - Ennadai Lake
    - Fort Reliance
    - Artillery Lake
    - Colville Lake
    - Inuvik
6. Low-latitude arctic tundra
    - Dubawnt Lake
    - Clinton-Colden Lake
    - Aylmer Lake
    - Reindeer Station
7. Northern coastal tundra
    - Coppermine
8. Higher-latitude continental arctic tundra
    - Repulse Bay
    - Curtis Lake
    - Snow Bunting Lake
    - Pelly Lake
9. Alpine tundra of the east slope of the Cordillera
    - Trout Lake (Babbagè River)
    - Canoe Lake
    - Florence Lake
    - Carcajou Lake

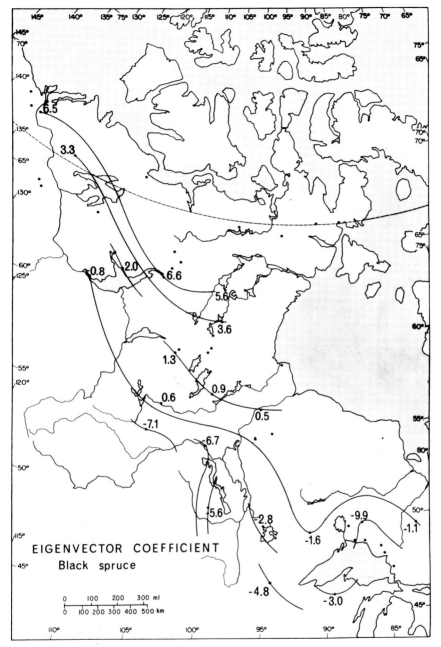

Fig. 28. Objective regionalization of vegetational, climatological, and geological features. The technique groups together those areas most similar in vegetational and physical characteristics. Note the general northwest–southwest trend of the areas falling together into the regions.

TABLE 22

Dominant species in Black Spruce (Forest) Understory Communities[a]

	Region 2			Region 3						Region 4		
	Raven lake	West Hawk	Clear Lake	Remi Lake	Klotz Lake	Rocky Lake	Waskesiu	Otter Lake	God's Lake	Ilford (area)	Lynn Lake	Wapata Lake
Maianthemum canadense	11	20		6	26	21	18	1				
Rubus pubescens		5	50	14	78	27	10	1				1
Mitella nuda		5	5	8	67	14	38		10	2		
Smilacina trifolia	20	20	15	40	15		5	33		3	6	7
Cornus canadensis	26	28	8	38	74	37	32	21	15	6	3	7
Linnaea borealis		10	10	16	51	32	47	2	40	6	2	2
Chamaedaphne calyculata	40	33		18	6	7		40	10	7	18	17
Picea mariana (seedlings)	9	10	5	50	13	5	10	41	15	42	31	36
Ledum groenlandicum	62	88	37	85	40	66	47	84	75	80	71	86
Vaccinium vitis-idaea	2	23	7		3	33	37	88	40	75	81	97
Rubus chamaemorus						8		32	15	14	30	33
Equisetum scirpoides					8	6	30			25		14
Empetrum nigrum										9	5	3
Betula glandulosa												
Arctostaphylos alpina (incl. *rubra*)										8		
Andromeda polifolia												
Vaccinium uliginosum										1		
Ledum decumbens										2		
Dryas integrifolia												
Cassiope tetragona												

[a] The species are arranged by study area and region. Values are average frequencies in stands in each study area. Species listed occur as wideranging dominants and are found in more than six study areas, with the exception of the two additional species in region 9.

using parameters that describe rather gross features of soils, topography, and climate.

This regionalization pattern is also apparent in the composition of the vegetational communities. Tables 22 and 23 indicate that the dominant species in the black spruce and the white spruce communities change in a striking manner along the climatic gradient from south to north. Species dominant in the communities of the more southern areas are

Regional Environmental Analysis

								Region 5							Region 6			Region 9
Black Lake	Yellowknife	Ft. Providence	Keller Lake	Kasmere Lake	Kasba Lake	Winter Lake	Ennadai Lake	Ft. Reliance	Pike's Portage	Artillery Lake	Colville Lake	Inuvik	Kazan River	Dubawnt Lake	Clinton-Colden	Florence-Carcajou
1																
		4														
8	8	8	2													
21		1	5													
5		22														
22	7	1	2	70		25	3									
31	44	31	50	30	40	38	21	20		24	45	20	53	35	25	
90	83	77	80	100	95	100	64	71		67	83	67	78	95	40	60	32
90	96	91	80	100	91	100	85	77		97	98	93	78	95	43	100	66
36	24	13	33	100	44	73	39	9		17	67	75	3	70	75	10	
4		45						1		37		12	93		3	
8		25	18		58	20	48	41		67	78	32	35	65	43	100	28
	2	1	26		20	26	49	7		52	53	47	48	85	65	90	49
		8	27		50		3	55		18	3	22	68				61
3		13	37		2	37		35		2		12	50				33
		25	45		65	45	75	60		78	63	68	90	65	10	90	66
	11	17	75	5	1	80	24	8		37	42	38	15	65	25	100	17
								6				20					76
																	44

absent from the communities of northern areas. There is a gradual transition in community structure from south to north.

The predictive capability of these techniques to discern and quantify the climate–vegetation interrelationships in the regions under consideration has yet to be tested in detail. Prior to one field season in the latter portion of the study, however, a prediction was made concerning the frequencies of dominant species that, it was anticipated, would be found

TABLE 23
Dominant Species in White Spruce Forest Understory Communities by Study Area and Region[a]

Species	Region 2	Region 3					Region 4			Region 5		
	Clear Lake	Remi Lake	Klotz Lake	Waskesiu	Otter Lake	Black Lake	Yellowknife	Ft. Providence	Ennadai	Ft. Reliance	Artillery Lake	Inuvik
Aralia nudicaulis	35	38	60									
Maianthenum canadense	45	34	20	20								
Mertensia paniculata	43	8	10		10							
Petasites palmatus	48	1	10	40		10						
Abies balsamea (seedlings)		53	20		70							
Arctostaphylos uva-ursi	35	38	60		35							
Clintonia borealis		43	40									
Mitella nuda	23	83	80					7				

Species	S1	S2	S3	S4	S5	S6	S7	S8	S9	S10	S11	S12
Viburnum edule	15						3	9				1
Rubus pubescens	15	89		4				3				6
Cornus canadensis	38	50	40	95	5			18				2
Rosa acicularis	43	3					50	26		13		
Linnaea borealis	5	38		85		65	30	46	13	7	1	
Epilobium angustifolium	30			55		60	35	9	10	8	3	3
Pyrola secunda	8		10	55		15	5	8	3			
Shepherdia canadensis	3					5	45	21		18	4	4
Picea mariana (seedlings)		8			15	5	25	35		2	3	
Vaccinium vitis-idaea				10		80	20	71	68	65	84	58
Geocaulon lividum						20	45	50		20	3	17
Betula glandulosa					5		10	4	67	3	55	21
Ledum groenlandicum				10			5	26	58	9	44	57
Rubus acaulis						5	5	12	37			
Empetrum nigrum							10	7	63	32	58	39
Arctostaphylos alpina							25	11		19	40	89
Vaccinium uliginosum								1	43	13	68	42
Dryas integrifolia										20	16	40

[a] Values are average frequencies in stands in each study area. Species listed occur in two or more study areas with a value of 35 or more in at least one.

in an area hither-to never sampled, the Kasba Lake area south of Lake Ennadai. Listed below are the predicted values and the actual values found in the stands sampled:

Species	Predicted	Actual
Betula glandulosa	34	20
Empetrum nigrum	38	42
Ledum decumbens	18	1
Ledum groenlandicum	70–80	95
Salix glauca	5	0
Vaccinium uliginosum	60	65
Vaccinium vitis-idaea	85	100

It can be seen that reasonably accurate predictions can be made in regard to the frequencies of species showing both high dominance and a high order of correlation with air mass frequencies. While the sample involved here was small, it nevertheless indicates that some encouragement might be accorded efforts to develop refined techniques for predicting climate from vegetational characteristics in boreal areas, or, conversely, for predicting the vegetational characteristics from climatic conditions known to prevail.

FUTURE POSSIBILITIES

The northern boreal forest and adjacent tundra even today remain virtually a *terra incognita* in terms of plant ecology but, although few studies have been conducted in the past, it can be anticipated that an increasing number will begin to appear. To provide an introductory groundwork for those who will undertake northern plant ecological studies, there are now a number of studies in which there is fairly extensive descriptive treatment of the vegetation of the northern portions of the forest and the forest–tundra ecotone. In the southern boreal forest zone, where forestry has a somewhat greater economic potential, studies have been appearing in print at an increasingly rapid rate. Because of the limited amount of time that most workers are willing to spend in the more northern or more remote southern areas, however, most studies tend to be in the style of a reconnaissance. Most have not been conducted by means of quantitative techniques, without which it is impossible to make meaningful comparisons between areas or between different stages of development along successional series. Suffice it to say that, while the field is opening, there remains a great deal to be done before we can say that we have begun to understand not only the ecological mechanisms at work in the north but even the kinds of plant communities that exist in these regions.

At the present time it is not possible to relate the physiological tolerances and characteristics of the species making up the flora directly to their performance in terms of frequency, density, and dominance. This is due largely to the almost total absence of information on the detailed physiological characteristics and tolerances of the species themselves. It is also due, of course, to the fact that even performance and characteristic behavior in an ecological sense are also largely—or totally—unknown. In the case of tree species, on the other hand, sufficient research into autecology has been conducted, principally because of economic value, to provide at least the outlines of ecological relationships. The same is true in the case of a few of the shrubs associated with the dominant trees.

It is now, of course, of the greatest ecological interest to attempt to discern what these environmental factors are that are so clearly distinguishable by the plants themselves but not by our conventional instrumentation or by the scattered meteorological data available for the region under consideration. That we do not understand these characteristics in no way reduces their significance. It is, in fact, ultimately to permit us to describe them and to understand the entire web of relationships among species that we pursue the course that we do. With this ultimate goal in mind, it seems reasonable that we should now attempt to obtain some rather basic information on the ecological performance of at least the more common species. Perhaps the greatest contribution of the study reported on these pages will have been to identify those species that would lend themselves most productively to future research along these lines.

SUMMARY

A rather large variety of material has been brought together here in support of the theme that, essentially, the vegetational communities of the boreal forest and tundra, in the area under consideration, possess variations in structure from place to place that can be shown to be correlated with regional differences in climate. The assumptions underlying this assertion, all of which require detailed consideration, are that the species of plants making up the vegetation have, by physiological and morphological adaptation, become adjusted to the specific local environment in which they are now found; that there are no widespread migrations along a broad front currently underway; that the vegetation in this region has not been markedly disturbed by human technological activities, hence represents essentially the native condition; and that the factors of geological substrate and soil formation are sufficiently similar over large areas so that we are relieved of the burden of taking soils into account as major determinants of the kind of vegetation found in each area and of differences found from place to place.

We are thus accomplishing three tasks: (1) describing the vegetation of areas for which few or no data have previously been available, (2) furnishing evidence as to the general validity of the continuum theory of vegetation on a broad regional basis, and (3) demonstrating that in this large area of northern central North America vegetational differences from one place to another are related to climatic differences, a relationship long suspected but regarding which, as Raup pointed out long ago (1941) there has been "no well defined bases for generalization."

In a discussion of North American forest flora and vegetation, Maycock (1963a) points out*:

> The sampling of a large number of forest stands throughout an extensive region enhances the value of the study since it will not only provide information relating to community organization and species relationships within the region but will produce a statistical summary that can be related to other extensive areas within the formation. It is only by exhaustive studies of this type that the problems of formation structure and variation will be eventually solved, and until information of this sort is available the ecological relationships of individual species within the formation and the factors governing their distributions will not be fully comprehended.
>
> Unfortunately studies of this type on a broad scale are very few in number. Many local studies are available but these include data for very few stands and they are limited in scope. Earlier workers who had been impressed by Clementian climax philosophy did not perceive the importance of sampling large numbers of stands but assumed that intensive studies of a few more mature stands suffice for a picture of the whole. A lack of recognition of the individualistic ideas of Gleason (1926) resulted in the paucity of quantitative information that is available for vast areas of the deciduous forest formation. Economics and a lack of suitable methods have also exerted an influence in this regard.

For all boreal regions there exists a great need for accurate and definitive ecological and phytosociological studies of the vegetation. For purposes of wise management, as population pressure and economic exploitation increase, there is now an overwhelming requirement that the studies be conducted as quickly as possible. It has become almost axiomatic in studies of regional plant ecology that results of broad significance almost inevitably emerge from so simple a technique as following changes in species populations and community characteristics along an environmental gradient (Whittaker, 1967, 1973).

*Reproduced by permission of the National Research Council of Canada.

REFERENCES

Anderson, A. J. B. (1971). Ordination methods in ecology. *J. Ecol.* **59**, 713–726.

Argus, G. W. (1964). Plant collections from Carswell Lake and Beartooth Island, northwestern Saskatchewan, Canada. *Can. Field Nat.* **78**, 139–149.

Argus, G. W. (1966). Botanical investigations in northwestern Saskatchewan: The subarctic Patterson–Hasbala lakes region. *Can. Field Nat.* **80**, 119–143.

"Atlas of Canada." (1957). Dept. Mines Tech. Surv. Geogr. Br. Ottawa, Canada.

Back, G. (1836). "Narrative of the Arctic Land Expedition to the Mouth of the Great Fish River and along the Shores of the Arctic Ocean, in the Years 1833, 1834, and 1835." London.

Baker, F. S. (1950). "Principles of Silviculture." McGraw-Hill, New York.

Bakuzis, E. V., and Hansen, H. L. (1965). "Balsam Fir." Univ. Minn. Press, Minneapolis.

Baldwin, W. K. W. (1958). "Plants of the Clay Belt of Northern Ontario and Quebec." *Nat. Mus. Can. Bull.* **156**, 1–324.

Beals, E. W. (1973). Ordination: Mathematical elegance and ecological naivete. *J. Ecol.* **61**, 23–35.

Bostock, H. S. (1964). "A Provisional Physiographic Map of Canada." Pap. 64-35, Geol. Surv. Can. Ottawa.

Bryson, R. A. (1966). Airmasses, streamlines, and the boreal forest. *Geogr. Bull. (Canada),* **8**, 228–269.

Carleton, T. J., and Maycock, P. F. (1978). Dynamics of boreal forest south of James Bay. *Can. J. Bot.* **56**, 1157–1173.

Christensen, E. M., Clausen, J. J., and Curtis, J. T. (1959). Phytosociology of the lowland forests of northern Wisconsin. *Am. Midl. Nat.* **62**, 232–247.

Clausen, J. J. (1957). A phytosociological ordination of the conifer swamps of Wisconsin. *Ecology* **38**, 638–646.

Cody, W. J., and Chillcott, J. G. (1955). Plant collections from Matthews and Muskox Lakes, Mackenzie District, N.W.T. *Can. Field Nat.* **69**, 153–162.

Cottam, Grant, and Curtis, J. T. (1956). The use of distance measures in phytosociological sampling. *Ecology* **37**, 451–460.

Craig, B. G. (1960). "Surficial Geology of North-Central District of Mackenzie, N.W.T." *Geol. Surv. Can. Pap.* **60–18**, 1–8.

Craig, B. G. (1961). "Surficial Geology of Northern District of Keewatin, N.W.T. *Geol. Surv. Can. Pap.* **61–5**, 1–8.

Craig, B. G. (1964). "Surficial Geology of East-Central District of Mackenzie." *Geol. Surv. Can. Bull.* **99**, 1–41.

Curtis, J. T. (1959). "The Vegetation of Wisconsin." Univ. of Wisconsin Press, Madison.

Damman, A. W. H. (1964). Some forest types of central Newfoundland and their relation to environmental factors. Can. Dep. For., For. Sci. Monogr. 8, pp. 1–62.

Damman, A. W. H. (1965). The distribution patterns of northern and southern elements on the flora of Newfoundland. *Rhodora* **67**, 363–392.

Dickinson, R. E. (1970). "Regional Ecology." Wiley, New York.

Gauch, H. G., Jr. (1973). A quantitative evaluation of the Bray–Curtis ordination. *Ecology* **54**, 829–836.

Gauch, H. G., Jr., and Whittaker, R. H. (1972). Comparison of ordination techniques. *Ecology* **53**, 858–875.

Gauch, H. G., Jr., Whittaker, R. H., and Wentworth, T. R. (1977). A comparative study of reciprocal averaging and other ordination techniques. *J. Ecol.* **65**, 157–174.

Gleason, H. A. (1926). The individualistic concept of the plant association. *Bull. Torrey Bot. Club* **53,** 7–26.

Graham, S. A. (1954). Scoring tolerance of forest trees. *Mich. For.* **4,** 2.

Greig-Smith, P. (1964). "Quantitative Plant Ecology." Butterworths, London.

Grigal, D. F., and Goldstein, R. A. (1971). An integrated ordination–classification analysis of an intensively sampled oak-hickory forest. *J. Ecol.* **59,** 481–492.

Hare, F. K. (1950). Climate and zonal divisions of the boreal forest formations in eastern Canada. *Geogr. Rev.* **40,** 615–635.

Hare, F. K. (1959). A photo-reconnaissance survey of Labrador–Ungava. Can. Dep. Mines Tech. Surv. Geogr. Br. Mem. 6, pp. 1–83.

Hare, F. K., and Taylor, R. G. (1956). The position of certain forest boundaries in southern Labrador-Ungava. *Geogr. Bull.* **8,** 51–73.

Heinselman, M. L. (1970). Landscape evolution, peatland types, and environment in the Lake Agassiz Peatlands Natural Area, Minnesota. *Ecol. Monogr.* **40,** 235–261.

Hopkins, D. M. (1959). Some characteristics of the climate in forest and tundra regions in Alaska. *Arctic* **12,** 215–220.

Jeglum, J. K., Wehrmann, C. F., and Swan, J. M. A. (1971). Comparisons of environmental ordinations with principal component vegetational ordinations for sets of data having different degrees of complexity. *Can. J. For.* **1,** 99–112.

Johnson, A. W., Viereck, L. A., Johnson, R. E., and Melchoir, H. (1966). Vegetation and flora. *In* "Environment of the Cape Thompson Region, Alaska." (N. J. Wilimovsky and J. N. Wolfe, eds.), pp. 277–354. U.S. Atomic Energy Commission, Oak Ridge, Tennessee.

Johnson, P. L., and Vogel, T. C. (1966). "Vegetation of the Yukon Flats Region, Alaska." U.S. Army Mat. Command, Cold Reg. Res. Eng. Lab., Res. Rep. 209. Hanover, New Hampshire.

Kendeigh, S. C. (1954). History and evaluation of various concepts of plant and animal communities in North America. *Ecology* **35,** 152–171.

Krebs, J. S., and Barry, R. G. (1970). The arctic front and the tundra-taiga boundary in Eurasia. *Geogr. Rev.* **60,** 548–554.

La Roi, G. H. (1967). Ecological studies in the boreal spruce–fir forests of the North American taiga. I. Analysis of the vascular flora. *Ecol. Monogr.* **37,** 229–253.

Larsen, J. A. (1965). The vegetation of the Ennadai Lake area, N.W.T.: Studies in subarctic and arctic bioclimatology. I. *Ecol. Monogr.* **35,** 37–59.

Larsen, J. A. (1971a). Vegetation of the Fort Reliance and Artillery Lake areas, N.W.T. *Can. Field Nat.* **85,** 147–167.

Larsen, J. A. (1971b). Vegetation and airmasses: Boreal forest and tundra. *Arctic* **24,** 177–194.

Larsen, J. A. (1972a). The vegetation of northern Keewatin. *Can. Field Nat.* **86,** 45–72.

Larsen, J. A. (1972b). Growth of spruce at Dubawnt Lake, Northwest Territories. *Arctic* **25,** 59.

Larsen, J. A. (1972c). Observations of well-developed podzols on tundra and of patterned ground within forested boreal regions. *Arctic* **25,** 153–154.

Larsen, J. A. (1972d). "Vegetation and Terrain (Environment): Canadian Boreal Forest and Tundra." Final Rep. UW-G1128, Center Clim. Res. Univ. of Wisconsin, Madison.

Larsen, J. A. (1973). Plant communities north of the forest border, Keewatin, Northwest Territories. *Can. Field Nat.* **87,** 241–248.

Larsen, J. A. (1974). Ecology of the northern continental forest border. *In* "Arctic and Alpine Environments" (J. D. Ives and R. G. Barry, eds.), pp. 341–369. Methuen, London.

Lee, H. A. (1959). Surficial Geology: District of Keewatin and Keewatin Ice Divide, Northwest Territories. *Geol. Surv. Can. Bull.* **51**, 1–42.
Linteau, A. (1955). Forest site classification of the northeastern coniferous section, boreal forest region, Quebec. Can. Dep. North. Affairs Nat. Res., For. Br. Bull. 118, pp. 1–85.
Lord. C. S. (1953a). "Geological Notes on Southern District of Keewatin, Northwest Territories." *Geol. Surv. Can. Pap.* **53–22**.
McFadden, J. D. (1965). "The Interrelationship of Lake Ice and Climate in Central Canada." Task NR 387-022 ONR Contr. No. 1202(07), Tech. Rep. 20, Dep. Meteorol., Univ. of Wisconsin, Madison.
McIntosh, R. P. (1967). The continuum concept of vegetation. *Bot. Rev.* **33**, 130–187.
Mackay, J. R. (1963). "The Mackenzie Delta Area, N.W.T., Canada." Can. Dep. Mines Tech. Surv., Geogr. Br. Mem. 8.
Maclean, S. W. (1960). "Some Aspects of the Aspen–Birch–Spruce–Fir Type in Ontario." Can. Dep. For., For. Res. Div., Tech. Note 94.
Maycock, P. F. (1957). "The phytosociology of Boreal Conifer–Hardwood Forests of the Great Lakes Region." Ph. D. Thesis, University of Wisconsin, Madison.
Maycock, P. F. (1963a). The phytosociology of the deciduous forests of extreme southern Ontario. *Can. J. Bot.* **41**, 379–438.
Maycock, P. F. (1963b). Plant records of the Ungava peninsula, new to Quebec. *Can. J. Bot.* **41**, 1277–1279.
Maycock, P. F., and Curtis, J. T. (1960). The phytosociology of boreal conifer–hardwood forests of the Great Lakes region. *Ecol. Monogr.* **30**, 1–35.
Maycock, P. F., and Matthews, B. (1966). An arctic forest in the tundra of northern Quebec. *Arctic* **19**, 114–144.
Mitchell, V. L. (1973). A theoretical tree-line in central Canada. *Ann. Assoc. Am. Geogr.* **63**, 296–301.
Nordenskjöld, O., and Mecking, L. (1928). The geography of the polar regions. *Am. Geogr. Soc. Spec. Publ.* **8**, 1–359.
Orloci, L. (1968). Information analysis in phytosociology: Partition, classification and prediction. *J. Theor. Biol.* **20**, 271–284.
Pielou, E. C. (1977). "Mathematical Ecology." Wiley, New York.
Porsild, A. E. (1943). Materials for a flora of the continental Northwest Territories of Canada. *Sargentia* **4**, 1–79.
Porsild, A. E. (1945). "The Alpine Flora of the East Slope of Mackenzie Mountains, N.W.T." *Natl. Mus. Can. Bull.* **101**, 1–35.
Porsild, A. E. (1950). Vascular plants of Nueltin Lake, Northwest Territories. *Natl. Mus. Can. Bull.* **118**, 72–83.
Porsild, A. E. (1951). Botany of southeastern Yukon adjacent to Canol Road. *Natl. Mus. Can. Bull.* **121**, 1–400.
Porsild, A. E. (1955). The vascular plants of the western Canadian arctic archipelago. *Natl. Mus. Can. Bull.* **135**, 1–226.
Porsild, A. E. (1957). Illustrated flora of the Canadian arctic archipelago. *Natl. Mus. Can. Bull.* **146**, 1–209.
Porsild, A. E. (1966). Contributions to the flora of southwestern Yukon Territory. *Natl. Mus. Can. Contrib. Bot. IV, Bull.* **216**, 1–86.
Ramirez Diaz, L., Garcia Novo, F., and Merino, J. (1976). On the ecological interpretation of principal components in factor analysis. *Oecol. Plant.* **11**, 137–141.
Raup, H. M. (1941). Botanical problems in boreal america. *Bot. Rev.* **7**, 148–248.

Raup, H. M. (1946). Phytogeographic studies in the Athabaska–Great Slave lake region, II. *J. Arnold Arbor. Harv. Univ.* **27,** 1–86.
Raup, H. M. (1947). The botany of southwestern Mackenzie. *Sargentia* **6,** 1–275.
Ritchie, J. C. (1956). The vegetation of northern Manitoba. I. Studies in the southern spruce forest zone. *Can. J. Bot.* **34,** 523–561.
Ritchie, J. C. (1958). The vegetation of northern Manitoba. III. Studies in the subarctic. Tech. Pap. Arct. Inst. 3, pp. 1–56.
Ritchie, M. C. (1960a). The vegetation of northern Manitoba, IV. The Caribou Lake region. *Can. J. Bot.* **38,** 185–197.
Ritchie, J. C. (1960b). The vegetation of northern Manitoba. V. Establishing the major zonation. *Arctic* **13,** 211–229.
Ritchie, J. C. (1960c). The vegetation of northern Manitoba. VI. The lower Hayes River region. *Can. J. Bot.* **38,** 769–788.
Ritchie, J. C. (1962). A geobotanical survey of northern Manitoba. Arct. Inst. North Am. Tech. Pap. 9.
Rowe, J. S. (1956). Uses of undergrowth plant species in forestry. *Ecology* **37,** 462–473.
Savile, D. B. O., (1956). Known dispersal rates and migratory potentials as clues to the origin of North American biota. *Am. Midl. Nat.* **56,** 434–453.
Savile, D. B. O. (1963). Factors limiting the advance of spruce at Great Whale River, Quebec. *Can. Field Nat.* **77,** 95–97.
Schimper, A. F. W. (1903). "Plant Geography upon a Physiological Basis." Clarendon, Oxford.
von Humboldt, A. (1807). "Ideen zu einer Geographie der Pflanzen nebst einem, Naturgemalde der Tropenlander." Tübingen.
Warming, E. (1909). "Oecology of Plants." Oxford Univ. Press. London and New York. (Translated by P. Groom and I. B. Balfour from *Plantesamfund,* Danish ed., 1895; German editions, 1896, 1902).
Whittaker, R. H. (1954). Plant populations and the basis of plant indication. *Angew. Pflanzensociol. Festschr. Aichinger,* **1,** 183–206.
Whittaker, R. H. (1967). Gradient analysis of vegetation. *Biol. Rev.* **42,** 207–264.
Whittaker, R. H. (1973). Direct gradient analysis. *In* "Ordination and Classification of Communities" (R. H. Whittaker, ed.), pp. 7–74. D. W. Junk, The Hague.
Wilde, S. A. (1946). "Forest Soils and Forest Growth." Chronica Botanica, Waltham, Massachusetts.
Wilde, S. A. (1958). "Forest Soils." Ronald, New York.
Wilton, W. C. (1964). The forests of Labrador. Can. Dep. For., For. Res. Br., Contrib. 610, pp. 1–72.
Wright, G. M. (1952). Second preliminary map, Fort Reliance, N.W.T. (map & descriptive notes). *Geol. Surv. Can. Pap.* **51–26.**
Wright, G. M. (1955). Geological notes on central district of Keewatin, N.W.T. *Geol. Surv. Can. Pap.* **55–11.**
Wright, G. M. (1957). Geological notes on eastern district of Mackenzie, N.W.T. *Geol. Surv. Can., Pap.* **56–10.**

7 Boreal Communities and Ecosystems: Local Variation

Simply stated, a plant community is an aggregation of individuals, representing one or more species, growing together. What is meant by "together" is open to interpretation—together on a cliff face, at a pond edge, in field, forest, prairie, or steppe, muskeg or meadow, or on any conceivable site where plants are found. A community may be called transitional, intermediate between any of the communities considered common or widespread in a region; it is meaningful to speak of transitional or ecotonal communities in terms such as forest–tundra and treed muskeg. Many terms, including aspen–parkland, oak opening, and lichen woodland are employed to identify ecotonal communities, i.e., communities transitional between aspen forest and prairie, oak forest and prairie, grassland and forest, and coniferous forest and tundra.

It has, indeed, been demonstrated that virtually all communities are transitional—that is, intermediate in species composition between communities found to the north or south, east or west, at higher or lower elevations, on different soils or topography, or under differing microclimatic conditions. Communities are also transitional in time, each community on any given plot of ground being linked along an ecotone with what came before and what will come after. Each individual plant association, in any given area, has a unique history of development, for no two plots of ground are ever quite the same, nor will chance events of seed dispersal, germination, and a multitude of other circumstances affecting the vegetation ever be identical. Thus, the previous chapter was devoted to discussion of the variation in plant community composition (the species) and structure (relative proportions of each species) over distance as measured in tens or hundreds of miles. In this chapter, the view of the landscape will be more myopic—concerned not with differences in communities on comparable sites over great distances, but with differences that occur between communities on different sites located close together and perhaps adjacent to one another.

One of the more instructive controversies in the history of ecology has developed over the interpretation of what constitutes the essential nature of local variation in plant communities—controversy over whether variation in communities found growing close together is continuous or discontinuous, whether, in short, the continuum is a valid concept for characterizing local plant community variation or whether, on the other hand, there are consistently recognizable discrete community types. The continuum theory of plant community structure has aroused a surprising amount of spirited discourse; the concluding argument in the debate has probably not yet been presented, but a few major contenders on both sides have delivered forceful arguments (Curtis, 1959; Whittaker, 1962, 1967, 1973; Daubenmire, 1966; Vogl, 1966; Cottam and McIntosh, 1966; McIntosh, 1967, 1968; Vasilevich, 1968; Cottam et al., 1973).

Traditionally, botanists classified vegetation according to types on the basis of the observation that closely similar plant assemblages recur where closely equivalent environments can be recognized. These were then grouped into abstract community classes, each with what was held to be consistent distinguishing characteristics. This concept of vegetation has led not to a history of rapid progress but instead to a variety of concepts, terms, and methods confusing even to specialists. Important to this view is the rarity of representative stands of typal communities in sufficiently undisturbed condition to afford botanists with data on what the vegetation type was like before disruption by human development. Disruption is most extensive in Europe, where the concept of community types was originally developed, but it is held that in all parts of the world the distinctiveness of types has been greatly weakened under man's pervasive influence. It is the traditionalist's belief, moreover, that the new methods of continuum analysis are biased—that they lead inevitably to the accumulation of data supporting the continuum hypothesis when actually continuua do not exist in nature. Traditionalists maintain that significant discontinuua exist in North American vegetational communities and that the discontinuities can be revealed whenever vegetation is studied following the European methods. It is the traditionalist view that with any of the several methods used in continuum analysis it is possible to demonstrate a continuum anywhere (Daubenmire, 1966).

In the traditionalist view there is, indeed, a continuum on the broad regional and global scales—the consequences of floristic change over distance. On a local scale, however, it is held that typal communities should be recognized as real and not imaginary. It is the flora that presents a continuum—the species in one area giving way imperceptibly to other species through contiguous zones. The argument asserts that typal communities exhibit minor compositional and structural gradients, separated abruptly by sharp ecotonal communities with steep gradients. Communities representing the types are sufficiently abundant to warrant being designated classes. In this view, human disturbance of vegetation has so disrupted these natural community types that it is now difficult or impossible to describe them; the result, too, is a mosaic of disturbed plant associations that yield data appearing to support the continuum concept.

Those who advocate the newer methods of plant community analysis regard this traditional approach as an oversimplification of reality. In their view, it is unlikely that the traditional methods are capable of discerning the real nature of what are exceedingly variable vegetational communities; traditional methods are incapable of handling the complexities of the interactions between vegetation and the environment; the traditional approach, moreover, cannot designate typal plant communities since "the consistent, distinguishing characters needed to recognize an association type vanish upon close examination" and "it is most unlikely that the environmental gradients operating concurrently... will be amenable to the subjective methods which have characterized the traditional approach" (Cottam and McIntosh, 1966).

There is not space here for a detailed marshaling of the evidence related to the controversy; it should be stated, however, that much effort has now been devoted to the development of quantitative methods of community and environmental analysis and description, and that the methods have been productive in elucidating relationships between plants and the environmental variables that influence community composition and population dynamics. More effective methods continue to evolve as the work progresses. One of the more cogent summaries of the value of continuum analysis in studies of environment and plant associations is that of Vasilevich (1968):

> The basic advantage of the continuum concept is the possibility of studying vegetation through a wide and comprehensive approach. Here the existence of discrete community-types is not postulated *a priori*, but if they really exist they can be defined in the course of analysis. On the other hand, existing discontinuities cannot be discovered if one classifies the vegetation following the community-types concept.... A continuous variation does not exclude the classification of vegetation.... If the degree of vegetation continuity is very high, the boundaries between natural types of plant communities will necessarily be very faint. In this situation we must decide whether classification will be of practical value. The basic difference in the approaches to vegetation study lies in the degree to which a classification is held to demonstrate entities which are consistently represented in the vegetation.... The

ecological individuality of a species is not an obstacle to the existence of discrete types of plant communities if there are factors connecting the species into groups.... The ecological individuality of species necessitates a continuum.... The adherents of a traditional community-types concept are apt to have "*a priori* notions of a type," McIntosh writes. This is, I believe, a very important point.... In particular it becomes clear why the adherents of such an approach to vegetation proved to be unable to elaborate satisfactory objective methods for vegetation classification.... Now the main effort of ecologists should be directed to the examination of how often one meets with a vegetational continuum in nature and with an estimation of degree of continuity of concrete samples....

The view adopted in the discussion of boreal communities in the following pages is that the concept of individual species response to the environment most closely approximates the conditions found to exist, that, in short, the continuum hypothesis is the one that, so far at least, most accurately represents the data taken in undisturbed boreal plant communities.

This latter point perhaps also requires some amplification; it has been stated repeatedly in recent ecological literature that no undisturbed plant communities exist today anywhere on the globe. This perhaps is true, but probably nowhere over the globe are plant communities as undisturbed as those in northern Canada. It is the writer's experience that a mile or two beyond the end of the road, the railroad, or the mining or fishing camp, the influence of civilization dwindles notably. At a distance of 100 or 500 mi from places of dense human habitation—here one does, indeed, find conditions essentially unchanged from prehistoric times. Those who claim this is not the case have not availed themselves of the somewhat inconvenient, uncomfortable, and expensive opportunities to obtain data in areas miles from human habitation. Many areas are now, in fact, less frequently visited by travelers of any sort than was the case in the days when fur and gold were a lure that drew adventurers into regions now rarely visited. In the past, many forest-tundra borderlands in northern Canada also were regularly occupied by hunting Eskimos or northern Indians; these areas today go for long periods of time without human visit of any kind—if one excepts aircraft that pass overhead at altitudes of a mile or more. In these areas there exist today unparalleled opportunities for research under pristine environmental conditions. But these areas are only for individuals whose sense of adventure and scientific accomplishment overcomes the discomforts involved—the isolation, the onslaught of insect pests, and the expense and difficulty of travel by air or canoe. Yet it is in some of these areas that the most valuable ecological studies can today be conducted, and the value of the knowledge to be gained in such studies far outweighs the temporary physical discomfort involved. It can, indeed, be said that the discomforts are today in no way as debilitating as in times past—good insect repellents and preserved foods are now available. Moreover, the dangers of travel were greater not many decades ago; the author cannot resist recording here the existence of the remains of a smashed wooden canoe of ancient vintage at the foot of a rapids along the Kazan River in Keewatin, the remnant of an exploration, who knows how many years ago, that ended abruptly far beyond the edge of civilization. If such places today bear the imprint of civilization, the imprint is ephemeral.

ENVIRONMENTAL GRADIENTS

It is the purpose of the remainder of this chapter to describe some of the events and circumstances of the boreal environment that influence the vegetational communities therein—summarized as factors related to moisture supplies, nutrients, and effects of permafrost and fire.

Boreal vegetation embraces a wide variety of communities, each representing the sum total of adaptations of the individual species to one another and to local environmental influences. There are, thus, two complex and, one might say, collective variables involved, the plant community on the one hand and the environment on the other, both multifactorial in character. The community is multifactorial in the sense that each species is a variable, affected by other species and by environmental factors; the environment is multifactorial in the sense that a number of conventional climatic and edaphic factors are involved, such as pH, nutrients, temperature, and moisture, each having an influence upon individual plants and collectively on the entire plant community.

There are, thus, two routes of entry into the ecology of a plant community: The first is through the composition, structure, and dynamics of the plant community itself, and the second is through the environment as expressed by various physical parameters. The two, ideally, should be combined so as to reveal the influence of each upon the other. In practice, the work usually is somewhat simplified by emphasizing one or the other initially and then inferring causal relationships between the two. The difference between the two approaches can be understood most clearly in pragmatic terms: The study of plant community response to known environmental gradients is called direct gradient analysis; indirect gradient analysis is the study of plant community variation from place to place, from which environmental gradients are inferred. In the simplest application of gradient analysis, plant community sampling is carried out at equal intervals along a transect that traverses a known environmental gradient, from wet to dry soil conditions, for example, or up a mountain slope. In this kind of analysis, the existence of the environmental gradient is assumed, preferably demonstrated, and documented with data on one or more physical parameters describing the gradient. The clearest results are obtained when several samples represent each interval and data are combined to average out sample-to-sample variation. With the direct method, measurements of the environment are used for arranging samples along a gradient, and the vegetational response to the gradient is expressed in the composition of the samples taken in the plant community. It is, thus, possible to observe the environmental change and the plant community response to it, even though many relationships between plants and environment revealed in the direct gradient analysis will be expressed in correlations for which causation is unknown (Whittaker, 1967).

Indirect gradient analysis employs variations in plant communities to reveal environmental gradients; the basic assumption is that relationships between species populations and environmental factors result in a coincidence between variations in environmental factors and variations in community composition and structure, and that the latter can be employed to reveal the former. In indirect gradient analysis, data usually are subjected to the statistical treatment known as ordination, in which the degree of similarity between the sampled communities is determined and the data displayed as an array; ordinations are based on similarity coefficients or, in the case of principal components analysis, on a complex statistical treatment of a correlation matrix. The purpose, in any case, is to reveal the existence of environmental gradients from the variation in plant community composition and structure; it remains for the analyst to infer the nature of the environmental gradients involved from whatever circumstantial evidence may be at hand. Measurements of environmental parameters can then be undertaken to quantify the relationships between the plant communities and environmental parameters.

As Whittaker (1967) points out, within each of the approaches a variety of statistical techniques can be used, and the choice of method depends on the decision of the ecologist, who should experiment with more than one technique to determine which is most suitable to the circumstances. Direct gradient analysis has the advantage of clarity and interpretability. In many situations, however, environmental gradients are not known, and it is the purpose of analysis to reveal them. Indirect gradient analysis employing comparative ordination can be applied in these situations in which direct gradient analysis is not appropriate. One or more hidden ecoclines may be discerned as a result of the variation revealed in the plant communities. Two or more gradients define an environmental hyperspace, representing in abstract form the complex of environmental variation. Groups of species close together in the hyperspace are related to one another by similarity of physiological tolerance limits to environmental factors, and gradient analysis provides the means of revealing their relationships to the ecoclines. Whittaker adds that the relationships should be expressed quantitatively, but it is evident that the complexity of the relationships studied requires methods that differ somewhat from those employed in most of the physical sciences. Gradient analysis, despite its limitations, has given biologists a better understanding of vegetational communities, as well as enhanced insights into the interconnections between plants and the environment.

ENVIRONMENTAL GRADIENTS: MOISTURE

Variations in a boreal community along a gradient of soil moisture from dry to wet can be arranged by vegetational strata (i.e., the level at which foliage occurs—trees are highest, shrubs intermediate, herbs lowest) and by the soil moisture regime in which each is found to be most abundant (Rowe, 1956). As Rowe points out:

> The position of each species relative to the horizontal moisture-preference scale is not meant to be interpreted in a hard and fast way. Plants of wide tolerance may be found in all sorts of communities, from dry to wet sites. However, most of the species exhibit a preference for one moisture regime (their "apparent optimum") over others, reflecting this preference in greater abundance and vigor, and it is expedient here to classify each according to the moisture column of closest fit. Intensive studies would make possible the subdivision of the moisture-preference classes as the finer responses of species relative to one another were distinguished, but for present purposes a five-point division of the moisture gradient is considered adequate.

The delineation of plant species along the moisture gradient is shown in Table 24. The table also indicates something of the successional trends in the forest communities. The common line of succession from poplar to pine to spruce or spruce-fir is marked by a diminution of light on the forest floor. Pioneer communities are characterized by an open canopy which allows tall shrubs and tall herbs to flourish, provided moisture is adequate. More advanced communities—forests of white spruce, spruce-fir, or black spruce—with closely spaced canopies make conditions suitable for the survival of more shade-tolerant low herbs and mosses. With some exceptions, the top half of the table lists species characteristic of coniferous types further along the successional series. The exceptions occur mainly in the "Dry" and "Wet" columns, and

TABLE 24
Moisture Preference of Common Selected Species of the Southern Boreal Forest[a]

	Very dry and dry forest, xerophytic species	Fresh forest, xeromesophytic	Moist forest, mesophytic	Very moist forest, mesohydrophytic	Wet forest, hydrophytic species
Vegetation strata		Pinus banksiana			Larix laricina
		Picea glauca			
		Betula papyrifera			
				Populus balsamifera	
				Picea mariana	
		Populus tremuloides			
				Abies balsamea	
Tall shrubs	Alnus crispa Shepherdia canadensis	Amelanchier alnifolia Corylus cornuta	Acer spicatum Viburnum trilobum	Acer negundo Cornus stolonifera	Alnus rugosa Salix petiolaris

Stratum						
Medium shrubs	Hudsonia tomentosa Juniperus communis	Diervilla lonicera Vaccinium myrtilloides		Lonicera dioica	Ledum groenlandicum Ribes triste	Andromeda polifolia Betula glandulosa Chamaedaphne calyculata
Tall herbs and grasses (3 dm plus)	Hedysarum alpinum Lathyrus venosus	Apocynum androsaemifolium Lathyrus ochroleucus	Aquilegia canadensis Osmorhiza longistylis		Anemone canadensis Calamagrostis canadensis	Aster junciformis Aster puniceus Glyceria borealis Petasites sagittatus
Medium herbs and grasses (1–3 dm)	Achillea millefolium Aster laevis	Corallorhiza maculata Prenanthes alba Vicia americana Viola rugulosa	Mertensia paniculata Petasites palmatus		Dryopteris spp. Geocaulon lividum Habenaria hyperborea	Caltha palustris Equisetum arvense Mentha arvensis
Low herbs, grasses and shrubs (less than 1 dm tall)	Antennaria species Arctostaphylos uva-ursi Juniperus horizontalis Lycopodium complanatum Polygala paucifolia Potentilla tridentata	Anemone quinquefolia Lycopodium obscurum Pyrola asarifolia Pyrola secunda	Coptis groenlandica Cornus canadensis Fragaria vesca Linnaea borealis Rubus pubescens Lycopodium annotinum Pyrola virens Trientalis borealis Vaccinium vitis-idaea		Equisetum scirpoides Galium triflorum Gaultheria hispidula Habenaria obtusata Mitella nuda	Carex species Drosera rotundifolia Galium trifidum Rubus acaulis Rubus chamaemorus Smilacina trifolia Vaccinium oxycoccus

[a] The species are arranged by strata in the vertical dimension and by moisture preference in the horizontal dimension. Condensed from Rowe (1956).

many of the species that prefer one or the other of the extremes appear to be intolerant of much shade. For example, *Arctostaphylos uva-ursi* and *Potentilla tridentata*, both low, shrubby plants, often form a ground carpet on dry, open pine stands, but they are absent or at least very rare wherever shade is intense. Other exceptions to generalization are illustrated by the tolerance to shade of some medium shrubs (*Viburnum edule* and *Rosa acicularis*) and of some tall and medium herbs (*Aralia nudicaulis* and *Petasites palmatus*). Some low-growing species occur in communities dominated by poplar or pine, where they grow under the taller shrubs and herbs, and they retain prominence in denser coniferous stands after the shading out of taller shade-intolerant species which concealed them in earlier successional communities. Feather mosses are among the most shade-tolerant of all species, and successional development from the more open pioneer stands to the closed coniferous forest is marked by an increase in importance of the feather mosses in the lowest layer of vegetation.

Differences in the rooting habits of each species are another complicating factor. Small plants are influenced primarily by soil-surface conditions. Large plants are influenced by the deeper, as well as the shallow, soil horizons. Small herbs on strongly leached dry soil may encounter more xeric conditions than tall shrubs and herbs, but the moisture conditions may be reversed if much wet, decayed wood lies on the ground.

The distributions of plant species do not always show a fine degree of correlation with gradients in the intensity of particular factors, because compensatory influences in the environment and variable ecological tolerances often tend to conceal these relationships. Nevertheless, as Rowe's study shows, detectable responses of plants to environmental variation do occur, and careful study with appropriate techniques will disclose the relationships between vegetation and the environment.

The studies by Maycock and Curtis (1960) established the significance of moisture as a determining factor in the composition and structure of the plant communities found in boreal forest regions. These studies, carried out in boreal forest stands along the southern fringe in central North America, are too detailed to describe fully here, but the species, both of trees and of herbs and shrubs of the ground layer were shown to be distributed in a manner that was readily correlated with the moisture regime. The factors taken into account in relating plant community composition to the environment were topography, soil conditions, light penetration, drainage, atmospheric moisture, and soil moisture. They reported as follows*:

*Reproduced by permission. Copyright 1960 by the Ecological Society of America.

Differences in floristic composition seemed generally to correspond with changes in these environmental factors and appeared most notably evident, the greater the contrast in moisture conditions. The species complement most frequently associated with coarse-textured, well-drained, sandy soils in partially open situations, was markedly distinct from the group of species characteristic of the poorly drained clays or organic soils in densely shaded habitats. Again, the composition of stands on mesic sites, which can be considered intermediate between these extremes, also differed markedly from those stands on the dry and wet sites. As data for many stands in a wide variety of environmental situations were accumulated, it became increasingly evident that, in spite of these differences between stands on the 3 extreme sites, they were not so markedly different when considered with groups of stands on sites intermediate between the extremes. In short, a gradational series occurred from the one extreme to the other, and the possibility of sharp differentiation on the basis of any single factor was impossible.

In a subsequent study carried out in northern central Ontario, Carleton and Maycock (1978) discerned that tree species rooted in an inorganic substrate could be seen to be distributed along a moisture and soil particle size gradient; the lowland conifer bog forests overlying deep, wet peat appeared to represent a different community with separate behavior patterns, and pH, conductivity, and nutrient availability seemed to be important. The assumption appears to be that, when moisture is continually abundant, other factors become more influential in establishing the distribution patterns of the tree species.

Another study on the relationships between moisture supply and the species composition of vegetation, this one in peatlands, was carried out at Candle Lake, Saskatchewan (Jeglum, 1973). By using depth to water table as an index of availability of moisture, it was possible to distribute the species found in the vegetational communities of the peatlands (Table 25) in much the same manner as had been done with the upland communities along a dry-to-wet gradient in the studies described above.

How characteristics of vegetation and soils vary along a hillslope from ridgetop to stream valley is clearly revealed in a study conducted in interior Alaska by Dyrness and Grigal (1979). Along a transect 3 km in length and spanning a difference in elevation of about 225 m, the vegetation and soils could be characterized from ridgetop to valley as follows:

At the summit of the ridge was an open black spruce/feathermoss–*Cladonia* zone, with an understory dominated by by *Alnus crispa, Ledum groenlandicum, Vaccinium uliginosum,* and the dominant moss was *Pleurozium schreberi*. There was no permafrost in this upper zone. Next in order downslope was a closed black spruce/feathermoss zone in which herbs were more abundant than on the ridge and the feathermoss *Hylocomium splendens* was the dominant moss species. Permafrost was present. This zone then graded into an open black spruce/

TABLE 25

Distribution of Selected Species by Categories of Water Level in Peatlands[a]

Category in which species attains a maximum value[b]	Depth to water level below surface (cm)					Depth to water level above surface (cm)			
	1 80	2 79–60	3 59–40	4 39–20	5 19–0	6 1–19	7 20–39	8 40–59	9 60 and more
Moist (water depth 80 cm or more)									
Cornus stolonifera	11–33[d]	9–10		2–14	1–7				
Salix bebbiana	34–67	3–20	3–54	10–48	1–17				
Vaccinium myrtilloides	43–33	10–40	23–15						
Calamagrostis canadensis	75–83	18–50	51–47	55–67	47–44	12–21		7–10	
Fragaria virginiana	26–50	6–20	3–8	10–14	20–3				
Galium triflorum	37–17	20–20	30–15	25–24	5–7				
Maianthemum canadense	20–17	20–10		3–5					
Pyrola asarifolia	23–50	7–10	3–15	8–33	9–14				
Very moist (water depth 79–60 cm)									
Picea mariana (≥10 cm dbh)[c]	117–17[e]	245–60	93–54	45–24	32–7				
P. mariana (3–10 cm dbh)	14–33[e]	1406–70	1056–54	213–33	49–10				
P. mariana (seedlings)	1015–50[e]	2280–80	3601–69	1277–39	133–10				
Alnus rugosa	0.1–17	24–30	305–23	29–14	7–10				
Ledum groenlandicum	38–50	58–70	52–69	12–57	8–10				
Salix myrtillifolia	1–17	3–50	3–38	1–10	0.4–3				
Equisetum arvense	38–50	42–70	26–47	20–42	8–17				
E. scirpoides		58–70	38–31	7–10					
Mitella nuda	13–33	25–60	14–47	29–33	21–10				
Petasites palmatus	15–50	20–50	3–15	5–19	15–7				
Rubus chamaemorus		22–60	35–38	25–10					
R. pubescens	14–50	24–30	10–8	14–14					
Vaccinium vitis-idaea	58–33	86–70	65–69	9–48	15–7				
Wet (water depth 59–40 cm)									
Chamaedaphne calyculata			91–15	65–5					

Species									
Carex tenuiflora				21-23	16-28	13-14			
Eriophorum spissum		1-10		23-15	7-5	7-3			
Oxycoccus microcarpus	3-17	30-70		57-61	48-42	11-10			
Petasites sagittatus	7-50			59-15	20-14	17-7			
Smilacina trifolia	10-17	16-80		34-76	40-61	28-31			
Viola renifolia	6-50	7-20		44-15	12-14				
Wet (water depth 39-20 cm)									
Larix laricina (≥10 cm dbh)			8-30[e]	64-31	60-29	70-14			
L. laricina (saplings)	80-17[e]	23-40		61-46	465-24	474-14			
L. laricina (seedlings)	133-17[e]	98-20		1128-54	1133-43	192-3			
Salix discolor		7-10		1-23	24-5				
Caltha palustris	27-17	30-10		43-31	23-67	37-24	3-7		
Carex canescens				24-15	22-19	22-10			
C. disperma	33-17	11-60		28-46	36-48	13-14			
Rubus acaulis	46-50	8-50		30-47	35-67	17-38			
Surface water level (water depth 19-0 cm)									
Agrostis scabra		30-10			6-23	13-24	25-48		
Aster junciformis	20-33			11-38	12-61	23-48	3-7		
Bidens cernua		3-10			15-10	36-7	15-14	10-14	13-10
Carex aquatilis	14-67	21-70		43-92	46-85	64-83	52-79	21-71	26-50	3-33
C. chordorrhiza		3-10		52-31	69-42	83-44	56-21		3-10
C. diandra-prairea		27-10		14-31	27-33	21-48	10-14		10-10
C. lasiocarpa		47-20		29-47	54-52	81-66	56-57		52-20	13-33
C. limosa				37-15	43-24	58-38	93-7		
Menyanthes trifoliata				8-15	48-28	40-34	7-7		
Scheuchzeria palustris				10-8	64-5	48-14	40-7		
Surface water level (water depth −1 to −19 cm)[f]									
Carex lacustris	25-33	60-10		24-15	28-19	51-17	65-43	3-14	37-90	36-100
C. rostrata	11-50	87-20		20-38	27-80	47-75	64-79	55-71	13-20	13-33
Glyceria grandis					26-14	5-7	52-74	20-14	3-20	27-33
Lemna trisulca					10-5	60-7	69-28	100-14		

(*continued*)

TABLE 25—Continued

Category in which species attains a maximum value[b]	Depth to water level below surface (cm)					Depth to water level above surface (cm)			
	1 80	2 79-60	3 59-40	4 39-20	5 19-0	6 1-19	7 20-39	8 40-59	9 60 and more
Shallow water (water depth −20 to −39 cm)									
Lemna minor					12-10	34-36	60-42	47-40	
Scolochloa festucacea		3-10		23-10	69-10	27-36	76-42	13-10	
Stellaria crassifolia				15-5	32-7	43-7	40-14		
Utricularia vulgaris					7-14	31-36	46-71	46-60	31-100
Shallow water (water depth −40 to −59 cm)									
Acorus calamus				3-5		52-14	17-57	66-40	24-67
Glyceria borealis				10-10	27-3	38-21	40-14	55-20	27-33
Phragmites communis					54-7	97-7	37-42	93-10	
Potamogeton vaginatus					3-3	40-7		90-20	
Scirpus acutus				99-5	14-7	13-21	67-29	44-50	15-67
Utricularia minor					3-3	15-14	6-42	74-20	3-33
Medium water (water depth −60 cm and more)									
Eleocharis palustris		3-10		9-14	60-17	45-50	33-71	47-60	38-100
Equisetum fluviatile			26-38	32-57	29-66	24-43	13-42	39-60	70-67
Hippuris vulgaris		3-10		35-57	12-7	7-21	7-14	28-40	93-33
Myriophyllum verticillatum					7-3	25-14	42-57	40-60	42-100
Ranunculus gmelini					5-10	10-28	23-14	33-20	77-33
Utricularia intermedia					36-21	35-50	33-29	67-20	67-33

[a] Species listed are selected from Jeglum (1971).
[b] Maximum value takes both frequency (or cover) and presence into account; for details see original paper.
[c] dbh refers to diameter at breast height.
[d] The first number refers to cover in the case of shrubs and percentage frequency in the case of semishrubs and herbs. The second number is percentage presence. Shrubs and trees are listed first in each category.
[e] Values for *Picea* and *Larix* refer to numbers per acre.
[f] Depth of water above surface: negative values refer to water depth above the ground.

feathermoss–*Cladonia* intermediate community in which *Cladonia* and *Sphagnum* occur in a closely mixed small scale mosaic. At the lower, more gently sloping portion of the transect, open black spruce/*Sphagnum* occupies the area, there is a shallow perched water table on the surface of the permafrost in summer, the tree canopy cover is only about 25%, and dominant understory species are *Ledum groenlandicum, Vaccinium uliginosum, Betula glandulosa, Calamagrostis canadensis, Vaccinium vitis-idaea,* and *Rubus chamaemorus.* The distinguishing feature is a virtually continuous cover of *Sphagnum* moss species.

At the low end of the transect, there is an intergrade community classified as an open black spruce/*Sphagnum* woodland *Eriophorum* intermixture. Permafrost is shallow, with upper levels at about 39 cm in later summer, and *Eriophorum* is an important species in the community. The area next to the stream and adjacent to streamside vegetation is dominated by black spruce woodland/*Eriophorum* in which the black spruce trees are scattered and stunted. There are tussocks of *Eriophorum*, *Ledum groenlandicum* is absent, and the most common shrub species is *Ledum decumbens. Vaccinium vitis-idaea* and *Rubus chamaemorus* are important, and the mosses are principally *Sphagnum* species. Immediately adjacent to the stream is a community made up of white spruce, alder, and *Calamagrostis*, with some common additional species including *Rosa acicularis, Vaccinium vitis-idaea,* and *Rubus chamaemorus.*

Various species reach peak abundance at certain portions of the hillslope. The changes downslope in the soil characteristics include a slight decrease in pH in the upper horizon and a slight increase in the pH of the next lower horizon; this is attributed to the immobilization of calcium in the lower surface horizon. Concentrations of nitrogen and magnesium increase downslope in both organic horizons; potassium and phosphorus decrease downslope. The mineral soils along the transect are uniformly poorly drained silt loam. Permafrost is present everywhere excepting the ridge summit.

ENVIRONMENTAL GRADIENTS: LIGHT AND SHADE

Plant species tolerant of shade possess physiological adaptations characterized by light saturation of the photosynthetic apparatus at low light intensities and higher rates of photosynthesis in shade than is the case with shade-intolerant species. Rowe (1956) notes that, while soil moisture regimes appear to be the most important factor in establishing the composition of the forest community, light conditions are a primary contributing factor in the composition and structure of the forest understory community, i.e., the species found growing there and the relative abundance with which they occur:

> A poplar canopy allows much more light to penetrate to the forest floor than does a spruce canopy, consequently tall shrubs are able to flourish under the former but not under the latter forest. In a comparable way, an open stratum of tall shrubs may allow tall herbs to survive, while a dense stratum eliminates all but the most tolerant. It is a fairly sound generalization that the relative height of the forest undergrowth species bears an inverse relationship to tolerance, since plants of low stature must be adapted to shade in order to survive. For example, the series: *Corylus cornuta* → *Aralia nudicaulis* → *Cornus canadensis* → *Calliergonella schreberi* (tall shrub → tall herb → low herb → moss) is clearly one of increasing tolerance to shade.

It should be noted that moisture and light conditions are usually correlated; dry sites are often more sparsely covered with trees than mesic sites, and it is not easy to discern whether moisture or light is the factor affecting the understory community to the greatest extent. Indeed, neither can be said to be the most important, since both are crucially involved in the conditions prevailing in the forest floor environment. The more open canopy of a sparsely treed forest permits more rainfall to reach the ground layer, but it also allows more rapid evaporation. A dense canopy will intercept precipitation, preventing it from reaching the forest floor, but it will tend to reduce evaporation from the soil surface. Evapotranspiration, however, will most likely be greater than in an open forest in which there are fewer trees and a smaller total leaf area. In a dense forest, on the other hand, understory development may be limited by the reduced availability of water as a result of competition from the roots of the trees. The relationships between light and moisture are, thus, interdependent and multivariate and cannot be expressed in a simple equation. There appears, however, to be a general tendency in northern communities for open-canopied forests to have an increased cover of shrubs and herbaceous species and for the deeply shaded ground surface beneath a closed canopy to be occupied by mosses. The latter are well adapted to the shaded conditions and to the intermittent nature of the supply of moisture to the surface of the ground.

The ordination studies of Maycock and Curtis (1960) on boreal forest stands in the Great Lakes region indicate that the quantitative values for species in the understory community are correlated with soil moisture; they point out, however, that light conditions are closely related to moisture and are also important:

> There seems little doubt that light is also an important environmental influence along this primary axis with a tendency for light values to be higher in the open stands of the lower end, lower in the mesic stands of the central portion, and very low in the close-canopied cedar swamps at the upper end. The influence of light is not, how-

ever, so simple since there is little doubt that light relationships are interwoven with other important environmental influences along the secondary axis.

Discussing these relationships, Maycock and Curtis add that each species seems to be integrated in terms of physiological adaptations to survive and compete within a range of conditions more narrow than those existing in the wide complex of communities found in the region. Each species, in other words, occupies a preferred niche. They point out*:

> These species are arranged in an orderly but independent manner along a series of environmental gradients which tend to converge upon a situation somewhat representative of mature conditions of a mesic nature, presenting an environmental complex of moderate moisture, low light and high amounts of incorporated organic matter. This situation is gradually displaced by conditions of increased moisture, less light and higher accumulations of organic matter in one direction: and in another by conditions of relatively decreased moisture, abundant light and impoverished amounts of incorporated organic matter. These relationships result in a three-sided environmental pattern. This situation bears no direct or linear relationship to successional tendencies but succession undoubtedly is an active agent in the complex.
>
> The entire assemblage of species can be said to represent a vegetational continuum, in which all species interact and position themselves in environmental situations most suited to their continued existence. Effects produced by or resulting from the interactions of species themselves are probably as influential as any separate or combined environmental factors, which similarly fluctuate in gradational series.

In a study on boreal forest stands at Candle Lake, Saskatchewan, Swan and Dix (1966) found that forest structure was best described as continuously varying, with individual species achieving greatest abundance in different ranges of the continuum. Balsam fir and black spruce dominate stands having a dense moss understory; paper birch, white spruce, jack pine, and trembling aspen dominate stands in which mosses are infrequent and vascular species are abundant. In this study, no relationships could be established between vegetation structure and three environmental measures thought to reflect available soil moisture (drainage regime, moisture content, water-retaining capacity), but the shade cast by the forest canopy appeared to be particularly important to understory character. Black spruce and balsam fir were generally associated with the deepest shade; the understory was characterized by the smallest number of vascular species of any community and by the highest percentage of moss cover. The soil acidity was high in the black spruce stands and low in the balsam fir stands, suggesting that some factor other than soil conditions was involved in the differences in the understory between these stands and the stands dominated by other species. The inference is that shade density is the controlling factor. Aspen, paper birch, jack pine, and (to a lesser extent) white spruce have

*Reproduced by permission. Copyright 1960 by the Ecological Society of America.

relatively thin canopies and support a well-developed understory of herbs and shrubs and a scant moss cover.

Rating the shade tolerance of forest trees, Zon and Graves (1911) rated balsam fir as the most tolerant species in the forests of eastern North America. Zon (1914) rated red spruce and hemlock more shade-tolerant than balsam fir; Hutchinson (1918) considered balsam fir less shade-tolerant than black spruce and a number of other species. The differences in assessing shade tolerance obviously result from subjective judgments that differ from one person to another. Graham (1954) tried to systematize the process by standardizing the criteria employed, and his shade-tolerance scheme is probably reasonably accurate. He rated the species as follows: most tolerant—balsam fir; highly tolerant—white spruce, black spruce, white pine; intolerant—red pine; most intolerant—jack pine, paper birch, tamarack, trembling aspen. Bakuzis and Hansen (1965) point out, however, that other factors are also involved in the distribution of a given species, including tolerance to heat and cold, soil acidity and alkalinity, wind, and mineral nutrient content of soil, to all of which there may be regional differences in the response of a given species. Bakuzis concludes, however, that, as stated by Rowe, the principal factor controlling the structure of the undergrowth in the forest is light; although temperature, root competition, litter accumulation, and soil characteristics exert a subsidiary effect, their main influence is upon composition rather than structure, i.e., upon the abundance of species rather than which species are present and which absent from the community.

ENVIRONMENTAL GRADIENTS: NUTRIENTS

By using depth to water table as an index of availability of moisture, it was possible for Jeglum (1971, 1973) to discern that the species found in the vegetational communities of the peatlands were distributed along a wet-to-dry gradient. The relationship between the vegetational composition of the communities and the pH of the upper soil horizon is shown in Table 26. This index was employed as a measure of soil fertility, the more acid soils commonly being lower in essential nutrients. The conclusion reached by Jeglum on the basis of the study on the Candle Lake peatlands was that moisture and soil fertility were the two major factors determining the distribution of species in the communities. The species demonstrated a wide variety of distributional patterns; different species attained the greatest abundance in every pH and moisture category and demonstrated tolerance amplitudes from narrow to broad.

Soil nutrient regimes in soil type series have been used by a number of investigators for purposes of site classification. In much reported work, however, the concept of fertility is poorly defined, and there is still an absence of adequate knowledge of the optimum nutrient requirements of trees and other plants upon which to base ecological studies. Foliar analysis, for example, is employed to measure the amounts of various nutrient elements in foliage and other plant parts on the assumption that a deficiency in the tissues represents a deficiency in the soil. A correlation often exists among soil nutrient supply, internal nutrient concentration, and growth. Although variations in light intensity and water supply, for example, tend to cause complicating and still largely inexplicable variations in these relationships, some reasonably accurate general observations have been made using the technique. For example, nitrogen deficiency-tolerant species have the capacity to build internal nitrogen concentrations greater in relation to the optimum than nitrogen-demanding species.

Ingestad (1962) prepared a table summarizing levels of foliar nutrient element content in *Picea, Pinus,* and *Betula* plants showing deficiency symptoms (Table 27). A subsequent study, (Gerloff et al., 1966) revealed that, at least in the light of this table, *Picea* and *Pinus* growing in natural communities in some areas show foliar concentrations of nitrogen and phosphorus consistently near or below deficiency levels. Such other elements as sulfur, calcium, and potassium are frequently below optimum concentrations and are often at deficiency levels. In Gerloff's analysis of more than 200 different native species, interesting differences among species are revealed in the ability to accumulate nutrient minerals on the same site. There are also differences on the part of individuals of the same species to accumulate nutrients on different sites.

Concentrations of nutrient elements in the foliage of evergreen bog plants were compared with those in plants growing on adjacent sites by Small (1972a), who found that the foliage of the bog species had a lower nitrogen and phosphorus content than the foliage of plants of other habitats, reflecting the paucity of available nitrogen and phosphorus in the bog substrate. Small (1972b) also found that bog plants showed no evidence of drought stress, pointing out that the concept of physiological drought as a cause of xeromorphic adaptations in bog species is more plausibly accounted for by other factors; nutritionally deficient environments, low temperatures, and short growing seasons favor sclerophyllous, evergreen leaves.

There are difficulties, however, in devising schemes for relating nutrient gradients to vegetation. This was shown by a study in which markedly different growth of aspen and pine stands was observed on sites

TABLE 26

Distribution of Selected Species by Categories of pH[a,b]

Class	1 pH 3.0–3.9	2 pH 4.0–4.9	3 pH 5.0–5.9	4 pH 6.0–6.9	5 pH 7.0–7.9
1: Very oligotraphic (pH 3.0–3.9)					
Chamaedaphne calyculata	65-20	86-10	97-5		
Ledum groenlandicum	66-100	45-60	44-42	13-24	
Vaccinium myrtilloides	39-60	3-20	10-5		
Carex tenuiflora	50-20	28-20	7-11	11-14	
Oxycoccus microcarpus	49-80	37-60	72-32	28-17	
Rubus chamaemorus	55-60	17-50	17-16	3-2	
Smilacina trifolia	52-80	25-60	30-42	34-32	5-13
Vaccinium vitis-idaea	68-100	84-50	58-37	12-19	
2: Oligotrophic (pH 4.0–4.9)					
Picea mariana (trees ≥10 cm dbh)[c]	138-80[d]	270-50	60-26	47-12	
P. mariana (saplings)	435-80	1659-50	1546-26	106-17	
P. mariana (seedlings)	1116-100	2559-60	5929-26	848-19	33-7
Andromeda polifolia	10-40	60-10	31-16	9-12	40-7
Agrostis scabra	3-20	41-30	19-32	15-19	7-7
Carex chordorrhiza	62-40	91-50	70-53	57-20	42-13
C. limosa	47-20	86-40	50-32	39-12	40-7
Equisetum scirpoides		55-40	48-26	7-3	30-7
Potentilla palustris	37-20	51-50	3-68	22-47	27-7
3: Mesotrophic (pH 5.0–5.9)					
Calamagrostis canadensis	3-40	49-60	47-42	47-42	72-13

Species					
Carex lasiocarpa	22-20	79-50	87-58	49-47	68-20
Equisetum arvense	31-20	30-40	47-26	19-24	10-13
4: Eutrophic (pH 6.0-6.9)					
Larix laricina (trees)	67-40[d]	7-10	14-11	67-19	6-7
L. laricina (saplings)	50-20	18-40	41-21	485-17	11-7
L. laricina (seedlings)	306-60	286-30	1420-21	1248-15	76-7
Carex aquatilis	39-60	14-60	51-74	52-83	37-67
C. rostrata	3-20	37-50	43-53	45-81	39-60
5: Very eutrophic (pH 7.0-7.9)					
Alnus rugosa	0.3-20	3-10	35-5	21-14	71-7
Galium trifidum		70-10	33-11	47-17	49-20
Glyceria striata		10-10		23-15	63-7
Hippuris vulgaris				19-17	50-13
Lemna minor				36-17	44-33
Mertensia paniculata		10-10	7-5	8-10	37-7
Mitella nuda	10-40	22-20	6-16	27-20	45-13
Rubus pubescens	3-20	7-10		13-7	63-7
Scirpus acutus				21-17	82-27
Viola palustris		3-10		20-2	57-7

[a] Table condensed from original data in Jeglum (1971). Reproduced by permission of the National Research Council of Canada.
[b] Mean values of abundance are followed by the percentage presence (number of stands in which the species occurred as a percentage of the number of stands in the category). A species was assigned to the category in which it attained the highest value of its mean quantity times its frequency in the class.
[c] dbh refers to diameter at breast height.
[d] Trees expressed in numbers per acre.

TABLE 27

Element Concentration in Leaves of Seedlings (as Percentage of Oven-Dry Weight) Corresponding to 50–90% of Maximum Growth (Range of Moderate Deficiency) and to 90–100% of Maximum Growth (Range of Optimum Growth)[a,b]

Element	Pine		Spruce		Birch	
	Range of moderate deficiency	Range of optimum growth	Range of moderate deficiency	Range of optimum growth	Range of moderate deficiency	Range of optimum growth
Nitrogen	1.1–2.4	2.4–3.0	0.9–1.8	1.8–2.4	2.4–3.4	3.4–4.0
Phosphorus	0.08–0.15	0.15–0.4[c]	0.07–0.10	0.10–>0.3	0.1[b]–0.2	0.2–0.4
Potassium	0.44–0.9	0.9–1.6	0.3–0.7	0.7–1.1	0.5–1.5	1.5–3.1
Calcium	0.03[c]–0.04	0.04–0.3	0.02[b]–0.09	0.09–0.6	0.06–0.16	0.16–0.6
Magnesium	0.05–0.12	0.12–0.18	0.02–0.09	0.09–0.16	0.10–0.17	0.17–0.5
Sulfur	0.06–0.20	0.20–0.25	0.09–0.13	0.13–0.18	0.22–0.29	0.29–0.32

[a] From Ingestad (1962).
[b] The species are *Pinus silvestris*, *Picea abies*, and *Betula verrucosa*. The values are similar to other known values for species of these genera such as *Pinus banksiana*, *Picea glauca*, and *Betula papyrifera*.
[c] Extrapolated values.

with seemingly identical conditions of soil profile and water table. Investigation disclosed that the unexpectedly high rate of tree growth in some areas was due to the presence of groundwater enriched with nutrients through contact with lenses of lacustrine clay. From this study, it is obvious that only chemical analysis of soil and groundwater can give a realistic picture of the actual chemical and physical environment of a forest. Poor growth of upland trees may be correlated with groundwater deficient in oxygen and mineral nutrients; this is not the case, however, with swamp species, such as white cedar, balsam fir, black spruce, and alder. It is, thus, apparent that not only local soil conditions but also the species of plants found there must be taken into account in efforts to describe nutrient conditions existing for plant growth. The effects of differences in groundwater on tree growth rates on different sites supporting black spruce forest were revealed in an early study conducted by Pierce (1953).

Native plants often grow in unusual nutritional environments, and many species have extraordinary capacities for selective accumulation or exclusion of elements. Adaptations to very high acidity and low fertility are characteristic of many plants. So well adapted are they to these unusual environments that they grow better there than under more average conditions.

There is a close relationship between the nitrogen in black spruce trees and the nitrogen content of the top 30 in. of mineral soils, although no correlation between nitrogen and black spruce is evident on sphagnum soils (Heilman, 1966). Nitrogen in mineral soils is found predominantly in the upper soil layers; the nitrogen of sphagnum soils is concentrated in the coldest part of the soil where mineralization of organic nitrogen is slow. Comparatively few black spruce were growing on the sphagnum bogs sampled by Heilman, and their rate of growth and nitrogen content were extremely low. The study indicated that, as the moss layer became thicker, most nitrogen was buried in the deep, cold strata of the soil.

A decrease in available nitrogen may be a necessary condition for forest paludification in cold regions, a process by which lowland spruce forest is progressively replaced by an open, treeless community dominated by mosses, low shrubs, and herbaceous species. The vigor of the trees is progressively reduced and, because most sphagnum species have high light requirements, the opening of the forest canopy favors invasion by *Sphagnum* mosses. As the mosses grow, they intensify the unfavorable characteristics with respect to mineralization of nitrogen and further tree growth. Release of this immobilized nitrogen explains the improvement in nitrogen availability and productivity following the burning of sphagnum-dominated forests.

The effect on forest succession of the incursion of mosses and the reduction of nitrogen and phosphorus supplies to trees is discussed later in this chapter; the work of Heilman (1966, 1968) has shown, for example, that in parts of Alaska there is a cyclic succession pattern brought about by paludification and subsequent disintegration of the forest community. The organic soils that develop under closed forest are greatly different in nutrient characteristics from soils on the organic substrate of the paludified areas.

Organic soils contain much more nitrogen per unit weight in the lower parts of the profile than mineral soils. Page (1971) shows that sombric brunisols and peats, for example, contain 6750–7850 kg/ha (6000–7000 lb/acre) of nitrogen in the upper 30 cm (12 in.) of soil, while podzols contain only 3350–4500 kg/ha (3000–4000 lb/acre). In most cases less than 1% of the total nitrogen supply is in available form (Table 28).

In sombric brunisols and peats, total phosphorus content in the upper 30 cm averages from 1000 to 1450 kg/ha (900–1300 lb/acre), but only about 17 kg/ha (15 lb/acre) of this total is available for plants.

Available calcium and magnesium are usually more abundant in poorly drained than in freely drained soils. Total calcium in the upper 30 cm averages 3900 kg/ha (3500 lb/acre) in soils with noncalcareous parent materials and up to 22,500 kg/ha (20,000 lb/acre) in soils influenced by calcareous rocks. Available calcium in the upper 30 cm ranges from about 225 to 11,200 kg/ha (200–10,000 lb/acre) or more. There are usually between 4500 and 9000 kg/ha (4000 and 8000 lb/acre) of magnesium in the upper 30 cm; an average of about 110–220 kg/ha (100–200 lb/acre) is in available form.

Total potassium contents average between 11,200 and 13,500 kg/ha (10,000 and 12,000 lb/acre) in available form.

Total soil nutrient contents reported by Page are similar to those given for soils in other parts of northern North America and Europe. Available nitrogen and phosphorus are low in all soil types, with freely drained soils having lower amounts than poorly drained soils. Available nutrients in the poorly drained soils are adequate to maintain the slow growth characteristic of plants subjected to unfavorable physical soil conditions. Where physical soil conditions are more favorable, however, the supply of available nutrients may limit tree growth.

After logging or burning, balsam fir forests in Newfoundland are often replaced by a *Kalmia* heath which appears capable of maintaining itself indefinitely. Low decomposition and mineralization rates are associated with the change from balsam fir or black spruce forest to *Kalmia* heath; Damman (1971) found that nutrients were immobilized to an appreciable extent under the *Kalmia*. Total nitrogen under the heath was

TABLE 28

Average Contents of Total Nitrogen, Phosphorus, Calcium, Magnesium, and Potassium in Humus and Mineral Soil and Relationships between Total and Available Nutrients

Nutrient element	Soil medium	Sample area	Average reading (all soil types)	Available nutrient as percentage of total present
Total Nitrogen	Humus	Avalon peninsula	1.2%	0.9
		Western Nfld.	1.2%	0.5
	Mineral soil	Avalon peninsula	0.2%	0.7
		Western Nfld.	0.1%	1.5
Total phosphorus[a]	Humus	Avalon peninsula	3.6 mEq/100 g	3.2
		Western Nfld.	2.6 mEq/100 g	5.6
	Mineral soil	Avalon peninsula	2.2 mEq/100 g	0.2
		Western Nfld.	1.5 mEq/100 g	0.5
Total calcium[a]	Humus	Avalon peninsula	12.1 mEq/100 g	89.5
		Western Nfld.	71.4 mEq/100 g	68.7
	Mineral soil	Avalon peninsula	11.4 mEq/100 g	1.4
		Western Nfld.	42.3 mEq/100 g	11.9
Total magnesium[a]	Humus	Avalon peninsula	9.1 mEq/100 g	58.2
		Western Nfld.	12.1 mEq/100 g	61.4
	Mineral soil	Avalon peninsula	32.1 mEq/100 g	0.9
		Western Nfld.	58.4 mEq/100 g	2.4
Total potassium[a]	Humus	Avalon peninsula	3.7 mEq/100 g	25.3
		Western Nfld.	2.7 mEq/100 g	42.7
	Mineral soil	Avalon peninsula	18.0 mEq/100 g	0.5
		Western Nfld.	27.6 mEq/100 g	0.5

[a] Data from Page (1971). Reproduced by permission of the National Research Council of Canada.
[b] Average based on 10% subsample only.

greater than under balsam fir or spruce, but it was not in a form available nutritionally to the plants. Phosphorus increased slightly, but there were large losses of potassium and perhaps some loss of calcium in the heath ecosystem. The differences between the fir and the spruce stands were less than those between these stands and the *Kalmia* community.

There are some revealing studies on peatland vegetation that describe the moisture and fertility regimes accounting for the remarkable variety of boreal plant communities found on low, flat, land surfaces. In northwestern Minnesota, where one such study was carried out, several million hectares of forested peatland occupy the bed of former glacial Lake Agassiz. Black spruce is the most abundant tree. These Lake Agassiz peatlands possess significant surface slopes. Areas located downslope from mineral soil islands in the peatlands support a rich swamp forest which often contains tall black spruce, tamarack, and northern white cedar, a shrub understory dominated by speckled alder, and an assemblage of mosses, herbs, grasses, and sedges. These areas contrast sharply with adjacent stands of muskeg dominated by dwarf black spruce, *Sphagnum* mosses, and ericaceous shrubs (Heinselman, 1963, 1970).

There is circumstantial evidence that forests located downslope from islands are productive and floristically rich because these sites are fertilized by mineral-bearing groundwater. The adjacent muskegs are unproductive and floristically poor because they have been isolated from the mineral-bearing waters by the accumulation of sphagnum peat. Where a mineral soil occurs within a peatland there is commonly a linear band of rich swamp forest downslope from the island. Such bands of floristically different vegetation have been termed water tracks because they seem to mark the course of mineral-bearing groundwater.

Along these water tracks, spruce foliage contains much higher levels of the nutrient elements nitrogen and phosphorus than the foliage of trees some distance away. The higher levels of nitrogen on good sites may be attributed to nitrogen fixation by speckled alder, abundant on good sites but absent on poorer sites. More rapid decomposition of litter (indicated by a lesser depth of peat accumulation), movement of enriched water, and proximity to mineral subsoil may influence the level of both nitrogen and phosphorus as well as that of other mineral nutrient elements in the water track.

How the levels of nutrient materials in soils are related to the local and wide-ranging geographical distribution of native species remains one of the least understood aspects of ecology, despite a relatively long history of research into the physiology of the mineral nutrition of plants. Most physiological work, however, has been carried out using plants of economic value (crop plants) and, whereas some of the available information is useful in studies on the distribution of native plants, there has so far been little application of the knowledge of wild native species. Research so far indicates only that it should be possible eventually to enhance greatly understanding of the basic physiological relationships of boreal species. Information on the morphological, physical, and chemical characteristics of many common boreal forest soils has accumulated rapidly in recent years, but more information is needed for an adequate understanding of plant–soil relationships.

The concentration of an element in plant tissues is in many instances a genetically determined trait; mineral content is genetically determined, and measurement of nutrient mineral content, thus, can be inadequate for describing the mineral content of soil. It is necessary to know the response of a plant species in mineral accumulation or exclusion (Gerloff et al., 1966; Garten, 1978). Garten concludes that the nutrient composition of the substrate sometimes fails to reflect the manner in which the plant species utilizes nutrient resources, because of physiological mechanisms that differentially absorb or exclude elements. The mineral element composition of plants is, in short, affected both by genotypic characteristics and the environment, and interactions between the two determine a plant's mineral content; there is a range of nutrient element concentration that a plant species is adapted to, and within this range the genetic characteristics of the plant and the concentration of the element in the soil both affect the concentration in plant tissues.

THE DYNAMICS OF COMPETITION: ADAPTATION

Boreal forest plant species follow the universal rule that each species, by evolutionary processes, has become adapted to a given range of environmental conditions within which it is capable of growing and competing with other species for moisture, light, and nutrients. Plants growing together in a vegetational community—a sedge meadow, a spruce forest, an open muskeg, for example—might at first glance be assumed to occupy the same habitat. On closer inspection this is seen not to be the case. The habitat of individual plants is a more narrowly circumscribed set of conditions than can be defined in such broad terms and, in truth, every meadow, forest, muskeg—every community—is an aggregation of many habitats occupied by plants of many different species.

Many species create conditions suitable for growth of individuals of other species. Thus, the shade beneath a spruce tree is usually occupied by several species of shade-tolerant herbs, shrubs, and mosses. This is in contrast to the spaces between trees where direct sunlight hits the forest floor. These spaces are occupied by fewer mosses and a greater abundance of herbaceous species requiring higher light intensities. If trees in a dense stand of mature spruce are destroyed by fire or wind, the area will soon be occupied by a community made up of species quite different from those occupying the forest in a mature stand. If any of the same species remain growing in the area, they are usually present in markedly fewer numbers than those they attain in a mature forest.

This is true most evidently in southern parts of the boreal forest where a rich complement of species exists, and is less apparent northward where species of mature forest are those that pioneer opened areas, largely perhaps as a consequence of absence of potential competitors for the newly available pioneer habitats (Black and Bliss, 1978).

In response to changes in light, moisture, and nutrient availability brought about by the changing structure of the community, a given species may, over a period of several years or decades, encounter at first increasingly favorable conditions and then conditions less and less favorable; it responds to the favorable conditions by vigorous growth and reproduction, becoming visibly more abundant, and then declines in number as conditions become unfavorable. Because many species are always rarities, apparently incapable of ever attaining great abundance, successional processes that involve species capable of attaining great abundance are most readily discernible and have thus received the most study. Observation of the more abundant species can thus be justified on the basis that the importance of a species in a given ecosystem is proportional to the amount of energy flowing through it relative to the amount flowing through the entire ecosystem. A species through which a high proportion of energy flows will be by definition a dominant species in the system.

The plants of any community are in competition for basic elements needed for life and growth. In the case of trees, this can be demonstrated by measuring accelerated diameter growth rates when trees are given additional growing space or, in contrast, by measuring suppression of growth and increased mortality in a dense forest. Shrubs and herbaceous plants also respond to strong competition by reduced growth rates, lowered reproductive success, and increased mortality. The mechanisms by which competition influences plants are not well understood. Plants always contend for moisture, nutrients, and light, but they may also compete in a multitude of other ways. Some produce antibiotic compounds that inhibit the growth of other plants, some are better able to resist attack by microorganisms and insect pests, some by reason of unpalatability are seldom eaten by rodents or larger herbivores, and some are incapable of growing in soils of high acidity and give way to other species when soil acidity increases during podzolization. Both the plant root and the shoot (or crown in the case of trees) are involved in competition, but the particular part each plays in competitive relationships with other plants has so far been difficult to determine or describe accurately.

It should be noted that the response of a species in its natural habitat may not be the same as its response to the conditions that prevail when it is grown in a laboratory or greenhouse. If relieved of competition, many plant species will grow much better under conditions other than those found where they grow naturally. But the effects of competition should not be overemphasized; even strongly competing species are influenced profoundly by the environment. At Candle Lake, for example, four factors significantly influence the peatlands on an annual overall basis (Swan and Dix, 1966): the long, harsh winters, the relatively dry subhumid climate, the moderately calcareous groundwaters enriched from marine shale bedrock, and a poor floristic and vegetational diversity compared to some peatlands in eastern Canada. These factors are so influential that competition is less significant in many habitats than sheer capacity to endure a variety of conditions. As a consequence, laboratory or greenhouse studies on competitive relationships may not furnish a realistic picture of what actually happens in the natural community.

An experiment performed by Ellenberg (Evans, 1963) illustrates that the response of a species to natural conditions may differ not only from that of single plant in the laboratory but from that of plants in pure stands. Ellenberg's experiment was conducted to determine the effect of depth to water table on the growth of six grasses. In pure stands, the optimum depth for all species was 20–35 cm. In mixed stands, however, this depth was not optimal for any of the species or even the community as a whole. The optimum for the whole community was 65 cm, while individual species growing as part of the community had optima ranging from 5 to 110 cm. These, in Ellenberg's terms, are "ecological" optima as against the "physiological" optima revealed in the pure stands. The point is that the response of plants to an environmental factor changes radically in the presence of competing individuals of other species. Physiological optima may reveal little of the performance of plants in the field in competition with other plants. Greenhouse conditions greatly simplify the relationships existing in nature.

The reasons for differences between greenhouse and natural conditions are undoubtedly numerous. A few examples may be revealing.

Stiell (1970), for example, found that the degree of competition between individual red pine trees could be discerned by measuring the amount of light received by each individual (some were in the shade of others). The differences in light available to each tree, however, failed to account entirely for the differences in growth rates among trees. Some other factors obviously were involved. When he washed the soil away from the roots of a large number of trees, Stiell found that the lateral root systems of individual trees were widely and irregularly dispersed. Root competition was widespread and diffuse. Moreover, it appeared to be unpredictable with respect to given trees. Stiell summarized:

> Numerous other trees will share the same overall rooting zone, and whether or not any two roots are competing will depend on how close to each other they are growing, how many other roots occur in the vicinity, and what conditions of soil moisture prevail. In these circumstances, a tree may have many competitors in its rooting zone, but which trees provide this competition cannot be determined from above ground.

Thus, it was obvious that, although some trees had shaded others out, their competitive advantage in this respect may have been outweighed by less favorable competitive relationships underground.

The above example illustrates that strong competition may exist among individuals of the same species. Competitive interactions among species are usually even more complex. An example of individual adaptations in species that compete with one another has been provided by Mueller-Dombois (1964). In an experiment, seedlings of jack pine, red pine, white spruce, and black spruce were grown from seed in rows along sloping beds of sand and loamy sand in large tanks in a greenhouse. Thus, although the conditions were only an approximation to what occurs in nature, the results revealed different physiological optima for each species. At the end of a year, the white spruce showed two peaks in height growth, and the other three species showed individually different peak responses. As Mueller-Dombois points out, early growth of white spruce is affected more strongly by water held in the upper parts of the soil than is growth in the other species. This may explain the preference shown by white spruce for very moist groundwater soils in certain portions of its range.

The mortality of black spruce on the upper slope of the sandy soil after watering had been discontinued can be explained by the inability of black spruce to form deep root systems, particularly on nutritionally poor soils. This may be the reason for the inability of black spruce to grow on sandy terrain with a deep water table. The survival of jack pine on the wet soil (in nature it is usually confined to drier sites) can perhaps be explained by its ability to form long lateral roots reaching upslope into less moist areas. The pronounced increase in jack pine height with decreasing depth to water table up to a certain optimum supports the result of field observations that the growth of jack pine is strongly influenced by depth to the water table. Red pine responds similarly to water table depths and has a greater mortality in moist soils because of its inability to form long lateral roots.

These are but two examples of adaptations that confer upon a species its unique ability to compete successfully in a given habitat—to grow and reproduce without being excluded by other contenders for the essentials of life. Lichen and moss species also show regional differences in photosynthesis and respiration rates, resistance to temperature stress, and adaptation to light and moisture regimes (Lechowicz, 1978; Busby *et al.*, 1978; Horton *et al.*, 1979; Hicklenton and Oechel, 1976a,b). These by no means exhaust the roster of competitive adaptations; many species have individually interesting adaptations, both physiological and morphological, which will be the subject of research for many years to come.

Climatic influences expressed in terms of temperature, air mass regimes, and other parameters, have distinct correlations with certain characteristics of species distribution. Mean summer temperatures and duration of the growing season determine total possible growth. Extremes in temperature, particularly during the growing season, may be lethal for individual plants. There are, moreover, some aspects of climate that unquestionably have an influence but produce effects more difficult to assess as to importance. Among these are winds, glaze damage, winter drying, and summer frosts. The strong winds of the Arctic are a cause of winter injury to dwarfed spruces found growing in protected sites north of the forest border. Wind-driven snow particles are an effective abrasive agent, and dwarfed trees at the edge of the barrens are given an annual trimming of those parts of the stem and branches protruding above the snow. Flying snow crystals eventually remove needles and bark, and it is characteristic of spruce clones that they produce either no growth above the winter snow line or possess spikes of bare, snow-blasted stem that jut upward from lateral branches. Such spruce clumps almost invariably grow in the protection of landscape features that provide shelter from prevailing winds, a windbreak behind which snow accumulates to a depth of a meter or more. Where conditions are not too severe, an occasional stem will survive and grow above the level at which the blowing snow is most destructive, creating the so-called candelabra effect in which a few bare stems, topped by a tuft of branches, rise from a dense cluster of lower laterals. Clumps of dwarfed spruce cause sufficient eddying of winds, which provides mutual protection, and a small, compact clone may be able to expand, at least for a short distance, beyond the protected site. It has been postulated that the abrasion mechanism may account for the abrupt nature of the northern tree line in some regions (Savile, 1963, 1972).

Trees subjected to snow abrasion also undergo desiccation, since water lost from the stem and branches in winter and spring cannot be replaced until the ground starts to thaw. Over the main extent of the boreal forest proper, however, snow abrasion is of little or no ecological significance, since the forest canopy serves as an effective windbreak.

Glaze damage and severe frosts in late spring may kill buds and affect growth. Unusually warm temperatures and strong winds during the winter months will cause winter drying and this, too, can have a marked influence on growth during the following summer (Cayford and Haig, 1961; Cayford et al., 1959). None of these extremes of weather can be considered unusual, since they recur at intervals. The effect may be apparent for only a season or may persist for years. Seedlings are particularly vulnerable to environmental extremes during germination and early growth. The frequency of adverse weather undoubtedly leaves a mark upon the community by affecting reproduction and the rate of succession (Elliott, 1979a,b).

It seems likely that such meteorological hazards as high winds have been influential in the evolutionary development of such adaptations as layering, which accounts for a large proportion of black spruce reproduction in most northern areas. High winds and other extremes of weather often have an adverse affect upon flowering, seed set, and seed maturation, but mild winds, on the other hand, increase the availability of carbon dioxide required for photosynthesis. Even high winds may have some advantage in that they aid seed dispersal, dislodging seeds and blowing them great distances over the surface of the snow. The effects of all these secondary physical factors upon growth and reproduction of plants in boreal forest regions merit a great deal more study.

Since plants are influenced by a number of interrelated factors, the result of much field observation will be difficult to interpret. All environmental factors may be suboptimal, in which case a shift in any one will change the responses of the plants. There is the danger, too, of considering the most obvious factor the one controlling distribution when it may be involved only in a subsidiary manner. Thus, a plant with a wide range of tolerance for moisture conditions may decline in abundance at the margin of a slight declivity in the terrain as a consequence of an inability to tolerate low night temperatures, but be thought to be intolerant of moist soils.

Adaptations to temperature have been discerned in both

morphological and physiological characteristics of a number of wide-ranging northern species. Some possess broad, villous leaves which expand relatively late in spring, apparently taking advantage of a short frost-free season for rapid completion of the life cycle. Other species are evergreen and grow slowly, but endure late spring frosts without damage. Probably most possess seeds requiring long periods of continuous low-temperature stratification for germination. In wide-ranging species there are differences in photosynthetic and respiratory acclimation responses, as demonstrated in *Ledum groenlandicum* (Smith and Hadley, 1974), that can probably be considered ecotypic characteristics. Species not wide-ranging, on the other hand, often do not show ecotypic variation in the individuals from northern and southern extremes of the range. *Trientalis borealis* is an example (Anderson and Loucks, 1973). Vowinckle *et al.* (1975), on the other hand, showed that maximum photosynthetic rates in black spruce growing near the northern forest border were considerably lower than the rates in spruce of more temperate regions. The temperature optimum for northern trees is 13°–15°C, and rates of 30% of maximum were recorded at 0°C. The trees also had a relatively high light saturation level. Black spruce, however, appears not to have well-developed edaphic ecotypes differentially adapted to mineral soils on uplands and organic soils in lowlands, even though there are considerable genetic differences in the spruces of the regions in seed size, germination characteristics, seedling size, and height growth (Fowler and Mullin, 1977). The markedly reduced growth rates of spruce at the northern edge of the range, compared to growth in more southern regions, provides an opportunity for detailed studies of the variation in many morphological and physiological characteristics along a north–south environmental gradient (Larsen, 1965; Mitchell, 1967, 1973; Kay, 1978).

If any general conclusion can be drawn from these studies, it appears to be that northern plants are adapted not only to a harsh but also a variable environment by means of limits of tolerance and characteristics of acclimation that allow them to survive abrupt changes in conditions from day to day and from one year to another. They tolerate changes that plants more narrowly adapted would not survive, responding either by migrating to more favorable regions or suffering the ultimate penalty of extirpation. The northern species and ecotypes are thus, in general terms, more genetically variable and more capable of acclimation than temperate or southern counterparts; only species rich in biotypes survive (A. W. Johnson and Packer, 1965; Parker and McLachlan, 1978; Lechowicz, 1978; Smith and Hadley, 1974; Kershaw, 1973; Mosquin, 1966; Hicklenton and Oechel, 1976a,b).

SUCCESSION

The historical development of the concept of succession in vegetational communities, as well as of the inevitable corollary, the concept of the climax community, is certainly one of the better examples of the kind of long-enduring scientific (if one can call it that) controversy resulting when adequate data are unavailable and many speculative schemes, based on insufficient evidence, are vigorously defended by their respective proponents. Little understanding has been obtained even yet on successional processes in northern plant communities. This is particularly the case with boreal communities, largely because of the time required for successional development, inability to find sequential development stages for field study, and often the difficulty of access to areas where logging or other disturbances have not destroyed the natural forest. Despite the relative simplicity of the structure of boreal communities, with few dominants and a narrow range of floristic diversity, there has been little opportunity either to disprove or to confirm a view that the ultimate stage in boreal succession is some form of closed spruce forest; for example, it was often accepted that in central Canada the spruce–fir association represented a climax, but succession rarely progresses to that stage because of fire, insect infestation, or windfall.

Fortunately, within recent years this has begun to change; useful techniques have been developed, and research has helped at least to sharpen the ability of ecologists to elucidate the structure of boreal communities and the nature of successional processes in boreal regions. We noted in a previous chapter that early observers considered the spruce–feather moss forest to be as close to a climax community as can be imagined; but it seems also to be the case that mature spruce forests are not stable communities capable of maintaining indefinitely the closed canopy and the deeply shaded forest floor. The prevalence of fire almost inevitably renders the point moot, but even without a disturbance that reinstates pioneer successional stages, it has become generally accepted that, as Rowe (1961) states: "Decadent forests do not return through an inevitable cycle to youth and an optimim phase but rather tend to remain open, ragged, and frequently brush-filled, awaiting the rejuvenating touch of fire, flood, or windfall-ploughing of the soil." Evidence of conversion of upland black spruce forest to wet sphagnum bog forest and, ultimately, to bog can be seen in some regions, notably Alaska (Heilman, 1966, 1968). The progression is accompanied by the replacement of feather mosses by *Sphagnum* mosses that accumulate on the soil surface. The upper level of permafrost rises, nitrogen levels in the substrate decline, and growth rate of the trees decreases. In the absence of fire, the process continues until only a scattering of a few stunted spruce remains. Spruce-sphagnum bogs that have been destroyed by fire repeat the successional cycle, starting with birch and other pioneer species, then going into a stage in which both birch and spruce are present, and, finally, to the spruce stage at which the paludification process begins. As Heilman points out, abrupt boundaries often exist between contrasting vegetation on burned and unburned areas.

Surface horizons of sphagnum soils are acid (average pH is between 3.85 and 4.0), contrasting with the higher pH of the organic surface layers of the mineral soils under the early successional forest. Some studies indicate decomposition is dependent on a fairly high pH, but Heilman indicates that even at pH 3.7 the decomposition rate is 85% of the rate at pH 5.0 to 8.0, indicating that pH is of little significance in reducing the rate of decomposition of organic matter or the availability of nitrogen as succession proceeds.

In Heilman's studies in central Alaska, mineral soils averaged more than twice as much total nitrogen (to a depth of 30 in.) as the sphagnum soils. Nitrogen of the mineral soil is located mostly in the upper layers of the soil, the result of rapid mineralization of organic nitrogen in the warmer portions of the soil. Most nitrogen is in the lower stratum of the sphagnum soils, probably the result of an opposite effect—low rates of mineralization in cold strata that render the nitrogen unavailable for uptake by plant roots. Heilman found that the nitrogen content of black spruce foliage declines progressively through the successional stages. Only those trees growing with birch and alder averaged above the critical level of about 1.0% nitrogen in oven-dry foliage. Low levels of nitrogen in the spruce of the sphagnum bogs was evidently the cause of the low growth rate of trees and the chlorotic condition of the needles.

There were also significant declines in the concentration of phosphorus and potassium in the black spruce foliage at later successional stages. Phosphorus in foliage is adequate in early successional stages but declines to deficiency levels in advanced stages; potassium in the trees on sphagnum soils in advanced stages of succession is also in the deficiency range.

The increasing unavailability of nitrogen as paludification proceeds may be essential to the transition from forest to bog. *Sphagnum* species are intolerant of shade, and reduction of density of the forest canopy may be needed for *Sphagnum* to become established. The invasion of sphagnum then accelerates the process by making nitrogen increasingly unavailable as a result of lowered rates of mineralization and raising of the upper level of permafrost. The latter then traps nitrogen in the permanently frozen substrata below the active layer of the sphagnum soil.

To conceive climax as applicable to boreal communities, it is necessary to modify classic concepts in at least one important way; climax is not a stand of spruce-dominated forest for every topographic site but is rather a climax pattern, as envisioned by Whittaker (1973), ranging from fen to closed-canopy forest along a topographic and microenvironmental gradient or continuum. Thus, in boreal regions, climax is not necessarily a forest community; climax in the neoclassic view is basically cyclical, encompassing all phases of cyclical regeneration in every community of the climax pattern. It can be thought of as a dynamic balancing in the communities of all the various phases represented by the individual stands of vegetation. As Sprugel (1976) states: "The ecosystem as a whole may be in a steady-state even if the individual stands are not."

In this light, the description of boreal communities becomes simpler and conceptually more tractable than was the case under older concepts of succession and climax; the basic data become quantitative tabulations of the structure and composition of associations as found in areas not greatly disturbed by human activity. It is, additionally, a description of the sequence of stages in cyclical regeneration in each segment of the climax pattern. Environmental continuua will have segments in which different associations dominate, and each will have its regeneration cycle, modified by circumstances and fluctuating environmental conditions. In this scheme, no thought is given to the possibility that primary succession from fen or muskeg to mesic forest will occur; the possibility that it may occur in boreal regions in the conventional sense, or over a time period that can be meaningful in terms of human generations, is negligible.

This appears to be a general rule in all regions of the forest, at least in Canada, although the differences in species composition of the forest communities from east to west and north to south create regional differences in successional patterns. In the boreal mixed wood region of southern Manitoba and Saskatchewan there are eight tree species, and these give rise to a large number of different dominance combinations (Rowe, 1956; Swan and Dix, 1966; Dix and Swan, 1971). In these areas, both white and black spruce are first-generation trees after fire, but they continue to regenerate in the forest for only a short time thereafter. As Dix and Swan (1971) affirm, different strategies of germination and growth for each tree species seem to explain composition and structural characteristics of the forest; light intensity at the forest floor and associated effects appear to offer explanations for many aspects of reproductive behavior. Low light intensity may be the chief factor limiting black spruce reproduction beneath a canopy of existing trees.

Discussing the successional relationships of the tree species in Saskatchewan, Dix and Swan (1971) point out that aspen, jack pine, balsam poplar, and white birch are all pioneer species unable to reproduce in an established forest. Seedlings and saplings of white spruce and black spruce in an established stand represent an initial incursion following a disturbance, rather than a continuing ability to invade beneath a closed canopy. Balsam fir, on the other hand, may occur in advanced stages of the forest, and it is usually not found immediately after a disturbance. Balsam fir is the only long-term invader, but it is not important in succession since it is abundant only rarely in Saskatchewan where it approaches the western limit of its geographical range.

Fire is without question the most important cause of the initiation of pioneer stages of succession. It is so ubiquitous that Dix and Swan could make no conjecture as to the nature of the forest that might develop were the area free of fire for a protracted period of time. They point out: "For Candle Lake forests as a whole, succession does not seem to be important. It seems more important that the area has probably undergone an infinite number of vegetational readjustments." They cite Rowe (1961) and confirm his opinion that the western boreal forest is a disturbance forest, one that does not fit classic concepts: "In view of these observations, any attempt to fit the vegetation into the mold of a climax concept would be unreal and, in our opinion, unjustified."

The relatively greater richness of the understory flora in southern regions broadens somewhat the possible patterns of succession in southern as compared to northern regions as far as the understory is concerned. Thus, Black and Bliss (1978) point out that in the black spruce stands near Inuvik, close to the northern limit of forest in the lower Mackenzie River region, no successional replacement of dominant vascular plants is found, as most shrubs and herbs simply resprout after burning and soon achieve preburn prominence.

In contrast, an orderly change of cryptogamic species is found, suggesting a continuous succession. Four stages in this sequence are described by Black and Bliss based on growth-related physiognomic changes and species changes with time since burning. Vegetation in general is homogeneous across the tree line in the *Picea mariana – Vaccinium uliginosum* woodland, with no measurable differences in species within and often between stages for most vascular plants. Only *Ledum palustre, Rubus chamaemorus, Eriophorum vaginatum,* and *Alnus crispa,* and mosses of the genus *Sphagnum* are found to increase in cover and frequency near the forest line and in the forest–tundra. *Cetraria* lichens are also more frequent in northern regions, but these also increase in frequency in older more southern stands. Most shrub species (i.e., *Salix glauca, S. pulchra,* and *Ledum groenlandicum*) decrease in cover but are generally present in northern stands. Only *Alnus crispa* dramatically increases in forest–tundra stands. *Betula glandulosa* is generally present but is a minor component of this woodland type. Other tree species (*Picea glauca, Betula resinifera,* and *Larix laricina*) are occasionally present in sampled stands but are absent by 150–180 years after a burn. *Larix laricina,* a minor component (less than 1% stems) of southern stands south of lat. 68°10′N), is not found in northern stands. Both *Picea glauca* and *Betula resinifera* are found in stands south of the forest line, but only *Picea glauca* is found north of the forest line and not in sampled stands (Black and Bliss, 1978).

In contrast to the relatively minor effects of fire on the vascular plants of the understory vegetation, the effect on black spruce is profound. Black and Bliss (1978) state that fire may play a role in creating shrub tundra in the Inuvik area on sites where black spruce forest has burned. In the absence of fire, there is maintenance of black spruce in the forested areas by vegetative layering. The ultimate destruction of black spruce stands is brought about by burning or cutting for firewood. In the Ennadai Lake area, at the northern edge of the forest–tundra transition, there is evidence that cutting of spruce for firewood by Eskimos has resulted in the establishment of tundra on areas once occupied by spruce clones (Larsen, 1965).

Observations by Strang (1973) on the uplands of the lower Mackenzie River valley, Northwest Territories, between Fort Good Hope and Point Separation, suggest that here the open spruce forest eventually dies out if fire is excluded, to be replaced by an almost tundralike vegetation. In areas where fire has been absent for 150 years or more there exists a dense growth of lichens, mainly *Cetraria nivalis* and *Cladonia* species, few or no tree seedlings, an open, unhealthy spruce stand, and a permafrost table close to the soil surface. In burned areas, the ground vegetation is initially the liverwort *Marchantia polymorpha,* followed within a few years by herbs, mosses, and low ericaceous shrubs, and then by a dense and vigorous growth of trees. Because tree seedlings evidently fail to become established in dense lichen mats, it appears that, without fire, the stand would disintegrate and disappear.

By way of contrast, in more southern boreal regions there is both a somewhat wider range of successional pathways and a much richer complement of understory species. There are greater differences among stands in the composition and structure of the understory association;

the flora is richer, there is a larger number of tree species capable of attaining a dominant role, light and soil conditions afford a wider range of possible combinations, and, in general, the understory associations are more varied, tend to differ in species composition beneath different associations of the dominant trees, and are subject to changes in composition as the canopy above changes as a result of whatever cause.

In the Chena River region of Alaska, an area where balsam poplar is commonly the pioneer on burned sites, succession involves ecesis by white spruce and, later, black spruce. As Viereck (1970) notes, seedlings of white spruce become established at some point in the development of the balsam poplar stands. As spruce litter accumulates, a thick moss mat develops in the stand. Thawing becomes progressively slower, and a permafrost layer is formed. This prevents drainage and creates a wet substrate on which *Sphagnum* mosses grow. These conditions favor black spruce over white spruce, and the former slowly attains dominance. As Viereck describes the transition*:

> These conditions favor black spruce, and the former becomes established in the stand. As succession continues, the permafrost layer rises closer to the surface. If the organic layer has developed sufficiently, the mineral layer may be entirely frozen and all tree roots become located in the organic layer. Nutrients in the organic layer are depleted, and some elements such as phosphorus, nitrogen, and manganese may become limiting. Soils are cold throughout the summer and waterlogged when not frozen. Black spruce grow slowly with an understory of ericaceous shrubs. Sphagnum mosses find optimum conditions for growth, thus creating an even thicker moss layer in a waterlogged condition.

Density and cover of willow and alder are high in the early pioneer stages and greatly decreased in the later stages of succession. *Salix alaxensis, S. pseudocordata,* and *Alnus incana* ssp. *tenuifolia* are all pioneer species that are intolerant of the shade under a canopy of balsam poplar and spruce. *Rosa acicularis* and *Viburnum edule* become established in the willow stage but reach maximum development in the balsam poplar stage where *R. acicularis* forms a nearly continuous shrub layer. In the black spruce–sphagnum forest, *Spiraea beauverdiana, Ledum groenlandicum, L. decumbens, Vaccinium uliginosum,* and *V. vitis-idaea* form a nearly continuous layer.

Herbaceous species, in the stands studied by Viereck, show great differences among stands in the successional sequence. Of the herbaceous species found in the willow stage, only one, *Calamagrostis canadensis* occurs in all other stands. *Linnaea borealis, Cornus canadensis,* and *Pyrola secunda* also occur in the white spruce–black spruce stage, but only one,

---

*Reproduced with permission of the Regents of the University of Colorado from *Arctic and Alpine Research.*

C. canadensis, occurs in what Viereck considers the climax. The only herb unique to the black spruce–sphagnum stage is *Eriophorum vaginatum.*

Three pioneer mosses, *Rhacomitrium canescens, Polytrichum juniperinum,* and *Ceratodon purpureus,* and three lichens, *Stereocaulon rivulorum, Cladonia macilenta,* and *C. pyxidata,* are not found in other stands. *Rhytidiadelphus triquetrus, Eurhynchium pulchellum,* and *Hylocomium splendens* form a nearly continuous layer in the white spruce and the white spruce–black spruce communities. Patches of *Sphagnum* occur in wet depressions. In all the spruce stands, lichens occur as isolated clumps in the moss mat. In the black spruce stands, *Sphagnum* species form thick mats. Forest structure, thus, appears to be best described as a complex of continuous variation.

In different regions, thus, individual species achieve optimal performance. In areas of southern central Canada, balsam fir and black spruce dominate stands having a dense moss cover and a poorly developed vascular understory. Paper birch, white spruce, jack pine, and trembling aspen dominate stands having a rich growth of vascular plants and only few mosses. In the southern boreal areas of both central Canada and Alaska, some of the understory species have distribution patterns specifically related to individual tree species. To the east, in the area around and immediately north of the Great Lakes, Maycock and Curtis (1960) have found that the hypothetical climax has little meaning in areas continually subjected to fire, windthrow, and budworm infestation. These and other disturbances continually cause a reversal of successional development and permit the continuous entrance of species suited to pioneer conditions. In the spruce stands of the peatlands of northern Minnesota, Heinselman (1970) has found essentially the same situation. He states*:

> If any direction is apparent, it is namely a trend toward elaboration of landscape types in the Lake Agassiz Peatlands Natural Area. Mostly the impression is one of ceaseless and almost random change, initiated by the innumerable local or regional events. The plant communities and peatland types do change predictably if the full sequence is known, but what occurs in any given area depends on an extraordinary complex series of interactions. Such a course of events can hardly be viewed as a fixed succession.

It is of interest that the performance of balsam fir tends to become more vigorous in the eastern sections of the region. MacLean (1960) describes the mixed wood community in the boreal forest of Ontario as commonly dominated by various mixtures of trembling aspen, white birch, black spruce, white spruce, and balsam fir. On shallow soils, black spruce is the common softwood component and persists for long

*Reproduced by permission. Copyright 1970 by the Ecological Society of America.

periods of time without replacement. The black spruce reproduces both by layering in patches of moss and by seedlings that become established on mineral soil. Mature or overmature stands are usually destroyed by fire under natural conditions; repeated burning may be disastrous, since the thin humus is destroyed and little but bare rock remains. Most stands have originated, however, from fire and, when it does not occur, disintegrate with overmaturity; balsam fir seems to be the only species that improves its position in such cases. Balsam fir reproduces in moderate shade and is not particularly demanding as to seedbed requirements. White birch seedlings may be abundant under the resulting mixed wood stands, but aspen and mature birch are rare, since both are intolerant of shade. It has been noted that fir seedlings usually begin to appear in an aspen–birch–spruce stand when the dominant trees are about 50 years old, providing there are sufficient balsam trees present to furnish ample seed. In another 20 years the fir seedlings become numerous. There is a tendency for balsam fir to take a more aggressive role in the central and eastern portions of its range, and Carleton and Maycock (1978) note that, in the absence of external catastrophe, some upland primary boreal forests seem to develop into secondary fir and/or spruce forests. Also apparent, however, are old, decadent spruce stands typified by complete absence of subsequent development or by considerably delayed regeneration; balsam fir only rarely attains dominance in a forest, and its success is due mostly to its ability to establish itself beneath the canopies of a wide range of other forest species.

Carleton and Maycock add that deforested sites in boreal regions become reforested far more rapidly and extensively than is the case in southern forests, and boreal tree species are highly productive in their youth so that they can gain a competitive advantage of height over shrub vegetation. The rapid reestablishment of highly vigorous tree populations following reforestation can be viewed as a consequence of selection for survival and competitive success in an unpredictable environment.

NEOCLASSIC SUCCESSION CONCEPTS

Early misconceptions regarding succession in northern forests may have resulted from the observation of inadequate numbers of stands at different stages of development. Ecologists assumed that one or a few observations of stages in succession were representative of a predetermined course of successional events. That their observations did not necessarily represent the range and variety of possible successional pathways was not at once apparent. The more realistic view, which holds that a much more varied and diversified number of pathways and stages in succession is possible is now replacing the old classic concepts. With advanced techniques it is possible to express the interaction of a multiple species complex with the environment, and successional changes occurring with time can be expressed as a vector, for example, by employing the technique of Goff and Zedler (1972) who state: "The concept of ordination is thereby enriched so that the array of vegetation within the hyperspace is not conceived as a static pattern but rather as a dynamic field."

In the James Bay region of Ontario where boreal forest stands were studied by Carleton and Maycock (1978)*, there appeared to be little evidence of succession:

> In general, the boreal species succession vectors tend to be short, circular, and divergent in their placement rather than long, linear, and convergent; that is, species do not show any tendency to progress toward a single climax type. An exception which does show a continuous linear trend is *Abies balsamea*. ... *A. balsamea* does not show any convergence toward other, more climax, species in the boreal models. Nor does *A. balsamea* seem to persist as a self-regenerating climax type....
>
> As seen with *Abies balsamea*, the longer a succession vector the greater the proclivity for association with other species at the various phases of growth. The shortness of most of the species vectors indicates that interspecies association is neither strong nor does it change very much during stand history.... *Pinus banksiana* exhibits a short, circular vector which is isolated from other species. Extremely monospecific forests of jack pine are a common feature of the region....
>
> The other species with short vectors seem to pair off on the ordination. Thus *Populus tremuloides* and *Betula papyrifera* lie close to one another, and although monospecific forests of these species are common, some admixture between them is not infrequent in the region. Similarly *Picea glauca* and *Populus balsamifera*, while growing mostly in monoculture on the higher and lower fluvial silt terraces of river bottoms, do frequently occur together in stands. *Picea mariana* and *Larix laricina* also tend to grow in association on those few wet, peaty sites in which tamarack occurs.... On certain deforested sites, *Picea mariana* can act as a primary colonist in a similar manner to *Pinus banksiana*. This seems to be the case in the more northern sections of the boreal forest region.... Postfire sites of the Cochrane clay belt in which patches of *Sphagnum* moss have survived provide the most favourable conditions for primary colonization by black spruce from seed. While jack pine is well known as a species possessing serotinous cones, releasing seed only after fire, black spruce is reported to be semiserotinous (Ahlgren 1974), i.e., some cones open at maturity to release seed while others are retained in a closed state next to the main axis of the tree crown.
>
> *Picea mariana* is also seen to behave as a shade-tolerant species which can invade beneath other tree canopies. However, this pattern is less pronounced than the similar pattern of *Abies balsamea*.
>
> The pattern which emerges here for jack pine forest is rapid establishment as a monoculture over fire-prone sandy or rock sites (cf. Cayford et al., 1967), with subsequent invasion and establishment in most cases by black spruce and (or) balsam fir. In approximately one-quarter of the older stands, however, no secondary species

*Reproduced by permission of the National Research Council of Canada.

establish in the understory and these pine forests show signs of opening up to become savannahlike. The reasons for this might be either the destruction of an established seedling and sapling population by, for example, a ground fire or the inability of black spruce seedlings to establish in certain pine forests.

As with the jack pine forests... the examination of structure in the trembling aspen stands largely confirms the pattern shown in the succession vector schemes. The preponderance of stands of younger to moderate ages suggests the recurrence of frequent catastrophic cycles in which trembling aspen establishes as a monospecific forest. Although other species may subsequently invade and establish as saplings, few stands appear in a state of transition from aspen dominance to dominance by either spruce or fir.

Conditions of light and shade in the boreal forest have been accorded much importance in establishing the structural patterns and reproductive success of the various species. The shade beneath a closed-canopy forest, thus, has been invoked as the cause of reproductive failure in the case of a number of boreal tree species, but Karpow (1961, 1962) has demonstrated that, at least in Eurasian forests, competition from the roots of other plants is also important. He points out that light is usually considered the limiting factor in the development of undergrowth plants and, presumably, in the development of tree seedlings as well, but when roots were cut to a depth of 50 cm around a test plot, plants of the herb layer developed much better in spite of unchanged light conditions. The limiting factor was the amount of available nitrogen. Plants on control plots exhibited clear symptoms of nitrogen deficiency. In contrast, the leaves on plants on experimental plots were deep green. The experiment shows that the poorly developed herb layer in boreal spruce forests is primarily the result of a low nitrogen supply, the consequence of competition from tree roots. Mosses receive nutrients through drip water and are able to grow vigorously. Tree seedling growth is also inhibited by root competition from older trees. Spruce seedlings grow slowly in birch forests in spite of favorable light conditions. They grow much better when competition is eliminated by cutting birch roots. When control and experimental plots were uniformly fertilized with ^{32}P, radioactivity was five to six times greater in seedlings growing in experimental plots than in seedlings in control plots. No inhibiting substances or root secretions from the trees were found that could have had any influence on the controls (Walter, 1971).

When all these studies are taken into consideration, it is apparent that many questions remain concerning the nature of succession in boreal regions. There are apparently many regional differences that can be elucidated only when additional detailed studies are undertaken. There is still much to be learned: Many important aspects of boreal community dynamics are yet to be revealed. The field is one in which exciting developments are imminent.

PERMAFROST AND SUCCESSION

A feature rendering the boreal forest distinct from all other forest types is the presence, in the northern parts of the forest, of permafrost, a permanently frozen zone underlying the upper or "active layer" of soil that thaws during the spring and summer months. The presence of permafrost in northern boreal lands creates conditions found in no other region except the arctic tundra, and it appears reasonable to assume that it is the presence of permafrost, along with other climatic influences such as the short, cool growing season, that creates environmental conditions inimical to the growth of trees north of the zone known as the northern forest border or the northern forest–tundra ecotone.

In the Far Northern lands, permafrost is continuous and in many regions extends to depths of hundreds of meters. Permafrost is continuously present beyond depths of 20–50 cm on the average over most of the arctic regions. In the northern sub-Arctic, in such areas as Churchill, Manitoba, at lat. 56°N, the active layer varies in depth with the type of substrate material, ranging from 200–250 cm for sandy soil to 100–200 cm for clay and 90–170 cm in areas of tussocks and hummocks. In contrast, at Resolute Bay, Northwest Territories, at lat. 75°N, the depth of the active layer is about 5 cm in clay and 8–12 cm in sand and gravel. Southward the permafrost becomes discontinuous, existing only on sites where the temperature regime of the ground is such that summer heating is insufficient to warm the deeper soils above the melting point of ice.

Permafrost in peatlands near the southern limit of discontinuous permafrost is found usually in raised peat landforms called palsas and peat plateaus. Palsas are mounds generally much less than 100 m in diameter, varying in height between 1 and 5 m (R. J. E. Brown, 1969, 1970; Zoltai and Tarnocai, 1971; Zoltai and Pettapiece, 1974). Peat plateaus are flat, elevated peatlands seldom exceeding 120 cm in height but covering large areas. Both have a perennially frozen core. Coalesced palsas may form peat plateaus that reach a height of 4–5 m. The low form of peat plateau occurs throughout the discontinuous permafrost zone, covering larger areas of peatland toward the north. In Canada, palsas are most frequently found in the Hudson Bay region. The vegetation cover varies from a low shrubby community to stunted open stands of black spruce, and, near the southern fringe, palsas and peat plateaus may be heavily wooded with black spruce or white birch. The palsas and peat plateaus of subarctic Canada are morphological variations of the same process, with variations chiefly due to differences in the availability of water.

On the southern fringe of the permafrost zone, vegetation plays an important role in the development of peat plateaus. Thin layers of permafrost develop under dense clumps of black spruce growing in a fen or bog. Snow cover typically is very thin under these dense stands, so that frost penetrates deeper than in the surrounding fen. The shade retards melting in summer. Because of the greater volume of water in the frozen state, and because of the lower relative weight of ice, doming takes place. The raised peat is dry, and its insulating effect protects the permafrost in summer; the thickness of the permafrost increases and is maintained eventually at equilibrium. Finally, wind, disease, fire, flooding, or cutting of trees induces decay of the plateau or palsa and it disappears.

In peat plateaus the permafrost is located exclusively within the peat. In palsas, although covered by peat, permafrost extends into the underlying mineral soil, firmly anchoring the palsa by the frozen core. The palsa, raised above the water table, offers an environment that is not too wet for the growth of black spruce trees, and the thickness of the active layer may average 60 cm under forest vegetation; it is least under a dense cover of trees and greatest under openings in the forest.

The physical action of permafrost upon root systems has been described by Benninghoff (1952): "In the high latitudes roots are not only encased in frozen material for a great part of the year, but by repeated freezing and thawing, especially during the autumn freeze-up, they are heaved, torn, split by forces of great strength." Moreover, the influence of increased plant cover upon the environment, contrary to the increased mesophytism commonly observed in the temperate zones, is far from salutary from the point of view of continued existence of the vegetational community.

> Plant succession in temperate regions tends to establish more mesophytic conditions in which drainage relations are less extreme. But in regions of severe frost climate, plants commonly generate conditions of extreme lack of drainage and greatly intensified soil frost; in short, the plants frequently destroy the very environmental conditions that favor their growth.

Vegetation increases the rate of accumulation of a cold reserve in the ground; also it decreases the velocity of air currents at ground level and thus permits temporary increases in temperature of aboveground plant parts well over that of the surrounding atmosphere during periods of high insulation. Thus, although conditions below ground surface favor low temperatures, aboveground parts often have temperatures appreciably higher than the surrounding air. Conifer needles may transpire during periods of warm weather while roots are still encased in ice; the result is "winter kill" in which the needles become desiccated and die from lack of moisture.

It seems intuitively reasonable to expect strong and readily apparent influences of permafrost upon vegetational associations in boreal regions, and studies that have been undertaken indicate that this is, indeed, the case. But the number of studies in which these interactions are revealed has so far been few. Dingman and Koutz (1974), however, have accomplished detailed mapping of vegetation, permafrost, and an index of solar radiation in the discontinuous permafrost zone of central Alaska and show that permafrost occurs where the average annual solar radiation is less than about 265 cal/cm^2/day. Where radiation is greater, a birch–white spruce forest has developed; the boundary of this forest corresponds closely to the permafrost boundary. Where permafrost is present, usually on north slopes, a black spruce forest with a thick moss cover is found.

In the studies on succession carried out by Viereck (1970) in central Alaska, changes brought about by the development of a thick organic layer result in the development of permafrost. Succession proceeds from the mesic balsam poplar and white spruce stages to a hydric slow-growing black spruce stage with a thick saturated sphagnum mat on a permafrost table only 20–30 cm below the moss surface.

A detailed study by Zoltai (1975) in the valley of the Mackenzie River, extending from lat. 61°30'N to the tree line near the Arctic Ocean, where most of the area is in the discontinuous permafrost zone, describes in detail the earth hummocks that have been observed to develop on fine-grained soils underlain by permafrost (Larsen, 1972; Zoltai and Tarnocai, 1974). These hummocks are small (40–50 cm high), permanent earth mounds, circular in outline, with a diameter of 1–2 m. Their internal structure shows buried organic layers in streaks and lumps, indicating intensive mixing. The hummocks are believed to be formed by an upward displacement of mineral soil in the center and a concurrent downward movement at the borders under the influence of frost action. In the Mackenzie Valley north of the Arctic Circle over 80% of the mineral soil surfaces have hummocks (Zoltai and Tarnocai, 1974).

The vegetation in the region reflects climatic zonation as modified by local factors. Southward, in the Fort Simpson area closed-crown forests of white and black spruce with a feather moss carpet are common. Further north, near Wrigley, the trees (mainly black spruce) are shorter and widely spaced, with reindeer lichen and moss ground cover. This trend toward shorter trees and wider spacing increases northward. North of Inuvik, small stands of spruce are separated by treeless tundra. The hummocky microrelief has a marked effect on the distribution of trees. Far fewer trees grow on the tops of hummocks than on the sides or in interhummock troughs. The proportion of trees growing on the sides and in troughs is about equal both in the south and in the north. The surface layers of hummocky terrain are subject to frost heaving, and trees are subject to tilting. The effect is especially pronounced in uneven-aged, undisturbed stands. Older even-aged stands also show tilted trees, but trees in younger stands still grow dominantly upright.

Zoltai has found that the distribution of lesser vegetation also reflects the moisture regime of the microtopography. Fruticose lichens dominate dry hummock tops, and mosses dominate the troughs. Feather mosses tend to cover the tops of the hummocks if left undisturbed for a long time, resulting in a complete moss cover.

Fires play a dominant role in the vegetational succession on hummocky ground. Immediately after a fire, liverworts and mosses, along with fireweed (*Epilobium angustifolium*) occupy the ground. Almost immediate establishment of black spruce occurs, and white birch may also appear. A century after a fire the ground vegetation and the tilted trees give the fire-originated black spruce-lichen forest the same appearance as undisturbed stands.

Zoltai asserts that not all spruce-lichen woodlands originated after fires. The trees were established in existing woodlands, indicating that spruce-lichen woodlands can perpetuate themselves without disturbances and, under some circumstances, may be climax communities.

SUCCESSION IN PEATLANDS

The long-term successional trend in peatlands south of the zone of permafrost appears to be a continual accumulation and thickening of the peat, accompanied by a rise in the water table and nutrient depauperization (Heinselman, 1963, 1970; Small, 1972a,b; Jeglum, 1972, 1973; Moore and Bellamy, 1974). As peat accumulates, it restricts drainage. Nutrient materials become tied up in undecomposed plant detritus. The main supply of moisture is from precipitation; the water found in surface horizons is consequently low in mineral nutrients unless it is in contact at some point with mineral soil. Because of slow decomposition, peat deposits persist for long periods of time, and bogs and muskegs are among the most stable types of vegetation in southern boreal regions. The driest stage of development, according to Jeglum, is the stage at which the muskeg community is dominated by black spruce. Fire and drought then cause a retrogression to fens or swamps. Without such a disturbance, however, it seems probable that muskegs and treed bogs, possessing relatively high water tables and a high degree of nutrient depauperization, represent a stable and advanced stage in the developmental trend from bogs and fens.

Peatlands have developmental patterns that appear to have an end point not greatly at variance from that of uplands. From a bare soil surface on uplands, the successional trend is toward increasing accumulations of plant materials which are slow to decompose and which prevent moisture in subsurface strata of the soil from evaporating. Eventually, peat accumulates to a depth of several feet, mineral nutrients become tied up in undecomposed plant materials, and mineral soil horizons can no longer be reached by the roots of most or all plants. Beneath deep peat accumulations, the mineral soil is persistently cold although only in the more northern areas is it encased in permafrost. Water from precipitation percolates downward and runs off in groundwater; as a consequence, there is little movement of nutrient materials upward, and a large proportion of the nutrients released by decomposition are washed away unless taken up immediately by plant roots. In upland areas, the trend toward development of thick peat deposits low in nutrients is not often carried to the extreme point, however, largely because of the great prevalence of fire. On islands or other uplands protected from fire, forests in advanced stages of development can occasionally be seen; forests on such sites do not differ greatly from forests found on muskegs possessing deep peat accumulations. In the forests on both sites, the conditions of peat development, nutrient availability, and moisture supply are very similar, despite the wide disparity in underlying topography and the different ways in which the forests developed. The accumulation of peat creates conditions, however, that do not conform to the classic concept stating that as succession proceeds there is an increase in mesophytism in the conditions prevailing in the forest—in short, that temperature extremes are moderated, moisture conditions tend to become neither too wet nor too dry, and nutrients become increasingly available. Instead, in boreal regions, peat accumulation creates conditions that tend to cause deterioration in the forest stand, resulting evidently in eventual replacement of the forest trees with open muskeg or bog. The conclusion in regard to boreal regions is that the classic concepts of climax do not apply.

As the peat thickens, the upper strata become isolated from nutrients, spruce growth becomes progressively dwarfed, and sphagnum mosses finally dominate. The peatlands are in a state of continual change, however slow, and the concept of climax seems meaningless. While the existence of succession cannot be questioned—if succession is taken to mean progressive change of one kind or another—the classic concept of succession as leading to a single, enduring "climax" community characteristic of a region cannot be considered tenable in boreal community dynamics. Some general outlines of the successional stages a given community is likely to pass through are now emerging, however, at least for communities occupying the more prevalent landform types; and this is indeed significant, for the knowledge will help immensely in understanding such important aspects of ecosystem dynamics as nutrient cycling and productivity.

THE ECOLOGY OF FIRE

Of the environmental events that disturb a forest stand, fire is the most widespread, frequent, and pervasive in its influence. Viewed from the air, expanses of natural forest present a complex mosaic of successional stages. Only when the effect of fire is recognized can the apparently haphazard pattern be explained. The pattern may not be due exclusively to fire; there are the added influences of topography, depth to water table, soil development, and permafrost. But when the mosaic appears not to conform in any discernible pattern to topography, streams, soils or permafrost, it is likely that fire history can be invoked as the cause. Sharp boundaries between stands of spruce and stands of aspen, birch, or other pioneer species are usually the edges of burns. Small, isolated stands of spruce, long stringers of mature trees flanked by streams, widely scattered large trees—all are often relicts of stands destroyed by fire (Rowe, 1971; Rowe and Scotter, 1973; Johnson and Rowe, 1975).

The widely distributed boreal broadleaf trees and conifers are well adapted to fire. Aspen, balsam poplar, white birch, and jack pine are usual pioneers after fire. Spruces, both white and black, are frequently early invaders. They are inconspicuous at first because of slower growth rates but later appear as components in mixed stands which they eventually will dominate. Only balsam fir seems poorly adapted to survival after fire, and balsam fir in abundance indicates a relatively moist, fireproof site.

The severity of fire has an important bearing on the success or failure of spruce regeneration. Light surface burns in early spring when humus is wet do not provide suitable conditions for spruce seedlings. The perennial parts of broadleaf trees, shrubs, and herbs are not destroyed by light fires, and such plants are stimulated by the opening of the canopy and the release of nutrients to the soil that result from fire. Severe fires that consume the humus layer are effective in eliminating broadleaf competition with spruces. The exposed or thinly covered mineral soil provides a suitable seedbed for spruce germination and seedling survival (Fig. 29).

• A deep accumulation of fine, dry fuel, a prerequisite for severe fire, is found more often under spruce than under broadleaf trees. In mixed stands there is thus a tendency for severe burning to occur under spruce and for light burning to occur under aspen. Spruce seedlings tend to reappear in the former and aspen sprouts in the latter locations.

• White spruce is very susceptible to destruction by fire. Its thin, easily damaged bark and shallow roots are severely affected by surface fires; severe fire will burn living roots as large as 8–9 in. in diameter. Mounds of cone scales around the base of some trees provide fuel for hot and

Fig. 29. Fire has a devastating effect upon a stand of black spruce, but fires are nevertheless a natural occurrence to which black spruce has adapted. Here a young forest has been burned, but serotinous cones at the tops of the dead trees will result in reseeding of the area, and spruce regeneration will soon occur.

persistent fires; the mounds are cone caches and feeding stations for red squirrels. A heavy growth of beard lichens also favors the spread of fire. Fires are especially destructive to white spruce seed. Cones open at maturity and, following a fire, regeneration of white spruce must generally originate from seed blown in from unburned areas.

Black spruce growing on uplands is as susceptible to destruction by fire as white spruce, but in wet lowland areas the chance of complete destruction is much reduced. Most seedlings of black spruce in burned-over areas are from seeds in serotinous cones unopened at the time of the fire. The regeneration of black spruce is similar in this respect to that of the other "fire species"—jack pine, lodgepole pine, and pitch pine.

Young birch is readily killed by fire because of its thin bark. The bark later becomes thicker, but it is then more flammable. Regeneration by sprouting from dormant buds around the base of a stump contributes to restocking, although regeneration from this source is not as important as seedling reproduction.

Quaking aspen is killed by hot fires, but in pure stands the accumulation of material on the forest floor is usually too light to carry a hot, persistent fire. Even if the tree is killed, new sprouts arise abundantly from the roots of the dead tree. Balsam poplar is probably the most resistant to destruction by fire of all the trees in the boreal forest. Bark thickness near ground level is often four or more inches on mature trees. Surface fires in balsam poplar stands also tend to be light because of the relatively thin accumulation of forest-floor material. Regeneration by root suckers is common in balsam poplar stands, and seedling reproduction can be abundant where mineral soil has been exposed and a source of seed is available. Both aspen and balsam poplar produce quantities of seeds widely disseminated by wind.

Shrubs are especially vulnerable to destruction by fire. Most of them, however, regenerate by means of sprouts arising from stem bases and roots. Willows produce minute, hairy seeds admirably adapted for wind dissemination, and they are usually well represented in the plant community in recently burned areas.

The effect of fire on herbaceous plants varies, because herbs differ greatly in life form, seed habits, and other characteristics. Regeneration in species appearing first after fire is often vegetative. Many if not most species invading burned areas are those with seeds easily disseminated by wind. Species producing large quantities of easily disseminated seed at an early age have the advantage. Germination of seed and survival of seedlings is highest where mineral soil has been exposed, and species reproducing vegetatively tend to prevail where the burn has been less intense. Mosses reproduce both by spores and vegetatively. Lichens are easily destroyed by a surface fire, and both soredia and fragmentation are involved in reproduction after fire. Lichen disseminules can be carried for long distances by wind blowing over the surface of snow in winter.

The density of vegetation recolonizing a recently burned area increases with the passage of time. Spruce, paper birch, aspen, and balsam poplar appear in burned areas within a year after fire. In burned areas of Alaska, for example, the average numbers of spruce, paper birch, and poplar seedlings and sprouts per acre were 2000, 8000, and 5000, respectively. In an area near Candle Lake, Saskatchewan, fire had been so frequent and pervasive in its influence that for the forest as a whole succession did not seem important. The area had apparently undergone many fires, followed by an equal number of postfire recolonizations. The Candle Lake area maintained highly productive forest populations near the starting point of succession; many pioneer tree species were represented (paper birch, jack pine, aspen, black spruce, white spruce, balsam poplar) but only one replacement species (balsam fir). Poor reproduction in many stands even suggested that a parkland might develop in the absence of disturbance (Dix and Swan, 1971).

Records of the frequency of fires in Canada show that their incidence generally decreases northward in the boreal forest zone (Johnson and Rowe, 1975). Observations by the author at the northern forest border in Keewatin indicate that, at least in this region, fire may be one mechanism by which the climatic boundary is maintained, but it is climate and not fire that explains why the forest has not invaded the tundra in recent times (Larsen, 1965; 1972). Buried fossil podzol soils and charcoal dated by carbon-14 methods show that forests extended north of Ennadai Lake as far as Dubawnt Lake at least once in postglacial time, presumably during a period of ameliorated climate. The clumps of black spruce presently found in the area north of the present forest border extending to Dubawnt Lake appear to represent relicts that survived the fires and now occupy rare favorable sites. The potential for spruce reproduction and spread of the forest over the landscape thus exists, and spruce forest would expand northward if environmental, primarily climatic, conditions were to become favorable. The restriction of spruce growth rates northward supports the view that spruce here reaches the climatic limit at which it can persist as a major component of the vegetation and as a forest tree, and that, in a relatively narrow zone north of the present-day forest border, it is capable of persisting by vegetative reproduction only in isolated clumps on rare favorable sites (Elliott, 1979a,b).

Raup observed long ago (1946) that very intense burning on dry sites, or repeated burning at short intervals on the same site, completely destroyed the plant cover. On the sandy uplands in the area he studied west of the Slave River this usually resulted in a dense pure stand of jack pine. A commonly observed forest type is one of young, vigorous spruces mixed with a scattered stand of tottering old jack pines. Thieret (1964) points out that, in southern Mackenzie, it is not at all unusual to find an island of unburned pine or spruce surrounded by burned forest in which a vigorous stand of young aspens is developing.

The role of fire in the northern coniferous forests of North America, thus, has been the subject of a large number of investigations. Fire is a recurring event that profoundly influences the structure and composition of the forests. Tree ring analyses, utilizing fire scars, as well as radiocarbon analysis of charcoal in soils, permit dating of the fire and initiation of succession so that postfire development of forests can be followed. Even so, very few generalizations can be made regarding postfire recovery sequences of the ground vegetation (Ahti, 1959; Bergerud, 1971; Shafi and Yarranton, 1973; Scotter, 1971). Maikawa and Kershaw (1976) found large differences between recovery sequences reported in the literature and concluded that it was clear that other factors in addition to the actual burning may complicate the primary effect of fire. Before generalizations can be made, they indicate, a knowledge of the consequences of fire in many different regions is necessary.

East of Great Slave Lake, for example, Maikawa and Kershaw sampled vegetation in burned areas of differing ages in lichen woodland in the Abitau–Dunvegan lakes area. They established four arbitrary phases of recovery: (1) the *Polytrichum* phase, an initial colonization period dominated by *Polytrichum piliferum* and lasting about 20 years; (2) the *Cladonia* phase, in which *Cladonia stellaris* and *C. uncialis* are particularly important, lasting to year 60; (3) the spruce–*Stereocaulon* phase, which lasts until about year 130, when increasing tree density and a closing canopy largely eliminate the lichen cover; and (4) the spruce–moss woodland phase, which develops with a ground cover dominated by *Hylocomium splendens*, *Pleurozium schreberi*, and *Vaccinium vitis-idaea*. The average reburn interval is less than 100 years and, in the absence of fire, the canopy closes and the lichen surface is replaced by a moss mat. It is evident, however, Maikawa and Kershaw say, that the sequence of phases leading to the final establishment of closed-canopy woodland contrasts with recovery sequences established for comparable woodlands in different regions of northern Canada. Four stages in postfire succession were also discerned by Shafi and Yarranton (1973) in the clay belt of Ontario. They found a different maturing of sites within areas burned at one time. The path of succession on different sites within areas burned at the same time also differs at certain stages of succession. The complex mosaic tends toward convergence to a few types at late successional stages. Shafi and Yarranton point out that evolutionary pressures in a repeatedly disturbed environment have resulted in species adapted for rapid growth and reproduction in the average time available before the recurrent incidence of fire, and for effective regeneration after fire (species with serotinous cones such as jack pine and black spruce). The species of the whole successional series must be regarded as adapted to survival of fire, rapid regeneration, and rapid reproduction. It is not surprising, therefore, they add, that most species are perennial, are capable of rapid vegetative reproduction, and either possess underground perennating organs or adaptations effective in rapid postfire propagule establishment. Carleton and Maycock (1978) have noted that deforested sites in boreal regions become reforested far more rapidly and extensively than those in southern forest regions. Seed production and dispersal occur at an early phase of tree growth before destruction recurs. Carleton and Maycock studied the closed boreal woodland of the Ontario clay belt, where white spruce and balsam poplar forests predominate and black spruce forests occupy a wide range of habitats from wet lowland bogs to very dry rock outcrops. Jack pine forms even-aged stands over outwash sand plains and, in contrast, a well-developed balsam fir forest is relatively infrequent, usually found on moist, sandy soil in an upland situation. This species, however, is abundant in a wide range of deciduous and mixed forests dominated by other tree species, mostly trembling aspen or paper birch.

Using a technique for following succession by means of vectors derived from ordinations of successive stages of forest regeneration after fire, a technique devised by Goff and Zedler (1972), Carleton and Maycock have found that, in general, the succession vectors of the dominant boreal species are short, circular, and divergent; that is, the species studied do not show any tendency to progress toward a single climax type. Balsam fir, unlike the other dominants, has a continuous linear trend, but it does not show any convergence toward other species in the boreal models nor does it persist as a self-regenerating climax type.

The pattern that emerges for jack pine in the study by Carleton and Maycock is rapid establishment as a monoculture over sandy or rocky sites, with subsequent invasion and establishment in most cases by black spruce and/or balsam fir. In some of the older stands, however, no secondary species had become established, and these forests showed signs of opening up to become savannalike.

Trembling aspen stands of moderate age with either a black spruce or balsam fir understory are very comparable in overall structure, Carleton and Maycock have found. Generally, a fir understory occurs beneath an aspen canopy in more southern stands on relatively calcareous substrates, while more northern stands on acid substrates favor the predominance of a black spruce sapling layer. Some black spruce and balsam fir stands have clearly been derived from a previous trembling aspen forest. A few older, decadent stands do occur in which sapling regeneration, although not completely suppressed, appears to be considerably arrested.

The structural classes for forests dominated by black spruce are the most varied of all the examples studied. Upland black spruce forest found on rocky knolls and well-drained sandy sites in the region seems to derive from primary establishment in association with a small amount of jack pine. As the black spruce trees become larger, jack pine mortality increases and balsam fir saplings invade. The lowland, relatively oligotrophic bog forests have either larger trees and a continuous tree canopy or a discontinuous canopy with very small trees, differences attributable to the fertility of the site and the depth of *Sphagnum* peat. Mixed lowland black spruce forests occur in marginal habitats subject to groundwater flushing; tamarack and black spruce may be primary colonizers followed by the influx of balsam fir and white cedar. In upland areas black spruce behaves much like jack pine, and competition for sites may occur between the two species. In the wet lowlands, however, black spruce forest divides into several distinctive types in which succession to balsam fir forest is indicated on the more fertile, and vegetative regeneration through layering enables black spruce to persist as a monoculture on more oligotrophic sites.

SAPLING ESTABLISHMENT

A study of the success of sapling establishment in forests of various ages dominated by the various tree species was conducted by Carleton and Maycock, revealing the degree of replacement success exhibited by each of the species. The scatter plot for *Picea mariana* indicates a broad relationship between saplings and trees, but variation around the central trend is very wide. It is apparent that *Picea mariana* can act as a primary colonist in a similar manner to *Pinus banksiana*. This seems to be the case

TABLE 29

Relative Frequency of Saplings of Different Species in Stands Dominated by the Tree Species Indicated[a]

Dominant tree species	Relative frequency (%) of saplings by species						No. of stands
	Black spruce	White spruce	Jack pine	Tamarack	White Birch	Other[b]	
Black spruce	89	2	0	3	6	0	56
White spruce	18	69	0	3	8	2	23
Jack pine	23	1	58	0	11	7	18

[a] Dominant trees have an importance index in stands of 150 or more (for definition of importance index see Curtis, 1959). These are stands in northern study areas only (for location of study areas see Fig. 8).
[b] Principally alder and willow species attaining sapling size, i.e., 1 in. diameter or more at about breast height.

TABLE 30

Relative Frequency of Saplings of Different Species in Stands Dominated by the Tree Species Indicated[a]

Dominant species	Relative frequency (%) of saplings by species								No. of stands
	Black spruce	White spruce	Jack pine	Aspen or balsam poplar	Balsam Fir	Tamarack	Other[b]	White birch	
Black spruce	76	4	1	4	4	7	3	1	25
White spruce	2	58	0	12	14	0	6	8	5
Jack pine	34	9	17	6	3	0	20	11	13
Aspen and/or balsam poplar	3	10	0	45	10	0	14	18	13
Balsam fir	8	1	0	1	20	0	58[c]	12	5

[a] Dominant trees have an importance index in the stands of 150 or more (for definition of importance index see Curtis, 1959). These are stands in southern areas only (for location of study areas see Fig. 8).
[b] Principally alder, willow, and other common shrub species attaining sapling size, i.e., 1-in. diameter or more at about breast height.
[c] *Acer rubrum* and/or *A. spicatum* in high proportion.

in the more northern sections of the boreal forest region. *Picea mariana* is also seen, however, to behave as a shade-tolerant species which can invade beneath other tree canopies; this pattern is, nevertheless, less pronounced than the similar behavior of *Abies balsamea*.

Of the other significant boreal tree species, *Larix laricina, Populus balsamifera, Populus tremuloides, Betula papyrifera,* and *Picea glauca* all showed tree-versus-sapling importance scatters of a pattern more or less similar to that for *Pinus banksiana*. This suggests that they are all species adapted to the colonization of open, nonforested sites. *Populus tremuloides* and *Betula papyrifera* can resprout fresh vegetative shoots following destruction by fire in addition to establishing on postfire sites by seed (Ahlgren, 1974). *Larix laricina* invades sedge meadows and alder thickets by growth from seed. *Populus balsamifera* and *Picea glauca* invade newly exposed fluvial silts and some upland sites. Unlike *Picea mariana, Picea glauca* sheds most of its seeds annually, and reproduction is usually associated with older trees in close proximity.

Information on the success of growth by saplings can be derived from data obtained by the author on forest structure over a large area of central Canada at the sites shown in Fig. 27. The data can be divided into two parts, one part representing the forest structure in the southeast portion of the region, near and in part encompassing the same clay belt region of Ontario in which Carleton and Maycock carried out their study. The second part takes in the more northern portions of the central Canadian forest extending to the northern forest border. Data from the northern portion include stands dominated solely by black spruce, white spruce, or jack pine, and the southeastern portion includes stands dominated by balsam fir and by trembling aspen and/or balsam poplar. The northern stands are beyond the region in which these latter species retain any degree of significance in the forest communities.

The data in Tables 29 and 30 demonstrate that in the northern stands each of the species—black spruce, white spruce, and jack pine—reproduce vigorously, with only a suggestion that black spruce may tend to invade the white spruce and jack pine stands; there is no suggestion that either white spruce or jack pine saplings are becoming established in any but forests dominated by trees of the same species. Southward, however, black spruce is reproducing more vigorously in jack pine stands than jack pine itself. Balsam fir reproduces relatively most vigorously in balsam fir, white spruce, and aspen stands. In general, the data for the southern stands confirm the finding of Carleton and Maycock in the clay belt region. Scatter diagrams of sapling presence in stands of various ages dominated by the major species confirm the work of Carleton and Maycock showing that stands of varied composition and a wide range of ages all possess high frequencies of black spruce saplings; there is little differentiation between northern and southern stands in this regard. White spruce possesses the capacity to reproduce in southern stands dominated both by these and other species; trembling aspen and jack pine show a marked reduction in sapling frequency in older stands dominated by either these or other species.

From these studies it becomes clear that, as Carleton and Maycock state, Clementsian monoclimax theory cannot adequately explain the scheme of succession vectors in the boreal forest. The emphasis, in summary, focuses heavily upon the importance of forest fire in boreal forest dynamics, and a typical forest community is dominated by one or more of a few highly productive species at or near the starting point of succession. Windthrow fire, hail storms, ice storms, and insect and fungal attack are all recognizable causes of major forest destruction in the region, but fire is the prime agent. In the absence of catastrophe, some upland primary boreal forests seem to develop into secondary fir and/or spruce forests, but old forest stands can also be found in which there is no regeneration. Dix and Swan (1971) comment upon this same characteristic feature in the uplands of Candle Lake, Saskatchewan, and as Rowe (1961) states, in such areas a parkland or a decadent brushland may eventually develop in the absence of disturbance. Environmental conditions tip the balance in favor of one process or the other. If the cause of the absence of advanced succession can be ascribed to the vegetation itself, then the causative influences must operate through the behavior evidenced by one or two of the most abundant species (Carleton and Maycock, 1978).

CONCLUSION

Variations in boreal forest communities occur in two dimensions. There is, first, the wide geographical dimension, in which regional differences are discernible in the species composition of communities and the frequencies with which species occur. There is, second, the local dimension, in which differences in species composition occur from one site to another as a consequence of variations in response of the different species to soils, topography, geology, and microclimate, or as a result of disturbance of forest stands by fire or other agency. In both dimensions, the variations in community structure are measurable in terms of absence or presence of species or in terms of abundance of individuals of each species—employing as a measure of abundance such statistics as frequency, density, dominance, or other numerical devices. In the boreal vegetation, as in many other vegetational zones throughout the world, it is difficult if not impossible to establish, on the basis of a given definition of a community, at what point regionally the community ends and another takes its place. Thus, from western to eastern Canada, the composition of the boreal forest undergoes a transition in terms of the dominant tree and understory species, but all transitions are gradual and no definite line of demarcation can be drawn separating one community from another. Even under a single dominant tree species—black spruce, for example—the understory vegetation changes in species presence and frequency (or density or dominance) from east to west and from north to south. Black spruce, over geographical distances, also shifts in its site preferences—it is found in lowlands in the south and northward increasingly occupies upland areas—and this has an effect upon the composition of the understory vegetation beneath the spruce canopy.

It is, thus, apparent that there are processes at work in the relationships among species to account for variations in composition and other characteristics of communities from one place to another, relationships that are defined, however loosely, as competition, successional interactions, response to fire, and so on. In all of this, one significant fact should not be overlooked. It is simply that description of the species compositional component, as Rowe (1961) points out, is but one aspect of the boreal ecosystem. It is an inventory of the biological material that carries out the cycling, the input and output, of energy and materials in the plants and soil of boreal regions.

The performance of different species is related in important ways to the performance of the ecosystem, but study of the ecosystem can be approached from the point of view of the whole as well as from the point of view of the parts—of individual species, interactions, and relationships. This means that study of the input and outgo of energy and nutrients in the plant–soil complex can be undertaken as a kind of summary of the interactions and interrelationships of the populations of individuals of the different species. Approached ultimately from both directions, it seems reasonable that the performance of the ecosystem in terms of the aggregated individuals can be described and measured. It is likely that some general principles will apply throughout the forest, but regional characteristics differ sufficiently so that an ecosystem model devised in one region will necessarily have to be modified before it will be applicable to another region. As Rowe (1961) points out, working principles for practical forestry established at regional or local levels cannot be expected to have circumpolar significance. It is necessary to work out the characteristics of ecosystem dynamics on a regional basis.

To return to the concept of the community, one of the key problems is simply the lack of a good statistical or mathematical construct adequate for representation or description of a given plant community; the parameters currently in use, such as frequency, density, dominance, or some expedient combination of these, leave much to be desired. As Colinvaux (1973) points out, the absence of many real data describing the fortunes of succession in the field make most hypotheses very general indeed. Each succession may actually have its own unique development and may create environmental conditions that modify the course of succession as it proceeds; there are, thus, a very large number of alternate pathways that succession may follow. It is possible to speculate that, in the future, equations or sets of equations will be available to implement research on community dynamics. At the present time, it can only be said that, given the state of development of our understanding of the boreal ecosystem, much remains to be learned at every level of boreal ecology.

REFERENCES

Ahlgren, C. E. (1974). Effects of fires on temperate forests. In "Fire and Ecosystems" (T. T. Kozlowski and C. E. Ahlgren, eds.), pp. 195–219. Academic Press. New York.

Ahti, T. (1959). Studies of the caribou lichen stands of Newfoundland. *Ann. Bot. Soc. "Vanamo"* **30**, 1–44.

Ahti, T. (1961). Taxonomic studies on reindeer lichens (*Cladonia*, subgenus *Cladina*). *Ann. Bot. Soc. Zool. Bot. "Vanamo"* **32**, 1.

Anderson, R. C., and Loucks, O. (1973). Aspects of the biology of *Trientalis borealis* Raf. *Ecology* **54**, 798–808.

Archibold, O. W. (1979). Buried viable propagules as a factor in postfire regeneration in northern Saskatchewan. *Can. J. Bot.* **57**, 54–58.

Auclair, A. N., and F. Glenn Goff, (1974). Intraspecific diameter differentiationas a measure of species replacement potential. *Can. J. For. Res.* **4**, 424–434.

Bakuzis, E. V., and Hansen, H. L. (1965). "Balsam Fir." Univ. of Minnesota Press, Minneapolis.

Barney, R. J., Van Cleve, K., and Schlentner, R. (1978). Biomass distribution and crown characteristics in two Alaskan *Picea mariana* ecosystems. *Can. J. For. Res.* **8**, 36–41.

Barney, R. J., and Van Cleve, K. (1973). Black spruce fuel weights and biomass in two interior Alaska stands. *Can. J. For. Res.* **3**, 304–311.

Beals, E. W. (1973). Ordination: Mathematical elegance and ecological naivete. *J. Ecol.* **61**, 23–35.

Bellefleur, P. and Auclair, A. N. (1972). Comparative ecology of Quebec boreal forests: A numerical approach to modelling. *Can. J. Bot.* **50**, 2357–2379.

Benninghoff, W. S. (1952). Interaction of vegetation and soil frost phenomena. *Arctic* **5**, 34–44.

Bergerud, Arthur T. (1971). Abundance of forage on the winter range of Newfoundland caribou. *Can. Field Nat.* **85**, 39–52.

Billings, W. D., and Mooney, H. A. (1959). An apparent frost hummock-sorted polygon in the alpine tundra of Wyoming. *Ecology* **40,** 16-20.

Billings, W. D., and Mooney, H. A. (1968). The ecology of arctic and alpine plants. *Biol. Rev.* **43,** 481-529.

Black, R. A., and Bliss, L. C. (1978). Recovery sequence of *Picea mariana-Vaccinium uliginosum* forests after burning near Inuvik, Northwest Territories, Canada. *Can. J. Bot.* **56,** 2020-2030.

Bliss, L. C. (1956). A comparison of plant development in microenvironments of arctic and alpine regions. *Ecol. Monogr.* **26,** 303-337.

Bliss, L. C. (1962). Adaptations of arctic and alpine plants to environmental conditions. *Arctic* **5,** 117-144.

Bliss, L. C., and Cantlon, J. E. (1957). Succession on river alluvium in northern Alaska. *Am. Midl. Nat.* **58,** 452-469.

Bliss, L. C., and Wein, R. W. (1972). Plant community responses to disturbances in the western Canadian Arctic. *Can. J. Bot.* **50,** 1097-1109.

Bond, G. (1967). Fixation of nitrogen by higher plants other than legumes. *Annu. Rev. Plant Physiol.* **18,** 107-126.

Braathe, P. (1974). Prescribed burning in Norway—effects on soil and regeneration. *Tall Timbers Fire Ecol. Conf. Proc., Tall Timbers Res. Sta., Tallahassee, Florida* pp. 211-222.

Brodo, I. M. (1973). Substrate ecology. *In* "The Lichens" (V. Ahmadjian and M. E. Hale, eds.), pp. 401-441. Academic Press, New York.

Brown, R. J. E. (1967). Permafrost in Canada. Map 1246A. Can. Geol. Surv., Ottawa.

Brown, R. J. E. (1969). Factors influencing discontinuous permafrost in Canada. *In* "The Periglacial Environment" (T. L. Péwé. ed.), pp. 11-54. McGill-Queen's Univ. Press, Montreal.

Brown, R. J. E. (1970a). Permafrost as an ecological factor in the Subarctic. *In* "Ecology of the Subarctic Regions." UNESCO, Paris.

Brown, R. J. E. (1970b). "Permafrost in Canada." Univ. of Toronto Press, Toronto.

Brown, R. T. (1967). Influence of naturally occurring compounds on germination and growth of jack pine. *Ecology* **48,** 542-546.

Brown, R. T., and Mikola, P. (1974). The influence of fruticose soil lichens upon the mycorrhizal and seedling growth of forest trees. *Acta For. Fenn.* **141,** 5-22.

Brown Beckel, D. K. (1957). Studies on seasonal changes in the temperature gradient of the active layer of soil at Fort Churchill, Manitoba. *Arctic* **19,** 151-183.

Busby, J. R., Bliss, L. C., and Hamilton, C. D. (1978). Microclimate control of growth rates and habitats of the boreal forest mosses, *Tomenthypnum nitens* and *Hyclocomium splendens*. *Ecol. Monogr.* **48,** 95-110.

Carleton, T. J., and Maycock, P. F. (1978). Dynamics of boreal forest south of James Bay. *Can. J. Bot.* **56,** 1157-1173.

Cayford, J. H. (1963). Some factors influencing jack pine regeneration after fire in southeastern Manitoba. *Can. For. Br. Dep. Publ.* **1016,** 1-15.

Cayford, J. H., and Haig, R. A. (1961). Glaze damage in forest stands in southeastern Manitoba. Can. Dep. For., For. Res. Br. Tech. Note 102. pp. 1-16.

Cayford, J. H., Hildahl, V., Nairn, L. D., and Wheaton, M. P. H. (1959). Injury to trees from winter drying and frost in Manitoba and Saskatchewan in 1958. *For. Chron.* **35,** 282-290.

Cayford, J. H., Chrosciewicz, S., and Sims, H. P. (1967). A review of silvicultural research in jack pine. *Can. For. Br. Dep. Publ.* **1173,** 1-255.

Chapman, S. B., ed. (1976). "Methods in Plant Ecology." Wiley, New York.

Churchill, E. D., and Hanson, H. C. (1958). The concept of climax in arctic and alpine vegetation. *Bot. Rev.* **24**, 127–191.
Clausen, J. J. (1957). A phytosociological ordination of the conifer swamps of Wisconsin. *Ecology* **38**, 638–645.
Colinvaux, P. A. (1973). "Introduction to Ecology." Wiley, New York.
Cormack, R. G. H. (1953). A survey of coniferous forest succession in the eastern Rockies. *For. Chron.* **29**, 218–232.
Cottam, G., and Curtis, J. T. (1956). The use of distance measure in phytosociological sampling. *Ecology* **37**, 451–460.
Cottam, G., and McIntosh, R. P. (1966). Vegetational continuum. *Science* **152**, 546–547.
Cottam, G., Goff, F. G., and Whittaker, R. H. (1973). Wisconsin comparative ordination. *In* "Ordination and Classification of Communities" (R. H. Whittaker, ed.), "Handbook of Vegetation Science," Vol. 5, pp. 193-222. D. W. Junk, The Hague.
Couper, W. S. (1939). A fourth expedition to Glacier Bay, Alaska. *Ecology* **20**, 130–55.
Crampton, C. B. (1977). A study of the dynamics of hummocky microrelief in the Canadian North. *Can. J. Earth Sci.* **14**, 639–649.
Creed, R., ed. (1971). "Ecological Genetics and Evolution." Blackwell, Oxford.
Curtis, J. T. (1959). "The Vegetation of Wisconsin: An Ordination of Plant Communities." Univ. of Wisconsin Press, Madison.
Dahl, E. (1951). On the relation between summer temperature and the distribution of alpine vascular plants in the lowlands of Fennoscandia. *Oikos* **3**, 22–52.
Damman, A. W. H. (1965). The distribution patterns of northern and southern elements in the flora of Newfoundland. *Rhodora* **67**, 363–392.
Damman, A. W. H. (1971). Effect of vegetation changes on the fertility of a Newfoundland forest site. *Ecol. Monogr.* **41**, 253–270.
Daubenmire, R. (1966). Vegetation: Identification of typal communities. *Science* **151**, 291–298.
Davis, R. B. (1966). Spruce-fir forests of the coast of Maine. *Ecol. Monogr.* **36**, 79–94.
Day, R. J. (1972). Stand structure, succession, and use of southern Alberta's Rocky Mountain forest. *Ecology* **53**, 472–478.
Decker, H. F. (1966). Plants. *In* "Soil Development and Ecological Succession in a Deglaciated Area of Muir Inlet, Southeast Alaska" (A. Mirsky, ed.), pp. 73–96. Inst. of Polar Studies Rep. 20, Ohio State Univ., Columbus.
Dingman, S. L., and Koutz, F. R. (1974). Relations among vegetation, permafrost, and potential insolation in central Alaska. *Arct. Alp. Res.* **6**, 37–42.
Dix, R. L., and Swan, J. M. A. (1971). The roles of disturbance and succession in upland forest at Candle Lake, Saskatchewan. *Can. J. Bot.* **49**, 657–676.
Dugle, J. R. (1966). A taxonomic study of western Canadian species in the genus *Betula*. *Can. J. Bot.* **44**, 929–1007.
Dyrness, C. T., and Grigal, D. F. (1979). Vegetation-soil relationships along a spruce forest transect in interior Alaska. *Can. J. Bot.* **57**, 2644–2656.
Elliott, D. L. (1979a). The stability of the northern Canadian tree limit: current regenerative capacity. Ph.D. thesis, Univ. of Colorado, Boulder.
Elliott, D. L. (1979b). The current regenerative capacity of the northern Canadian trees, Keewatin, N.W.T., Canada: Some preliminary observations. *Arct. Alp. Res.* **11**, 243–251.
Endler, J. A. (1977). "Geographic Variation, Speciation, and Clines." Princeton Univ. Press, Princeton, New Jersey.
Evans, L. T. (1963). Extrapolation from controlled environments to the field. *In* "Environ-

mental Control of Plant Growth" (L. T. Evans, ed.), pp. 421–437. Academic Press, New York.

Fowler, D. P., and Mullin, R. E. (1977). Upland–lowland ecotypes not well developed in black spruce in Northern Ontario. *Can. J. For. Res.* **7**, 35–40.

Garten, C. T., Jr. (1978). Multivariate perspectives on the ecology of plant mineral element composition. *Am. Nat.* **112**, 533–544.

Gates, D. M. (1968). Transpiration and leaf temperature. *Annu. Rev. Plant Physiol.* **19**, 211–238.

Gauch, H. G., Jr. (1973). A quantitative evaluation of the Bray–Curtis ordination. *Ecology* **54**, 829–836.

Gauch, H. G., Jr., and Whittaker, R. H. (1972). Comparison of ordination techniques. *Ecology* **53**, 868–875.

Gerloff, G. C., Moore, D. G., and Curtis, J. T. (1966). Selective absorption of mineral elements by native plants of Wisconsin. *Plant and Soil* **25**, 393–405.

Gill, D. (1973). Modification of northern alluvial habitats by river development. *Can. Geogr.* **17**, 138–153.

Goff, F. G., and Zedler, P. H. (1972). Derivation of species succession vectors. *Am. Midl. Nat.* **87**, 397–412.

Golley, F. B., ed. (1977). "Ecological Succession." Academic Press, New York.

Gordon, A. G. (1976). The taxonomy and genetics of *Picea rubens* and its relationship to *Picea mariana*. *Can. J. Bot.* **54**, 781–813.

Graham, S. A. (1954). Scoring tolerance of forest trees. *Mich. For.* **4**, 1–2.

Grandtner, M. M. (1963). The southern forests of Scandinavia and Quebec." Laval Univ. For. Res. Found., Contrib. 10 pp. 1–25.

Greig-Smith, P. (1964). "Quantitative Plant Ecology," 2nd ed. Butterworth, London.

Handley, W. R. C. (1963). Mycorrhizal associations and *Caluna* heathland afforestation. *Bull. For. Comm., London* **36**, 1–70.

Harborne, J. B., ed. (1972). "Phytochemical Ecology." Academic Press, London.

Harborne, J. B. (1977). "Introduction to Ecological Biochemistry." Academic Press, New York.

Hare, F. K. (1954). The boreal conifer zone. *Geogr. Stud.* **1**, 4–18.

Hare, F. K. (1959). A photo-reconnaissance survey of Labrador–Ungava. Can. Dep. Mines Tech. Surv., Geogr. Br. Mem. 6, 1–83.

Harper, F. (1964). "Plant and Animal Associations in the Interior of the Ungava Peninsula." Univ. of Kansas Mus. of Nat. Hist., Lawrence.

Heilman, P. E. (1966). Change in distribution and availability of nitrogen with forest succession on north slopes in interior Alaska. *Ecology* **47**, 825–831.

Heilman, P. E. (1968). Relationship of availability of phosphorus and cations to forest succession and bog formation in interior Alaska. *Ecology* **49**, 331–336.

Heinselman, M. L. (1957). Silvical characteristics of black spruce. Lake States For. Exp. Sta., Sta. Paper 45, pp. 1–30. For. Serv., U.S. Dep. Agric.

Heinselman, M. L. (1963). Forest sites, bog processes, and peatland types in the glacial Lake Agassiz region, Minnesota. *Ecol. Monogr.* **33**, 327–374.

Heinselman, M. L. (1970). Landscape evolution, peatland types, and the environment in the Lake Agassiz Peatlands Natural Area, Minnesota. *Ecol. Monogr.* **40**, 235–261.

Hicklenton, P. R., and Oechel, W. (1976a). Physiological aspects of the ecology of *Dicranum fuscescens* in the subarctic. I. Acclimation and acclimation potential of CO_2 exchange in relation to habitat, light, and temperature. *Can. J. Bot.* **54**, 1104–1119.

Hicklenton, P. R., and Oechel, W. C. (1976b). Physiological aspects of the ecology of

Dicranum fuscescens in the subarctic. II. Seasonal patterns of organic nutrient content. *Can. J. Bot.* **55**, 2168–2177.
Hopkins, D. M., and Sigafoos, R. S. (1950). Frost action and vegetation patterns on Seward Peninsula, Alaska. *U.S. Geol. Surv. Bull.* **974-C.** 101 pp.
Horton, D. G., Vitt, D. H., and Slack, N. G. (1979). Habitats of circumboreal-subarctic *Sphagna:* I. A. quantitative analysis and review of species in the Caribou Mountains, northern Alberta. *Can. J. Bot.* **57**, 2283–2317.
Hustich, I. (1954). On forests and tree growth in the Knob Lake area, Quebec–Labrador peninsula. *Acta Geogr.* **17**, 1–24.
Hustich, I. (1966). On the forest–tundra and the northern tree-lines. *Ann. Univ. Turku. Ser. A2* **36**, 7–47.
Hutchinson, A. H. (1918). Limiting factors in relation to specific ranges of tolerance of forest trees. *Bot. Gaz.* **66**, 465–493.
Ignatenko, I. V., Knorre, A. V., Lovelius, N. V., and Norin, B. N. (1973). Phytomass stock in typical plant communities in the "Ary-Mas" forest. *Sov. J. Ecol.* **4**, 213–217.
Ingestad, T. (1959). Studies on the nutrition of forest tree seedlings: Spruce. *Physiol. Plant.* **12**, 568–593.
Ingestad, T. (1962). Macro element nutrition of pine, spruce, and birch seedlings in nutrient solutions. *Medd. Statens Skogsforskningst.* **51**, 1–150.
Jeffree, E. P. (1960). A climatic pattern between latitudes 40° and 70°N and its probable influence on biological distributions. *Proc. Linna. Soc. London* **171**, 89–121.
Jeglum, J. K. (1971). Plant indicators of pH and water level in peatlands at Candle Lake, Saskatchewan. *Can. J. Bot.* **49**, 1661–1676.
Jeglum, J. K. (1972). Wetlands near Candle Lake, central Saskatchewan. I. Vegetation. *Musk Ox* **11**, 41–58.
Jeglum, J. K. (1973). Boreal forest wetlands near Candle Lake, central Saskatchewan. II. Relationships of vegetational variation to major environmental gradients. *Musk Ox* **12**, 32–48.
Johnson, A. W., and Packer, J. G. (1965). Polyploidy and environment in arctic Alaska. *Science* **148**, 237–239.
Johnson, E. A. (1975). Buried seed populations in the subarctic forest east of Great Slave Lake, Northwest Territories. *Can. J. Bot.* **53**, 2933–2941.
Johnson, E. A., and Rowe, J. S. (1975). Fire in the subarctic wintering ground of the Beverley caribou herd. *Am. Midl. Nat.* **94**, 1–14.
Karpow, W. G. (1961). On the influence of tree root competition on the photosynthetic activity of the herb layer in spruce forest. *Dokl. Acad. NAUK SSSR* **140**, 1205–1208. (in Russian).
Karpow, W. G. (1962). Some experimental results on the composition and structure of the lower strata of the *Vaccinium*-rich spruce forests. *Probl. Bot.* **6**, 258–276 (in Russian).
Kay, P. A. (1978). Dendroecology in Canada's forest–tundra transition zone. *Arct. Alp. Res.* **10**, 133–138.
Kershaw, K. A. (1973). "Quantitative and Dynamic Plant Ecology." Arnold, London.
Kjelvik, S., and Karenlampl, L. (1975). Plant biomass and primary production of Fennoscandian subarctic and subalpine forests and of alpine willow and heath ecosystems. *In* "Fennoscandian Tundra Ecosystems" (F. E. Wielgolaski, ed.), Part I, Plants and Microorganisms, pp. 111–120. Springer-Verlag, Berlin and New York.
Korchagin, A. A., and Karpov, V. G. (1974). Fluctuations in coniferous taiga communities. *In* "Vegetation Dynamics" (R. Knapp, ed.), Handbook of Vegetation Science, Part VIII, pp. 227–231. D. W. Junk, The Hague.

Kourtz, P. (1967). Lightning behavior and lightning fires in Canadian forests. *Can. For. Br. Dep. Publ.* **1179**, 1-33.
Kozlowski, T. T., and Ahlgren, C. E., eds. (1974). "Fire and Ecosystems." Academic Press, New York.
Kozlowski, T. T., and Borger, G. A. (1971). Effect of temperature and light intensity early in ontogeny on growth of *Pinus resinosa* seedlings. *Can. J. For. Res.* **1**, 57-65.
Larcher, W. (1975). "Physiological Plant Ecology." Springer-Verlag, Berlin and New York.
La Roi, G. H., and Dugle, J. R. (1968). A systematic and genecological study of *Picea glauca* and *P. engelmannii*, using paper chromatograms of needle extracts. *Can. J. Bot.* **46**, 649-687.
Larsen, J. A. (1965). The vegetation of the Ennadai Lake area N.W.T.: Studies in subarctic and arctic bioclimatology. *Ecol. Monogr.* **35**, 37-59.
Larsen, J. A. (1972). Vegetation and terrain (environment): Boreal forest and tundra. Rep. Center Clim. Res., Univ. Wisconsin, Madison, UW-G1128. 231 pp.
Larsen, J. A. (1973). Plant communities north of the forest border, Keewatin, Northwest Territories. *Can. Field Nat.* **87**, 241-248.
Larsen, J. A. (1974). Ecology of the northern Continental forest border. *In* "Arctic and Alpine Environments" (J. D. Ives and R. G. Barry, eds.), pp. 341-369. Methuen, London.
Lawrence, D. B. (1958). Glaciers and vegetation in southeast Alaska. *Am. Sci.* **46**, 89-122.
Lechowicz, M. J. (1978). Carbon dioxide exchange in *Cladina* lichens from subarctic and temperate habitats. *Oecologia* **32**, 225-237.
Lechowicz, M. J., and Adams, M. S. (1974a). Ecology of *Cladonia* lichens. I. Preliminary assessment of the ecology of terricolous lichen-moss communities in Ontario and Wisconsin. *Can. J. Bot.* **52**, 55-64.
Lechowicz, M. J., and Adams, M. S. (1974b). Ecology of *Cladonia* lichens. II. Comparative physiological ecology of *C. mitis*, *C. rangiferina*, and *C. uncialis*. *Can. J. Bot.* **52**, 411-422.
Lechowicz, M. J., Jordan, W. P., and Adams, M. S. (1974c). Ecology of *Cladonia* lichens. III. Comparison of *C. caroliniana*, endemic to southeastern North America, with three northern *Cladonia* species. *Can. J. Bot.* **52**, 565-573.
Logan, K. T. (1973). Growth of tree seedlings as affected by light intensity. *Can. For. Serv. Publ.* **1323**, 1-19.
Lutz, H. J. (1956). Ecological effects of forest fires in the interior of Alaska. *U.S. Dep. Agric. Tech. Bull.* **1133**, 1-121.
McCullough, H. A. (1948). Plant succession on fallen logs in a virgin spruce-fir forest. *Ecology* **29**, 508-513.
MacFarlane, J. D., and Kershaw, K. A. (1978). Thermal sensitivity in lichens. *Science* **201**, 739-741.
McIntosh, R. P. (1978). "Phytosociology." Academic Press, New York.
McIntosh, R. P. (1967). The continuum concept of vegetation. *Bot. Rev.* **33**, 130-187.
McIntosh, R. P. (1968). The continuum concept of vegetation: Responses. *Bot. Rev.* **34**, 253-314.
McIntosh, R. P. (1968). The continuum concept of vegetation: Responses; Reply. *Bot. Rev.* **34**, 315-332.
McIntosh, R. P., and Hurley, R. T. (1964). The spruce-fir forests of the Catskill Mountains. *Ecology* **45**, 314-326.
MacLean, D. W. (1960) Some aspects of the aspen-birch-spruce-fir type in Ontario. Can. Dep. For., For. Res. Div. Tech. Note 94.

McNaughton, S. J., and Wolf, L. L. (1970). Dominance and niche in ecological systems. *Science* **167**, 131–139.
Maikawa, E., and Kershaw, K. A. (1976). Studies on lichen-dominated systems. XIX. The postfire recovery sequence of black spruce–lichen woodland in the Abitau Lake region, N.W.T. *Can. J. Bot.* **54**, 2679–2687.
Major, J. (1974). Differences in duration of successional series. *In* "Vegetation Dynamics" (R. Knapp, ed.), Handbook of Vegetation Science, Part VIII, pp. 157–159. D. W. Junk, The Hague.
May, R. M. (1976). "Theoretical Ecology: Principles and Applications." Saunders, Philadelphia, Pennsylvania.
Maycock, P. F., and Curtis, J. T. (1960). The phytosociology of boreal conifer–hardwood forests of the Great Lakes region. *Ecol. Monogr.* **30**, 1–35.
Mitchell, V. L. (1967). An investigation of certain aspects of tree growth rates in relation to climate in the central Canadian boreal forest. Univ. of Wis., Dep. of Meteorology, Tech. Rept. 33, pp. 1–62.
Mitchell, V. L. (1973). A theoretical tree-line in central Canada. *Ann. Assoc. Am. Geogr.* **63**, 296–301.
Moore, P. D., and Bellamy, D. J. (1974). "Peatlands." Springer-Verlag, Berlin and New York.
Moore, T. R., and Verspoor, E. (1973). Above ground biomass of black spruce stands in subarctic Quebec. *Can. J. For. Res.* **3**, 596–598.
Mosquin, T. (1966). Reproductive specialization as a factor in the evolution of the Canadian flora. *In* "The Evolution of Canada's Flora" (R. L. Taylor and R. A. Ludwig, eds.), pp. 41–65. Univ. of Toronto Press, Toronto.
Mueller-Dombois, D. (1964). Effect of depth to water table on height growth of tree seedlings in a greenhouse. *For. Sci.* **10**, 306–316.
Mutch, R. W. (1970). Wildland fires and ecosystems—A hypothesis. *Ecology* **51**, 1046–1051.
Page, G. (1971). Properties of some common Newfoundland forest soils and their relation to forest growth. *Can. J. For. Res.* **1**, 175–192.
Parker, W. H., and McLachlan, D. G. (1978). Morphological variation in white and black spruce: Investigation of natural hybridization between *Picea glauca* and *P. mariana*. *Can. J. Bot.* **56**, 2512–2520.
Patten, B. C. (1961). Competitive exclusion. *Science* **134**, 1599–1601.
Peterson, E. B. (1965). Inhibition of black spruce primary roots by a water-soluble substance in *Kalmia angustifolia*. *For. Sci.* **11**, 473–479.
Pielou, E. C. (1969). "An Introduction to Mathematical Ecology." Wiley, New York.
Pierce, R. S. (1953). Oxidation–reduction potential and specific conductance of ground water: Their influence upon natural forest distribution. *Proc. Soil Sci. Soc. Am.* **17**, 61–65.
Place, I. C. M. (1955). The influence of seed bed conditions on the regeneration of spruce and balsam fir. *Can. For. Br. Bull.* **117**.
Raup, Hugh M. (1941). Botanical problems in boreal America. *Bot. Rev.* **7**, 147–248.
Raup, H. M. (1946). Phytogeographic studies in the Athabaska–Great Slave Lake region. *J. Arnold Arbor. Harv. Univ.* **27**, 2–78.
Rencz, A., and Auclair, A. N. D. (1978). Biomass distribution in a subarctic *Picea mariana*–*Cladonia alpestris* woodland. *Can. J. For. Res.* **8**, 168–176.
Rice, E. L. (1974). "Allelopathy." Academic Press, New York.
Roche, L. (1969). A genecological study of the genus *Picea* in British Columbia. *New Phytol.* **68**, 505–554.

Rodin, L. E., and Bazilevich, N. I. (1966). The biological productivity of the main vegetation types in the northern hemisphere of the Old World. *For. Abstr.* **27**, 369–372.

Rodin, L. E., and Bazilevich, N. I. (1968). "Production and Mineral Cycling in Terrestrial Vegetation." Oliver & Boyd, Edinburgh.

Rosswall, T., ed. (1971). "Systems Analysis in Northern Coniferous Forests, IBP Workshop." Kratte Mosugn, Sweden, Swed. Nat. Sci. Res. Counc., Oslo.

Rouse, W. R. (1976). Microclimatic changes accompanying burning in subarctic lichen woodland. *Arct. Alp. Res.* **8**, 357–376.

Rouse, W. R., and Kershaw, K. A. (1971). The effects of burning on the heat and water regimes of lichen-dominated subarctic surfaces. *Arct. Alp. Res.* **3**, 291–304.

Rowe, J. S. (1956). Uses of undergrowth plant species in forestry. *Ecology* **37**, 461–473.

Rowe, J. S. (1961). Critique of some vegetational concepts as applied to forests of northwestern Alberta. *Can. J. Bot.* **39**, 1007–1017.

Rowe, J. S. (1969). Plant community as a landscape feature. *Essays Plant Geogr. Ecol. Symp. Terrestr. Plant Ecol.* (K.N.H. Greenidge, ed.), pp. 63–81. Saint Francis Xavier Univ., Antigonish, Nova Scotia.

Rowe, J. S. (1971). Spruce and fire in northwest Canada and Alaska. *Tall Timbers Fire Ecol. Conf. Proc., Tall Timbers Res. Sta., Tallahassee, Florida* pp. 245–254.

Rowe, J. S., and Scotter, G. W. (1973). Fire in the boreal forest. *Quat. Res. (N.Y.)* **3**, 444–464.

Savile, D. B. O. (1963). Factors limiting the advance of spruce at Great Whale River, Quebec. *Can. Field Nat.* **77**, 95–97.

Savile, D. B. O. (1964). North Atlantic biota and their history (A review). *Arctic* **17**, 137–141.

Savile, D. B. O. (1972). "Arctic Adaptations in Plants." *Can. Dep. Agric., Monogr.* **6**, 1–81.

Scotter, George W. (1964). Effects of forest fires on the winter range of barren-ground caribou in northern Saskatchewan. *Wildl. Manage. Bull.* **1**, 1–111.

Scotter, G. W. (1971). Wildfires in relation to the habitat of barren-ground caribou in the taiga of northern Canada. *Tall Timbers Fire Ecol. Conf. Proc., Tall Timbers Res. Sta., Tallahassee, Florida* pp. 85–106.

Shafi, M. I., and Yarranton, G. A. (1973). Vegetational heterogeneity during a secondary (postfire) succession. *Can. J. Bot.* **51**, 73–90.

Shafi, M. I., and Yarranton, G. A. (1973). Diversity, floristic richness, and species evenness during a secondary (postfire) succession. *Ecology*, **54**, 897–902.

Siren, G. (1974). Some remarks on fire ecology in Finnish forestry. *Tall Timbers Fire Ecol. Conf. Proc., Tall Timbers Res. Sta., Tallahassee, Florida* pp. 191–210.

Small, E. (1972a). Ecological significance of four critical elements in plants of raised *Sphagnum* peat bogs. *Ecology* **53**, 498–502.

Small, E. (1972b). Water relations of plants in raised *Sphagnum* peat bogs. *Ecology* **53**, 726–728.

Smith, E. M., and Hadley, E. B. (1974). Photosynthetic and respiratory acclimation to temperature in *Ledum groenlandicum* populations. *Arct. Alp. Res.* **6**, 13–27.

Sprugel, D. G. (1976). Dynamic structure of wave-regenerated *Abies balsamea* forests in the north-eastern United States. *J. Ecol.* **64**, 889–912.

Stiell, W. M. (1970). Some competitive relations in a red pine plantation. *Can. For. Serv. Publ.* **1275**, 1–16.

Strain, B. R., and Billings, W. D. eds. (1974). "Vegetation and Environment." D. W. Junk, The Hague.

Strang, R. M. (1973). Succession in unburned subarctic woodlands. *Can. J. For. Res.* **3**, 140–142.

Sutton, R. F. (1969). Silvics of White Spruce. *Can. For. Br. Dep. Publ.* **1250**, 1–57.
Swan, J. M. A., and Dix, R. L. (1966). The phytosociological structure of upland forest at Candle Lake, Saskatchewan. *J. Ecol.* **54**, 13–40.
Thieret, J. W. (1964). Botanical survey along the Yellowknife Highway, N.W.T., Canada. II. Vegetation. *Sida* **1**, 187–239.
Tisdale, E. W., Fosberg, M. A., and Poulton, C. E. (1966). Vegetation and soil development of a recently glaciated area near Mount Robson, British Columbia. *Ecology* **47**, 517–23.
Thomson, J. W. (1972). Distribution patterns of American arctic lichens. *Can. J. Bot.* **50**, 1135–1156.
Thomson, J. W., Scotter, G. W., and Ahti, T. (1969). Lichens of the Great Slave Lake region, Northwest Territories, Canada. *Bryologist* **72**, 137–177.
Tranquillini, W. (1979). "Physiological Ecology of the Alpine Timberline." Springer-Verlag, Berlin and New York.
Ugolini, F. C. (1966). Soils. *In* "Soil Development and Ecological Succession in a Deglaciated Area of Muir Inlet, Southeast Alaska" (A. Mirsky, ed.), pp. 28–72. Ohio State Univ. Inst. of Polar Studies, Columbus.
Van Cleve, K., Viereck, L. A., and Schlentner, R. L. (1970). Accumulation of nitrogen in alder ecosystems developed on the Tanana River floor plain near Fairbanks, Alaska. *Arct. Alp. Res.* **3**, 101–114.
Van Cleve, K., Viereck, L. A., and Schlentner, R. L. (1972). Distribution of selected chemical elements in even-aged alder (*Alnus*) ecosystems near Fairbanks, Alaska. *Arct. Alp. Res.* **4**, 239–255.
van Groenewoud, H. (1965). Ordination and classification of Swiss and Canadian coniferous forest by various biometric and other methods. *Ber. Geobot. Inst. Eidg. Tech. Hochsch. Stift. Ruebel Zurich* **36**, 28–102.
Vasilevich, V. I. (1968). The continuum concept of vegetation: Responses. *Bot. Rev.* **34**, 312–314.
Viereck, L. A. (1970). Forest succession and soil development adjacent to the Chena River in interior Alaska. *Arct. Alp. Res.* **2**, 1–26.
Vincent, A. B. (1965). Black spruce: A review of its silvics, ecology and silviculture. *Can. For. Br. Dep. Publ.* **1100**, 1–79.
Vogl, R. J. (1966). Vegetational continuum. *Science* **152**, 546.
von Humboldt, A. (1807). "Ideen zu einer Geographie der Pflanzen neben einem, Naturgemalde der Tropenlander." Tübingen.
Vowinckel, T., Oechel, W. C., and Boll, W. G. (1975). The effect of climate on the photosyntheses of *Picea mariana* at the subarctic treeline. 1. Field measurements. *Can. J. Bot.* **53**, 604–620.
Walter, H. (1971). "Ecology of Tropical and Subtropical Vegetation." Oliver & Boyd, Edinburgh.
Washburn, A. L. (1969). "Weathering, frost action, and patterned ground in the Mesters Vig district, Northeast Greenland." *Medd. Gronl.* **176**, 1–305.
Wein, R. W., and Bliss, L. C. (1973a). Biological considerations for construction in the Canadian permafrost region (1). *In* "Permafrost: The North American Contribution to the Second International Conference," pp. 767–770. Natl. Acad. Sci., Washington, D.C.
Wein, R. W., and Bliss, L. C. (1973b). Experimental crude oil spills on arctic plant communities. *J. Appl. Ecol.* **10**, 671–682.
Whittaker, R. H. (1962). Classification of natural communities. *Bot. Rev.* **28**, 1–239.

Whittaker, R. H. (1966). Forest dimensions and production in the Great Smoky Mountains. *Ecology* **47,** 103–121.
Whittaker, R. H. (1967). Gradient analysis of vegetation. *Biol. Rev.* **42,** 207–264.
Whittaker, R. H. (1969). Evolution of diversity in plant communities. *In* "Diversity and Stability in Ecological Systems" (G. M. Woodwell and H. H. Smith, eds.), pp. 178–196. Brookhaven Natl. Lab., Upton, New York.
Whittaker, R. H. (1970a). "Communities and Ecosystems." Macmillan, New York.
Whittaker, R. H. (1970b). The biochemical ecology of higher plants. *In* "Chemical Ecology" (E. Sondheimer and J. B. Simeone, eds.), pp. 43–70. Academic Press, New York.
Whittaker, R. H. (1973). Direct gradient analysis: Techniques. Direct gradient analysis: Results. *In* "Ordination and Classification of Communities" (R. H. Whittaker, ed.), Handbook of Vegetation Science, Part V, pp. 7–52. D. W. Junk, The Hague.
Whittaker, R. H., and Gauch, H. G., Jr. (1973). Evaluation of ordination techniques. *In* "Ordination and Classification of Communities" (R. H. Whittaker, ed.), Handbook of Vegetation Science, Part V, pp. 287–322. D. W. Junk, The Hague.
Wolfe, J. A., and Leopold, E. B. (1967). Neogene and early Quaternary vegetation of northwestern North America and northeastern Asia. *In* "The Bering Land Bridge" (D. M. Hopkins, ed.), pp. 193–206. Stanford Univ. Press, Stanford, California.
Wright, H. E., Jr., and Heinselman, M. L. (1973). Ecological role of fire. *Quat. Res. (N.Y.)* **3,** 319–328.
Zoltai, S. C. (1972). Palsas and peat plateaus in central Manitoba and Saskatchewan. *Can. J. For. Res.* **2,** 291–302.
Zoltai, S. C. (1975). Structure of subarctic forests on hummocky permafrost terrain in northwestern Canada. *Can. J. For. Res.* **5,** 1–9.
Zoltai, S. C., and Pettapiece, W. W. (1974). Tree distribution on perennially frozen earth hummocks. *Arct. Alp. Res.* **6,** 403–411.
Zoltai, S. C., and Tarnocai, C. (1971). Properties of a wooded palsa in northern Manitoba. *Arct. Alp. Res.* **3,** 115–129.
Zon, R. (1914). Balsam fir. *Bull. U.S. For. Serv.* **55,** 1–68.
Zon, R., and Graves, H. S. (1911). Light relations and tree growth. *Bull. U.S. For. Serv.* 92. 59 pp.

8 Nutrient Cycling and Productivity

The biosphere is a regenerative system. Based on the utilization of incident solar energy, the capacity for renewal is an essential property. Life is maintained by the constant cycling of atmospheric gases and soil mineral nutrients and the ceaseless production of organic matter. The characteristics of biological systems have been shaped over the course of billions of years of evolution, and the cycling processes, it appears, are carried out with a great deal of efficiency, no doubt the consequence of evolutionary selection at work on the principal enzyme systems involved in the production of organic compounds. On a global basis, energy is seldom a limiting factor in the productivity of forest ecosystems; plant growth is more often critically influenced by the availability of nutrients and water. Nutrient elements occur in the atmosphere, the soil, rocks, and living and dead organic material. The atmosphere is the source of oxygen and carbon dioxide, as well as a major nutrient—nitrogen—occurring in gaseous form. Available mineral nutrients in soil are in the form of ions located either in humus or dissolved in the soil solution. There are minerals in inorganic materials and in rock, from both of which they are released by bacterial activity and by weathering processes. In the biogeochemical flux of elements, available soil and gaseous nutrients are taken up and assimilated by vegetation and microorganisms and made available again through respiration, biological decomposition, and leaching from living and dead organic matter. Minerals in soil and rock, converted to soluble available nutrients through weathering, enter the system and then are continually recycled. They enter an intrasystem cycle and tend to remain there, recycling, until finally they are washed away by the regional hydrological cycle, going into rivers and finally reaching the sea. The cycling of minerals occurs as a function of time; seasonal and daily fluctuations in the intensity of the cycling must be considered in measurements of the quantities involved.

There are, moreover, feedback control mechanisms, complex and including great numbers of interactions, which must be understood before an accurate conception of the system can be achieved. During recent years there has been great interest in understanding the functioning of ecological systems, interest derived from the hope that such knowledge will yield an improved understanding of biological organisms, populations, and communities.

Modern ecology employs systems analysis as a tool to understand how communities of living organisms utilize the resources of the environment to maintain a dynamic, fluctuating, pulsating stability and to perpetuate each species generation after generation. Systems analysis employs numbers to describe, for example, the amount of energy flowing through an ecosystem, the biomass or weight of organisms living at a given time, the productivity or weight of new organisms produced over a given period of time, and the quantity of mineral elements cycling through organisms, soil, and atmosphere. These are the materials, the processes, and the forces at work in the ecosystem. Much of the effort expended by ecologists in analyzing an ecosystem is given over to making quantitative determinations of the energy, materials, mass of each kind of organism, and total productivity of the system. The remainder of this book considers some of the knowledge available on these matters as they relate to the boreal forest, its plants, and its animals.

Ecosystems can be said to possess two nutrient cycles, a biological cycle and a geological cycle. The biological cycle involves the exchange and recycling of nutrients among plants, animals, and soil. The geological cycle involves the fixation of nutrients from the gases of the atmosphere and the release of nutrient minerals by the weathering of rocks. Forests often attain high levels of production on soils too poor for agriculture. This is possible because nutrients recycle continuously between the soil and the plants. Nutrients in the organic matter of living plants and soil humus represent much of the nutrient capital available in the forest ecosystem.

Boreal soils range from dry to consistently moist, but since precipitation exceeds evaporation in boreal regions, even the soils on dry uplands are frequently moistened. Often boreal soils are decidedly anaerobic in lower horizons, acidic because of the nature of anaerobic respiration and because levels of calcium and magnesium are low. These soils usually have a high proportion of partly decomposed organic matter. Nutrient elements may be present in abundance, yet unavailable for utilization by plants because they are tied up in unassimilable organic compounds. Conditions are unfavorable for rapid bacterial decomposition of organic material but, when nutrient release does occur, the nutrients are quickly incorporated again into living plants. This process contrasts with those in plants and soils in temperate and tropical zones where there is an annual cycling of elements from leaves to litter to soils and back to leaves again. In boreal soils, the large proportion of nutrients tied up in undecomposed material literally creates nutritional deficiencies in the midst of abundance. In the soils of northern boreal regions where permafrost is present, accumulating organic material in litter continually raises the surface of the permanently frozen zone, and nutrients trapped below the freeze line are lost to the ecosystem for long periods of time, at least until some disturbance results in melting of the upper layers of permafrost—an event that may not occur for thousands of years.

NUTRIENT CYCLING AND THE BOREAL ECOSYSTEM

The cycling of elements vital to life—carbon, hydrogen, oxygen, nitrogen, and so on—is a natural process that modern techniques of systems analysis have been devised to emulate. As presently developed, a systems model is at best a crude approximation to the "reality" of nature; the dilemma is that a model simple enough to be manageable does not accurately describe the natural ecosystem, and a model complex enough to describe nature accurately is too complex for existing computational methods. As mathematical techniques and computer capabilities improve, more accurate ecosystem models will be devised. Then it will become vitally important to have accurate conceptions of the events that transpire in cycling processes. This will require detailed knowledge of the quantities of the various materials involved in cycling. It will require also a detailed understanding of the physical and biochemical events involved—such physiological and biochemical processes as assimilation of mineral nutrients, respiration, photosynthesis, nitrogen fixation, denitrification, nitrification, synthesis of proteins and fats, and translocation of organic foodstuffs. With these considerations in mind, we can present only a bare description of the kinds of cycling processes that will presumably be encountered when more intensive analyses of boreal ecosystems are undertaken. Let us, then, outline the general cycling of the major elements involved in boreal forests.

THE NATURE OF CYCLING PROCESSES

The soils of the sub-Arctic have few if any properties unique to the region. Perhaps if a single pedogenic property can be singled out as being strongly characteristic of boreal soils, it would be the pronounced tendency for a thick surface organic mat to accumulate under the subarctic conditions. Nearly all the plant material produced by the forest community eventually falls to the ground or exists as root material already within the soil. This organic material is attacked and decomposed by a wide variety of fungi, bacteria, and subterranean animals. These organisms mineralize the humus of the soil, returning nutrient elements to the soil in chemical forms in which the elements again can be assimilated and used by growing plants. Though boreal soils often contain ample reserves of nitrogen, phosphorus, and other nutrient mineral elements, most are in organic compounds not assimilable by angiosperms and gymnosperms. The rate at which they are mineralized often regulates the rate at which the plants making up the forest cover can grow.

TABLE 31

Major Chemical Reactions in Soils

Respiration
$$C_6H_{12}O_6 + 6O_2 \rightarrow 6CO_2 + 6H_2O$$
Carbon dioxide   Water

Ammonification
$$CH_2NH_2COOH + 1\tfrac{1}{2}O_2 \rightarrow 2CO_2 + H_2O + NH_3$$
Glycine   Oxygen   Ammonia

Nitrification
$$NH_3 + 1\tfrac{1}{2}O_2 \rightarrow HNO_2 + H_2O$$
Nitrous acid

$$KNO_2 + \tfrac{1}{2}O_2 \rightarrow KNO_3$$
Potassium nitrite

Denitrification
$$C_6H_{12}O_6 + 6KNO_3 \rightarrow 6CO_2 + 3H_2O + 6KOH + 3N_2O$$
Glucose   Potassium nitrate   Potassium hydroxide   Nitrous oxide

$$5C_6H_{12}O_6 + 24KNO_3 \rightarrow 30CO_2 + 18H_2O + 24KOH + 12N_2$$

Nitrogen fixation
$$N_2 \rightarrow 2N \text{ ``activation'' of nitrogen}$$
$$2N + 3H_2 \rightarrow 2NH_3$$

In mineralization, organic compounds are decomposed, and the nutrient elements are released in such inorganic materials as carbon dioxide, ammonium, nitrate, orthophosphate, sulfate, soluble silicates, and other substances. These are the products that are formed most abundantly under aerobic conditions. Anaerobically, nitrate and sulfate are not formed abundantly, but methane, nitrogen, ammonium, and sulfide appear instead (Table 31). The pool of elements in natural ecosystems is relatively constant because of the continuing tendency toward steady-state conditions in which depletion is balanced by an influx of new material from rocks or the atmosphere. Despite wide distribution, little is known about humic substances. The fulvic acids that form complexes with metal ions are important, although very little is known about the specific reactions of fulvic acid in boreal environments.

CHEMICAL CHANGES DURING DECOMPOSITION

Chemical breakdown of litter has so far been described only in the most general terms. Water-soluble components disappear first, followed successively by alcohol and ether-soluble fractions and then by the hemicellulose, cellulose, and lignin fractions. As decomposition proceeds, the organic plant material serves as food for the growth of microorganisms which in turn synthesize proteins which persist in the soil as constituents of the microbial biomass. Some plant materials and microbial cell structures protected by colloid absorption can persist for long periods of time. It has been found, for example, that organic matter in podzolic B horizons can have a residence time ranging from 300 to 1200 years, longer periods being characteristic of more northern regions. Plant materials high in nutrients and low in lignin are the most rapidly decomposed, depending on moisture content, aeration, and soil temperature. It is only under cool, anaerobic conditions such as are found in boreal regions that undecomposed plant material accumulates and peat is formed.

Ecosystems retain nutrients by cycling within the system. If the cycle is broken by destruction of the vegetation, loss of nutrients is rapid because of the absence of plants, increased water drainage, better aeration, and accelerated decay processes.

NUTRIENT CYCLES THROUGH SOIL POPULATIONS

The control agents of boreal forest ecosystems are microorganisms. Though they are individually small, their total biomass is great, and they exert a large effect through the turnover of chemical elements in biogeochemical cycles. Little biological study has been attempted on boreal forest soils, and limited data on numbers and kinds of microorganisms are available.

Studies by Christensen and Cook (1970) on three significantly different humified peat bogs in central Alberta—a fibrisol (the least decomposed), a mesisol, and a well-decomposed humisol—revealed high numbers of bacteria in humic horizons and generally also at greater depths. Large numbers of aerobes, up to nearly 10 million per gram oven-dry weight of peat, were notable at the surface of each profile. Up to 10 million psychrophiles, 12 million mesophiles, and 75,000 thermophiles were found in deep layers of mesisol and humisol. Counts were highest in August and September.

Mesophilic and psychrophilic fungi were moderately abundant in the top 30 cm, but their numbers dropped off rapidly with depth, probably because of anaerobic conditions. Actinomycetes were virtually absent from all samples, and algae were present down to 46 cm in the mesisol, but only in small numbers (fewer than 2000 per gram). Euglenoids, *Chlorococcum* types, and members of the Ulotrichaceae were also found.

In these soils, tests for nitrification and nitrogen fixation were negative. Only the first stage of denitrification took place, and organisms possessing the ability to convert nitrate to nitrite occurred in the greatest numbers below the surface layers; there were far fewer in the cold fibrisol. Bacteria capable of ammonification of protein were present in significant numbers. The counts for iron-reducing bacteria showed no consistent pattern but were considerably (over 100 times) higher in peat soils than for comparable layers in mineral soils. From 2000 to 17 million iron reducers per gram oven-dry peat were recorded.

The bacteria found at depth in the mesisol and the humisol could be facultative anaerobes, residual spores of *Bacillus* species active in these layers when originally at the surface, or bacteria and spores leached down the profile. However, the wide range of bacteria present in the lower layers suggests that residual spores are not the only possibility, and leaching is perhaps not important in soils with a high water content. Study of the anaerobic population might reveal whether facultative anaerobes are responsible for the marked increase in bacterial numbers at depth. The limited numbers of bacteria in the deeper layers of the fibrisol may be due to the less humified nature of the material, or to the fact that the lower layers of this particular example were at a relatively low temperature, only just above 0°C.

Fungi are mostly strict aerobes, and this was demonstrated by their virtual restriction to the surface layers. In the humisol, only the upper 25 cm had significant numbers of fungi. They were relatively abundant in the mesisol down to 54 cm, although it was less aerobic than the humisol at this depth and the pH was about 4.0. The fibrisol had few fungi below 30 cm, although its water content and degree of decomposition were similar to those of the mesisol, and its pH was an acid 3.5. Either very high acidity or other factors, possibly temperature and degree of aeration, limited survival in the lower layers.

The absence of actinomycetes was not surprising in view of the acid nature of the soils. The populations of algae were low. Those occurring beneath the surface must exist heterotrophically, and their rate of growth is probably slow.

THE CARBON CYCLE

Carbon rapidly cycles between inorganic and organic states. A large percentage of the atmospheric carbon captured by green plants, representing much of the organic carbon in living matter, is returned each year to the atmosphere. By decomposing the dead organic matter, bacteria play an essential role, replenishing the atmospheric carbon dioxide available for photosynthesis—about 30 tons/ha of carbon dioxide on a global basis. The major source of atmospheric carbon dioxide is microbiological decomposition of dead plants and animals, although plants and animals also produce large amounts of carbon dioxide from respiration. Microorganisms involved in the carbon cycle are mainly mesophilic heterotrophs.

In temperate and tropical upland forests, organic materials are decomposed rapidly, so that carbon inflow equals the rate of loss from the system. In soil, the amount of carbon lost yearly is essentially equivalent to the quantity of organic carbon added. This is not true of bogs or marshes, however, and is generally not true of much of the boreal forest, where an accumulation of undecomposed plant detritus exists in the form of peat (Fig. 30). Soils that differ only slightly will often have completely different species dominating decomposition. Once cellulose, lignin, and other substances have been degraded to simpler materials, they are attacked by a wide range of soil organisms. The decay of a complex substance such as lignin involves a succession of microorganisms, each of which carries out a relatively small number of biochemical steps.

THE NITROGEN CYCLE

In the decay of nitrogen-containing materials, proteins are hydrolyzed to amino acids which are then deaminated to liberate ammonium. Nitrogen in the purine and pyrimidine bases of nucleic acids is also liber-

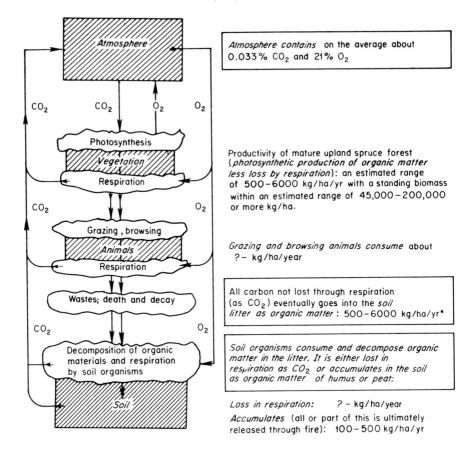

Fig. 30. The carbon cycle. The values are generalized estimates derived from available data. Definitive values for different regions and habitats will become available as additional studies are conducted. References consulted in compiling the figure include Rodin and Basilevič (1968); Reader and Stewart (1972), Moore and Verspoor (1973), Barney and Van Cleve (1973), Reader (1978), Alban et al. (1978), Rencz and Auclair (1978), and Barney et al. (1978).

ated as ammonium. The resistance of nitrogen compounds to microbial attack is such that only a small proportion of the nitrogen reservoir of soil is mineralized in each growing season (Fig. 31). Proteins are adsorbed by clays and may be rendered less available to microbial attack in this state. A lignin–protein complex has been observed experimentally; mixing protein with lignin produces a material even more resistant to mineralization than clay–protein complexes.

Degradation of protein is carried out largely by fungi; bacteria, in general, do not attack native protein. Peptones and peptides, however, support bacterial growth and, once the protein has been broken down into these compounds, bacteria continue the process. The enzymes proteinase and peptidase are pH sensitive and are effective only when conditions are favorable.

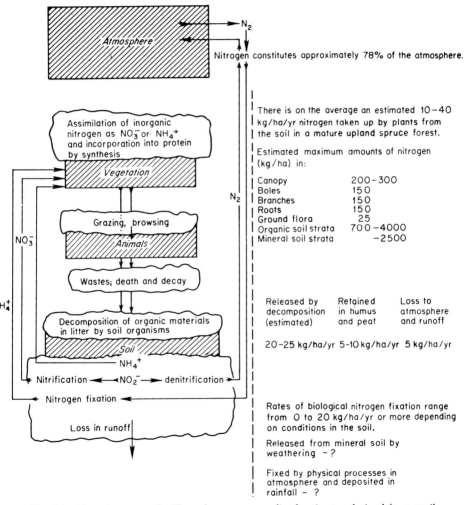

Fig. 31. The nitrogen cycle. The values are generalized estimates derived from available data. Definitive values for different regions and habitats will become available as additional studies are conducted. References consulted in compiling the figure include Heilman (1966, 1968), Nihlgård (1970, 1972), Page (1971), Damman (1971), Weetman and Webber (1972), Alban et al. (1978), Larsen et al. (1978), and Crittenden and Kershaw (1979).

Nitrification and Denitrification

Dead tissues of plants and animals are thus decomposed in a series of steps to simpler compounds, and in the degradation of protein one end product is ammonium. A fraction of this ammonium is used directly by higher plants, but a large portion is assimilated by microorganisms and oxidized to yield nitrate as the final product. Nitrate is readily assimilated by plants but is also easily washed away into the groundwater. Nitrification is of minor importance in acid podzol soils.

There is accumulating evidence that certain vegetational communities, particularly those approaching mature or climax forest stages, inhibit nitrification. This is evidently an adaptation for circumventing the last stage of nitrification, in which ammonia is converted to nitrate, a step requiring the expenditure of considerable amounts of energy. The energy is retained in the system and the cycling is expedited when this stage is eliminated, and the plants reabsorb the nitrogen as ammonium ions.

Denitrification results in a net loss of nitrogen from soil to atmosphere. Counts in excess of a million denitrifying microorganisms per gram are not uncommon. Anaerobic conditions in water-logged soil lead to denitrification and a loss of gaseous nitrogen and nitrous oxide. Denitrification organisms utilize nitrate in place of oxygen under anaerobic conditions, but organic compounds are required as an energy source and denitrification will not occur unless organic substrates are available. In general, the release of nitrogen from soils in boreal regions is influenced by pH (the production of both gaseous nitrogen and inorganic nitrogen as ammonium and as nitrate is relatively slow in acid soils) and by cool temperatures which slow the rate of mineralization and, as a result, nitrogen accumulates in the acid soils of the boreal forest in forms unavailable to plants.

Nitrogen Fixation

The rate of annual depletion of nitrogen from boreal soils by denitrification or by deposition in peat is not known. To maintain a steady state, however, losses must equal gains, so the deficit must be made up by organisms that fix nitrogen.

Several genera of nonleguminous northern plants possess root nodules, notably *Alnus*, the alders; *Myrica*, bog myrtle or sweet gale; *Hippophae*, distributed in large areas of Europe and Asia; *Elaeagnus*, of widespread distribution in North America, Europe, and Asia; and *Shepherdia*, soapberry, found abundantly in Canada. In northern environments, species representing these genera are often much more abundant than legumes. Their capacity to fix nitrogen is assumed to be of considerable ecological significance. Of these plants, the alders have been the most intensively studied and have been found to assimilate nitrogen at a rate that appears adequate to permit rapid enrichment of nitrogen-deficient soils. Nitrogen fixation is also accomplished by a variety of free-living microorganisms.

Iron salts have been implicated in the nitrogen metabolism of *Azotobacter, Clostridium,* algae, *Aerobacter,* and *Achromobacter,* but the specific requirement is difficult to establish. As a rule, soils more acid than pH 6.0 contain very few *Azotobacter* cells. It seems likely that significant nitrogen fixation in boreal forest is accomplished by free-living *Clostridium.* When the organic content is high and the pH favorable, nitrogen fixation is probably frequent under the anaerobic conditions prevalent in most of the boreal soil types studied.

Nitrogen-fixing bacteria serve as control agents of the nitrogen economy of the whole ecosystem; colonization of bare ground by nitrogen-fixing plants leads to production of organic matter, nonsymbiotic nitrogen fixers can then grow, and a balanced ecosystem can be established. Crocker and Major (1955) studied changes in organic nitrogen of ground in Alaska laid bare by a receding glacier. During the alder thicket stage (0 to 50 years) nitrogen accumulated in the mineral soil alone at the rate of 1.5 g/m^2/year and in the forest floor mineral soil system at 4.9 g/m^2/year. In another study, they reported nitrogen accumulated at a steady rate of 4.0 g/m^2/year in mineral forest soil. Other studies demonstrated that alder can add an average annual total nitrogen of 360 kg/ha or more at 20 years (Van Cleve et al., 1970).

THE NITROGEN CYCLE IN FOREST STANDS

Of particular interest are the quantities of nitrogen involved in nitrogen cycling between vegetation and soil. The unavailability of nitrogen in boreal soils is related to the widespread formation of the type of humus known as mor. Mor humus, in which the decomposition of organic matter is incomplete, often contains large quantities of nitrogen (1–2%). Little of the nutrient content of mor humus is available for plants and marked nitrogen deficiencies often exist in mor soils. Mor humus is also characterized by a high carbon to nitrogen ratio, a lack of ligno-protein substances, a low pH, a very active fungal population, few or no nitrifying bacteria, mycorrhizal infection of roots and possibly increased efficiency of nitrogen, potassium, and phosphorus uptake by trees and other plants, a high moisture content, low soil temperatures, poor aeration, and the presence of abundant mosses that rely on rain and tree drip for their mineral nutrients. In mor humus, with storage of nitrogen at the soil surface, the turnover of nitrogen is slow. Turnover may even gradually decline with time until it results in markedly decreased fertility of mature forest soils, with a noticeable effect upon the vigor and growth of trees and other vegetation and a general disintegration of the stand.

The following is an estimate of the quantities of nitrogen cycled in a black spruce stand of about 65 years of age [data from Weetman and Webber (1972) and other sources]:

Net annual uptake by trees	15 kg/ha/year
Annual return in litter fall	5–10 kg/ha/year
Total accumulation (available and unavailable) in the humus layer	212 kg/ha
Total accumulation available in the humus layer	15 kg/ha
Total (available and unavailable) in humus and mineral soil	1510 kg/ha
Total available in humus and mineral soil	22 kg/ha

The nitrogen content of litter fall is difficult to measure accurately unless data are taken over a period of years, since annual litter fall has been shown to be quite variable in amount. The return of nitrogen to the soil humus may be greater than the 5–10 kg/ha/year indicated above.

The figures indicate an accumulation of nitrogen in the humus layer in excess of the difference between the amount in litter fall and the amount taken up by the trees. The difference is attributable to other sources of nitrogen, which could include the following:

Nitrogen compounds in precipitation	2–15 kg/ha/year
Symbiotic fixation by alder and/or other shrubs early in the development of a stand (such shrubs are sporadic in occurrence in mature stands)	10+ (?) kg/ha/year
Possible accumulation of nitrogen by *Hylocomium* mosses and other ground vegetation, exudation of nitrogen compounds by roots, possible absorption of nitrogen by organic material in vegetation and humus layers	(?) kg/ha/year

There is also a loss of nitrogen from the soil by various means, and this could occur as the result of one or more of the following:

Leaching losses; small since the nitrate nitrogen content of humus and soil is very low	0–1 kg/ha/year

Losses as nitrous oxide and nitrogen gas by bacterial denitrification; low on well-drained sites with good aeration	(?) kg/ha/year
Losses as ammonia; small because of low decomposition rate. Loss by chemical action is usually insignificant	(?) kg/ha/year

Estimates of the loss of nitrogen from closed stands are difficult to obtain, but it seems unlikely that losses brought about by these factors are very significant. Weetman and Webber (1972) indicate that the input of nitrogen through precipitation (both as nitrogen compounds dissolved in water and as dust) is an important factor in the mineral nutrition of plants. There is, moreover, a great store of potentially decomposable organic nitrogen in the humus. Nitrogen deficiencies are due to blockage of the nitrogen cycle by conditions in the humus; Weetman and Webber conclude that nitrogen appears to be a major limiting factor for tree growth, at least in eastern Canada, and that its availability is largely associated with the rate and type of humus layer decomposition. The supply of the other major nutrients, phosphorus, potassium, magnesium, and calcium, is related to the mineralogical composition of soils on upland sites. It appears that most soils developed on glacial till in the Canadian shield and the Appalachian region are mineralogically rich enough to support nutrient losses associated with one full-tree logging operation every 50 years. However, this cannot be applied to forests growing on coarse, waterlain deposits with a low cation-exchange capacity or to forests growing on organic accumulations with shallow rooting and no input of nutrients from lateral water movement.

Annual macroelement transfer from jack pine trees was studied by Foster (1974), who found that tree litter was the most important source of nitrogen, phosphorus, calcium, and magnesium for the forest floor (51–69% of the total depending on the element), whereas throughfall supplied most potassium (54% of the total). Ground vegetation litter contributed significant amounts of nutrients (7–23% of the total depending on the element), but stemflow added little (1–8% of the total). Potassium in the throughfall was derived mainly from leaf wash, whereas nitrogen, phosphorus, calcium, and magnesium in throughfall were derived primarily from precipitation entering the ecosystem. The jack pine forest floor received an annual total of 30 kg/ha of nitrogen, 22 kg/ha of calcium, 19 kg/ha of potassium, 3 kg/ha of magnesium, and 2 kg/ha of phosphorus from the processes studied. Most of the nutrients in these totals were returning to the forest floor from the vegetation.

Since nutrient concentrations in ground vegetation litter were higher than those in tree litter, the contribution of ground vegetation to the weight of nutrients in litter was proportionally greater than its contribution to the total litter biomass. The considerable quantities of nutrients cycled in the ground vegetation should not be overlooked when computing forest ecosystem nutrient budgets and cycles, Foster points out.

Not all nutrients in leaf wash are derived from leaching aerial components of the forest. Some come from dry fallout outside the ecosystem; thus the quantity of nutrients leached from the vegetation may be overestimated. Nihlgard (1970) also has demonstrated that foliar leaching contributes significant quantities of nutrients to leaf washing in spruce forests. The forest floor also receives nutrients in leaf wash, which are leached from the ground vegetation. Potassium, which occurs in an inorganic form in plants and can be readily extracted with water, is the element leached from the canopy in the greatest amounts. Nihlgard states that the higher amounts of nutrients in the rainwater of a spruce forest, compared with those of a beech forest in his study, are probably explained by the presence of live, dying, and dead needles on the branches throughout the year.

Foster and Gessel (1972) found marked seasonal differences in the rate of litter fall and cycling of nutrients in jack pine forest. The production of litter was very low in July and August but much greater in September and October. An autumn supply of carbon and nutrients, thus, provide a favorable substrate for a spring flush of decomposers, thereby enhancing the release of nutrients to plants during the period of most rapid growth.

The high variability in amounts and concentrations of elements encountered when sampling throughfall, stemflow, and litter fall processes should be appreciated when making measurements or interpreting data, Foster and Gessel caution. For example, coefficients of variation of amounts in a single collection for these three processes were 24, 65, and 26%, respectively. Nutrient quantities in each process are also subject to fluctuation controlled by climatic conditions; e.g., above-normal precipitation could increase the amounts of nutrients entering the forest. Additional research on the cycling of nutrients in boreal forests would be most valuable, they point out.

OTHER ELEMENTS

Studies of the cycling of other nutrient elements in boreal soils—sodium and sulfur, for example—are few or lacking altogether and can be discussed only in general terms (see Fig. 32). Sodium and much of the potassium are washed out from freshly fallen litter within a few weeks. Phosphate and magnesium content also falls rapidly. About half of the

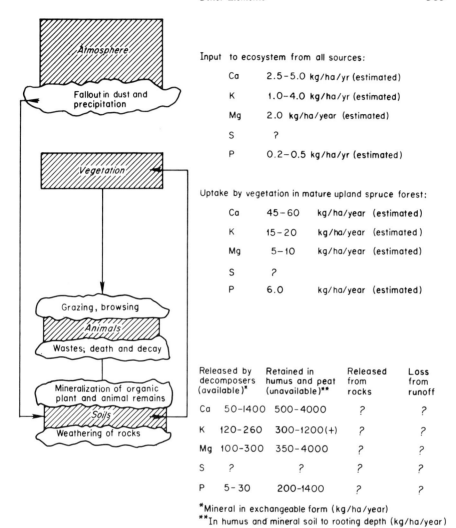

Fig. 32. Cycling of mineral elements. The values are generalized estimated derived from available data. Definitive values for different regions and habitats will become available as additional studies are conducted. References consulted in compiling the figure include Heilman (1968), Weetman and Webber (1972), Page (1971), and Damman (1971).

calcium is fairly rapidly mobilized. The autotrophic *Thiobacillus thiooxidans* and related organisms may be instrumental in producing calcium sulfate and phosphoric acid from calcium phosphate. Organically bound phosphorus can also be used by some algae.

The biogeochemical conversions of sulfur have an effect on the reactions of other elements. Biologically generated hydrogen sulfide and sulfuric acid are derived from organic sulfur of cellular compounds such as amino acids, assimilation of inorganic sulfur compounds by microbial organisms, oxidation of inorganic or amino acid sulfur, and reduction of sulfate and sulfur to sulfide.

Atmospheric sulfur is usually present as the sulfide of hydrogen or sulfur dioxide. These two compounds originate from the biological decomposition of dead plant and animal materials. Hydrogen sulfide washed from the atmosphere in rain and snow is often deposited as calcium sulfate. Oxidation of hydrogen sulfide is accomplished by sulfur bacteria and photosynthetic bacteria. Some organisms can also oxidize elemental sulfur.

The standing-crop weights of each nutrient in forest stands indicate that the forest requires relatively modest amounts of most of these major nutrients, a large portion of which are recycled in litter fall, root decay, tree mortality, and washing of nutrients from leaves by rain. The quantities annually immobilized in woody tissues are much less significant and, actually, the values for precipitation and dust input approach or exceed the values for nutrients contained in trees. At the high end of the range they exceed the values for all major nutrients, except phosphorus, contained in full trees. These make up for logging losses of major nutrients without considering inputs from soil weathering. However, the values so far obtained show a wide variation and are from widely scattered sources. Actual values need to be determined for various forest regions before firm conclusions can be achieved. The significance of leaching losses is more difficult to judge, and there is at present only incomplete knowledge of the abilities of trees to extract nutrients from unweathered minerals.

EFFECTS OF FIRE ON CYCLING

Destruction of trees, ground vegetation, and soil organic matter by fire exposes the surface of the mineral substrate, and the results in terms of cycling properties of the soil are considerable. A study on soil nutrients following fires in northern Saskatchewan was carried out by Scotter (1964) who found considerable variability in the characteristics of the various stands studied but was able to make some generalizations from the data (Table 32).

Total exchange capacity decreased in three of the four burned-over areas. A 5-year-old burn showed a slight increase in total exchange capacity. Of the individual cations, exchangeable calcium increased on

TABLE 32

Chemical Soil Properties on Burned-Over and Unburned Soils[a]

Site description	Exchangeable cations in mEq per 100 g of dry soil					Sum of cations	Total exchange capacity	Total nitrogen (ppm)	Available phosphorus (ppm)
	H	Na	K	Ca	Mg				
5-year-old burn	4.1	0.1	T[b]	0.4	0.2	4.8	7.3	50	8.5
Mature black spruce	7.0	0.1	T	0.2	0.4	7.7	11.8	100	1.0
13-year-old burn	3.4	0.2	T	0.2	0.1	3.9	4.8	40	26.5
Mature black spruce	3.9	0.1	T	0.2	0.1	4.3	4.9	40	14.5
5-year-old burn	1.9	0.1	T	0.4	0.1	2.5	4.4	50	13.5
Mature jack pine–black spruce	2.4	0.2	T	0.2	0.1	2.9	3.2	30	T
22-year-old burn	1.2	0.1	T	0.4	0.1	1.8	2.9	20	37.0
Mature jack pine–black spruce	2.4	0.2	T	0.2	0.1	2.9	3.2	30	T

[a] From Scotter (1964), reproduced with permission.
[b] T, trace.

three sites and remained the same on the fourth; little change was noted in exchangeable potassium, magnesium, or sodium. Total nitrogen abundance in mineral soils did not follow any apparent trend. It increased on one burn, was reduced on two, and remained the same on the fourth. On each of the four burns, available phosphorus was more abundant than in the adjacent unburned forests.

According to Scotter, infrequent fire may not be disastrous to the fertility of mineral soils. Greater postfire soil temperature extremes and reduced soil acidity, however, may stimulate or inhibit the germination and growth of plants depending on seedling requirements. Studies are needed to determine which plants are favorably or unfavorably influenced by soil temperature extremes and reduced acidity in recently burned areas.

PRODUCTIVITY: COMMUNITY COMPARISONS

The productivity of the earth's vegetation represents most of the total productivity of the ecosystems found on the earth's land surface. The greatest mass of organic matter is in forests. In the tropical rain forest the amount of organic matter is on the order of 500 tons/ha. In broad-leaved forest it is about 370–410 tons/ha. In northern forests it is considerably less, being on the order of 22–60 tons/ha. The influence of climate upon growth, hence upon productivity and the rate of nutrient cycling, has thus been clearly demonstrated on a global basis (Leith and Whittaker, 1975). It can also be demonstrated between northern and southern regions within the boreal forest. In the southern parts of the boreal forest, black spruce is found in a wide range of habitat types, with the consequence that growth rates vary markedly from stand to stand and range over a wide scale of values. A narrowing in average growth rates can be discerned along the latitudinal gradient northward. At the northern forest border, the rate of growth on optimum sites does not greatly exceed the rate on the poorest sites, indicating that only the most narrow range of habitable environment is available at the northern edge of the forest (Figs. 33, 34). The best available sites permit only the minimum growth consistent with continued survival; these sites have growth rates comparable to those shown by trees only on the poorest sites in regions to the south. When a line is drawn through the maximum growth rates found at each latitude, the intersect in central Canada (long. 101°W) occurs at approximately lat. 65°N, about 4° beyond the range of the spruce forest at this longitude, but 2° or less beyond the range of spruce as a species (if one excepts the extension along the Thelon River which

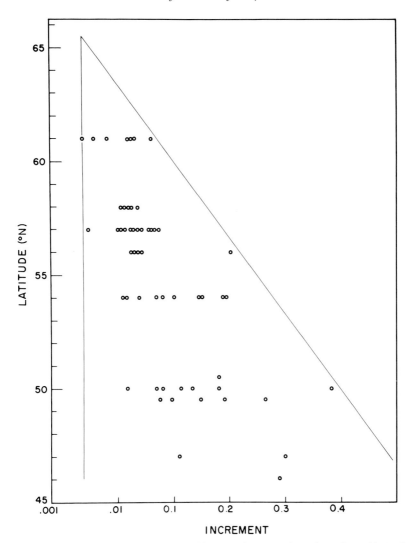

Fig. 33. Relative growth rates averaged for trees from stands at the indicated latitudes; data from 10–20 randomly selected trees in sampled stands. The widest variation in growth rate from one stand to another is found in the southern latitudes; the maximum rate declines northward. The increment is a relative value, calculated by obtaining the total volume of a cylinder (multiplying basal area at breast height of the tree by the height of the tree) divided by the age of the tree. Values for all trees in a stand were then averaged. Note change in increment scale. (Data from Larsen, 1965; Mitchell, 1973.)

Fig. 34. Cone production by spruce at the northern edge of the forest–tundra ecotone may be surprisingly heavy. The seeds, however, evidently are mostly infertile, and the little reproduction that occurs is at least principally by the process of layering. Note the snow abrasion on the stem at the right, indicating that the surface of the snow was just below this point for a time during winter when high winds prevailed and abrasion from driven snow resulted. The middle stem shows little abrasion, and the stem at left little if any; the stem at the right served as a windbreak and protection for the other two.

may well have a special microclimate because of the environmental effects of the river).

Plants of different species accumulate different concentrations of nutrients and differ in ability to produce organic matter, so the distribution of nutrients in forest ecosystems depends upon the species of plants present. The total production or the net primary productivity is usually considered a good measure for ranking species in an ecosystem according to their significance. Studies on primary production of dry matter in peatland vegetation have been carried out in Canada and have proved to be of considerable importance. Up to 400,000 km^2 of the Canadian land surface is estimated to be peat covered and occupied by lowland forest of great potential economic value.

Reader and Stewart (1971) found that a peatland in central Canada had a total net primary production value of 1943 g/m² (19,430 kg/ha²). In the treeless portion of the peatland, the subsurface biomass was 4.6 times the value of the aerial biomass. With so much of the total biomass found below the peatland's surface, estimated subsurface production in the area (1461 g/m²) far outweighed its aerial counterpart. The large subsurface production value is in part a result of the continuous development of new adventitious roots as aerial stems are buried by accumulating peat.

The annual net production for all nonvascular species (55.4 g/m²) was small compared with the total for vascular species (1777.5 g/m²), with 82% of the vascular production associated with subsurface plant components. This total annual net production of bog vegetation is significantly higher than the values obtained by other workers. It reflects a larger value for subsurface biomass in the net production of peatland vegetation than had previously been assumed would be the case.

Measurements of the aboveground biomass of black spruce stands in Quebec gave values ranging from about 10,000 kg/ha for lichen woodland to 163,000 kg/ha (16,300 g/m²) for a spruce–moss forest containing 5000 trees averaging more than 36 m²/ha breast-height stem area (Moore and Verspoor, 1973). Total biomass was not measured in this study; a summary of the measurements obtained is given in Table 33, as well as a comparison with two spruce stands in northern Eurasia and one in northern Quebec.

The total live biomass in a lichen woodland in northern Quebec was measured by Rencz and Auclair (1978). The community was dominated by black spruce and *Cladonia alpestris*; of the total biomass, these two species contributed 75%. The only other species was *Betula glandulosa*, with 12% of the total. The biomass was distributed among the tree, shrub, and ground-layer species in about the ratio 3:1:1, with lichen and leaf tissue representing one-third of the total biomass; stems and roots constituted 40 and 28% of the total, respectively, and black spruce contained 76% of all stem tissue. Shrub root mass (39% of shrub dry weight) was high in comparison with tree root mass (34% of black spruce weight).

The comparisons of biomass levels and tissue distributions presented in Table 33 indicate that the lichen woodlands and other forest communities, as well as the closed-canopy forests in the areas shown, are generally comparable and without major regional differences. As Rencz and Auclair point out, the lichen woodland has a much lower biomass

TABLE 33

Biomass of Selected Forest Ecosystems for Comparisons of Regional Productivity[a]

				Above ground			Lichens and mosses	Below ground
	Community[b]	Region	Total	Trees	Shrubs	Herbs		
1.	Spruce–lichen woodland	No. Quebec		9,590				
2.	Spruce–lichen woodland	No. Quebec		21,060				
3.	Spruce–lichen woodland	No. Quebec		29,250				
4.	Spruce–moss forest	No. Quebec		82,270				
5.	Spruce–moss forest	No. Quebec		163,360				
6.	Spruce–moss forest	No. Quebec		78,360				
7.	Northern spruce forest	USSR		78,000				
8.	Central spruce forest	USSR		200,000				
9.	Spruce forest, lowland	Alaska		16,136				
10.	Spruce forest, upland	Alaska		24,056				
11.	*Larix dahurica* forest	USSR	14,025	3,335	3,250	0	7,440	14,330
12.	Spruce–lichen woodland	No. Quebec	28,825	19,970	1,375	0	8,480	
13.	Spruce–lichen woodland	No. Quebec	33,372	18,637	5,040	53	9,642	12,919
14.	*Pinus sylvestris* forest	Finland	33,367	31,326	1,921	0	1,024	7,673
15.	*Picea rubens–Abies*	Tennessee	341,580	340,000	960	220	400	

[a] Units are kilograms per hectare.
[b] Sources: 1–6: Moore and Verspoor (1973); 7–8: Rodin and Bazilevich (1966), cited in Moore and Verspoor (1973); 9–10: Barney, *et al.* (1978); 11–15 cited in Renze and auclair (1978) as follows—11: Ignatenko *et al.* (1973); 12: Moore and Verspoor (1973); 13: Rencz and Auclair (1978); 14: Kjelvik and Karenlampi (1975); Whittaker (1966).

compared with boreal closed-canopy forests and with temperate deciduous forests. In comparison with these latter forest communities, the lichen woodland has higher root/shoot ratios and a higher shrub and ground-layer biomass than the other forests. The data presented for this area in Quebec are generally consistent with those presented for the biomass distribution in two Alaskan forests by Barney *et al.* (1978).

Biomass and productivity of an alder swamp in northern Michigan can also be employed for comparison. Parker and Schneider (1975) determined total aboveground dry weight biomass and annual production for two sites of different soil texture in an alder swamp of Michigan's upper peninsula. The more poorly drained site averaged 5300 and 640 g/m^2 per year, while the better-drained site averaged 3100 and 570 g/m^2 per year. The smaller standing-crop biomass on the better-drained site was due to a greater abundance of *Alnus rugosa*. The tree stratum constituted 97 and 93% of the total biomass and 84 and 80% of the total production on the site. The understory strata constituted the remaining 3 and 16% of the biomass and production on the first site and 7 and 20% on the second.

In another study, aboveground dry matter production in three stands of trembling aspen in Canada was measured (Pollard, 1972). Aboveground biomass and annual production were estimated for several years in aspen stands aged 6, 15, and 52 years old. Based on regressions of dry weight on stem diameter, biomass (stems and branches) estimates were 21,500 kg/ha in the juvenile stand, 51,200 kg/ha in the intermediate stand, and 91,800 kg/ha in the mature stand. Net annual aboveground production (stems and branches) for these stands was 6900, 7000, and 1340 kg/ha, respectively. Foliage amounted to 2600, 2600, and 1500 kg/ha.

THE LICHEN ASSOCIATES OF THE COMMUNITIES

Lichens have long been known to be of great importance in the ecology of the northern regions, principally because of their great abundance in many habitats and the significance of their role in the understory community of open forest. They are also of importance in the diet of caribou, although the extent to which they are utilized under natural conditions is not yet fully established (Scotter, 1964; Bergerud, 1974).

Despite the general recognition of their importance, only recently have quantitative indications of their significance in the ecology of the northern regions been forthcoming. In a study of the effects of forest fires on the winter range of barren-ground caribou in northern Saskatchewan, Scotter (1964) has shown that, in terms of pounds of organic substance per unit area, lichen weight in the northern forests is often equal in importance to the weights of grasses, herbs, and shrubs combined and may exceed them at certain stages of forest development. For example, black spruce, white birch, and jack pine forests of six age classes all possessed lichens—both ground-inhabiting forms and species growing usually on the bark and twigs of trees—equal in forage yield weights to herbs and shrubs (Table 34).

As stands increase in age, the importance of the lichens increases accordingly. Whether destruction by fire greatly reduces the available forage for caribou, as indicated by Scotter, or whether fire actually enhances lichen availability after a relatively brief regeneration period, is a problem yet not entirely clarified. Scotter holds that fire greatly reduces the forage available to caribou, a situation reversed only after one or more decades of regenerative growth. Bergerud (1974), on the other

TABLE 34

Weights of Winter Forage Available to Barren-Ground Caribou in Northern Saskatchewan[a]

	Age class of the trees (years)					
	1–10	11–30	31–50	51–75	76–120	120+
Forests dominated by black spruce and white birch						
Grasses[b]	63	9	1	1	4	2
Herbs	122	5	7	9	1	3
Shrubs	96	403	403	348	470	455
Lichens[c]	5	75	193	283	355	482
Other	4					
Total	290	492	604	641	830	942
Forests dominated by jack pine[d]						
Grasses	21	18	5	0	0	0
Herbs	33	66	1	Trace	Trace	4
Shrubs	31	195	386	611	256	373
Lichens	2	203	86	181	277	130
Other	Trace					
Total	87	482	473	792	533	507

[a] Pounds per acre of air-dry forage. From Scotter (1964), reproduced with permission.
[b] Including grasslike plants.
[c] Ground lichens. In mature coniferous forests, arboreal lichens may contribute to the winter diet of barren-ground caribou, particularly under severe weather conditions. Accessible arboreal lichens per acre amounted to 628 lb in the black spruce stands and 375 lb in the jack pine stands.
[d] There were six jack pine stands sampled and 32 black spruce–white birch stands.

hand, citing the results of his own field experience and that of other workers, has concluded that caribou do not require lichens and that lichens actually make up a small proportion of the food usually consumed by these animals, with the consequence that fires that are destructive of lichen cover are not for this reason damaging to the caribou if other food sources persist.

Whatever their significance in the nutrition of caribou, lichens constitute a major part of the organic material of the northern forests, and Scotter (1964) demonstrated their importance not only as a ground-inhabiting element in the community but also as an often overlooked portion of the arboreal part of the community: A considerable weight of lichens is found growing on the bark or trunk and branches of black spruce and jack pine trees. An average specimen of the former supports a total growth (in air-dry weight) of about 2488 g and the latter a total of 3584 g. On a per-acre basis, this amounts to about 1069 lb of lichens per acre of black spruce forest and 1830 lb/acre in mature jack pine forest. These lichens, Scotter points out, may be of considerable significance in the ecology of caribou, since they can be utilized for food at least whenever snow depths make ground lichens and other foods difficult to secure.

These figures illustrate the importance of lichen species in the ecology of the northern forest and should inspire an increased interest in lichens as components of the boreal ecosystems. Since there must be a close interaction, as yet little understood, between forest trees and the composition of the forest understory, such a high proportion of lichens makes these plants of considerable significance in studies of seed germination and growth of seedlings. The fact that lichens are not only a major component of northern forest communities in terms of weight but are also among the nitrogen-fixing species of the forest contributes even more to their importance.

That the nitrogen fixing of lichens is of significance in the nitrogen nutrition of higher plants, at least under some conditions, is brought out by a study on *Stereocaulon paschale* (Crittenden and Kershaw, 1979) showing that this species is important in the postfire recovery sequence of succession in spruce woodlands of northern regions. *Stereocaulon* is an important component of the ground vegetation in the spruce woodlands in later stages of succession, at a time when nitrogen deficiencies are apt to be most severe for spruce, since much available nitrogen can be tied up in vegetation or is inaccessible in slowly decomposing humus or beneath a rising permafrost table under a moss and lichen ground cover. Estimates by Crittenden and Kershaw suggest that substantial quantities of nitrogen are fixed by *Stereocaulon* and that the nitrogen compounds enter the upper layers of soils by leaching during rainfall, by decomposition of the lichens, and after release from organic compounds as a consequence of fire. They state that quantitative information concerning these pathways is so far lacking, but it should be forthcoming. The value of lichens as sources of nitrogen in the northern forests is enhanced by their ability to survive long periods of severe environmental stress and begin active metabolism once conditions have become favorable. As Millbank and Kershaw (1973) state: "Metabolism of lichens is essentially opportunist; when conditions are right, absorption, assimilation, synthesis, or whatever the process may be, becomes active for as long as possible and then reverts to a state of quiescence when conditions deteriorate."

Nitrogen fixation by lichens under natural conditions is widespread, notably in *Peltigera, Leptogium, Nephroma, Sticta,* and *Stereocaulon,* as well as in other species in which a filamentous blue-green alga is a phycobiont in lichen symbiosis. Reported findings suggest that all lichens with these algae are capable of fixing nitrogen (Hitch and Stewart, 1973; Hitch and Millbank, 1975a,b; Millbank and Kershaw, 1973; Millbank, 1976), and common decay fungi associated with rotting wood are also capable of nitrogen fixation (Larsen *et al.,* 1978).

On a worldwide basis, only 8% of the 18,000 lichen species have blue-green algae as phycobionts, and in general lichens containing green algae are more widely distributed, but nitrogenase activity in the fixing species can occur at a high rate and these lichens are capable of releasing relatively large amounts of nitrogen to the soil. Free-living algae also fix nitrogen in the soil, and it seems that under some conditions either the nitrogen-fixing algae or the lichens can contribute significantly to soil nitrogen fertility.

ECOSYSTEM NUTRIENTS AND MANAGEMENT

Understanding of the mineral nutrient cycling in boreal ecosystems, as in ecosystems of all kinds, is dependent upon accurate measurements of a variety of parameters. It is impossible to quantify the mineral cycle without data on biomass and net primary productivity, on the mineral nutrient content of plant tissues contributed annually to the detritus of the forest floor, on minerals leached from living leaves and twigs, and on the contribution of nutrient elements from rainfall, subsurface water, atmospheric aerosols, weathering of rocks and soil particles, and all other sources. Involved also are the characteristics of the community at various successional stages, diversity profiles of tree, shrub, and herbaceous strata, and the mechanisms controlling interactions among individuals of the same and of different species and between plants and the abiotic environment.

Summing all this into a scheme of input–output is one of the goals of any effort to design an ecosystem model; the model then serves both as an aid to understanding and as a practical guide in establishing forest management and regional conservation policies. It will most likely be a long time before models are developed to the point where they are capable of incorporating all these factors into their design in sufficient detail to be of interest, for example, to physiologists or soil physicists, or of translating the dynamic interrelationships of producers, consumers, and decomposers into quantitative terms. The art of modeling, however, has progressed rapidly, and it is to be expected that improvements in technique will continue to appear, especially when it is realized more widely that models are useful and inexpensive adjuncts to more traditional research methodology. There are economic reasons for studying boreal forest ecosystems and for employing modeling as a means of establishing practices that enhance the productivity and lasting value of commercial forestry operations. Although forest management may seem a goal far removed from many efforts expended to describe and model ecosystems, particularly ecosystems as geographically remote from major human settlement as are many regions of the boreal forest, almost all the work will have a bearing on management policy in some way; it may improve understanding of the global-scale ecological relationships among life, land, water, and atmosphere, or it may enhance an individual forester's ability to predict the future productivity and biomass of a modest stand of spruce. Whatever the ultimate use to which a model will be put, the aims of a modeling program are description, prediction, and development of the ability to control the ecosystem studied.

REFERENCES

Alban, D. H., Perala, D. A., and Schlaegel, B. E. (1978). Biomass and nutrient distribution in aspen, pine, and spruce stands on the same soil type in Minnesota. *Can. J. For. Res.* **8**, 290–299.

Alexandrova, V. D. (1970). The vegetation of the tundra zones in the U.S.S.R. and data about its productivity. *In* "Productivity and Conservation in Northern Circumpolar Lands" (W. A. Fuller and P. G. Kevan, eds.), pp. 93–114. Int. Union Conserv. Nature and Natural Resources, Morges, Switzerland.

Andreyev, V. N. (1966). Peculiarities of zonal distribution of the aerial and underground phytomass in the East European Far North. *Bot. Zh. (Leningrad)* **51**, 1401–1411.

Barney, R. J., and Van Cleve, K. (1973). Black spruce fuel weights and biomass in two interior Alaska stands. *Can. J. For. Res.* **3**, 304–311.

Barney, R. J., Van Cleve, K., and Schlentner, R. (1978). Biomass distribution and crown characteristics in two Alaskan *Picea mariana* ecosystems. *Can. J. For. Res.* **8**, 36–41.
Bazilevich, N. I., and Rodin, L. E. (1971). Geographical regularities in productivity and the circulation of chemical elements in the earth's main vegetation types. *Sov. Geogr.* **12**, 24–53.
Bergerud, A. T. (1971). Abundance of forage on the winter range of Newfoundland caribou. *Can. Field Nat.* **85**, 39–52.
Bergerud, A. T. (1974). Decline of caribou in North America following settlement. *J. Wildl. Manage.* **38**, 757–770.
Bormann, F. H. (1971). The nutrient cycles of an ecosystem. *Sci. Am.* **223**, 92–101.
Burges, A. (1967). The decomposition of organic matter in the soil. *In* "Soil Biology" (A. Burges and F. Raw eds.), pp. 479–492. Academic Press, New York.
Christensen, P. J., and Cook, F. D. (1970). The microbiology of Alberta muskeg. *Can. J. Soil Sci.* **50**, 171–178.
Corns, J. G., and La Roi, G. H. (1976). A comparison of mature with recently clear-cut and scarified lodgepole pine forests in the lower foothills of Alberta. *Can. J. For. Res.* **6**, 20–32.
Crittenden, P. D., and K. A. Kershaw. (1979). Studies on lichen-dominated systems. XXII. The environmental control of nitrogenase activity in *Stereocaulon paschale* in spruce–lichen woodland. *Can. J. Bot.* **57**, 236–254.
Crocker, R. L., and Major, J. (1955). Soils development in relation to vegetation and surface age at Glacier Bay, Alaska. *J. Ecol.* **43**, 427–448.
Damman, A. W. H. (1971). Effect of vegetation changes on the fertility of a Newfoundland forest site. *Ecol. Monogr.* **4**, 253–269.
Epstein, E. (1972). "Mineral Nutrition of Plants: Principles and Perspectives." Wiley, New York.
Foster, N. W. (1974). Annual macroelement transfer from *Pinus banksiana* Lamb. forest to soil. *Can. J. For. Res.* **4**, 470–476.
Foster, N. W., and Gessel, S. P. (1972). The natural addition of nitrogen, potassium and calcium to a *Pinus banksiana* Lamb. forest floor. *Can. J. For. Res.* **2**, 448–455.
Gordon, A. G. (1975). Productivity and nutrient cycling by site in spruce forest ecosystems. *In* "Energy Flow—Its Biological Dimensions: A Summary of the IBP in Canada 1964–1974." Can. IBP Prog., Natl. Res. Counc., Ottawa.
Hegyi, F. (1972). Dry matter distribution in jack pine stands in northern Ontario. *For. Chron.* **48**, 193–198.
Heilman, P. E. (1966). Changes in distribution and availability of nitrogen with forest succession on north slopes in interior Alaska. *Ecology* **47**, 825–831.
Heilman, P. E. (1968). Relationship of availability phosphorus and cations to forest succession and bog formation in interior Alaska. *Ecology* **49**, 331–336.
Hitch, C. J. B., and Millbank, J. W. (1975a). Nitrogen metabolism in lichens. *New Phytol.* **75**, 239–244.
Hitch, C. J. B., and Millbank, J. W. (1975b). Nitrogen metabolism in lichens. *New Phytol.* **74**, 473–476.
Hitch, C. J. B., and Stewart, W. D. P. (1973). Nitrogen fixation by lichens in Scotland. *New Phytol.* **72**, 509–524.
Ingestad, T. (1959). Studies on the nutrition of forest tree seedlings: Spruce. *Physiol. Plant.* **12**, 568–593.
Kershaw, K. A. (1972). The relationship between moisture content and net assimilation rate of lichen thalli and its ecological significance. *Can. J. Bot.* **50**, 543–555.

Kershaw, K. A. (1977). Studies on lichen-dominated systems. XX. An examination of some aspects of the northern boreal lichen woodlands in Canada. *Can. J. Bot.* **55**, 393–410.

Kozlowski, T. T. (1979). "Tree Growth and Environmental Stress." Univ. of Wash. Press, Seattle.

Kramer, P. J., and Kozlowski, T. T. (1979). "Physiology of Woody Plants." Academic Press, New York.

Larsen, M. J., Jurgensen, M. F., and Harvey, A. E. (1978). N_2 fixation associated with woody decay by some common fungi in western Montana. *Can. J. For. Res.* **8**, 341–345.

Lieth, H., and Whittaker, R. H., eds. (1975). "Primary Productivity and the Biosphere." Springer-Verlag, Berlin and New York.

Lucas, R. E., and Davis, J. F. (1962). Relationships between pH values of organic soils and availabilities of 12 plant nutrients. *Soil Sci.* **92**, 177–182.

Lutz, H. J. (1956). Ecological Affects of Forest Fire in Alaska. *U. S. Dep. Agric. Tech. Bull.* **1133**, 1–121.

Mahendrappa, M. K., and Weetman, G. F. (1973). Nitrogen concentrations in the current foliage and in fresh litter of fertilized black spruce stands. *Can. J. For. Res.* **3**, 333–337.

Mikola, P. (1971). Man's modification of taiga ecosystems. *In* "Systems Analysis in Northern Coniferous Forest, IBP Workshop, Kratte Masugn, Sweden," pp. 161–165. Swed. Nat. Sci. Res. Counc., Oslo.

Millbank, J. W. (1976). Aspects of nitrogen metabolism in lichens. *In* "Lichenology: Progress and Problems," (D. H. Brown, D. L. Hawksworth, and R. H. Bailey, eds.), pp. 441–456. Academic Press, New York.

Millbank, J. W., and Kershaw, K. A. (1973). Nitrogen metabolism. *In* "The Lichens," (V. Ahmadjian and M. E. Hale, eds.), pp. 289–307. Academic Press, New York.

Miller, W. S., and Auclair, A. N. (1974). Factor analytic models of bioclimate for Canadian forest regions. *Can. J. For. Res.* **4**, 536–548.

Mitchell, V. L. (1973). A theoretical tree-line in central Canada. *Ann. Assoc. Am. Geogr.* **63**, 296–301.

Moore, T. R., and Verspoor, E. (1973). Aboveground biomass of black spruce stands in subarctic Quebec. *Can. J. For. Res.* **3**, 596–598.

Morrison, I. K. (1973). Distribution of elements in aerial components of several natural jack pine stands in northern Ontario. *Can. J. For. Res.* **3**, 170–179.

Morrison, I. K. (1974). Dry-matter element content of roots of several natural stands of *Pinus banksiana* Lamb. in northern Ontario. *Can. J. For. Res.* **4**, 61–64.

Nihlgard, B. (1970). Precipitation, its chemical compositional effect on soil water in a beech and a spruce forest in south Sweden. *Oikos* **21**, 208–217.

Nihlgard, B. (1972). Plant biomass, primary production and distribution of chemical elements in a beech and a planted spruce forest in south Sweden. *Oikos* **23**, 69–81.

Olson, J. S. (1970). Carbon cycles and temperate woodlands. *In* "Analysis of Temperate Forest Ecosystems," (D. E. Reichle, ed.), pp. 226–246. Springer-Verlag, Berlin and New York.

Page, G. (1971). Properties of some common Newfoundland forest soils and their relation to forest growth. *Can. J. For. Res.* **1**, 174–192.

Parker, G. R., and Schneider, G. (1975). Biomass and productivity of an alder swamp in northern Michigan. *Can. J. For. Res.* **5**, 403–409.

Patrick, W. H., Jr., and Khalid, R. A. (1974). Phosphate release and sorption by soils and sediments: effect of aerobic and anaerobic conditions. *Science* **186**, 53–55.

Pollard, D. F. (1972). Above-ground dry matter production in three stands of trembling aspen. *Can. J. For. Res.* **2,** 27–33.
Reader, R. J. (1978). Primary production in northern bog marshes. *In* "Freshwater Wetlands," (R. E. Good, D. F. Whigham, R. L. Simpson, and C. G. Jackson, eds.), pp. 53–62. Academic Press, New York.
Reader, R. J. and Stewart, J. M. (1971). Net primary productivity of bog vegetation in southeastern Manitoba. *Can. J. Bot.* **49,** 1471–1477.
Rencz, A. N., and Auclair, A. D. (1978). Biomass distribution in a subarctic *Picea mariana–Cladonia alpestris* woodland. *Can. J. For. Res.* **8,** 168–176.
Rice, E. L., and Pancholy, S. K. (1972). Inhibition of nitrification by climax ecosystems. *Am. J. Bot.* **59,** 1033–1040.
Rice, E. L., and Pancholy, S. K. (1973). Inhibition of nitrification by climax ecosystems. II. Additional evidence and possible role of tannins. *Am. J. Bot.* **60,** 691–702.
Rodin, L. E., and Bazilevich, N. I. (1966a). The biological productivity of the main vegetation types in the Northern Hemisphere of the Old World. *For. Abstr.* **27,** 369–372.
Rodin, L. E., and Bazilevich, N. I. (1966b). "Production and Mineral Cycling in Terrestrial Vegetation." Oliver & Boyd (Translated from the Russian).
Rosswall, T., ed. (1971). "Systems Analysis in Northern Coniferous Forests, IBP Workshop, Kratte Mosugn, Sweden." Swed. Nat. Sci. Res. Counc., Oslo.
Rouse, W. R., and Kershaw, K. A. (1971). The effects of burning on the heat and water regimes of lichen-dominated subarctic surfaces. *Arct. Alp. Res.* **3,** 291–304.
Scotter, G. W. (1962). Productivity of arboreal lichens and their possible importance to barren-ground caribou (*Rangifer arcticus*). *Arch. Soc. Zool. Bot. Fenn. "Vanamo"* **16,** 156–161.
Scotter, G. W. (1963). Growth rates of *Cladonia alpestris, C. mitis,* and *C. rangiferina* in the Taltson River region, N.W.T. *Can. J. Bot.* **41,** 1199–1202.
Scotter, G. W. (1964). Effects of forest fires on the winter range of barren-ground caribou in northern Saskatchewan. *Wildl. Manage. Bull.* **1,** 1–111.
Scotter, G. W., and Miltmore, J. E. (1973). Mineral content of forage plants from the reindeer preserve, Northwest Territories. *Can. J. Plant Sci.* **53,** 263–268.
Small, E. (1972). Ecological significance of four critical elements in plants of raised sphagnum peat bogs. *Ecology* **53,** 498–503.
Thomson, J. W., Scotter, G. W., and Ahti, T. (1969). Lichens of the Great Slave Lake region, Northwest Territories, Canada. *Bryologist* **72,** 137–177.
Van Cleve, K., Viereck, L. A., and Schlentner, R. L. (1971). Accumulation of nitrogen in alder (*Alnus*) ecosystems near Fairbanks, Alaska. *Arct. Alp. Res.* **3,** 101–114.
Van Cleve, K., Viereck, L. A., and Schlentner, R. L. (1972). Distribution of selected chemical elements in even-aged alder (*Alnus*) ecosystems near Fairbanks, Alaska. *Arct. Alp. Res.* **4,** 239–255.
van Groenewoud, H. (1975). Microrelief, water level fluctuations, and diameter growth in wet-site stands of red and black spruce in New Brunswick. *Can. J. For. Res.* **5,** 359–366.
Watts, R. F. (1965). Foliar nitrogen and phosphorus level related to site quality in a northern Minnesota spruce bog. *Ecology* **46,** 357–361.
Weetman, G. F., and Algar, D. (1974). Jack pine nitrogen fertilization and nutrition studies: Three year results. *Can. J. For. Res.* **4,** 381–398.
Weetman, G. F., and Webber, B. (1972). The influence of wood harvesting on the nutrient status of two spruce stands. *Can. J. For. Res.* **2,** 351–369.
Zoltai, S. C. (1975). Structure of subarctic forests on hummocky permafrost terrain in northwestern Canada. *Can. J. For. Res.* **5,** 1–9.

9 The Trophic Pyramid: Animal Populations

Energy flow in biotic communities involves energy transfer and biomass changes that result when animal consumers eat plants or prey. Energy is transferred between individuals; there is a transfer of nutrients into or within the animal community; there is uptake, retention (in the biomass), and eventual loss to the forest floor when animals die and decay.

There is an extensive literature on the physiological mechanisms involved in energy intake and expenditure in mammals, fish, birds, and insects. Models of physiological systems have been constructed that describe the ecological bioenergetics of living organisms, and these studies clearly demonstrate that it is essential to measure production both seasonally and from year to year. Daily and seasonal changes are characteristic of arctic and subarctic ecosystems, and very often there are high rates of primary production for short periods of time, followed by long (cold) periods in which accumulated organic and nutrient matter cannot be utilized at other trophic levels. In biomass studies of mammals, for example, it is important to distinguish between seasonal changes and cyclical changes involving more than 1 year, since the biomass of the animals usually fluctuates greatly throughout any 12-month period. Animal census figures, moreover, can suggest a stable population from one year to the next, but this does not distinguish whether the population is young with high turnover or old with low turnover; accurate measurement of biomass production requires an accounting of population changes throughout the year. Wild animal respiration rates are presently not measurable, so that estimates of net rather than gross production are the best that can be had with available research methods. There are also great differences among species in their relationships between biomass and production, usually a function of the size of the animal. Voles and deer mice, for example, represent one extreme; they have an astonishing capacity to produce new tissue, but they accumulate little biomass because of high energy requirements; they have little reserve food supply, and they are susceptible to a wide range of predators. All these factors are essentially the opposite in the case of a large predator such as the wolf or even such prey species as the moose which, because of size and defense abilities, is vulnerable to predation only when very young, sick, or old. These are all factors that must be considered in describing relationships between biomass and production in animal populations; low population densities may indicate low biomass but high production rates; rapid growth of such a population is often blocked at some point by limited environmental resources, and rapid growth is followed by a rapid decline in number. Accurate description of the ecological relationships of the organisms involved must always take into account the cyclical nature of the population and the nature and limits of the environmental response to population changes of any kind.

PLANTS AND ANIMALS

The plant life of any region constitutes by far the greater bulk of the biomass existing at any given time, but people—both primitive and urbanized—usually have found in animals a more constant source of interest. Not only are animals inherently fascinating because of their obvious biological kinship to humans, but they are also a source of food and clothing, and some are beasts of burden, partners in the hunt, guardians, or companions. Many animals in various parts of the world have been extensively studied, and the literature available on them is voluminous. Many other animals have yet to be studied in detail. The discussion here will provide only brief descriptions of some of the more important species found in the boreal forest and will take as examples the few that have so far been studied as components of the trophic pyramid, a concept now familiar to ecologists. In this concept, animals are grouped according to feeding habits—herbivores, carnivores—and are studied as though they were transducers along a circuit through which a current of energy is carried.

Boreal animals are abundant in many regions of the northern forests. In spring and summer, millions of birds wing north to breed and nest—including the nuthatch, ruby-crowned kinglet, raven, broad-winged hawk, redwinged blackbird, myrtle warbler, purple finch, rose-breasted grosbeak, northern shrike, ovenbird, rusty blackbird, lesser yellowlegs, nighthawk, and solitary sandpiper. Several species of thrush, warblers, and flycatchers also abound. Waterfowl of many species nest along the shores, notably the green-winged teal, grebe, loon, osprey, and great blue heron, as well as species of gulls.

Mammals present include the red squirrel, nocturnal flying squirrel, woodchuck, porcupine, muskrat, beaver, snowshoe hare, moose, black bear and more rarely the grizzly, as well as mice, voles, squirrels, and rarely in some areas elk and woodland and barren-ground caribou. Carnivores include the coyote, timber wolf, lynx, marten, fisher, weasel, mink, otter, and wolverine. Many amphibians and insects are also found and, in summer, clouds of mosquitoes swarm around any animal—including humans. Flies known locally as bulldogs drive beasts into wild and apparently aimless dashes through the brush. Of particular annoyance are the swarms of blackflies.

Of the more than 300 species of birds known in central Canada only about 30 species, or one-tenth, remain in the boreal woods during winter. During the winter months foxes and coyotes eat mice, voles, and shrews, caribou eat lichens and other plants, grouse eat buds and catkins, the snowshoe hare feeds on bark and shrubs, and owls eat voles as well as hares, squirrels, and small birds.

Although the winter landscape seems stark and bare and nearly devoid of life, there is, nevertheless, a complex pattern of energy flow through the ecosystem, an invisible web possessing strength and stability sufficient to bring ample numbers of animals through the harshest winter to ensure survival of breeding populations. Then, in spring, the numbers of animals over the landscape seem to increase almost instantaneously. Migrants arrive from the south. One catches glimpses of young moose. By fall, juvenile animals are a few months of age and have gained strength and vigor. The scene is again set for the annual migration or for the annual preparation for a long winter in the forest. Every species has its own pattern of survival, its own preference for food and habitat type during each season of the year. The literature on food preferences and behavioral characteristics is now fairly extensive for many boreal animal species and cannot be repeated here, but a few salient points can be mentioned for some of the more abundant boreal species, reserving discussion for later pages in the chapter.

Moose

A full-grown moose consumes large quantities of food, 20–30 kg/day of buds, stems, bark, and leaves of birch, poplar, hazel, alder, and other shrubs, in addition to aquatic vegetation when it is available in spring and summer in the shoreline shallows of lakes and ponds.

At peak densities, moose may be as numerous as 10 or more animals per square mile. Although individual moose have been known to move considerable distances, they do so rather slowly and it is not unusual for an individual to remain in one locality for long periods of time if ample food and cover are available.

The role of fire and logging in improving moose range has been studied fairly intensively, and a few general conclusions can be drawn from the work. Both fire and logging can improve moose range, but there is a need for scattered stands of large timber for suitable escape and winter cover; in the case of logging, this implies avoidance of selection or group selection silvicultural methods for areas where high moose populations are to be maintained.

Moose are considered the least gregarious of North American cervids and are generally widely dispersed in summer, but in winter in some areas they are found to congregate in favorable habitats at densities of from 2 to as many as 25 moose per square mile. As many as a dozen may be found wintering together in a relatively small area such as a lowland along a stream with abundant willows. The aggregating behavior is related to the annual cycle of breeding, forage availability, topography, seasonal conditions related to snow, and so on (Peek et al., 1974).

Studies in western Quebec show that, in summer, moose spend about 14 hr/day feeding, an hour of which is occupied eating aquatic plants and the rest given over to eating terrestrial species, mostly willows and paper birch. Wintering moose in the area consume less birch and more mountain maple (Joyal, 1976; Joyal and Scherrer, 1978).

Studies on food consumption by moose in coniferous broad-leaved forests of eastern Russia are discussed by Sukachev and Dylis (1964), who state that the animals consume 8% of the biomass of their food species in the wintering grounds annually; of this, 37% is used in the expenditure of metabolic energy and 63% is returned to the soil surface in altered organic and mineral compounds. The moose first eat the new growth of poplar in clearings of the spruce forest, and overbrowsing of the poplar can result in cessation of growth and slowing of succession toward a forest community dominated by spruce, the result of insufficient shade for ecesis by the spruce.

Barren-Ground Caribou and Woodland Caribou

Caribou resort to practically every type of terrestrial and aquatic habitat. They frequent ridges, dwarf birch thickets, sedge bogs, peat bogs, areas of upland spruce and tamarack, wooded muskegs, and willow thickets. Their winter forest habitat is characterized by sparse to dense and tall timber.

Woodland caribou are most common where forest cover is present on less than one-fourth of the land area, the remaining portion being covered with large tracts of muskeg, barrens, and water. The species partially overlaps the range of moose, but its distribution generally lies further north. A change in ungulate dominance over the past 100 years from caribou to moose in some areas has been attributed generally to fires and logging.

Barren-ground caribou are unique in the volume of lichens consumed in all seasons of the year, although grasses and sedges make up a major portion of their diet when they are available in summer. Willow and dwarf birch catkins, twigs, and leaves are also eaten. Barren-ground caribou move about over the landscape a great deal and may travel 20 or more miles a day even during the periods between the annual spring and fall migrations. This seems to be an adaptation to the relatively low productivity of the vegetation in northern areas and the consequent need for the animals to keep moving into a fresh range.

That fire destroys caribou range is an established belief. On the other hand, another study (Bergerud, 1971) showed that *Cladonia* lichens were common on some sites 10 years after a forest fire, along with *Vaccinium angustifolium, Rhododendron canadense, Rubus idaeus,* and *Cornus canadensis*. In one area of the study, which was carried out in Newfoundland, a site burned 22 years previously had developed lichen woodland rather than spruce forest, apparently because of poor spruce seed germination, and in this case the fire had increased the potential pasture for caribou. The air-dry weights of lichen in pounds per acre ranged from 200 to 600 on forest study sites and from 300 to 12,000 in lichen woodland. The weight of lichens did not exceed 1000 lb/acre until some time after 40 years following the fire. In general, Bergerud states, many fires benefit caribou, since they destroy stands of mature closed-canopy forest and permit lichen and shrub growth; a severe fire causes the animals to vacate an area, but succession is soon underway, and evergreen shrubs become abundant in a few years, followed by lichens. Both are important in the winter diet of the caribou. The relative importance of each, and the effect of fire remains to be established with certainty; there are perhaps regional differences in its effect upon caribou range (Scotter, 1964; Bergerud, 1974).

Black Bear and Grizzly Bear

The black bear is a large animal, ranging from 125 to 150 kg on the average, but large animals weighing more than 250 kg have been recorded. It has no shoulder hump like the grizzly. Mating takes place in June or July after the animals reach 3 years of age.

The black bear is omnivorous and will eat a wide variety of food, from insects and dead animals to leaves, berries, nuts, and seeds. Bears girdle both pole- and timber-sized trees and locally the damage can be quite severe.

Studies on the summer diet of black bears in Quebec indicate that about 70% of the total is green vegetation, with the rest made up of a variety of animal remains and fish (Juniper, 1978).

The grizzly bear inhabits remote areas of the Rocky Mountains and, more rarely, the northern edge of the interior forest (the "barren-ground" grizzly). A rugged animal of immense proportions and great strength, it has a shoulder hump and a massive head. Average individuals range from 200 to 300 kg, and weights up to 450 kg have been recorded.

Grizzlies mate every 2 or 3 years. Both the black bear and the grizzly hibernate throughout most of the winter, the length of the hibernation depending upon the weather and the amount of food available in the fall when fat stores accumulate.

The grizzly is omnivorous, eating roots, green vegetable materials, berries, small animals, fish, and any larger animals alive or dead that they find available.

Snowshoe Hare

The snowshoe hare is the characteristic lagomorph of the northern coniferous forest. Coniferous swamps, willow–alder swamps, poplar–birch second growth, cedar, and spruce swamps are preferred. Hares are seldom found in cover not having sufficient density to shield from detection from above, and conifers evidently must be present. The hare eats buds, bark, roots, and stems of woody as well as herbaceous plants.

Canada Lynx

The Canada lynx is larger than the bobcat and has long hair, broad feet, and long ears tufted with long hairs. Its legs are long, with heavily furred feet. For food, the lynx depends mainly on snowshoe hares but, when hares are scarce, it will eat mice, squirrels, foxes, and birds.

The lynx is widespread and well-known throughout the more remote forested regions, and the cyclical behavior of the lynx population is probably the best documented example available. This is due to the fact that it is trapped for its fur, and the records of fur-buying agencies show clearly that lynx densities increase and decline with cyclical regularity. Since snowshoe hares constitute by far the major proportion of its food, the lynx cycle is a response to the cyclical behavior of the population density of the hares. The natural enemies of the lynx, other than humans, are negligible; it is a top predator.

Beaver and Porcupine

Few actual data are available on the extent to which beavers affect the vegetation of boreal regions, but they have been reported to cause, by flooding, 14% of all mortality in a number of spruce–fir stands studied over a 7-year period.

The beaver is entirely vegetarian, eating the bark and wood of twigs, branches, and trunks of aspen, willow, birch, hazel, and a wide variety of other species.

There has been considerable study of the damage done by porcupines to forests; the foods preferred by these animals include spruce, larch, balsam, fir, hemlock, birch, and pine. Bark is the main source of winter food. Herbaceous growth constitutes the main diet in spring and summer. The entire stem of young trees is frequently stripped, but bark feeding may also occur in patches.

The winter food of porcupines in spruce forests of New Brunswick is predominantly the bark of coniferous trees, principally spruce, although the bark of pines and larch are also favored (Speer and Dilworth, 1978).

Mink and Weasel

A number of species are here discussed together, although much is known about each individually and they constitute an interesting and valuable group. In terms of the food chain and trophic pyramid, however, they can perhaps here be considered as possessing similar food habits and natural enemies. At least 50% of the food of weasels consists of mice and voles, reaching an even higher proportion in winter, and the remainder is made up of a wide variety of animal material. The prey of mink includes a wide variety of animals including muskrats and hares; in summer, crayfish may make up a large proportion of their diet. Predators of weasels and mink include hawks, owls, foxes, and probably wolves and lynx on occasion. The fur of the mink and of one or two of the weasel species (ermine) is of commercial value.

Marten

The marten is weasellike in appearance and about three-fourths the size of a small house cat. It may attain densities of about one per square mile of forest over many areas, and there is evidence that the population is subject to cycles with population peaks on the average of about every

10 years. Food includes squirrels, snowshoe hares, mice, grouse, small birds, and not infrequently nuts and berries. One study has revealed that on an annual basis, mice comprise more than two-thirds of the marten diet, in winter 80% and in summer 60%. Natural predators of the marten are the fisher, wolf, and lynx, as well as large owls. It is trapped for fur, and humans have consistently been its primary predator in recent times.

Fisher

The fisher is similar in appearance to the marten but is nearly twice as large and nearly four times as heavy, obtaining average weights of 3–6 kg for adult males and 2–3 kg for adult females. Individuals evidently have a home range 15–30 km in diameter and may wander much farther in winter. It feeds primarily on small mammals such as snowshoe hares, mice, squirrels, and porcupines and occasionally captures grouse. During summer and autumn it consumes berries and is omnivorous. Pelts are valuable in the fur trade and, like the marten, the fisher's most dangerous predator is the commercial trapper.

Red Squirrels

Red squirrels prefer to feed on the male flower buds of spruce and balsam fir both in summer and through fall and winter. They may somewhat influence the establishment of balsam fir at the expense of spruce, since balsam fir seed is released early when there is an abundance of other food available and spruce seed is released later and is used to a greater extent.

Between 1 and 3 million red squirrels are commonly harvested each winter in Canada. The economic value of this fur harvest usually exceeds $1 million, placing the red squirrel among the country's top furbearers.

The natural regulation of red squirrel populations is still poorly understood, but recent work has implicated both territoriality and food supply as important limiting factors. Squirrels are probably territorial in winter, at least to the extent that they defend a food cache, without which starvation appears to be inevitable.

Mice, Voles, and Shrews

The variety and number of small mammals present above, on, and under the forest floor is always surprising to the uninitiated, and there is not space here to list all the species found in the boreal forest. Small mammals, however, often number 50–200 or more per 2.47 ha (1 acre), often varying greatly from year to year. They are 10–20 times as numerous as birds over a given area and, since they are not migratory, are residents throughout the year. In terms of food consumption, individuals of the various species eat from one-fourth to three times their own body weight a day; shrews are insectivorous, mice and voles eat seeds, fruits, roots, and tubers, and some species also include insects and other animal foods in their diet.

Grouse

Of the North American "grouse," two are birds of forest habitat—the spruce grouse and the ruffed grouse.

The spruce grouse is associated with northern conifer associations; it is a bird of the wilderness, feeding on buds, tips, and needles of spruce, balsam fir, and larch for its winter diet. Elimination from its southern range is a result of hunting or disturbance by logging and settlement.

The ruffed grouse has an optimum range along the southern fringe of the boreal regions, although its total range extends considerably both north and south of this zone. Thickets are important for providing fall, winter, and early spring cover and are apt to harbor the birds when near openings or clumps of such shrubs as bearberry, the fruit of which is used as food when uncovered in the spring. Ruffed grouse are primarily browsers, eating foliage, twigs, catkins, and buds of woody plants in wide variety.

Other Birds

Small, forest-inhabiting birds are certainly of ecological significance, but their relationships in the forest ecosystem have not been studied extensively. Seed-eating birds may affect the regeneration of certain tree species, seed germination may be assisted by the passage of seeds through birds, and insect eaters may help maintain populations of certain insects at endemic levels. Birds may carry spores of forest pathogens.

The total breeding population may exceed 300 pairs per 247 ha (100 acres) in forests and 150 pairs in areas regenerating after fire or logging. Fairly distinct bird communities can be recognized as being characteristic for bogs, spruce forest, and other vegetational communities. One spruce forest with 233 territorial males per 247 ha included 32 different species, a greater variety than any other vegetational community. However, in total numbers there were more birds per 247 ha in bog communities.

WILDLIFE: A BIOTIC FACTOR

The ecological relationships among forest trees, as well as among shrubs, vines, herbs, fungi, and lower animals and wildlife, are still far from adequately understood. Herbivores sometimes control development of the vegetation, influencing plant succession and determining which species become dominants of the communities. Carnivorous animals such as lynx, fisher, and wolf eat little plant food, and they relate to the vegetation, as far as the food chain is concerned, only through the prey species.

Many of the problems with which animal ecologists must deal are held in common with plant ecologists. This is a result of the dependence of animals upon the plant community. Since community classification has been largely the work of plant ecologists, these classifications have often been used—even though they may not be very useful—by individuals dealing with problems of animal ecology. Problems peculiar to animal ecology result from the mobility of animals. Daily and seasonal travels may take many animals to several different kinds of plant communities.

Population fluctuations are intense in young ecosystems, and they appear to select for the more prolific species. Constancy in number evidently leads, on the other hand, to a reduced number of offspring, a greater degree of parental care, and more intense territorial behavior. Species adapted to fluctuating systems have a high rate of potential increase. The flow of energy through such populations is high, and such species are expensive from a thermodynamic point of view.

In every ecosystem, all the animals are under the control of a large variety of processes. A bad crop of seeds. for example, will mean that seed eaters must wander far in search of food. The abundance of taiga animals fluctuates widely as a response to the years of abundance that alternate with periods of great scarcity. Weather conditions play a direct role in many of these fluctuations.

BIOMASS OF THE AVIFAUNA

The birds found nesting in undisturbed northern boreal forest habitats in an area at the west end of Great Slave Lake were studied by Carbyn (1971). A list of the birds found on five plots is included in Table 35. The vegetation on these plots is described as follows: plot 1, jackpine and trembling aspen, plot 2, black spruce trees, scattered thickets within a feather moss–black spruce association; plot 3, black spruce dominant, lichens forming the main ground cover; plot 4, similar to that of plot 2; plot 5, open black spruce–sphagnum bog.

TABLE 35
Biomass of Birds in Grams per 100 Acres (40 ha)[a,b]

Family	Plot 1[c] Biomass	(%)	Plot 2[c] Biomass	(%)	Plot 3[c] Biomass	(%)	Plot 4[c] Biomass	(%)	Plot 5[c] Biomass	(%)
Fringillidae (gramnivorous)	1328	25.1	2472	45.0	684	21.2	1524	49.2	2260	66.9
Turdidae (omnivorous)	2028	38.3	1696	31.0	1656	50.9	1200	38.7	488	14.4
Parulidae (insectivorous)	984	18.5	808	15.0	448	13.7	240	7.7	632	18.7
Sylviidae (insectivorous)	24	0.5	288	5.0			48	1.5		
Paridae (insectivorous)			88	2.0			88	2.9		
Bombycillidae (omnivorous)	928	17.5			464	14.2				
Tyrannidae (insectivorous)			88	2.0						
Total biomass	5292		5496		3252		3100		3380	

[a] From Carbyn (1971), reprinted with permission.
[b] Calculations are for each family of passerines nesting on the study plots. Percentages of biomass for each family per plot are listed.
[c] Stand types are as follows: plot 1, semiopen stand of jack pine and trembling aspen; plot 2, black spruce–feather moss with black spruce thickets; plot 3, black spruce with lichens as the main ground cover; plot 4, black spruce–feather moss; plot 5, open black spruce–sphagnum bog.

The total biomass of breeding passerine birds varied considerably for the different plots, but gramnivores and omnivores contributed the highest percentages to the total avian biomass. Food habits of birds often change seasonally; hence it is quite likely that birds categorized as omnivorous and gramnivorous here may feed to a large extent on insects when available.

The spruce-fir forests of eastern North America are periodically devastated by the eastern spruce budworm (*Choristoneura fumiferana*) which may defoliate thousands of acres of white spruce and balsam fir, resulting in widespread tree mortality. Several species of birds respond numerically to increasing spruce budworm densities, although they do

TABLE 36

Number of Singing Males Recorded Each Year[a,b]

	Plot 1			Plot 2		
Species	1966	1967	1968	1966	1967	1968
Broad-winged hawk	½	½	½	½	½	0
Ruffed grouse	0	½	½	½	0	1
Flicker	1	1	1	1	1	1
Pileated woodpecker	0	½	0	0	½	0
Yellow-bellied sapsucker	1	1	1	1	1	1
Hairy woodpecker	½	½	1	½	½	1
Yellow-bellied flycatcher	0	0	1	2	2	2
Least flycatcher	1	1	2	3	2	2
Wood peewee	0	1	1	0	0	0
Olive-sided flycatcher	0	0	a	0	0	0
Canada jay	½	½	½	½	½	½
Blue jay	½	½	½	½	½	½
Black-capped chickadee	0	0	0	0	1	½
Boreal chickadee	0	0	0	0	0	1
Red-breasted nuthatch	½	½	½	0	0	½
Brown creeper	1	1	½	0	0	0
Winter wren	1	1	1	½	1	½
Robin	0	0	0	0	t	½
Hermit thrush	a	½	a	a	a	½
Swainson's thrush	1	1	1	1	½	½
Veery	0	0	0	a	0	0
Golden-crowned kinglet	1	1	0	1	1	1
Ruby-crowned kinglet	0	½	0	1	1½	0
Red-eyed vireo	2	2	2	1	½	1
Black and white warbler	2	1	1½	2	2	1
Tennessee warbler	0	a	0	0	0	0

TABLE 36—*Continued*

	Plot 1			Plot 2		
Species	1966	1967	1968	1966	1967	1968
Nashville warbler	1	a	½	2½	2½	3½
Parula warbler	0	½	0	0	0	0
Magnolia warbler	2½	1	2	4	4	3½
Black-throated blue warbler	a	0	0	0	0	0
Myrtle warbler	0	t	0	1½	1	0
Black-throated green warbler	2	2	2½	0	0	0
Blackburnian warbler	a	0	0	0	0	0
Chestnut-sided warbler	½	½	½	1½	1	1
Bay-breasted warbler	0	½	0	0	0	0
Ovenbird	2½	3	3½	2	3	4
Mourning warbler	½	½	½	½	½	½
Canada warbler	3½	2	2	t	0	0
Scarlet tanager	a	½	0	0	t	0
Rose-breasted grosbeak	a	a	0	0	a	0
Evening grosbeak	t	t	t	t	t	0
Purple finch	1	1	1	½	½	½
Pine siskin	0	0	0	0	0	t
Slate-colored junco	0	t	0	0	0	0
White-throated sparrow	1	2½	1	4	4	4½

[a] Data from Sanders (1970), reproduced with permission.
[b] ½ indicates a singing male with a territory either considerably larger than the plot or with a territory lying approximately one-half inside and one-half outside the plot; t (transient) indicates a male singing occasionally, but not consistently, inside the plot; a (adjacent) indicates the presence of a species not recorded inside the plot, but which was apparently breeding in the same forest type adjacent to the plot.

not contribute significantly to the regulation of the spruce budworm at high densities. Flocks of vagrant species may cause appreciable mortality in declining budworm populations, however, and it is possible that the role of birds in regulating budworm numbers may be very important at low budworm densities.

Sanders (1970) carried out ecological investigations near Black Sturgeon Lake, northwestern Ontario, to determine the status of the population of breeding birds during an endemic phase of the spruce budworm population cycle. The numbers of singing males recorded each year are shown in Table 36. A value of ½ was assigned in instances where the apparent territory of the male was partly inside and partly outside the

TABLE 37
Populations and Energy Removed by Birds Detected Regularly in Two Areas of Taiga Woods near Fairbanks, Alaska[a]

Species	Population in period[b] no/100ha			Energy removed in period[b] (Mcal/100 ha)[c]			99-day summer total (Mcal/100 ha)[c]
	1	2	3	1	2	3	
Trail A							
Dark-eyed junco	24.8	70.6	62.9	14.2	32.9	28.3	75.4
White-crowned sparrow	17.3	55.7	33.9	13.2	34.6	20.4	68.2
Fox sparrow	25.2	31.4	24.7	20.0	20.3	15.5	55.8
Common redpoll	37.8	64.3	35.8	17.6	24.4	13.1	55.1
American robin	24.5	31.4	16.0	21.1	22.1	10.9	54.1
Swainson's thrush	37.8	29.0	6.2	24.1	15.1	3.1	42.3
Yellow-rumped warbler	33.8	15.7	14.8	16.5	6.3	5.7	28.5
Yellow warbler	24.8	28.2	12.9	9.2	8.5	3.8	21.5
Ruby-crowned kinglet	10.4	9.4	16.0	3.7	2.8	4.5	11.0
Varied thrush	3.9	6.3	4.3	3.4	4.4	2.9	10.7
Orange-crowned warbler	13.3	9.4	6.8	5.3	3.1	2.2	10.6
Hammond's flycatcher	7.3	13.7	5.5	2.8	4.2	1.7	8.7
Rusty blackbird	4.9	1.1	6.2	3.7	0.7	3.7	8.1
Black-capped chickadee	2.5	7.8	10.5	1.1	2.9	3.8	7.8

Species						
Gray-cheeked thrush	5.6	7.1	0.6	3.5	3.6	0.3	7.4
Lesser yellowlegs	1.1	5.5	0	1.2	4.9	0	6.1
Alder flycatcher	4.9	9.1	3.7	1.9	2.8	1.1	5.8
Townsend's warbler	5.2	6.3	4.3	2.0	2.0	1.3	5.3
Wilson's warbler	3.4	3.2	5.0	1.3	1.0	1.5	3.8
Northern phalarope	3.9	1.2	0	2.8	0.7	0	3.5
Savannah sparrow	3.6	2.4	0.6	2.1	1.1	0.3	3.5
Lincoln's sparrow	2.5	0.8	0.9	1.3	0.3	0.4	2.0
Total	298.5	409.6	271.6	172.0	198.7	124.5	495.2
Percent				34.7	40.1	25.1	99.9
Trail B							
Dark-eyed junco	7.6	47.6	55.1	4.3	22.2	24.8	51.3
Yellow-rumped warbler	17.3	26.9	19.5	8.4	10.7	7.5	26.6
Swainson's thrush	22.4	16.4	7.5	14.3	8.5	3.8	26.6
Towsend's warbler	28.5	2.2	1.5	10.8	6.8	4.5	22.1
American robin	2.4	13.5	7.5	2.1	9.5	5.1	16.7
Common redpoll	12.4	6.0	12.0	5.8	2.3	4.4	12.5
Varied thrush	3.6	2.9	6.0	3.1	2.0	4.1	9.2
Orange-crowned warbler	7.6	8.9	4.5	3.1	2.9	1.4	7.4
Ruby-crowned kinglet	1.2	8.9	6.0	0.4	2.6	1.7	4.7
Alder flycatcher	3.0	2.9	2.9	1.1	0.9	0.9	2.9
Total	106.0	136.2	122.5	53.4	68.4	58.2	180.0
Percent				29.7	38.0	32.3	100.0

[a] From West and DeWolfe (1974), reproduced from *The Auk* **91**, 757–775.
[b] Period 1 = 38 days (24 May–30 June); 2 = 31 days (1 July–31 July); 3 = 30 days (1 August–30 August).
[c] Mcal = megacalorie = 10^6 calories.

study plot. It was assumed that each singing male indicated the presence of a breeding pair. The total number of breeding pairs on two plots remained virtually constant from year to year, 115 pairs per 247 ha/year on one plot and an average of 131 pairs per 247 ha/year on the other. The relative constancy of the total number of breeding pairs, and of the species composition of each plot from year to year, suggest that these were reliable estimates of the normal breeding populations.

For most regions, estimates of the amount of energy removed from the vegetational trophic level of the ecosystem are not yet available. For one area in Alaska studies made under the International Biological Program, however, estimates are now available and are given in Table 37 in which data obtained by West and DeWolfe (1974) are presented.

THE TROPHIC LEVELS

The trophic levels can be arranged in linear sequence and, since the transfer of energy between two successive levels is never very efficient, there is a progressive diminution of energy content along the food chain. The trophic structure of the community is based on the standing crop of vegetation, and the apex is the top carnivore level. The energy content of the trophic levels is designated according to the energy content of the (1) primary producers, (2) herbivores and detritivores, and (3) carnivores.

As McCullough (1970), however, has pointed out:

> Ecologists speak of "energy flow" through ecosystems in contrast to the cycling and reutilization of nutrients. In reality, energy does not "flow" in ecosystems; in the form of food it is located, captured and digested by organisms at the expense of a considerable performance of work. Far from flowing, energy (i.e., food) is dragged forcefully from one trophic level to the next.

All ecosystems possess resiliency to a remarkable degree, and the fact that animal populations tend usually to fluctuate between higher and lower levels of density does not mean that the ecosystem is in a state of instability. Indeed, as pointed out by McCullough, any system functioning within its capacity to adjust may be said to be in a state of stability. Successional cycles may be repeated indefinitely without harm. It is of importance to recognize that population sizes within an ecosystem undergo changes without disrupting the stability of the system as a whole. The biomass of populations in a system, thus, also varies considerably from one time to another. The concept of carrying capacity has often resulted in thinking in terms of a single level, while in actuality a whole range of levels is within the normal range of the system. The moose-versus-wolf example used in the discussion below illustrates the changes in biomass and density that are possible within a trophic level. Energy consumption by herbivores can also change over time. A large animal has a lower maintenance requirement per unit biomass than a smaller one, and it can survive by consuming plant biomass with a lower average caloric and nutrient content. The biomass of the different species at a trophic level thus cannot be combined without considering the characteristics of the various species in that level.

THE PLANT–MOOSE–WOLF SUBSYSTEM

Since studies of moose have been more numerous than those of the other mammals, they constitute the best available description of the trophic relationships existing between an animal species and the other components in the boreal ecosystem. The moose is both a herbivore and a prey species and, as such, vividly illustrates the flow of energy that continually occurs in the boreal community of plants and animals.

There are seasonal cycles in the utilization of food plants by the moose population, presumably as a result of seasonal variations in availability of the species that, for one reason or another, are the most palatable or otherwise desirable. In Alaska, for example, food habit studies showed that willow (*Salix* species), birch (*Betula papyrifera* var. *kenaica*), cottonwood (*Populus trichocarpa*), and aspen (*Populus tremuloides*) constituted practically all the winter food of moose. Alder (*Alnus fruticosa*), rose (*Rosa acicularis*), highbush cranberry (*Viburnum edule*), and other plants were rarely consumed. During spring, twigs of willow and birch were most frequently eaten. Other foods were fruit and dry leaves of willow, hardened aspen twigs, green willow leaves, hardened alder twigs, sedges, hardened *Vaccinium* twigs, and succulent willow twigs. Summer food included green willow leaves, willow twigs, and *Equisetum*, as well as green shrub–birch leaves, grass, and birch twigs. Fall food included twigs of willow and birch, followed in descending order by birch fruit, dry and green willow leaves, aspen twigs, green birch leaves, alder twigs, and *Equisetum* (Cushwa and Coady, 1976).

Feeding activities influence the quantity and quality of primary production, in areas of high moose populations reducing the production of young plant growth by as much as 50% or more, the greater part of which is due to the decrease in growth rates of damaged plants. Moose damage mainly the best specimens of young trees, resulting in less robust trees becoming dominant. The consumption of a given weight of vegetation by moose gives an erroneous impression of their influence on primary forest production; other factors are also an important influence on the rates of primary production and energy flow. Plant succession on areas burned by wildfire in southern central Alaska followed a variety of patterns, resulting in the creation of a useful moose winter range for a period of at least 50 years. Under average conditions, stands appeared to furnish good forage for 15–20 years after the fire.

Preliminary analyses of the plant–moose–wolf food chain of Isle Royale National Park show that the weight of vegetation consumed by the herd of moose was about 3000 tons annually and that their wolf predators consumed about 45 tons of moose. From the bottom to the top of the pyramid in terms of energy content, the following was estimated: plants, $11{,}000{,}000 \times 10^3$ kcal; moose, $46{,}000 \times 10^3$ kcal; wolves, 780×10^3 kcal.

Jordan et al. (1971) analyzed the population and biomass dynamics of the moose of Isle Royale by constructing a simple model from which calculations were carried out by digital computer. The model assumed that the annual pattern of population and biomass dynamics remained constant throughout the 10-year study period. They concluded that, strictly in terms of the quantity of transfer of flesh from moose to wolves, this represents an inefficient cropping system, because a large portion of biomass is maintained for many years before it is harvested. The observed low ratio of wolf population biomass to moose population biomass is related in part to the slowness of turnover in the moose. While this model did not test the impact of environmental changes or feedback processes, it appears to be a useful tool for summarizing data and for evaluating assumptions about the ecosystem (Table 38).

TABLE 38

Weights of Consumed Vegetation and of the Moose and Wolf Populations of Isle Royale National Park[a]

Area of Isle Royale	544 km²
Wolf population weight (in midwinter)	812 kg (24 average wolves)
Moose population weight (in midwinter)	367,000 kg (1000 average moose)
Weight of moose killed by wolves per year	61,200 kg (taking as a maximum estimate; 90% of moose dying per year are killed or eaten by wolves)
Weight of vegetation eaten by moose per year	22,000,000 kg
Weight of moose killed by each wolf per year	2550 kg (7 kg/day; not all of which is consumed since a large portion—bones, etc.—is inedible)
Portion of vegetation going into wolf population per year	0.003%

[a] Data from Jordan et al. (1971).

THE WOLVERINE AND WOLF COMPARED

An example of a predator that has probably a minimal influence on the vegetation, even indirectly, perhaps largely because of its low population density, is the wolverine, one of the most poorly known of the larger carnivores of Canada, a fact reflected by the dearth of information in the literature on the species (van Zyll de Jong, 1975). The species is reported to still occur in Labrador, Quebec, western Canada, and Ontario, but there have been no recent reports of wolverines from Labrador or Newfoundland.

Wolverines normally occur at relatively lower densities than other carnivores of comparable size; data on densities are not easily obtained, but estimates for Scandinavia vary from one wolverine per 200–500 km^2 (approximately 193 mi^2). For comparison, estimates of wolf densities in North America vary up to one wolf per 26 km^2 (10 mi^2) for high-density areas. These estimates indicate that wolverines tend to be less abundant than wolves even in optimum habitats. There seems to be a low rate of natural increase incapable of compensating for losses due to human hunting and trapping.

Studies of food habits show that the wolverine is an omnivore in summer, feeding on a large variety of food including carrion, small mammals, insect larvae, eggs, and berries. It is a carnivore in winter. Herbivores, especially caribou, the most numerous large herbivore over most of the wolverine's range, are the most important winter food. Most of the large herbivores eaten are thought to be carrion, resulting from predation by other carnivores or death from other causes.

The wolverine can be regarded as a seasonal scavenger on the fringe of the food web. The distribution and abundance of wolverines in Manitoba coincides with that of the barren-ground caribou in winter (van Zyll de Jong, 1975). In the mountain ranges of Alberta, British Columbia, and the Yukon, wolverines are still common, and large and diverse ungulate populations occur.

Wolves are large and formidable predators. The weight of individual animals varies from 30 to 60 kg. The adult male takes an active part in the rearing and training of the young. Although in summer the wolf supplements its diet with vegetable matter such as grass, roots, and berries, in winter, hunting is the source of by far the largest proportion of the food supply, and several families may join together in a single hunting pack.

The diet of wolves on the winter range of the barren-ground caribou in open spruce forest of southern Mackenzie was estimated by Kuyt (1972) to be made up of about 75% caribou, the staple food, with the remainder consisting of otter, wolf, wolverine, mink, fish, and other

TABLE 39

The Diet of Wolves: Data from Various Sources

Region	Source	Percentage of diet[a]	
		Ungulate prey	Other
Mt. McKinley, Alaska	Murie (1944)	69 (caribou)	31
Rocky Mts., Canada	Cowan (1947)	80 (big game)	20 (rodents)
Northern Wisconsin	Thompson (1952)	90 (white-tailed deer)	10 (showshoe hare)
Northern Minnesota	Stenlund (1955)	90 (white-tailed deer-	10 (small mammals)
Algoma Park, Ontario	Pimlott et al. (1969)	Mostly white-tailed deer	Secondly beaver
Southern Alaska	Merriam (1964)	95 (black-tailed deer)	5
Northwest Territories	Kelly (1954); Banfield (1954); Kelsall (1960)	Ca. 100 (caribou)	
Mackenzie forest, winter	Kuyt (1972)	75 (caribou)	25 (see text)
Mackenzie tundra, summer	Kuyt (1972)	40 (caribou)	60 (see text)

[a] Most important item when noted in report.

substances. In summer, their diet is somewhat lower in caribou, about 40%, with more birds, rodents, fish, squirrels, and microtine rodents being consumed. Kuyt calculated that a wild wolf will eat about 7 lb of meat a day, on the average, which comes to about 23 average caribou a year. This, on the basis of reports from other sources compiled by Kuyt, can be compared with the diet of wolves of other areas presented in Table 39. Wolf densities of as high as one wolf per 7 mi^2 occur locally at times in winter in areas where caribou are concentrated.

In the open forest caribou range in northwestern Manitoba, in inhabited areas the density of caribou increased from 14 to 68 animals per square mile from January to April; this occurred as the animals congregated in smaller areas. The highest wolf density was about one wolf per 8 mi^2, although the average density was about one per 14 mi^2; the latter density was in January, when the animals were dispersed over about 3600 mi^2, and a lowered wolf density was found in April when the caribou range had contracted to about 680 mi^2. It seemed that the lowered wolf density in late winter, when the caribou were most concentrated, was the result of the wolves being forced to avoid the forest because of deep snow (and stay on lakes or in areas that had been blown clear of snow) or to leave the area to seek denning sites in preparation for the whelping of cubs (Parker, 1973).

ROLE OF INSECTS

At the other extreme from the large carnivores in terms of individual size, but probably far more significant in terms of biomass, are the insect populations of the boreal ecosystem. Insects damage plants in a variety of ways and reduce their photosynthetic capacity, hence net primary production, by sap consumption, consumption of tissues, and damage to reproductive organs, all of which produce effects that are cumulative and lasting, particularly if the insect populations are high and the damage is appreciable. Fortunately, however, except for a relatively few forest pests, most insects occur in comparatively small numbers and have little effect upon the growth of plants. Large numbers of defoliators injure trees by reducing photosynthesis and interfere with transpiration and translocation. In conifers, photosynthesis is affected by the absence of foliage for up to 3 or 4 years. On the other hand, it appears that the rate of normal insect grazing usually does not impair annual plant (primary) production; it may even accelerate growth. Although outbreaks do reduce plant production temporarily, they commonly occur in individual plants or in whole forest systems that have passed peak efficiencies in biomass production (Mattson and Addy, 1975).

The spruce budworm is the most widely distributed destructive forest insect in North America. Severe outbreaks have occurred in northeastern spruce–fir forests at intervals during the past 150 years. The most recent series of outbreaks date to the mid-1930s in Ontario, with successively later outbreak periods eastward through Quebec and into the Atlantic region. Six outbreaks of spruce budworm in Quebec, as shown by radial-growth studies, occurred at intervals of 44, 60, 26, 76, and 37 years. The moths appear in July or August and deposit eggs in masses on the undersides of needles. The eggs hatch in 10 days, and the larvae spin hibernation shelters under bud and bark scales and crevices. They molt and remain until spring, emerging before vegetative buds expand and mining into old needles, unopened buds, or staminate flowers. Eventually they move to vegetative buds and feed under protective loose webs among the needles. They feed on new foliage and on old foliage after the new needles are all gone. Severe timber mortality paves the way for future outbreaks by creating extensive even-aged forests. Much research has been conducted on the spruce budworm in various parts of North America; much more is needed to provide the information required to cope with recurrent epidemics in different regions.

Insecticides are needed to prevent serious timber mortality but, owing to hazards to other forms of life, continued research is needed on minimum dosages and on the influence of insecticide applications on wildlife and aquatic fauna as well on natural enemies of the budworm.

In eastern spruce–fir forests, severe defoliation by the spruce budworm results in the rapid growth of great numbers of balsam fir seedlings, many of which had been present in a suppressed state in the understory of the forest for many years. Ghent (1958) points out that balsam fir accumulates a regeneration potential, in the form of seedlings, for many years under conditions that prevent the seedlings from replacing the overstory until the latter is destroyed by defoliation or other cause. Flowering balsam fir is prone to budworm infestation, and when the balsam fir of the overstory, mixed often with white spruce, reaches this stage of development, a budworm outbreak is often imminent. The spruce–fir forest seems capable of perpetuating itself without the intervention of periodic destruction by the budworm, according to Ghent, but under natural conditions infestation is in most instances inevitable after a period of time unless fire intervenes.

Observations on deciduous forest in Russia revealed that insectivorous birds reduced the numbers of forest insect pests, such as the gypsy moth and brown-tailed moth, by 40–70%, and that experiments in which the birds were kept away from the trees by nets resulted in the reduction of tree growth by half (Sukachev and Dylis, 1964); whether this is the case in spruce forest is not noted, but severe infestations are probably beyond the ability of the bird population to control.

In summary, even during periods of unusually high numbers, animals consume only a small proportion of the organic material synthesized by the forest vegetation, even though the amount of food consumed is many times greater than the biomass of the consumer animals. For this reason, the typical diagrammatic trophic pyramid is misleading, since the upper tiers are actually much smaller than can be depicted in any kind of geometric relationship; the biomass of unutilized carbon compounds in the boles of forest trees, for example, far outweighs that of the forest animals dependent upon vegetation for food and is another order of magnitude removed from the weight of the predators dependent upon the herbivores.

EFFECT OF FIRE ON ANIMAL POPULATIONS

One of the important resources of the boreal forest is its wildlife, especially the furbearers and the big game. From it both the white population and the natives derive very substantial economic benefits. Fur-bearing animals are adversely affected by severe fires, particularly animals unable to take refuge in water. The marten, one of the most valuable fur-bearing animals in northern forests, is more menaced by forest fires than any other furbearer.

The opinion is frequently expressed that forest fires create more favorable conditions for moose, but it is not invariably certain that the species composition of the vegetation following a wildfire will be useful to moose. In 1953, Leopold and Darling, writing of Alaskan conditions, observed:

> The mere passage of a fire through timberland does not necessarily create optimum conditions for moose. Some burns produce a grassland stage; others come back in pure spruce; many produce aspen with little birch or willow which are the most palatable and productive browse plants.

It should also be recognized the regrowth of browse species following severe fires is not immediate; years may pass before burned areas again support an appreciable amount of food for moose. As Lutz pointed out, it is doubtful that moose can reproduce rapidly enough to utilize fully the browse in extensive burns before it grows out of their reach. This would be a problem with tree species such as aspen, birch, and the larger willows (Lutz, 1956).

Recent studies by other wildlife ecologists have led them to the conclusion that it is desirable for moose to have access to older stands bearing well-grown coniferous trees in order that they may obtain certain nutrients. The research points to the desirability of a winter range for browsing ungulates where there is a variety of palatable species predominantly in an early stage of growth but where there is also an intermixture of stands of other ages, including areas bearing submature or mature associations with some palatable coniferous species. The most desirable winter range for moose will be one well diversified as to species composition and age of stands but predominantly of new growth following deforestation.

Throughout its range, the moose prefers pioneer plant communities or vegetation representing early stages of successional development, but the barren-ground caribou lives in plant communities characterized by tundra and forest–tundra transition zones. Fruticose lichens of the *Cladonia* group, *Cetraria* species, and *Stereocaulon* species, together with certain species of *Usnea* and *Alectoria* growing on trees, form the principal winter food of caribou. These plants are all killed by forest fires, and recovery is slow. A conservative estimate of the length of time required for recovery is 40–50 years, but in some instances in the Far Northern regions it may be much more. As pointed out previously, the northern range of the caribou in the central and western regions of Canada may differ from that of Newfoundland or elsewhere, for example, since in the latter region fire appears to improve the range for caribou as much as for moose (Bergerud, 1974).

POPULATION CYCLES

Many boreal species undergo regular cycles in abundance that are still inexplicable. The main cyclical species are either snow-adapted or snow-dependent, namely, the snowshoe hare, ruffed grouse, sharp-tailed and spruce grouse, and voles. Lynx, fox, mink, and marten may also show periodicity in population densities. Theories on the causes of cycles range from the notion that they are a side effect of the sunspot cycle to the possibility that internal physiological mechanisms act to restore balance when a population gets too large or too small. There have been successful attempts to correlate population changes with weather factors, particularly the winter conditions involving the physical characteristics of the snow, and with a number of other environmental factors such as fire and food availability, as well as such factors as disease and parasite infestation. Diseased animals often appear during the decline of the population, but no single disease shows up consistently, and it seems probable that disease and increased parasitism occur as a side effect. Some studies show that predators are not responsible for prey cycles, since the predators do not appear to remove the prey fast enough to control a population peak, but other studies show that, at least with some species, it seems likely that the prey controls the population of the predator.

Keith and colleagues (see References) have studied the population cycle in the snowshoe hare and its predators and have been unable to detect a dominant influence of any one weather variable, except that cold winter weather may have an effect since adult survival decreases during very cold, snowy winters. Increased litter size the following spring constitutes a compensatory response to the increased mortality among adults over the winter.

The lynx is a significant factor in the overwinter mortality of snowshoe hares, accounting for an average of about 20% of the loss. It is not known what impact this mortality has on the densities of the prey species, but annual lynx mortality, occurring primarily in winter, appears to be directly correlated with the hare population; the number of lynx born each year that survive through the first winter appears to be directly correlated with the number of showshoe hares. There appears to be a minimum density of hares below which a female lynx is not capable of successfully rearing kittens.

A study on the degree to which weather variations were associated with changes in the ruffed grouse population over a period of 20 years in northern Minnesota showed that warm days in spring and summer tended to be associated with a high grouse population the following April, and that warm days during winter tended to be associated with a low grouse population the following April. The influence of weather was interpreted as follows: Warm spring weather allows early nesting, and warm summer weather ensures a plentiful supply of insect food for the young grouse. Warm winter weather encourages crusting of snow from rain or sleet and prevents grouse from gaining access to the protective blanket of snow for roosting, thus exposing them to cold, to predators, and thus to increasing mortality. Both of these effects are substantiated by the known behavior of ruffed grouse and the effect of the environment upon reproductive behavior and winter survival (Larsen and Lahey, 1958).

It seems that the events affecting the grouse population must be related to weather, food, predation, or disease, perhaps in some combination or differing combinations from one time to another, since the cyclical fluctuations in population are synchronous throughout most of the ruffed grouse range on many known occasions.

A severe decline in the population of ruffed grouse in Manitoba and

Saskatchewan in 1971–1973 was recorded by Rusch et al. (1978), who found that the fall ratios of young to adult grouse did not change significantly during the decline. The implication was that reproductive failure was not an important factor in the decline, a conclusion supported by Ransom (1965) on the basis of the observation of a 1961 decline in the Turtle Mountain area of Manitoba. The evidence of predation by raptors led the investigators to conclude that most of the grouse mortality was caused by predators and that the cyclical declines in grouse numbers were caused by a shift in predator food habits during and after a die-off of snowshoe hares. They suggested that high rates of predation were sufficient to cause grouse declines but were not necessarily involved in all declines; a high mortality of young in spring caused by adverse weather or food shortages may be involved in cyclic declines at other times and places. Rusch et al. (1978) point out: "No one explanation of grouse declines is likely to be widely accepted without field experiments and long-term studies of the prey and predator species involved in the 10-year cycle."

Rusch (1976) explains the cycle as follows: Hares increase to peak density under conditions in which food in the form of woody browse species is abundant, and then the population declines as hares consume the available food and begin to starve. Lynx, horned owls, and hawks dependent upon hares increase with the hares and, when the hare population declines, they must turn to other species for food. Grouse populations build up during the period of hare abundance, because predators feed mainly on hares. When hare populations are decimated, the wide-ranging predators seek out the grouse. The population of grouse, however, is unable to sustain the high number of predators, and both decline rapidly. The stage is set for the recovery of both prey and predator populations, since the woody vegetation can now regenerate and hares can once again increase; the cycle is repeated. Within recent years evidence has accumulated that the population fluctuations in all the species involved follow this cyclical sequence (Rusch and Keith, 1971; Meslow and Keith, 1971; Rusch et al., 1972; Adamcik et al., 1978; Adamcik and Keith, 1978; Rusch et al., 1978).

A suggestion for modification of this sequence of cyclical events has been put forth by Fox (1978), who finds correlations between the periodic increase in forest and brush fires across Canada and the snowshoe hare and lynx cycle. His analysis of fire data for six Canadian provinces indicates that fire clearly appears to be a periodic phenomenon, complexly related to weather conditions, and that the hare—lynx-owl-hawk—grouse cycle seems to be a forced oscillation rather than a predator–prey, parasite–host, or hare–vegetation relationship. Conditions in the forest as a result of different snowfall from one year to the next, Fox indicates, seem to account for the variation not explained by fire. In Fox's view, the abundance of woody browse in the years following fire maxima permits a rapid increase in the hare population, followed by an increase in the predator population, but this high level can be maintained only for a few years, while the vegetation is in early successional stages. After an initial increase in browse availability after fire, the trend is downward until fire again occurs and renews the cycle. Fox points out that this view of the cause of the northern wildlife cycle is supported by the fact that fluctuations in wildlife populations have been damped and the periodicity disturbed in areas, such as the Great Lakes region, where fire protection has been effective. Fox states: "The main significance of demonstrating periodicity in fire and weather data is that it gives us good cause to suspect the hare-lynx cycle and other game cycles of being 'forced oscillations' driven by environmental fluctuations." This circumstance, along with the abundant evidence for postlogging and postfire increases in animals with dietary preferences for plant species typical of early successional stages, supports the view that the cyclical behavior of northern populations of grouse, hares, hawks, owls, and other animals has its origin in events related to weather and fire rather than intrinsic cyclical tendencies in the animal populations. It is conceivable, however, that both are involved, with one or the other predominant in importance at different times.

As Fox points out, there are many substantial data on the browse preference of wildlife species in the boreal forest; there is also evidence that habitat use by the animals is dependent upon the presence of these species. The order of preference for snowshoe hares in winter, for example, is (1) willow, aspen, birch, hazel, and other second-growth woody species (2) white cedar; (3) pines; (4) firs; and (5) spruces. Moose follow much the same sequence of preference. Ruffed grouse, however, are more limited and prefer winter buds of aspen. Herbage within reach of a hare is in short supply in a late-successional stage forest but is very abundant soon after fire or logging.

The populations of several species of small mammals commonly fluctuate synchronously over broad regions in the boreal forest, and weather factors may play a major role in keeping these populations fluctuating in phase. One study shows that the relationship between the red squirrel population and its food supply is controlled by weather events. In red squirrel populations there are statistically significant correlations among population fluctuations, as shown in the fur returns, white spruce flower bud and cone production, and weather factors affecting bud and cone production. Kemp and Keith (1970) have shown that flower bud differentiation during a preceding summer leads to substantial cone crops. The winter diet provided by flower buds stimulates squirrel reproduction. Increased rates of reproduction immediately before a heavy cone crop then trigger a marked response to increased cone production the following year. The anticipatory rise in reproduction coupled later with good winter survival maximizes the red squirrel's ability to utilize a widely fluctuating food supply.

Ehrlich and his colleagues have pointed out that, while there is still considerable disagreement as to whether or not there is an optimum population density—i.e., a density at which the probability of the population persisting for another generation is maximized—it is clear that, when a population goes above a certain size, resource shortages, predator pressures, interspecific competition, disease, or other factors that operate in a density-dependent manner will tend to depress numbers (Ehrlich *et al.*, 1972).

Ehrlich found in one alpine study area, however, that a late June snow wiped out many populations, both among insects and small mammals. It was clear from the wide variety of animals affected that the downward trends in population size were almost certainly caused by the random environmental event and were not a density-dependent response. Ehrlich concluded that a clear picture still is not available concerning the kinds of factors controlling population size in different kinds of organisms and that, in ecologically distinct areas, the task is clear: The dynamics of taxonomically and geographically diverse samples of many populations must be followed in great detail over a long period of time.

REFERENCES

Adamcik, R. S., and Keith, L. B. (1978). Regional movements and mortality of great horned owls in relation to snowshoe hare fluctuations. *Can. Field Nat.* **92**, 228–234.

Adamcik, R. S., Todd, A. W., and Keith, L. B. (1978). Demographic and dietary responses of great horned owls during a snowshoe hare cycle. *Can. Field Nat.* **92**, 156–166.

Adamcik, R. S., Todd, A. W., and Keith, L. B. (1979). Demographic and dietary responses of red-tailed hawks during a snowshoe hare fluctuation. *Can. Field Nat.* **93**, 16–27.

Banfield, A. W. F. (1954). Preliminary investigation of the barren ground caribou. *Can. Wildl. Serv. Wildl. Manage. Bull., Ser.* **1**, No. 10A & 10b, Ottawa.

Banfield, A. W. F. (1974). "Mammals of Canada." Univ. of Toronto Press, Toronto.

Bergerud, A. T. (1971). Abundance of forage on the winter range of Newfoundland caribou. *Can. Field Nat.* **85**, 39–52.

Bergerud, A. T. (1974). Decline of caribou in North America following settlement. *J. Wildl. Manage.* **38**, 757–770.

Bergerud, A. T. (1971). Abundance of forage on the winter range of Newfoundland caribou. *Can. Field Nat.* **85**, 39–52.

Carbyn, L. N. (1971). Densities and biomass relationships of birds nesting in boreal forest habitats. *Arctic* **24,** 51–61.

Cowan, I. M. (1947). The timber wolf in the Rocky Mountain national parks of Canada. *Can. J. Res. Dev.* **25,** 139–174.

Cushwa, C. T., and Coady, J. (1976). Food habits of moose, *Alces alces*, in Alaska: A preliminary study using rumen analysis. *Can. Field Nat.* **90,** 11–16.

Davidson, A. G., and Prentice, R. M. eds. (1967). "Important Forest Insects and Diseases of Mutual Concern to Canada, the United States and Mexico." Dep. of Forestry and Rural Development, Ottawa.

Doerr, P. D., Keith, L. B., Rusch, D. H., and Fischer, C. A. (1974). Characteristics of winter feeding aggregations of ruffed grouse in Alberta. *J. Wildl. Manage.* **38,** 601–615.

Ehrlich, P. R., Breedlove, D. E., Brussard, P. F., and Sharp, M. A. (1972). Weather and the "regulation" of subalpine populations. *Ecology* **53,** 243–247.

Erskine, A. J. (1977). Birds in boreal Canada: Communities, densities, and adaptations. *Can. Wildlf. Serv., Rep. Ser.* **41,** 1–73.

Fischer, C. A., and Keith, L. B. (1974). Population responses of central Alberta ruffed grouse to hunting. *J. Wildl. Manage.* **38,** 585–600.

Fox, J. F. (1978). Forest fires and the snowshoe hare–Canada lynx cycle. *Oecologia* **31,** 349–374.

Ghent, A. W. (1958). Studies of regeneration in forest stands devastated by the spruce budworm. II. Age, height, growth, and related studies of balsam fir seedlings. *For. Sci.* **4,** 135–146.

Ghent, A. W., Fraser, D. A., and Thomas, J. B. (1957). Studies of regeneration in forest stands devastated by the spruce Budworm. I. Evidence of trends in forest succession during the first decade following budworm devastation. *For. Sci.* **3,** 184–208.

Gill, D. (1972). The evolution of a discrete beaver habitat in the Mackenzie River delta, Northwest Territories. *Can. Field Nat.* **86,** 233–239.

Gill, D. (1973). Ecological modifications caused by the removal of tree and shrub canopies in the Mackenzie Delta. *Arctic* **26,** 95–111.

Harper, F. (1956). "The Mammals of Keewatin." Univ. of Kansas Museum of Natural History, Lawrence.

Harper, F. (1961). "Land and Fresh-water Mammals of the Ungava Peninsula." Misc. Publ. No. 27, Univ. of Kansas Museum of Natural History, Lawrence.

Harper, F. (1964). "Plant and Animal Associations in the Interior of the Ungava Peninsula," pp. 1–58. Misc. Publ. No. 38, Univ. of Kansas Museum of Natural History, Lawrence.

Harper, F. (1964). "Caribou Eskimos of the Upper Kazan River, Keewatin," pp. 1–74. Misc. Publ. No. 36, Univ. of Kansas Museum of Natural History, Lawrence.

Harper, J. L. (1977). "Population Biology of Plants." Academic Press, New York.

Hatler, D. F. (1972). Food habits of black bears in interior Alaska. *Can. Field Nat.* **86,** 17–31.

Joyal, R. (1976). Winter foods of moose in La Verendrye Park, Quebec: An evaluation of two browse survey methods. *Can. J. Zool.* **54,** 1765–1790.

Joyal, R., and Scherrer, B. (1978). Summer movements and feeding by moose in western Quebec. *Can. Field Nat.* **92,** 252–258.

Jonkel, C. J., and Cowan, I. M. (1971). The Black Bear in the Spruce-Fir Forest. The Wildlife Soc. Monogr. 27, New York.

Jordan, P. A., Botkin, D. B., and Wolfe, M. L. (1971). Biomass dynamics in a moose population. *Ecology* **52,** 147–152.

Juniper, I. (1978). Morphology, diet, and parasitism in Quebec black bears. *Can. Field Nat.* **92,** 186–189.

Keith, L. B., (1966). Habitat vacancy during a snowshoe hare decline. *J. Wildl. Manage.* **30,** 828-832.
Keith, L. B., and Surrendi, D. C. (1971). Effects of fire on a snowshoe hare population. *J. Wildl. Manage.* **35,** 16-26.
Keith, L. B., and Windberg, L. A. (1978). A Demographic Analysis of the Snowshoe Hare Cycle. The Wildlife Soc. Mongr. 58, New York.
Kelly, M. W. (1954). Observations afield on Alaskan wolves. *Alaska Sci. Conf. Proc.* **5,** 1-8.
Kelsall, J. P. (1960). Cooperative studies of barren ground caribou, 1957-58. *Can. Wildl. Serv., Wildl. Manage. Bull., Ser.* **1,** No. 15.
Kemp, G. A., and Keith, L. B. (1970). Dynamics and regulation of red squirrel (*Tamiasciurus hudsonicus*) populations. *Ecology* **51,** 763-779.
Kuyt, E. (1972). Food habits of wolves on barren ground caribou range. *Can. Wildl. Serv. Rep. Ser.* **21.**
Larsen, J. A., and Lahey, J. F. (1958). Influence of weather upon a ruffed grouse population. *J. Wildl. Manage.* **22,** 63-70.
Leopold. A. S., and Darling, F. F. (1953). "Wildlife in Alaska." Ronald Press, New York.
Lutz, H. J. (1956). Ecological effects of forest fires in the interior of Alaska. *U.S. Dep. Agric., Tech. Bull.* **1133,** 1-121.
McCullough, D. R. (1970). Secondary production of birds and mammals. *In* "Analysis of Temperate Forest Ecosystems" (D. E. Reichle, ed.), pp. 107-130. Springer-Verlag, Berlin and New York.
Mattson, W. J., and Addy, N. D. (1975). Phytophagous insects as regulators of forest primary production. *Science* **190,** 515-521.
Merriam, H. R. (1964). The wolves of Coronation Island. *Alaska Sci. Conf. Proc.* **15,** 1-17.
Meslow, E. C., and Keith, L. B. (1971). A correlation analysis of weather versus snowshoe hare population parameters. *J. Wildl. Manage.* **35,** 1-15.
Miller, C. A., and Angus, T. A. (1971). *Choristoneura fumiferana* (Clemens), spruce budworm. *In* "Biological Control Programs against Insects and Weeds in Canada" (P. S. Corbet, ed.), pp. 127-130. Commonwealth Agric. Bur., Farmham Royal, Slough, England.
Morris, M. W. (1973). Great Bear Lake indians: A historical demography and human ecology. *Musk Ox* **11,** 3-27.
Murie, A. (1944). The wolves of Mount McKinley. *U.S. Dep. Interior, U.S. Nat. Park Serv., Fauna Ser.* **5,** 1-238.
Nellis, C. H., and Keith, L. B. (1968). Hunting activities and success of lynxes in Alberta. *J. Wildl. Manage.* **32,** 718-722.
Nelson, R. K. (1973). "Hunters of the Northern Forest." Univ. of Chicago Press, Chicago, Illinois.
Parker, G. R. (1972). Biology of the Kaminuriak population of barren ground caribou. Part. I. *Can. Wildl. Serv. Rep. Ser.* **20.**
Parker, G. R. (1973). Distribution and densities of wolves within barren ground caribou range in northern mainland Canada. *J. Mammal.* **54,** 341-348.
Pearson, A. M. (1975). The northern interior grizzly bear *Ursus arctos.* L. *Can. Wildl. Ser. Rep. Ser.* **34,** 1-86.
Peek, J. M., LeResche, R. E., and Stevens, D. R. (1974). Dynamics of moose aggregations in Alaska, Minnesota, and Montana. *J. Mammal.* **55,** 126-137.
Peterson, R. L. (1955). "North American Moose." Univ. of Toronto Press, Toronto.
Pimlott, D. H. (1959). Reproduction and productivity of Newfoundland moose. *J. Wildl. Manage.* **23,** 381-401.

Pimlott, D. H., Shannon, J. A., and Kolenosky, G. B. (1969). The ecology of the timber wolf in Algonquin Provincial Park. *Ontario Dep. Lands For. Res. Rep. (Wildl.)*, **97**.
Prebble. M. L., and Carolin, V. M. (1967). Spruce budworm. *In* "Important Forest Insects and Diseases of Mutual Concern to Canada, the United States and Mexico" (A. G. Davidson and R. M. Prentice, ed.), pp. 75–79. Dep. of For. Rural Dev., Ottawa.
Pruitt, W. O., Jr. (1967). "Animals of the North." Harper, New York.
Ransom, A. B. (1965). Observations of a ruffed grouse decline. *Can. Field Nat.* **79,** 128–130.
Rounds, R. C. (1978). Grouping characteristics of moose (*Alces alces*) in Riding Mountain National Park, Manitoba. *Can. Field Nat.* **92,** 223–227.
Rusch, D. H. (1976). The wildlife cycle in manitoba. Res. Br., Manitoba Dep. Natl. Res., Environ. Manage. Div., Inform. Bull. Ser., 11.
Rusch, D. H., and Keith, L. B. (1971). Seasonal and annual trends in numbers of Alberta ruffed grouse. *J. Wildl. Manage.* **35,** 803–822.
Rusch, D. H., and Keith, L. B. (1971). Ruffed grouse–vegetation relationships in central Alberta. *J. Wildl. Manage.* **35,** 417–429.
Rusch, D. H., Meslow, E. C., Keith, L. B., and Doerr, P. D. (1972). Response of great horned owl populations to changing prey densities. *J. Wildl. Manage.* **36,** 282–296.
Rusch, D. H., Gillespie, M. M. and McKay, D. I. (1978). Decline of a ruffed grouse population in Manitoba. *Can. Field Nat.* **92,** 123–127.
Sanders, C. J. (1970). Populations of breeding birds in the spruce–fir forests of northwestern Ontario. *Can. Field Nat.* **84,** 131–135.
Scotter, G. W. (1964). Effects of forest fires on the winter range of barren ground caribou in northern Saskatchewan. *Wildl. Manage. Bull.* **1,** 1–111.
Scotter, G. W. (1967). Effects of fire on barren-ground caribou and their forest habitat in northern Canada. *Proc. 32nd N. Am. Wildl. Conf.* pp. 246–259.
Sharp, H. S. (1978). Comparative ethnology of the wolf and the Chipewyan. *In* "Wolf and Man" (R. L. Hall and H. S. Sharp, eds.), pp. 55–80. Academic Press, New York.
Singer, F. J. (1978). Seasonal concentrations of grizzly bears, north fork of the Flathead River, Montana. *Can. Field Nat.* **92,** 283–286.
Smith, J. G. E. (1975). The ecological Basis of Chipewyan socio-territorial organization. *Proc. Northern Athapaskan Conf. Ottawa*, Paper 27, pp. 389–461.
Speer, R. J., and Dilworth, T. G. (1978). Porcupine winter foods and utilization in central New Brunswick. *Can. Field Nat.* **92,** 271–274.
Stenlund, M. H. (1955). A field study of the timber wolf (*Canis lupus*) on the Superior National Forest, Minnesota. *Minn. Dep. Conserv. Tech. Bull.* **4**.
Strang, R. M. (1973). Succession in unburned Arctic woodlands. *Can. J. For. Res.* **3,** 140–142.
Sukachev, V.. and Dylis, N. (1964). "Fundamentals of Forest Biogeocoenology" (translated from the Russian by J. M. MacLennan). Oliver & Boyd, Edinburgh.
Tamm, C. O., ed. (1976). Man and the boreal forest. *Proc. Meet., Stockholm, Swed. Nat. Sci. Res. Counc., Ecol. Bull.* **21.**
Thompson, D. Q. (1952). Travel, range, and food habits of the timber wolves in Wisconsin. *J. Mammal.* **33,** 429–442.
van Zyll de Jong, C. G. (1975). The distribution and abundance of the wolverine (*Gulo gulo*) in Canada. *Can. Field Nat.* **89,** 431–437.
West, G. C., and DeWolfe, B. (1974). Populations and energetics of taiga birds near Fairbanks, Alaska. *Auk* **91,** 757–775.

10 Boreal Ecology and the Forest Economy

There has been a lapse of decades since the last review of advances in fundamental knowledge of the ecology of the boreal forest and of known scientific problems that could be considered accessible to study with available techniques (Raup, 1941). During that time there have been not only major advances in ecological knowledge but also major changes in the manner in which the environment, and mankind's relation to it, are viewed. We have witnessed the development, moreover, of numerical techniques useful in work for describing ecosystems and for predicting their behavior under a variety of conditions, techniques that will give us a much more detailed and sophisticated view of the place of man in the world's biomes and, collectively, of mankind in the global ecosystem. There are techniques available that will enable people to manage the human economy in accord with global ecological realities, providing of course that human society can adapt quickly enough to the equally threatening realities of its own nature. There must be a finish to aggressive, shortsighted, primitive modes of conduct; if this is not accomplished, then they will put a finish to any expectations for human life that might be termed advanced, civilized, and enduring, as opposed to nasty, brutish, and short. A continuing flow of scientific papers on research conducted in central Canadian boreal forest and arctic tundra regions has indicated a continuing interest in northern ecology, although the rate of increase of published reports can hardly be said to presage a flood of literature such as may be anticipated on the ecology of biomes studied under the auspices of the International Biological Program.

The circumpolar and latitudinal extent of boreal (subarctic) and low-arctic tundra regions makes it readily apparent that there will be many aspects of the ecology of these regions that will be of widespread significance—to practicing foresters, game managers, land planners, environmental pollution agencies, and everyone concerned with wise use and preservation of a resource of tremendous significance to the future of northern countries—a resource with much unrealized potential for economic exploitation and with hopefully never-to-be realized potentials for ecological disaster. The responsibilities of the individuals charged with the wise use and management and preservation of northern renewable resources are very great indeed, and one can only encourage and support them in their efforts. The discussion here, however, is initially intended to deal more directly with some of the less tangible aspects of ecology, specifically with some of the underlying ideas that tend to interest ecologists and which, as in most natural sciences, may ultimately be seen to be either of very great fundamental importance or so totally irrelevant or inane as to make us wonder a decade hence what all the fuss was about.

The past history and current dynamics of the vegetation of the boreal and tundra regions present involved and fascinating problems that will keep ecologists occupied for many generations to come. Sufficient taxonomic groundwork has been laid to indicate that this field, particularly when taken in conjunction with the genetics of populations, remains virtually unexplored beyond the initial studies which reveal the existence of more questions than they answer.

Problems in the field of applied botany are also numerous and varied. Although a great deal of research has been carried out, much remains to be done before knowledge is sufficient to permit forests to be managed in a manner that will not violate, at least in some manner, the precepts of conservation. There is much to be discovered about aboriginal and even modern utilization of wild plants by indigenous peoples. Much undoubtedly can be done in the way of improvement of species by genetic selection for forestry purposes, for agriculture, and for grazing. As economic use intensifies, weed control and soil improvement and fertilization practices will almost certainly be refined. There are perhaps many opportunities for the utilization of native plant species in the production of drugs and other useful materials that have as yet only been hinted at. There is a need for more comprehensive floras of the boreal regions, although regional floras are now fairly numerous and most useful. Studies on variation within a given genus or species over a wide area, such as that conducted with *Salix glauca* by Argus (1965), will further enhance knowledge of the plants of the boreal regions, particularly in regard to development and perpetuation of subspecies, varieties, and ecotypes.

REGIONAL VARIATIONS

There seems to be a rather interesting paradox in the views of the nature of the boreal and arctic flora as described by the various botanists who have studied the plants of the northern regions. On the one hand, we have the impression that the flora of these regions is composed of genera and species bewildering in their variability. On the other hand, we have somewhat more recently been introduced to the concept that the boreal and arctic flora is, in actuality, a very narrowly variable one in which a high degree of genetic uniformity has been achieved and maintained to ensure perpetuation of rather special physiological and biochemical adaptations required for the maintenance of life in such harsh environments (Mosquin, 1966). To furnish evidence for such a low degree of variability, Mosquin cites as examples such genera as *Saxifraga, Antennaria, Arnica, Potentilla, Taraxacum, Koenigia, Corydalis, Epilobium, Dryas, Erysimum, Lesquerella, Cochlearia, Eutrema, Cardamine, Draba, Braya, Arabis, Descurainia, Oxytropis, Astragalus, Pedicularis*, and a number of others often represented in boreal and arctic regions by one or two or at most a few species. It, thus, seems apparent that we are dealing with two groups of plant species in these regions, those with a high degree of variability at least as expressed in morphological characteristics, and those with a high degree of physiological uniformity, at least from the convincing evidence presented by Mosquin. In the latter species, reproductive devices such as apomixis and self-pollination are so numerous as to validate the assumption that these species have achieved adaptation to severe habitats by means of mechanisms that narrow the possibility of variability and enhance the probability that physiological characteristics adapting the plants to a rather narrow spectrum of environmental conditions will be perpetuated. This is a subject that might profitably be investigated somewhat further.

The problem of the distribution of the various species or varieties in the *Salix glauca* complex in North America has been studied by Argus, who managed to discern considerable sense amid the confusion and who showed that with ample herbarium collections all degrees of intergradation could often be demonstrated between what were formerly considered distinct species. Thus, not only has the number of species been reduced, but the extent of variability within species has been broadened even more than had previously been thought to be the case. Each has, moreover, been identified with a refugium in which willow species must have survived the Pleistocene glaciations (Argus, 1965, 1973).

In his cautious interpretation of the *Salix glauca* complex and its relationships to Pleistocene history, Argus points out that the explanation may not be a satisfactory one for all time but may require modification as improved knowledge of polyploidy, hybridization, and population variation is acquired. The most important feature is the degree of intergradation demonstrated between the various components of the *Salix glauca* complex. It is conceivable that similar relationships may eventually be found among certain other *Salix* species.

It is perhaps of significance that the genera found to be complex and difficult in the boreal and arctic regions are those that often are wide-ranging throughout the North American continent. It is conceivable that their complexity is the direct consequence of having been forced into numerous refugia during the Pleistocene glaciations, in each of which they would have had an opportunity to evolve along separate pathways, later to be again brought together as they reoccupied the regions left open by the dwindling ice. Finally, at the end of the Pleistocene (if, indeed, it is the end and we are not today living in but another interglacial period) there remained a complex pattern of regional variations in the species of these genera. As range limits once again began to fuse, further mixing took place, continuing to the present day. It can easily be seen that such an occurrence would produce a most confusing admixture of characteristics and physiological traits, and it is no wonder that early botanists collecting in these regions were impressed by the variation within species.

On the other hand, species that survived the Pleistocene glaciations in the refugia of the Far North—the unglaciated portions of the arctic islands poleward of the main body of the Pleistocene ice—may have been subjected to an environment even harsher than that existing there today. They would have followed the course of evolution described by Mosquin, restricting variation to the narrow limits imposed by the unusually cold and wet conditions existing during the short growing season. Thus, in looking at the arctic and boreal flora, perhaps what we see are two groups, increased variation in species that survived south of the glacial and decreased variation in those that survived north of the ice. Admittedly, this hypothesis is put forth tentatively and with caution, but (as with so many other guesses that have seemed so logical in the past) at the moment it seems to be a reasonable explanation of at least some of the observed phenomena. This is a field in which we make many assumptions which cannot—and perhaps will never—be satisfactorily demonstrated to have an acceptable degree of truth, hence our efforts must always be categorized as highly speculative.

SUCCESSION

In regard to the problems of succession, competition, and other ecological relationships, Raup's (1941) apt summary of the state of knowledge applies in large part even today: "With so much confusion in the content of northern plant associations, their organization into developmental series, especially under the existing climatic and physiographic conditions, becomes equally obscure."

To Raup and others, however, it was apparent that the successions are not conditioned so much by biological factors as by climatic and edaphic ones. Competition assumes far less importance in the structure of communities in the sub-Arctic and Arctic than it does in more temperate regions. Raup cites Polunin's difficulty with the concept of seral development in arctic regions, of which Polunin states: "The vegetation is so poor as to suggest that hardly any successional advances or even marked changes (except in a few favored localities) can have taken place since the first colonization after the final ice retreat" (Polunin, 1935). Raup proposes a concept of climax in boreal regions, at least the more northern portions, stating that, while forests produced on the oldest surfaces may have attained the equilibrium of a climatic climax, other forest types over large regions might more properly be called subclimax. While the latter have attained a degree of long-lasting stability they might, given sufficient time, proceed further along a successional pathway, the nature of which can as yet only perhaps be hinted at. It is apparent that one of the great needs of ecological theory in northern regions is a refinement and clarification of the concept of climax, or a decision whether the concept can be said to apply at all in these regions of the world.

The effects of the physical action of frost upon soils and vegetation in boreal regions has not been adequately studied. The physical action of soil frost phenomena upon vegetation, particularly root systems, has often been noted. Benninghoff gives the following description (1952): "In the high latitudes roots are not only encased in frozen material for a great part of the year, but by repeated freezing and thawing, especially during the autumn freeze-up, they are heaved, torn, split by forces of great strength." Moreover, in the Arctic and northern sub-Arctic an increased plant cover does not commonly signify an increase in mesophytism as is usually observed in temperate zones. On the contrary, the influence of an increased plant cover may be far from salutary from the point of view of continued existence of the vegetational community, since the cover of vegetation usually decreases the depth of the active layer. The interaction of soil and vegetation apparently is at least partly responsible for many of the cyclical phenomena reported in areas possessing permafrost (Billings and Mooney, 1959; Hopkins and Sigafoos, 1951; Larsen, 1965). The influence of wind may be significant in the environment of boreal plant communities, and adaptations to persistently high winter wind velocities may be found in many species (Fig. 35). While it can be seen that the primary influence of vegetation is

Fig. 35. Layering is the principal mode of reproduction at the northern edge of the forest–tundra ecotone, and clumps of dwarfed spruce survive almost entirely by this method. The occasional vertical stem is eventually defeated by the snow-blasting effect of high winds in winter, which first destroy twigs and branches and eventually erode the outer bark.

to increase the evident rate of accumulation of cold reserves in the ground, vegetation also decreases the velocity of air currents within the stratum occupied by stems and leaves, thus tending to reduce heat loss by conduction and convection and often permitting a temporary increase in the temperature of aboveground plant parts well over that of the surrounding atmosphere during periods of high insolation. Thus, while conditions below the soil surface tend to favor low root temperatures, aboveground plant organs often have temperatures appreciably higher than air temperature even a few centimeters above the vegetational layer. In boreal plants, the temperature differential between aboveground and belowground parts is probably greater than that occurring in temperate zone vegetation, creating physiological stresses which have perhaps been overcome by special adaptations.

As one approaches the northern regions, low summer temperatures and the shortness of the growing season evidently are primarily factors in the reduced number of species present. The low temperatures and shortness of the growing season greatly restrict the number of degree-days above 32°F, and even this is a generous measure of the time available for growth, since only after the solstice does the last of the snow melt and growth begin.

THE FOREST BORDER

In North America, the continental forest border or transition zone where forest ends and tundra begins is surely one of the most fascinating biotic regions. At least in the case of tree species, the temperature-dependent relationship between assimilation rates and respiration rates may be significant in setting at least northern (or upper altitudinal) limits to geographical ranges. In trees, respiratory losses are proportionately greater than in herbaceous plants, apparently because of the need to maintain a trunk-and-branch system. About two-thirds of their total matter is devoted to supporting structures (Warren Wilson, 1967). As a result, probably more than one-half the energy budget of a tree is devoted to "defense expenditure" related solely to the need to shade out competitors. Whenever assimilation becomes insufficient to maintain this elaborate defense system, the trees give way to small species with lower fixed respiration costs.

This relationship probably is not solely responsible for the location of the forest border in all regions. Secondary influences at work include length of growing season (although this is related to energy accumulation), soil conditions, moisture supplies, snow and sand abrasion, seed dispersal, conditions for seedling survival, and diseases and insect infestation (Savile, 1963). At one place or another, all become of importance in limiting the extension of the range of northern forest trees. In trees approaching the limit established by respiration and assimilation, dwarfed stature and reduced vigor become readily apparent. For such trees the influence of nutritional deficiency and other physical and biological hazards are all the more disastrous when they occur (Fig. 36).

The ecology of spruce at the continental forest border has been studied in the Ennadai Lake region by the author (Larsen, 1965). Most striking are tree lines on hills at the south end of Lake Ennadai where the spruce community of the higher slopes grades upward into a treeless

Fig. 36. Maximum vertical growth of black spruce is not much more than 4–6 m at the northern edge of the forest–tundra ecotone, even on favorable upland sites such as this. The area shown is along the Pike's Portage route between the eastern arm of Great Slave and Artillery lakes. A medium-sized backpack leans against the rock in the center foreground.

summit. A narrowing of average growth rates can be discerned along the latitudinal gradient northward. At Ennadai, the rate of growth on optimum sites does not much exceed the growth rate on the poorest sites, indicating that only the most narrow range of habitable environment is available at the northern edge of the forest. The best available sites at Ennadai permit only minimum growth consistent with continued survival; these sites have growth rates comparable to those of only the poorest sites in regions to the south. When a line is drawn through the maximum growth rates found at each latitude, intersection with the line parallel with the lowest rates occurs at approximately lat. 65°N, about 4° beyond the range of spruce as a species (with the exception of the outlier of spruce forest along the Thelon River which may well have a special microclimate because of the river, or a special history). Subsequent to this study, Mitchell (1973) analyzed spruce growth rates in greater detail, coming to the similar conclusion that the location of the theoretical black spruce tree line coincides closely with the known northern limits of this species in the study region. He points out that the theoretical tree line does not directly coincide with the forest border, apparently because of fluctuations in the position of the arctic front in summer.

The most striking vegetational boundary controlled by climatic conditions in the boreal region is the northern forest border. The coincidence of the path of the summer cyclonic storms with the northern limit of forest in Canada was evidently first recognized by Stupart (1928), who pointed out that these storms passed southward of the mean summer position of the arctic frontal zone (see Chapter 3).

Hare and Ritchie (1972) have shown that global solar radiation per annum in Canada increases southward from values near 90 kly at the arctic tree line to about 110 kly at the boundary between open woodland and closed forest zones (i.e., the northern forest line). More significant is the absorbed solar radiation, which in the same span ranges from 50 to 55 kly at the tree line to about 80 kly at the forest line. Zonal divisions of the vegetation appear to correlate closely with mean net radiation; growing season net radiation is fairly constant over both the forest and tundra with a sharp drop between. The arctic tree line in Canada occurs near annual net radiations of 18–19 kly.

In summary, perhaps there could be no more suitable statement than that of Went (1950) who writes that the distribution of plants is not just a question of frost damage but is correlated with many specific temperature requirements which are met only in certain climates.

Cool-weather plants will die within a relatively short time in tropical or temperate environments even without the competition afforded by other plants. "An analysis of the genetical basis of climatic response may provide some interesting insights into the problem of evolution and migration of species, because it will indicate how many genes have to participate to allow invasion into a new climatic territory." There is no doubt, Went adds, that the genes controlling climatic response are the most important in the survival of a species.

SOILS AND PODZOLIZATION

The most intense podzolization occurs in boreal regions of continuous coniferous forest, in climates characterized by high precipitation and low rates of evaporation; on acid, sandy glacial deposits and unconsolidated sandstone material. There has long been active discussion of the degree to which parent materials contribute to the final character of soil in boreal regions where intensity of leaching and movement of materials in chelated form are unusually rapid. Under such conditions, upper mineral horizons in highly podzolized soils are often nearly depleted of all but organic and siliceous materials, resulting in a relatively homogeneous substrate for plant growth—at least for plants with shallow root systems—regardless of the nature of the parent material. Under conditions favoring high degrees of podzolization, mature podzols develop on a basic, acidic, or neutral substrate, with readily detectable differences occurring only in deeper layers. Regional variations in northern soils often occur most noticeably along a roughly latitudinal cline, grading northward into the often azonal tundra soils. Further studies are needed to elucidate in greater detail the relative importance of the various factors at work in soil genesis in the northern boreal and tundra regions. Arctic brown soil was not described in the scientific literature until 1955, yet in the author's experience it occurs extensively throughout at least the low-arctic regions. The reason for this apparent neglect may be that most soil research has been done on tundra soils in Alaska, where arctic brown soil is apparently a relatively rare soil type, although the first description of arctic brown soil is from an Alaska site as described by Tedrow and Hill (1955). The gradations between typically arctic soils and boreal podzols are undoubtedly a subject meriting further study. According to Hill and Tedrow (1961), podzolization processes are at work in arctic soils, but the effects are often difficult to discern on other than laboratory evidence. Well-developed podzols have been observed on tundra, and frost action occurs in spruce forest (Larsen, 1972). Diminution in the apparent intensity of podzolization northward may be due to the increase in the intensity of frost action which deforms soil profiles, but this is a point on which further research is needed. Differences in vegetation seem closely related to soil differences, and it is reasonable that the relationship must reflect differences in soil physicochemical characteristics and modes of genesis.

COMMUNITY STRUCTURE

Considering the forest in broad outline, readily detectable differences in floristic composition can be discerned at the eastern and western limits of the continent as well as at the northern and southern limits of the forest; the ground flora of eastern Canada differs markedly from that at the western continental limit, and northern boreal communities are distinct in composition from those found at southern limits along any given line of longitude. Similarity is inversely proportional to distance.

It is of significance that the differences in floristic composition in the boreal communities appear greater along a southwest–northeast axis for any given distance than they do along a northwest–southeast axis for an equal distance. Since climatic gradients are greater along the former axis, it can be inferred that the greater floristic variation for a given distance along this axis is due primarily to the climatic relationships prevailing in the region. In Quebec–Ungava, isopleths of climatic parameters more nearly parallel latitudinal lines, and here the vegetational parameters, particularly the forest border, tend more nearly eastward and westward, coinciding with the direction of the mean airflow.

A number of methods are now available for recording in great detail the composition and structure of plant communities. The method employed by the author (see Chapter 7) measures similarities between understory vegetational communities and was used to demonstrate that community differences between areas appear to represent a response to environmental differences and that they are strongly related to climatic gradients. Communities on study sites located geographically closest to one another generally demonstrate the greatest similarity; those farthest apart geographically are usually least similar in composition. In many communities, the sharp north–south variation in boreal forest community structure is readily apparent.

It is thus apparent that the regional differences in community structure possess a coherence demonstrable by these techniques, and that the quantitative analysis of community structure not only reveals a continuity in vegetational composition but also makes it possible to relate spatial changes in composition to variation from one region to another in the physical environmental parameters.

From the author's initial reconnaissance of the Canadian boreal and tundra communities it is apparent that (1) we have techniques available for studies relating environment and vegetation over rather broad geographical regions and (2) response of vegetation to environment is both comprehensible and ultimately predictable. Criticism has been raised that this approach to the study of plant community ecology is a curious recrudescence of determinist philosophy; in the author's view it is the substitution of scientific method for practices based largely upon intuition, subjective observation, and literary description. Only the future will tell, of course, whether one is superior to the other. Further improvement of the techniques and more intensive application to a wider range of communities is now needed and, it is hoped, will be undertaken by an increasing number of investigators in boreal and tundra regions. It is probably obvious that in my view this will represent a major advance in study of the ecology of these biomes.

THE FOREST ECONOMY: RESOURCES

Almost one-half of the total world reserve of exploited coniferous forest is found in the USSR, the remainder existing largely in North America and Europe (Table 40). Four-fifths of the forested area of the Soviet Union consists of coniferous species, found mainly in the taiga across the northern part of the country. Among the coniferous species, pine provides the timber most widely used for commercial purposes. The amount of pine in the forests of Siberia is considerable, making up about 25% of the forested area. Spruce grows with pine in most commercially useful forests, often in pure stands adjacent to pure stands of pine. Larch dominates the forests of eastern Siberia and the Soviet Far East. Large tracts of taiga are covered by birch, aspen, and alder, especially areas where mature stands of conifers have been logged. To the south of the taiga extend the mixed and deciduous forests of central Russia. In the future, the Soviet economy will continue to require a large volume of mature timber and, although the pulp and paper industry has begun use of a wider variety of timber, it will apparently continue to rely largely on spruce for some time to come.

In Canada, forestland occupies about 35% of the total land area. Forest suitable for regular harvest amounts to about 25% of the total area. With this resource, forest industries have long been a major factor

TABLE 40
Exploitation of World Timber Reserves[a]

Region	Percentage of total forest area surveyed	Timber reserves (million cubic meters)			Average reserves (million cubic meters per hectare)		
		Total	Coniferous	Nonconiferous	Total	Coniferous	Nonconiferous
Europe (not USSR)	96	10,780	7,120	3,660	80	90	65
North America	100	36,640	27,140	9,500	100	135	60
Central America	24	1,080	330	750	80	85	80
South America	20	8,300	1,060	7,240	115	135	110
Africa	71	5,620	70	5,550	45	30	45
Asia (not USSR)	96	22,020	4,620	17,400	100	120	95
Australia and Oceania	97	1,320	220	1,100	65	50	70
Subtotal		85,760	40,560	45,200			
USSR[b]	100	42,813	34,610	8,203	94	108	59
Total world		128,573	75,170	53,403	91	107	79

[a] Data from Gerasimov *et al.* (1971). Copyright © 1971 W. H. Freeman and Company.
[b] For comparability, the timber reserves were calculated in the same manner as those of other world regions per hectare of forest area, whereas usually in the USSR they are calculated per hectare of forest-covered area.

in the Canadian economy. Canada cuts more than 8% of the total annual world production of industrial wood, a harvest exceeded only by the Soviet Union and the United States.

Canada has put 191 million acres of remote and slow-growing forest unsuitable for regular harvest into a special category considered to have economic importance as an emergency timber reservoir but of primary value for recreation and as a vast catchbasin for the precipitation that flows into the nation's streams and rivers.

The timber existing in the productive forests of Canada is about 80% softwoods by volume, the most abundant species being black and white spruce, balsam fir, and jack pine, with other species predominant in the mountain and coastal forests of the western regions. The remaining 20% is hardwood, most common of which are poplars and white birch. About 55% of the wood harvest is utilized for saw logs and veneer, and 35% is utilized for pulpwood. The remainder goes for poles, pilings, mining timbers, and fuel. Raw materials used in the pulp and paper industry consist of 94% softwood and 6% hardwood. In sawmilling the same proportion holds, and the plywood and veneer industry consumes 80% softwood and 20% hardwood. Technological change probably will influence the species used in the production of wood pulp, with new installations utilizing increased proportions of hardwood.

On a regional basis there is considerable variation in Canada from full utilization of natural regenerative capability to vast unharvested surpluses. Although large areas are beyond existing transport systems, accessibility is improving, and many more areas soon will be under managed harvest.

Improved utilization of the forest resource is being achieved by better harvesting technology. If hardwoods now in accessible areas were utilized to a greater extent, a considerable portion of the projected demand could be satisfied without opening new forest areas. The unexploited forest areas offer excellent opportunities to plan for the future in ways that will satisfy both needs for timber and for activities associated with environmental appreciation. It should also be possible to increase the productive capacity of forests now being managed. Since 1900, more than 2 million acres of land in Canada have been afforested, reforested, and artificially regenerated. About 60% of this involved artificial regeneration of cutover and burned-over forest land by seeding with conifers. In early attempts, plantation management was often inadequate or neglected, but more recently most planted or otherwise regenerated forestlands are under skilled management, and the success of the efforts has increased markedly. Studies of the costs of research and the benefits of forest programs demonstrate that the programs pay for themselves many times over. In one research project, for example, the cost of producing genetically superior seed was more than offset by a 2–5% increase in yield of marketable timber.

MANAGING THE FOREST ECOSYSTEM

An ecologically sound land management plan must begin with a comprehensive ecosystem analysis. A resource survey that merely characterizes soils, minerals, terrain, demography, plants, and animals is inadequate. To be effective, a survey must provide a good analysis of energy flow, nutrient cycling, population dynamics, and species relationships. For a land management effort to be successful, for example, it is necessary to have knowledge available on the reproductive cycle of key plant species, conditions under which regeneration of desired species is successful, and response of the system to a wide variety of environmental conditions. The second step in land management programs is consideration of land use alternatives, based on historical and cultural characteristics of the region, natural resources, climate, and uses for products that can be harvested from the land. An assessment of ecological costs and benefits can then be made.

The regions occupied by boreal forest are endowed with the kinds of trees best suited to meet the needs of advanced industrialized countries, mainly the coniferous or softwood species—pine, spruce, hemlock, fir, cedar, larch. In Canada and other countries with forests of this type, wood-based industries promise to keep the economy in a favorable position for the foreseeable future.

It is obvious that the boreal forests of Canada, Sweden, Finland, and the USSR will continue to be influenced by people to an ever-increasing degree. Exploitation not only of the forest itself will increase, but also there will be increased utilization of the land for mining and hydroelectric power. It will be necessary under these circumstances to intensify land management practices so as to avoid the potentially disastrous effects of some forms of land use. Burning of lichen woodlands, for example, can have a pronounced effect on groundwater regimes, greatly accelerating evaporation and reducing the water from rainfall available for normal runoff into streams and rivers. The search for petroleum and other mineral resources in the Canadian north has resulted in the removal of a significant fraction of vegetation along winter roads and seismic lines, although these areas represent only a small fraction of the total land surface. In areas of permafrost with a high ice content disturbance can result in melting and settling of the surface. This can present a significant problem for both ecologists and engineers (Haag and Bliss, 1974). Land use planning in northern regions must give consideration to vegetational structure in assessing landscape suitability for any activity that would result in removal of vegetation and formation of the mudholes known as thermokarst (Hutchinson and Freedman, 1978; Jenkins et al., 1978).

It is essential that modern forestry and other land utilization practices be based on a sound understanding of the ecosystems involved, so that the effects of these practices will be known and predictable. So far, modification of the more northern portions of the taiga ecosystem by silvicultural and other forms of utilization has, at least in many areas, been relatively slight compared with the drastically altered forests of most temperate and subtropical zones. The taiga ecologically is still a stable ecosystem with many forested areas resembling those existing before human influence occurred. To retain them in this state, or in a condition even more productive than the original forest, and to do so on a permanent basis for decades and centuries to come, will require more thought and foresight than has been exercised in the past (Ritchie, 1977).

EXPLOITATION AND MANAGEMENT

Vegetation is a primary resource in boreal regions, and it is one that will become increasingly valuable as world populations increase. There will be, in future years, growing economic pressures to exploit the resources of boreal regions. Demands for wood will open what are now marginal lands for logging. Recreational uses such as hunting and fishing will increase the probability of disturbance in remote areas, particularly by fire. There will be increasing demands for more roads, more airfields, more summer fishing camps, and more fall hunting lodges. Mineral exploration will increase, new economically valuable deposits will be found, and new mining towns will be established, linked to southern regions by roads and railroads. All these events will tend to work great changes in the natural vegetation and wildlife of the northern regions, and in a real sense most of them will be destructive. It is imperative that good management principles and practices be elucidated quickly so the adverse effects of increasing exploitation can at least be minimized. It is fortunate that there is already an effort of some magnitude under way with this goal in mind.

Forest research stations are now active in many regions, many of them cooperative endeavors between governments and universities, and they support the research of many dedicated foresters, botanists, wildlife ecologists, and scientists specializing in other fields. Out of this effort will come many good techniques for managing the ecosystems of northern regions, and it is to be hoped that these will be utilized to protect the ecosystem from at least the more destructive impact of increasing economic exploitation.

THE BIOREGENERATIVE SYSTEM

This is not the place to review the knowledge already available concerning techniques that can be employed to maintain the northern ecosystems in a condition of dynamic stability and productivity. It will suffice to say that all the techniques and management practices are based on the concept of the ecosystem as a regenerative system; the life of the system is maintained by the constant renewal of organic matter which, utilizing the incoming energy of the sun, maintains itself, reproduces, and preserves a state of dynamic equilibrium despite disturbance in the form of variations in physical conditions that occur from time to time. The knowledge needed to carry out effective management of these renewable resources includes, then, a knowledge of the effect upon the ecosystem of the various disturbance factors and the manner in which adverse effects can be minimized. This, in essence, is the purpose of all management and conservation activities, though the ways to acquire the knowledge needed involve a wide variety of scientific fields ranging from soil science and entomology to molecular biology and genetics.

The value of forest vegetation in its natural state is often underestimated, simply because much of the natural forest still remaining in undisturbed condition is not economically exploitable, hence in the opinion of many individuals is essentially worthless. This is, of course, an invalid opinion and one that would quickly change should prices, for example, double or triple for wood and wood products. Often ignored, too, in this view is the fact that the vegetative cover acts to store mineral nutrients that would otherwise wash away and be lost to the ecosystem forever. The vegetative cover is also essential to wildlife and the more intangible aesthetic values of the natural landscape. It is the purpose of forest management and the various forestry operations to retain as much as possible the nutrient elements, the wildlife populations, and the aesthetic and recreational values of the forest even while it is being exploited for its resources. On lands that are now economically marginal, these resources can best be conserved by maintaining the forest in as undisturbed a condition as possible.

MANAGEMENT IN NORTHERN REGIONS

Management of northern coniferous forest is rendered easier than management of the natural vegetation of many other parts of the world because of one fact. Ecosystems adapted by evolution to an environment with wide natural fluctuations can resist human exploitation better than ecosystems adapted to stable and unvarying environments. Thus, the taiga is much more resistant to total collapse than is the tropical forest. But in the Far Northern taiga, improper exploitation can cause the more advanced community structure of a mature forest to collapse and may bring about substitution by a treeless vegetational community of little value economic or otherwise. In many areas where exploitation is to be undertaken, perhaps the best management practice is to preserve some areas in the natural state, thus creating a mosaic of exploited sites interspersed in a network of protected natural sites. This serves to prevent total destruction of the forests of any region and allows the forest communities to retain bioregenerative capabilities both in terms of vegetation and wildlife. It has been pointed out that the conservation of nature has utilitarian aspects that are even more compelling than aesthetic ones. In some of the Far Northern forested areas it may be necessary to protect the forest cover completely, because once removed the trees do not regenerate, at least for many decades, and cutting may reduce the vegetation to a more-or-less permanent tundra subclimax (Larsen, 1965; Strang, 1973; Gill, 1973).

The literature on forest management practices is now so large that no effort can be undertaken here to review it; there are now, moreover, management simulation models that accurately predict the effect of given management decisions for 50 or more years into the future. Thus, silvicultural treatments, width of strips that are clear-cut during harvest, age structure of the forest trees, flooding or changes in groundwater levels caused by dams, and a great variety of other environmental influences can now be analyzed in terms of their effect on the forest and on forest productivity (Hegyi and Tucher, 1974; Ek et al., 1976; Adams and Ek, 1974; Jeglum, 1975).

Future research will include development of methods for achieving lasting increases in forest productivity by improving microbiological and chemical activity and speeding up the cycling of nutrients. Several new techniques already appear promising, including mechanical mixing of surface organic material into soil and thinning of stands to increase surface soil temperatures during the growing season. Paludification of forests is a frequent initial phase in the conversion of forest to bog. In interior Alaska, sphagnum bogs have been reported on north slopes as steep as 50%. Burning might be effective for converting these stands to a productive mineral soil. The nitrogen content of black spruce foliage declines progressively through successional stages, but trees growing in conjunction with birch and alder average above the critical level, and management practices that encourage a mixture of spruce, birch, and alder might increase forest productivity.

WILDLIFE MANAGEMENT

Wildlife is part of the boreal regenerative system and, like vegetation, is capable of renewal if disturbance is not so great as to exceed the tolerance limits of the population. For many animal species, however, these tolerance limits are low and, while vegetation may recover fairly rapidly from disturbance, there are some forms of disturbance that seem to affect animal populations beyond recovery. Barren-ground caribou, for example, are known to have once wintered but are seen no more in areas that do not now differ vegetationally a great deal from presettlement times but are crossed by roads or railroads. Destruction of the forest cover by fire may have been an important cause of the abandonment of this former range by caribou, and hunting must also have been a factor, but the simple presence of humans and machines may have contributed more than we know or are willing to admit. With even more massive developments anticipated for the future, we can now only wonder what the effect on wildlife will be. It seems inevitable, however, that the changes will be for the worse and that great concern is justified over the fate of wildlife in the northern regions.

Not so many years ago, all human life in the north was based totally on wildlife. Today, wildlife is still the only local resource capable of sustaining human life. All support for technological development and population growth must come through corridors of transport to already heavily developed middle latitudes. Thus, to exploit resources of the northern regions, both renewable and nonrenewable, lands farther south must be exploited even more intensively than they are today. There is increasing danger that northern development will not only damage irreparably the renewable resources of the north but will increase the destructive pressures that now already bear upon environments of lands in middle latitudes.

The value of northern wildlife is fourfold; for many native peoples it is still important for subsistence, it is important economically as the source of fur, it is central to economic development based on recreational hunting and fishing, and it forms the foundation upon which much of the indigenous human culture is based. Without wildlife, the fabric in which the lives of the local peoples are wrapped becomes threadbare. It is even conceivable that survival of the human race depends upon the retention of ways of life based on close adaptation to the environment, retention of the knowledge still held by small groups of native peoples. It is also conceivable that survival may depend upon the existence of wildlife populations that can sustain human life. We are not inclined to look upon the existence of human life on earth as an experiment, but the fate of thousands of extinct species is evidence that every species is an experiment and that most of them have failed. The fate of technology is still in the balance, and it may well be that the less specialized native peoples will be those best equipped to survive through future millenia. This is another reason to justify the view that we are custodians of the world's resources, both renewable and nonrenewable, and that needless destruction is not only unjustifiable but reprehensible.

The life style and economy of the Far Northern regions are today an admixture of modern industrial and traditional domestic pursuits. The prevailing opinion in Canada is that traditional ways of life will decline and disappear, to the general benefit of all concerned. There are persistent hints, however, that not everyone shares this view, and hunting and trapping retain considerable significance in the lives of large groups of native people—who are in the majority in most parts of the north. It is becoming increasingly evident that there will be economic and political significance in accurate assessments of the degree to which northern plant and animal products can be exploited by the local human communities without damage to potentials for annual renewal. There is much that must be known before permissible annual harvests of caribou, moose, and fur-bearing animals for food and clothing, and of forest trees for buildings and firewood, for example, can be established. Studies on the northern economy and the traditional ways in which local resources are utilized have begun to appear, however, and, in addition to their inherent interest, have furnished a basis for future research which cannot escape being of great value (Usher, 1976).

A fascinating account of the uses to which the resources of the region are put by the native Kutchin people of central Alaska is to be found in "Hunters of the Northern Forest" by Nelson (1973). He points out, for example, that wood from spruce trees is used to build houses, caches, canoes, furniture, and a variety of other things, as well as for fuel in stoves and campfires. Beds in camps are made of spruce boughs, and spruce pitch is spread on cuts and sores. Birch is used for firewood and is known as the traditional material used for the construction of sleds, snowshoes, and the covering of canoes. A wide variety of other plant species are utilized for one purpose or another, from berries for food (notably *Vaccinium vitis-idaea* and *V. uliginosum*) to spruce roots for lashing birch bark to canoe frames and marsh sedges for dog bedding in winter months. For food, the Kutchin depend on moose, fish, snowshoe mare, muskrat, waterfowl, and caribou—all of which are obtained in a variety of ways. Changes in availability of the different species from year to year result from weather conditions, and variable population densities and fluctuations undoubtedly had an effect on populations of the aboriginal Kutchin, although they are not so significant today when food can be delivered by aircraft during times of scarcity. There are advantages to be had in the northern economy, as Usher (1976) points out: "The North may well be the only place where a poor man's table is laden with meat as a matter of course." He adds the warning that widespread destruction of an ecosystem capable of producing meat may have disastrous consequences:

> It would surely be the height of irresponsibility to impair the productivity of lands which can supply food only in the form of meat, at a time when the world may well be entering a period of food shortages—particularly meat shortages—of such proportions that we cannot now even imagine what they will mean to our daily lives and to our society.

It is of significance for the future utilization and management of boreal forest resources that interest in "game ranching" is increasing—the use of wild forest areas for growing large ungulate animals such as moose or caribou under close surveillance (Telfer and Scotter, 1975). The possibilities for successful game ranching seem good, largely as a result of increased ecological knowledge of niche specialization of multispecies assemblages and the need to use land resources to the best possible advantage, particularly if global red meat production reaches a point beyond which the potential for further increases is limited. There are at present not a large number of studies available that describe animal populations of boreal regions in an ecosystem context, the result perhaps of the fact that energy and nutrient totals cycling through the animal populations are relatively small. It is obvious that this insufficiency of adequate research should not be allowed to persist in view of the importance of animal protein in the diet of the peoples of northern regions. Initially there must be an assessment made of the biomass of populations of each species by sex and age class, followed by determination of the standing crop and annual permissible harvest. With increased knowledge, it should eventually be possible to model the animal population and its ecological relationships so that predictions can be made of the effect of increases in primary production upon the herbivore population. This knowledge could then be employed to make predictions of the effect of changes in primary production upon carnivore populations, as well as the effect of removal of predators upon the herbivore population. Only with increased knowledge, moreover, will the effect of logging on animal populations be predictable with reasonable accuracy. The effect of other human activities, as well as that of physical variables such as weather, will also, it is hoped, eventually be predictable as a consequence of the development of ecosystem models.

Some initial studies tend to confirm the belief that the effect of these environmental influences on animal populations can be measured. Such models would simulate plant production and succession, behavior of the animal population, the effect of forest management practices, the effect of various animal harvest policies, and the effect of fires, diseases, insect infestations, and storms. One such study (Walters *et al.*, 1975) has resulted in computer simulation of a barren-ground caribou population in which a caribou biomass is programmed to respond to hunting pressure as well as food availability, the latter through a foraging submodel that takes into consideration such factors as snow depth, seasonal migrations, and area of useful habitat.

PRODUCTION AND MANAGEMENT

The specific management techniques of value in the control and conservation of wildlife are different for each species, and it is not the purpose here to summarize what is already a very large scientific literature, one that grows rapidly year by year. There is still much to learn, however, in terms of the natural history and behavior of the various bird and mammal species and in terms of the effect upon populations of such innovations as pipelines, seismic lines, impoundments, and pollution. The management of wildlife populations even in the Far North has in recent years taken on dimensions that could not have been imagined only a few decades ago. Fortunately, as the pressures of new technology and economic development increase, there seems also to be increasing public awareness of the unique values of wildlife and wilderness, so we can have some hope that preservation measures will be undertaken at the same time economic development of remote regions continues to expand.

FOREST MANAGEMENT

The economic value of the northern forest regions will continue to increase, and it is to be anticipated that exploitation of forest resources, even those in what are now remote areas, will expand. The potential of the forest biomass for satisfying the economic demand for building material and newsprint is well known, but not as widely recognized is its potential as a source of hydrocarbon fuels and chemicals. It is estimated that the net photosynthetic productivity of the earth is 115×10^9 tons/year, of which forests account for 42% and croplands 6% of this total. The net productivity of forests exceeds the annual consumption of fossil fuels, but only a small proportion is currently utilized as an energy source; presently, only wood-based industries have the potential for rapid conversion to wood fuel. If this were to be accomplished, a total saving of 3% or more of the national energy budget (in the United States), for example, would result. The major difficulty in utilizing the present biomass production of forests for energy is that much of the forest is either inaccessible for all practical purposes or there is no technology existing to convert it into hydrocarbon fuels in the regions where these fuels are most needed—the deep tropics do not require household heating and presently have a relatively low demand for transportation fuels. In the far northern boreal regions, where both are required for almost all human activities, local demands for home heating can be supplied by wood, but there is as yet no existing technology for conversion of wood products into hydrocarbon fuels for internal combustion engines. So far, economic competition from traditional petroleum materials will preclude any major effort toward exploitation of wood products; as supplies of petroleum dwindle, however, new energy sources will be sought and, in northern regions, consideration will undoubtedly be given to the development of ways to convert wood into substitutes for petroleum products. Effective utilization will require more development effort than has been economical in the past, but changing conditions will very likely modify this situation in the future.

At the present time, the value to the Canadian economy of wood products obtained from the boreal forest is equaled by no other single resource. Growth of the economy in many provinces is dependent upon growth of the lumber industry. In British Columbia, for example, 44% of all industrial value added comes from forest industries, of which 30% is attributable to the lumber industry; this has led to the design of a number of econometric models of the softwood lumber industry which, like ecological models, are employed to forecast the effect of various policies upon future trends (Manning, 1975). The similarity of ecological and econometric models is often striking; ecology has, in fact, been described as biological economics. Canada's production of 16% of the annual world output of wood pulp and 38% of the world output of newsprint makes it one of the few major wood-producing and forest nations of the world. Roughly 90% of the total production of timber-based industries in the country is divided about equally among British Columbia, Ontario, and Quebec. Despite the importance of the wood industry, however, few studies have been made on the interrelationships among various wood products industries and among the wood industries and other industries in the provincial economies. Such studies are needed, and it is apparent that a start has been made (Raizada and Nautiyal, 1974). There are, moreover, studies underway on the regional climates most suited to forest economies (Miller and Auclair, 1974), as well as a wide variety of other factors that influence regional production and utilization of forest products. It is apparent that in the future there will be little or no economic distinction between croplands and forestlands; the trees of wilderness areas will be viewed as a crop to be managed just as croplands are managed today—for maximum productivity on a long-term basis.

THE OUTLOOK FOR THE FUTURE

Perhaps one of the most encouraging observations one could make in regard to northern development is that a prediction made by J. B. Tyrrell in 1897 has not been fulfilled. Speaking to the British Association for the Advancement of Science, this geologist and explorer of the interior plains of Canada spoke of the day soon to come when an electric railway powered by local hydroelectric plants would stretch from Great Slave Lake to Chesterfield Inlet, bringing countless happy homes to the interior Canadian barren grounds. The inhospitable aspects of the Canadian northern interior have postponed that development for a few decades—and one might add probably for many decades to come.

Many years and efforts by many workers will be required for anything resembling an adequate understanding of many of the relationships so lightly touched upon in this volume. But as the efforts to elucidate more fully the relationships between life and the environment expand, knowledge of vegetation will be particularly important, because vegetation is not only the primary producer of the world's complement of organic material, but is also a factor of considerable importance in maintaining the world's ecosystem in a condition that permits the continued existence of humanity.

In the distant—or perhaps not so distant—future it is quite conceivable that definite measures will have to be undertaken to preserve some degree of natural balance in the world ecosystem. When that time comes, knowledge of the vegetation of the various regions of the world will be of great importance, particularly of regions that remain in a relatively undisturbed condition. The vast oceans of the world, of course, are of great importance as buffer agents in maintaining the balance of the elements that compose the atmosphere, but the land areas are by no means of negligible import. Of the vegetated land surface of the earth, somewhat more than one-third is given over to arctic, alpine, and northern temperate vegetational types. It is these areas, along with a fairly large proportion of tropical vegetational types, that remain in a relatively undisturbed condition. The forests and other lands of the temperate zone are increasingly being given over to farming, residential, and industrial complexes. The few small areas of natural landscape that remain in a protected condition will, under these circumstances, become of increasing importance. Even if they are allowed to remain relatively unexploited, they may still be considered to possess the potential for becoming increasingly important, because they help to recharge the atmosphere with oxygen and extract carbon dioxide, they help maintain worldwide ecological balances, and they serve a function in the hydrological cycle. It may ultimately be found that these are the most important functions they perform.

Perhaps, as Aldo Leopold has inferred, they will also serve in a most significant way as wilderness: "Of what avail are forty freedoms without a blank spot on the map?"

While the day is already past when the northern lands were a blank spot on geological or hydrographic maps, perhaps they will remain so on demographic maps, at least in relative terms and in comparison with urban areas. It is the author's personal hope that they will do so. For there is a beauty here that urban areas, no matter how carefully planned or filled with cultural advantages, can never match. There is a peace and vast tranquility that can be found only "far from the madding crowd's ignoble strife...." One cannot help but wonder what the future holds for the human race. It is a great source of comfort to know there is still at least one place to which one can escape for a week or a month and find a semblance of wilderness.

REFERENCES

Adams, D. M., and Ek, A. R. (1974). Optimizing the management of uneven-aged forest stands. *Can. J. For. Res.* **4**, 274–287.
Adzhiev, M. E. (1976). Taming the Siberian taiga. *New Sci.* **81**, 382–384.
Alban, D. H., Perala, D. A., and Schlaegel, B. E. (1978). Biomass and nutrient distribution in aspen, pine, and spruce stands on the same soil type in Minnesota. *Can. J. For. Res.* **8**, 290–299.
Argus, G. W. (1965). The taxonomy of the *Salix glauca* complex in North America. *Contrib. Gray Herb. Harv. Univ.* **196**, 1–142.

Argus, G. W. (1973). The Genus *Salix* in Alaska and the Yukon. *Nat. Mus. Can. Publ. Bot.* **2**, 1–279.
Armstrong, T., Rogers, G., and Rowley, G. (1978). "The Circumpolar North." Methuen, London.
Barr, B. M. (1979). Soviet timber: Regional supply and demand, 1970–1990. *Arctic* **32**, 308–328.
Bazilevich, N. J., Rodin, L. E., and Rozov, N. N. (1970). Geographical aspects of biological productivity. *Sov. Geogr.* **12**, 293–317.
Benninghoff, W. S. (1952). Interaction of vegetation and soil frost phenomena. *Arctic* **5**, 34–44.
Billings, W. D., and Mooney, H. A. (1959). An apparent frost hummock-sorted polygon cycle in the alpine tundra of Wyoming. *Ecology* **40**, 16–20.
Day, R. J. (1970). Stand structure, succession, and use of southern Alberta's Rocky Mountain forest. *Ecology* **53**, 472–478.
Ek, A. R., Balsinger, J. W., Biging, G. S., and Payandeh, B. (1976). A model for determining optimal clear-cut strip width for black spruce harvest and regeneration. *Can. J. For. Res.* **6**, 382–388.
Endelman, F. J., Northup, M. L., Hughes, R. R., Keeney, D. R., and Boyle, J. R. (1973). Mathematical modeling of soil nitrogen transformations. *Am. Inst. Chem. Eng. Symp. Ser.* **70**, 83–90.
Euler, D. L., Snider, B. and Timmermann, H. R. (1976). Woodland caribou and plant communities on the Slate Islands, Lake Superior. *Can. Field Nat.* **90**, 17–21.
Foster, N. W., and Gessel, S. P. (1972). The natural addition of nitrogen, potassium, and calcium to a *Pinus banksiana* Lamb. forest floor. *Can. J. For. Res.* **2**, 448–455.
Fraser, D. A. (1962). Growth of spruce seedlings under long photoperiods. Can. Dep. For. For. Res. Div. Tech. Note 114. pp. 1–18.
Fraser, D. A. (1966). Vegetative and reproductive growth of black spruce [*Picea mariana* (Mill. BSP.)] at Chalk River, Ontario, Canada. *Can. J. Bot.* **44**, 567–580.
Gardiner, L. M. (1975). Insect attack and value loss and wind-damaged spruce and jack pine stands in northern Ontario. *Can. J. For. Res.* **5**, 387–398.
Gerasimov, I. P., Armand, D. L., and Tefran, K. M., eds. (1971). "Natural Resources of the Soviet Union." Freeman, San Francisco.
Gill, D. (1973). Modification of northern alluvial habitats by river development. *Can. Geogr.* **17**, 138–153.
Haag, R. W., and Bliss, L. C. (1974). Functional effects of vegetation on the radiant energy budget of boreal forest. *Can. Geotech. J.* **11**, 374–379.
Hare, F. K., and Ritchie, J. C. (1972). The boreal bioclimates. *Geogr. Rev.* **62**, 333–365.
Hegyi, F., and Tucker, T. L. (1974). Testing silvicultural treatments by computer simulation. In "Proc. Workshop For. Fertiliz." Can. Dep. Environ., Can. For. Serv., Great Lakes For. Res. Cent. Tech. Rep. 5, pp. 59–64.
Heinselman, M. L. (1957). Black Spruce. *U.S. Dep. Agric. For. Serv., Lakes States For. Exp. Sta. Pap. No.* **45**, 1–30.
Hill, D. E., and Tedrow, J. C. F. (1961). Weathering and soil formation in the arctic environment. *Am. J. Sci.* **259**, 84–101.
Hopkins, D. M. (1950). Frost action and vegetation patterns on Seward Peninsula, Alaska. *U.S. Geol. Surv. Bull.* **974-C**, 1–101.
Horton, K. W., and Lees, J. C. (1961). Black spruce in the foothills of Alberta. Dep. For., For. Res. Div., Tech. Note 110, pp. 5–54.
Hutchinson, T. C., and Freeman, W. (1978). Effects of experimental crude oil spills on subarctic boreal forest vegetation near Norman Wells, N.W.T., Canada. *Can. J. Bot.* **56**, 2424–2433.

Jarvenpa, R. (1979). Recent ethnographic research—Upper Churchill River drainage, Saskatchewan, Canada. *Arctic* **32,** 355-365.
Jeglum, J. K. (1975). Vegetation-habitat changes caused by damming a peatland drainageway in Northern Ontario. *Can. Field Nat.* **89,** 400-412.
Jenkins. T. F., Johnson, L. A., Collins, C. M., and McFadden, T. T. (1978). The physical, chemical, and biological effects of crude oil spills on black spruce forest. *Arctic* **31,** 305-323.
Larsen, J. A. (1965). The vegetation of the Ennadai Lake area, N.W.T.: Studies in arctic and subarctic bioclimatology. *Ecol. Monogr.* **35,** 37-59.
Larsen, J. A. (1972). Observations of well-developed podzols on tundra and of patterned ground within forested boreal regions. *Arctic* **25,** 153-154.
Larsen, J. A. (1974). Ecology of the northern continental forest border. *In* "Arctic and Alpine Environments" (J. D. Ives and R. G. Barry, eds.), pp. 341-369. Methuen, London.
Llano, G. A. (1956). Utilization of lichens in the arctic and subarctic. *Econ. Bot.* **10,** 367-392.
MacLean, D. W. (1960). Some aspects of the aspen-birch-spruce-fir type in Ontario. Dep. For., For. Res. Div., Tech. Note 94, pp. 5-24.
Maini, J. S., and Carlisle, A., eds. (1976). "Conservation in Canada—A Conspectus." Dep. Environ., Canada For. Ser., Ottawa.
Manning, G. H. (1975). The Canadian softwood lumber industry: A model. *Can. J. For. Res.* **5,** 345-351.
Manning, G. H., and Grinnell, H. R. (1971). Forest resources and utilization in Canada to the year 2000. *Can. For. Serv., Publ.* **1304,** 1-180.
Miller, W. S., and Auclair, A. N. (1974). Factor analytic models of bioclimate for Canadian forest regions. *Can. J. For. Res.* **4,** 536-548.
Mitchell, V. L. (1973). A theoretical tree-line in central Canada. *Ann. Assoc. Am. Geogr.* **63,** 296-301.
Mosquin, T. (1966). Reproductive specialization as a factor in the evolution of the Canadian flora. *In* "The Evolution of Canada's Flora" (R. L. Taylor and R. A. Ludwig, eds.), pp. 43-65. Univ. of Toronto Press, Toronto.
Nelson, R. K. (1973). "Hunters of the Northern Forest." Univ. of Chicago Press, Chicago, Illinois.
Polunin, N. (1935). The vegetation of Akpatok Island. II. *J. Ecol.* **23,** 161-209.
Prebble, M. L., ed. (1976). "Aerial Control of Forest Insects in Canada." Dep. Environ., Can., For. Serv., Ottawa.
Pruitt, William O., Jr. 1978. "Boreal Ecology." Edward Arnold, London.
Raizada, H. C., and Nautiyal, J. C. (1974). An input-output model of Ontario based industries. *Can. J. For. Res.* **4,** 372-380.
Rapport, D. J., and Turner, J. E. (1977). Economic models in ecology. *Science* **195,** 367-373.
Raup, H. M. (1941). Botanical problems in boreal America. *Bot. Rev.* **7,** 147-248.
Ritchie, J. C. (1977). Northern fiction—Northern homage. *Arctic* **31,** 69-74.
Rouse, W. R., and Kershaw, K. A. (1971). The effects of burning on the heat and water regimes of lichen-dominated subarctic surfaces. *Arct. Alp. Res.* **3,** 291-304.
Rouse, W. R., and Kershaw, K. A. (1971). The effects of burning on the heat and water regimes of lichen-dominated subarctic surfaces. *Arct. Alp. Res.* **3,** 291-304.
Rowe, J. S. (1972). "Forest Regions of Canada." *Can. For. Serv., Dep. Environ.* **1300,** 1-172.
Rowe, J. S. (1970). Spruce and fire in northwest Canada and Alaska. *Proc. Annu. Tall Timbers Fire Ecol. Conf., Tall Timbers Res. Sta., Tallahasse, Florida* 245-254.
Rowe, J. S. (1961). Critique of some vegetational concepts as applied to forests of northwestern Alberta. *Can. J. Bot.* **39,** 1007-1017.

Sater, J. E., Ronhovde, A. G., and Van Allen, L. C. (1971). "Arctic Environment and Resources." Arctic Inst. of N. Am., Montreal, Canada.
Savile, D. B. O. (1963). Factors limiting the advance of spruce at Great Whale River, Quebec. *Can. Field Nat.* **77**, 95–97.
Speiss, A. E. (1979). "Reindeer and Caribou Hunters." Academic Press, New York.
Strang, R. M. (1973). Succession in unburned sub-Arctic woodlands. *Can. J. For. Res.* **3**, 140–142.
Stupart, R. F. (1928). The influence of arctic meteorology on the climate of Canada especially. *In* "Problems of Polar Research" (W. L. G. Joerg, ed.), pp. 39–50. Am. Geogr. Soc. Spec. Publ. 7, New York.
Tamm, C. O., ed. (1975). Man and the boreal forest. *Swed. Nat. Res. Counc., Ecol. Bull.* **21**, 1–153.
Tedrow, J. C. F., and Hill, D. E. (1955). Arctic brown soil. *Soil Sci.* **80**, 265–275.
Telfer, E., and Scotter, G. W. (1975). Potential for game ranching in boreal aspen forests of western Canada. *J. Range Manage.* **28**, 172–180.
Usher, P. J. (1976). Evaluating country food in the northern native economy. *Arctic* **29**, 105–120.
van Groenewoud, H. (1971). Microrelief, water level fluctuations, and diameter growth in wet-site stands of red and black spruce in New Brunswick. *Can. J. For. Res.* **1**, 359–366.
Vasilyev, P. N. (1971). Forest resources and forest economy. *In* "Natural Resources of the Soviet Union" (I. P. Gerasimov, D. L. Armand, and K. M. Yefron, eds.), pp. 187–216. Freeman, San Francisco, California.
Vincent, A. B. (1965). Growth of black spruce and balsam fir reproduction under speckled alder. *Can. For. Br. Publ.* **1102**, 7–14.
Walters, C. J., Hilborn, R., and Peterman, R. (1975). Computer simulation of barren-ground caribou dynamics. *Ecol. Model.* **1**, 303–315.
Warren Wilson, J. (1967). Ecological data on dry-matter production by plants and plant communities. *In* "The Collection and Processing of Field Data" (E. F. Bradley and O. T. Denmead, eds.), pp. 77–127. Wiley (Interscience), New York.
Weetman, G. F., and Webber, B. (1972). The influence of wood harvesting on the nutrient status of two spruce stands. *Can. J. For. Res.* **2**, 351–369.
Went, F. W. (1950). The response of plants to climate. *Science* **112**, 489–494.

11 Epilogue

It is inevitable that the boreal forest will be subjected to accelerating forces of change within the next few decades, and it is of considerable importance that its ecology be understood in detail so that conservation and management measures can be effectively based on sound knowledge. It is conceivable that irreparable damage might be the consequence of ignorance, perhaps of what may now seem some minor and insignificant aspect of the relationship between vegetation and physical environment; conservation and rational land management programs always must be undertaken in the light of fundamental scientific principles, and the latter can be revealed and understood only by means of ample data and sound analysis. Fortunately, there is yet time to obtain much of the needed data on the boreal forest plant and animal communities, in decided contrast to the situation existing in many other parts of the world.

Within what is left of this century, Hare and Ritchie (1972) point out, the boreal forest of North America will almost certainly be subjected to massive economic invasion:

> Already the southern margin has suffered widespread cutting by loggers who practice their craft with the ruthlessness so typical of our continent's pioneers. The tundra to the north is threatened by the search for oil and gas. Communication lines increasingly thread the forest. The last great cohesive wilderness is about to be irreversibly altered. There is little time left to reveal its nature.

Ecosystem modeling has become an important tool in assessing the benefits, in terms of quality of human life, of continued industrialization. As Passmore (1974) writes, it is not absurd to hope that the Western industrialized nations will be able to find ways to solve the major environmental problems. It is encouraging to note that philosophers, such as Passmore, have begun to analyze the assumptions involved when value judgments are made concerning the consequences of disposal of wastes into sea and air, destruction of ecosystems, human population growth, resource depletion, and expansion of industrial activity. The issues are large, complex, and difficult, but awareness is the first step in identifying rational approaches to the problems; there is, it is encouraging to see, a general awareness that there is a point at which industrial growth reduces rather than enhances the quality of life in a region. The world's ecological problems can only be solved, Passmore points out, by "that old-fashioned procedure, thoughtful action." The intellectual and political freedom of the West, as well as traditional scientific technological ingenuity, are a decided advantage in the effort.

The task, however, is not a small one and, as Ritchie (1977) adds, it is compounded by many "treacherous questions of politics ... the loss of credibility and independence by the scientific community, particularly that segment concerned with the environment."

It is clearly apparent that much can be done to ameliorate the environmental injury caused by exploitation, providing sound research is done on the methods to be used and scientific rather than political judgments prevail in the planning and administration of the environmental management efforts (Bliss, 1978).

I was not able in the space available here to touch as much as I would have liked on the historical interaction between humans and the boreal forest. There are, however, increasing numbers of such studies forthcoming, and it is to be hoped that many more will appear in the future. There are also interesting and valuable studies available on the primitive prehistoric relationships between humans and nature in the north; those by Gordon (1977) and Bergerud (1974) and others, for example, describe the ecological relationships, if they can be so called, between humans and caribou, the social organizations of both, and the interactions between them. There will be much interesting and valuable knowledge to be had from studies of man and the environment in boreal regions before the advent of industry. The knowledge will bear directly on the more intensive management of land use that will be needed in the future. The scientific base for this view has been aptly summarized by Kellman (1975)*:

> The earth's plant cover is now seen to be composed of species populations whose structures are seldom uniform and whose evolutionary history is rarely as narrowly channeled as once supposed. Environmental conditions impinging upon these species populations have been shown to be intricate and species population ranges rarely stable. Vegetation history increasingly reveals varied patterns of change and the presence of unfamiliar species assemblages in the past. Past human influences upon the earth's plant cover now appear to have been complex. In sum, the bases of deductive systems have been eroded by the complexity of detail in form and process now revealed in the earth's plant cover.

*Reproduced by permission of the Methuen & Co. Ltd.

> Recent research on the earth's plant cover reflects this methodological transformation. The frontiers of the field no longer comprise descriptions of vegetation in formalized modes. Instead, the frontiers lie in detailed treatments of the species populations that constitute this plant cover, their patterns, control mechanisms and modes of integration into intricate and continually evolving assemblages. . . .

In writing the first chapter of this exploration into boreal systematics I had the temerity to reproduce an equation of Margalef's, indicating that it represented the goal of systems analysis—to integrate the community of individuals, each bounded in its behavior by the limits of its capacities as a member of its species, taking continually into consideration the range of environmental influences that continually are at work upon the community. I am now reminded of a conversation I had in regard to that equation with a graduate student, more mathematical than I, who expressed a great esthetic and philosophical appreciation of the equation—with which I was completely in accord. But then, in one simple statement, he put the whole picture in much better and more realistic perspective. "How," he asked, "do you put the numbers into it?"

It is obvious, of course, that putting the numbers into it remains to be done. I have devoted several chapters in this book, and two decades of summer fieldwork in boreal forest and tundra, to the kind of effort that one day will permit the putting of numbers into the equation. The goal conceptually is in sight, but it is distant and surely will not be reached in my lifetime. But it is, I believe, attainable, and it is, I believe, worth the effort. Mankind must sooner or later take a firm grip on its environment and its destiny and put hard thought and solid science behind its venture on earth. Soft and unrigorous attitudes toward man and nature, and this includes many traditional deeply ingrained, long-cherished beliefs and sentimental attitudes toward nature, no longer suffice; men and women are demonstrably capable of hard and rational thought, and they will ultimately prevail only if they seize upon the advantages that rational thought offers. The relationship of mankind to nature is, in the light of our traditions, ambiguous; we are part of nature, we say, and yet we can control nature. We can understand nature, and we can exploit, and devastate, and crush if we are wont to do so, or if we are simply unthinking and careless. We can build a long and enduring future, however, if we use knowledge in a rational way—to ask intelligent questions, seek answers in intelligent ways, and use intellectual capacities—as incorporated into the sciences—for the attainment of rational goals and a sensible balance of human life with the resources available.

Systems analysis will be of enormous assistance in this effort. It is, in a sense, the only method yet available for carrying out the advanced accounting of global resources. But the reader's perusal of these chapters will lead to the impression that the principles of that accounting have not yet been revealed in anything resembling the working body of knowledge that will be required. We do not yet know how to put numbers into the equations in anything resembling the detail that eventually will be required. The knowledge, however, will accumulate.

The goal is in sight, but to attain it will in some respects require traversing a labyrinth of ignorance and, perhaps even at times, superstition. There is, however, adventure and intellectual excitement in store for those who will embark upon the quest of seeking a path through the maze. It will be adventure of a scientific kind. The life will be that of the mind. The satisfaction will come from having added to the store of ecological knowledge. It is the new adventure—the life of the mind and a quest for understanding of life and of the nature of the universe.

REFERENCES

Bergerud, A. T. (1974). Decline of caribou in North America following settlement. *J. Wildl. Manage.* **38,** 757–770.

Bliss, L. C. (1978). The report of the Mackenzie Valley pipeline inquiry, Volume I, II: An environmental critique. *Musk-Ox* **21,** 28–38.

Gordon, B. C. (1977). Prehistoric Chipewyan harvesting at a barrenland caribou water crossing. *West. Can. J. Anthropol.* **7,** 69–83.

Hare, F. K., and Ritchie, J. C. (1972). The boreal bioclimates. *Geogr. Rev.* **62,** 334–365.

Kellman, M. C. (1975). "Plant Geography." St. Martin's, New York.

Passmore, J. (1974). "Man's Responsibility for Nature." Scribner's, New York.

Ritchie, J. C. (1977). Northern fiction–Northern homage. *Arctic* **31,** 69–74.

Appendix
I *Analysis of Boreal Soils*

TABLE I.1
Chemical Composition (%) of More Abundant Parent Rock Types[a,b]

Chemical composition	SiO_2	Al_2O_3	Fe_2O_3	FeO	CaO	MgO	Na_2O	K_2O	TiO_2	P_2O_5	MnO
1. Granodiorite	72.95	14.11	0.37	1.89	1.46	0.56	4.29	2.80	0.19	0.06	0.06
2. Granite	72.87	13.58	0.41	1.38	3.32	0.67	2.81	3.58	0.26	0.33	0.04
3. Granite	75.95	11.71	1.83	1.35	0.24	0.24	3.37	4.36	0.20	0.02	0.06
4. Quartz feldspar	73.09	14.01	0.32	0.89	2.05	0.37	4.15	1.63	0.20	0.05	0.02
5. Quartz monzonite	72.79	14.62	0.14	1.39	1.16	0.33	4.19	4.45	0.12	0.11	0.05
6. Greywacke	62.61	16.78	0.38	5.81	2.80	2.99	3.79	1.56	0.68	0.16	0.08
7. Conglomerate	61.96	17.20	1.42	4.49	1.00	3.27	5.27	2.04	0.60		0.10
8. Shale	60.79	16.79	1.05	5.67	0.96	2.76	2.80	3.77	0.80	0.12	0.04
9. Clay	58.90	26.63	1.40		0.56	0.16	0.42	0.31	1.25		0.01
10. Clay	48.58	14.10	4.94		9.38	4.82	1.67	2.78			
11. Argillite	58.55	18.95	0.51	5.16	2.70	2.70	1.85	4.07	0.53	0.17	
12. Greenstone	49.24	10.35	2.07	12.18	8.52	8.55	2.38	1.39	1.60		0.09
13. Schist	47.50	17.03	4.15	6.67	14.90	6.70	0.76	0.76	0.58	0.07	0.15
14. Gneiss	63.84	20.34	3.34	3.98	0.64	2.20	0.95	2.42	0.80		
15. Gneiss	53.36	16.40	1.02	6.49	12.57	3.17	1.51	1.40	0.56		0.26
16. Metabasalt	49.86	15.07	2.39	9.63	9.64	6.34	1.76	0.43	0.53	0.20	0.37

[a] Data from J. A. Maxwell et al. (1965), reproduced with permission from *Geol. Surv. Can. Bull.* **115**.
[b] Areas represented are 1, Yellowknife; 2, North Manitoba; 3, Labrador; 4, Yellowknife; 5, Quebec; 6, Yellowknife; 7, Ontario; 8, Ontario; 9, Ontario; 10, Ontario; 11, Manitoba; 12, Ontario; 13 Ontario; 14, Cochrane District; 15, Northern Manitoba; 16, Quebec. Numbers 1–5 are igneous rock types, 6–10 are sedimentary rocks, and 11–16 are metamorphic rocks.

TABLE I.2

Chemical Analysis of Mineral Fraction (%) of Representative Soils: Podzolic Soils

	Gray-brown podzolic[a]				Brown-wooded[b]				Gray-wooded[b]			
	A_1	A_2	B_1	B_2	C	A_2	B	C	A_2	B_{21}	B_{22}	C
SiO_2	74.0	77.8	76.1	69.8	53.7	67.8	68.6	62.6	79.3	73.5	72.7	65.7
Al_2O_3	8.4	9.2	11.0	14.2	10.0	13.6	13.2	9.3	8.9	12.5	12.2	8.4
Fe_2O_3	3.3	2.9	4.0	6.2	4.5	5.0	5.2	4.0	3.6	4.7	4.8	3.4
CaO	0.9	0.5	0.7	0.6	9.0	1.0	1.0	7.6	0.5	0.6	0.8	7.6
MgO	0.8	0.8	1.0	1.4	5.8	1.4	1.5	2.2	0.8	1.3	1.2	1.9
Na_2O	0.9	1.1	1.0	0.7	0.7	0.6	0.6	0.6	0.7	0.7	0.6	0.6
K_2O	1.8	2.0	2.0	1.7	1.9	2.3	2.2	1.9	1.8	2.2	2.1	1.5
TiO_2	0.8	0.7	0.7	0.6	0.5	0.7	0.7	0.6	0.6	0.6	0.6	0.5
P_2O_5	0.35	0.09	0.08	0.07	0.09	0.13	0.10	0.16	0.14	0.10	0.12	0.16
MnO	0.13	0.14	0.12	0.11	0.09							

[a] Tavernier and Smith (1957), with permission of Academic Press.
[b] Wright et al. (1959). Reprinted from *Can. J. Soil Sci.* **39**, 32–43, published by the Agricultural Institute of Canada.

TABLE I.3

Chemical Analyses of Mineral Fraction (%) of Representative Soils:[a] **Podzol Soils**

	Podzol[b]				Podzol[c]			Podzol[d]		
	A_1	A_2	B_1	B_2	C	A_1	B	C	A_2	B	C
SiO_2	52.9	83.3	69.6	72.7	77.9	64.5	71.9	73.5	79.0	69.0	75.5
Al_2O_3	7.0	6.7	9.6	10.3	10.0	11.2	12.7	12.0	9.8	14.0	11.1
Fe_2O_3	1.1	1.7	4.0	3.6	3.1	2.8	2.3	2.0	1.3	4.1	3.0
CaO	0.9	0.5	0.6	0.6	0.5	2.8	1.8	2.9	0.4	1.8	1.7
MgO	0.1	0.2	0.3	0.4	0.5	1.5	0.6	0.9	0.5	0.7	0.6
Na_2O	0.4	0.5	0.5	0.7	0.5						
K_2O	2.1	2.9	3.4	3.4	3.8	2.1	1.0	2.9	1.7	2.4	2.4
TiO_2	0.7	0.9	0.8	0.7	0.5	1.0	1.6	1.3	0.7	2.1	1.0
P_2O_5	0.13	0.04	0.08	0.08	0.08	0.29	0.14	0.32	0.01	0.64	0.20
MnO	0.01	0.01	0.01	0.02	0.03						

[a] Remaining nonmineral material is the organic fraction of the soil.
[b] Cited in Jenny (1941): data from C. F. Marbut (1935).
[c] Wilde (1946). Weakly podzolic loam with maple–basswood forest.
[d] Wilde (1946). Hardpan podzol with hemlock, yellow birch, balsam fir forest.

Appendix

II Broad Geographical Species Relationships

By means of a factor analysis the species in the understory of the black spruce communities of the various study areas in central Canada (shown in Fig. 3) were grouped according to the frequency with which they occur. The species making up these groups show similar performance in terms of frequency, i.e., attaining highest frequency together in certain of the study areas (and lowest in other study areas). Thus, there is a group of species (making up the first factor in the analysis) that occur with highest frequency in the southern and eastern portions of the region in which the study areas are located, hence positively correlated with air masses of eastern origin and negatively correlated with air masses originating in the northwest and moving eastward (see Chapter 7 and Larsen 1971). In the table presented here, the species in a number of the more important groups are listed, and the areas in which they attain highest frequency are named. The species are listed in order of declining magnitude of importance in the group; i.e., those with highest factor scores are given first. The groups are also listed in order of declining magnitude of importance, i.e., declining magnitude of variance explained.

Table II.1

Factor 1 (areas with largest number of species present and in highest frequencies—Klotz, West Hawk, Clear)

Aralia nudicaulis
Anemone quinquefolia
Fragaria virginiana
Mitella nuda
Aster macrophyllus
Coptis groenlandica
Streptopus roseus
Trientalis borealis
Maianthemum canadense
Galium triflorum
Rubus pubescens
Clintonia borealis

Factor 2 (Areas—Ilford, God's, Lynn)

Salix bebbiana
Viburnum edule
Epilobium angustifolium
Salix myrtillifolia

Factor 3 (Area—Ennadai)

Potentilla palustris
Carex brunnescens
Salix planifolia
Larix laricina (seedlings)
Vaccinium uliginosum

Factor 4 (Areas—Waskesiu, Clear)

Lathyrus ochroleucus
Pyrola asarifolia
Salix discolor
Aster ciliolatus
Linnaea borealis
Petasites palmatus

Factor 5 (Area—Klotz)

Diervilla lonicera
Viola renifolia
Oryzopsis asperifolia
Aster macrophyllus
Trientalis borealis
Streptopus roseus

Factor 6 (Areas—Rocky, Ilford, Lynn)

Galium boreale
Arctostaphylos rubra
Kalmia polifolia
Geocaulon lividum

TABLE II.1—*Continued*

Factor 7 (Areas—Trout, Bemidji)

Gaultheria procumbens
Smilacina trifolia

Factor 8 (Areas—Ilford, Lynn, Rocky, West Hawk, Klotz, Trout, Remi, Bemidji, Clear, God's)

Chamaedaphne calyculata
Oxycoccus microcarpus
Rubus chamaemorus
Kalmia polifolia

Appendix
III Community Composition

As described in the text (see Chapter 6) there are groups of understory species that tend to be found growing together in the communities of a given area. The tables that follow give the correlations between frequency values of species sampled in the study sites shown in Fig. 27 (the 44 study sites from Hudson Bay to the western Cordillera). The groups of species listed in the following tables are those found to have high correlation with an air mass frequency: species highly correlated positively with arctic air masses increase in frequency toward the northeast, those correlated positively with air masses of Alaska–Yukon origin increase in frequency toward the northwest, those correlated positively with Pacific air increase toward the southwest, and those correlated with southern air masses increase in frequency toward the southeast. Since the species in the groups are positively correlated with one or another of the air mass types, there are species groupings in which all the species are correlated with one another. The high degree of correlation between the species in these groups serves to demonstrate that vegetational community composition varies along a spatial continuum that, in the case of the species listed, appears to be a response to climate-related environmental gradients. For a discussion of the methods employed in the study, see Larsen, (1971). The data given here are previously unpublished.

TABLE III.1

Correlation between Frequency of Species[a]

	Aralia	Aster	Coptis	Cornus	Linnaea	Maianthemum	Mertensia	Mitella	Oryzopsis	Trientalis
Anemone quinquefolia	0.45	0.96	0.93	0.72	0.53	0.65	0.42	0.72	0.73	0.88
Aralia nudicaulis		0.43	0.42	0.67	0.77	0.76	0.83	0.88	0.28	0.73
Aster macrophyllus			0.89	0.67	0.51	0.71	0.38	0.67	0.70	0.86
Coptis groenlandica				0.83	0.53	0.66	0.36	0.67	0.66	0.87
Cornus canadensis					0.71	0.76	0.58	0.76	0.51	0.81
Linnaea borealis						0.64	0.71	0.87	0.40	0.73
Maianthemum canadense							0.61	0.72	0.42	0.73
Mertensia paniculata								0.76	0.55	0.63
Mitella nuda									0.66	0.63
Oryzopsis asperifolia										0.62
Trientalis borealis										

[a] Average 0.66. A number of species found in considerable abundance in the southeastern portion of the study region show similar behavior in the frequencies with which they appear in the black spruce communities of the various study areas. Eleven species showing the highest intercorrelation are given.

TABLE III.2
Intercorrelations among Those of a Group of Species in White Spruce Communities Characterized by High Correlations with Alaska–Yukon Air Masses[a]

	Carex	Carex	Dryas	Equisetum	Eriophorum	Larix	Saussurea
Arctostaphylos alpina[b]	0.91	0.90	0.92	0.78	0.87	0.55	0.88
Carex scirpoidea		0.81	0.88	0.60	0.99	0.50	0.99
Carex vaginata			0.82	0.69	0.74	0.36	0.98
Dryas integrifolia				0.52	0.83	0.31	0.89
Equisetum scirpoides					0.57	0.84	0.54
Eriophorum spissum						0.53	0.99
Larix laricina							0.50
Saussurea angustifolia							

[a] Correlations are given between frequencies of species showing the highest intercorrelations.
[b] Arctous alpina, some authors. Average 0.74.

TABLE III.3

Intercorrelations Among Those of a Group of Species in White Spruce Communities Characterized by High Correlations with Arctic Air Masses[a]

	Empetrum	Ledum	Ledum	Salix	Vaccinium	Vaccinium
Betula glandulosa	0.91	0.77	0.83	0.82	0.88	0.53
Empetrum nigrum		0.79	0.87	0.86	0.93	0.66
Ledum decumbens			0.67	0.51	0.94	0.56
Ledum groenlandicum				0.88	0.86	0.63
Salix glauca					0.76	0.47
Vaccinium uliginosum						0.61
Vaccinium vitis-idaea						

[a] Average 0.75. Shown are correlations between frequencies of species affording the highest degree of intercorrelation.

TABLE III.4

Intercorrelations among Those of a Group of Species in Jack Pine Communities Characterized by High Correlations with Southern Air Masses[a]

	Aster	Cornus	Diervilla	Galium	Gaultheria	Lycopodium	Maianthemum	Oryzopsis	Trientalis
Alnus rugosa	0.08	0.33	0.21	0.67	0.06	0.33	0.23	0.07	0.07
Aster ciliolatus		0.84	0.80	0.29	0.76	0.78	0.61	0.79	0.78
Cornus canadensis			0.51	0.30	0.44	0.52	0.47	0.48	0.49
Diervilla lonicera				0.41	0.92	0.98	0.76	0.98	0.97
Galium triflorum					0.02	0.58	0.43	0.21	0.22
Gaultheria procumbens						0.83	0.65	0.98	0.97
Lycopodium obscurum							0.65	0.92	0.92
Maianthemum canadense								0.71	0.73
Oryzopsis asperifolia									0.99
Trientalis borealis									

[a] Average 0.57. Shown are correlations between frequencies of species affording the highest degree of intercorrelations.

Appendix IV Frequency of Occurrence of Lichens

During the course of sampling of vegetational communities in northern Canada, the author made some studies on the frequency of occurrence of lichen species in a number of the communities. The following tables give the relative frequency with which the lichen species indicated occurred in the communities. The lichen identifications have been made by J. W. Thomson of the University of Wisconsin, and gratitude is hereby expressed for his contribution to the research.

TABLE IV.1

Ground Lichen species in Black Spruce Communities[a]

Species	Area in which sampled community is located											Percentage presence in stands
	Otter (20)[b]	Otter (25)	Otter (35)	Wapata (85)	Wapata (45)	Wapata (80)	Wapata (90)	Wapata (80)	Wapata (100)	Waskesiu (85)	Providence (100)	
Cetraria nivalis	x											9
Cladonia alpestris		x	x	x	x				x			45
C. amaurocraea												9
C. coccifera							x			x		18
C. cornuta							x	x			x	27
C. crispata				x		x	x					27
C. cristatella							x					9
C. deformis							x			x		18
C. gonecha				x				x				18
C. gracilis var. chordata							x					9

458

Species	1	2	3	4	5	6	7	8	9	10	11	%
C. gracilis var. dilata	x		x			x	x		x			45
C. mitis	x	x	x	x	x	x	x	x			x	90
C. mitis f. setigera				x								
C. rangiferina	x		x	x		x	x				x	45
C. sylvatica						x					x	9
C. uncialis			x	x		x					x	27
Icmadophila ericetorum												
Lecidea berengeriana				x								9
Peltigera aphthosa	x		x		x	x			x		x	45
P. canina									x			9
P. malacea	x		x						x			18
P. polydactyla		x								x		18
P. pulverulenta				x	x	x						18

[a] Tabulated are the species of ground lichens found in 11 mature stands dominated by black spruce located at Otter, Wapata, and Waskesiu lakes in northern Saskatchewan, as well as Fort Providence in the Northwest Territories of Canada.

[b] The percentage frequency with which lichens occurred in transects of 20 one-m² quadrats is given in parentheses. The percentage refers to the presence of the lichen species in one or more quadrats of the transect (an x indicates the species was present).

TABLE IV.2

Species of Ground Lichens Found in Six Mature Stands Dominated by Jack Pine, Located at Otter, Wapata, and Waskesiu Lakes in Northern Saskatchewan

Species	Otter (65)[a]	Wapata (85)	Wapata (100)	Wapata (95)	Black (70)	Waskesiu (80)	Percentage presence
Cetraria nivalis	x						16
Cladonia alpestris	x				x		33
C. amaurocraea	x		x		x		50
C. botrytes				x			16
C. coccifera					x		16
C. cornuta			x	x			33
C. crispata			x	x			33
C. cristatella				x			16
C. deformis				x			16
C. gonecha				x			16
C. gracilis var. *chordata*			x	x	x		33
C. gracilis var. *dilatata*			x	x			33
C. mitis	x	x		x	x	x	100
C. subulata			x				16
C. uncialis	x			x	x		50
C. verticillata				x			16
Lecidea berengeriana				x			16
Peltigera malacea		x			x		33
P. polydactyla		x					16
Stereocaulon tomentosum					x		16
S. paschale						x	16
Stereocaulon sp./spp.	x						16

[a] The percentage frequency with which lichens occurred in transects of 1-m² quadrats is given in parentheses. There were 20 quadrats to a transect. The percentage presence refers to presence of the lichen species in the transect at one or more quadrats.

TABLE IV.3

Corticolous Lichen Community Found in a Black Spruce Stand in the Lynn Lake Area[a]

Species[b]	Stand 1	Stand 2
Alectoria americana	13	
A. glabra	48	90
A. nidulifera	91	
Cetraria ciliaris	74	95
C. pinastri	4	80
C. sepincola		5
Crocynia neglecta	13	22
Evernia mesomorpha	91	95
Lecanora coilocarpa	4	
L. pinastri		5
L. varia		63
Lecidea vernalis	21	
Parmelia physodes	74	95
P. sulcata	39	42
Parmeliopsis aleurites	54	
P. ambigua	43	5
P. hyperopta	4	11
Usnea cavernosa	8	
U. comosa	26	11
U. glabrata	57	42
U. glabrescens	13	42
U. glabrescens subsp. *glabrella*	4	
U. hirta	30	
U. scabrata subsp. *nylanderiana*	4	
U. sorediifera		11

[a] Lichens were collected from spruce boles at breast height, from a foot-high cylinder around the tree. Percentage frequency in 23 quadrats in stand 1 and 19 quadrats in stand 2 is shown.

[b] Species present in one quadrat on jack pine present in the same stand were *Alectoria nidulifera, A. glabra, Cetraria ciliaris, Evernia mesomorpha, Lecanora varia, Parmelia physodes, P. sulcata, Parmeliopsis ambigua, P. aleurites,* and *Usnea glabrata.*

TABLE IV.4

Corticolous Lichen Community on Black Spruce in the Lynn Lake Area[a]

Species	Stand 1	Stand 2
Alectoria americana	15	
A. glabra	23	83
A. nidulifera	69	58
Bacidia chlorantha		9
Cetraria ciliaris	60	100
C. pinastri	7	75
Cyphelium tigillare	7	
Evernia mesomorpha	75	96
Lecanora coilocarpa	15	17
L. pinastri		9
L. varia		28
Lecidea euphorea		9
Parmelia physodes	46	100
P. sulcata	53	66
Parmeliopsis aleurites	30	9
Usnea comosa	53	17
U. glabrata	54	28
U. glabrescens	23	17
U. hirta	53	33

[a] The lichens were those occurring on black spruce twigs and were sampled from a single branch at breast height. There were 13 trees in stand 1 and 12 trees in stand 2.

Appendix V
Species Frequently in Black Spruce Communities

Table V.1

Species Occurring with Sufficient Frequency to be Recorded in Quadrat Data from Transects Run in Black Spruce Communities[a] by Various Workers[b]

A. Southeastern (southern Quebec, eastern Ontario, Maritimes)

Acer spicatum
Alnus rugosa
Amelanchier bartramiana
Anemone quinquefolia
Aquilegia canadensis
Aralia nudicaulis
Aster acuminatus
Aster macrophyllus
Carex eburnea
Carex pensylvanica
Carex trisperma
Carex woodii
Chrysosplenium americanum
Circaea alpina
Clintonia borealis
Coptis groenlandica
Corylus cornuta
Cypripedium reginae
Diervilla lonicera
Dryopteris cristata
Dryopteris thelypteris
Epigaea repens
Fragaria virginiana
Galium asprellum
Galium palustre
Galium triflorum
Gaultheria hispidula
Gaultheria procumbens

Geum rivale
Iris versicolor
Kalmia angustifolia
Lonicera canadensis
Lonicera dioica
Lonicera villosa
Lycopodium clavatum
Lycopodium sabinaefolium
Monotropa uniflora
Nemopanthus mucronatus
Oryzopsis asperifolia
Osmunda claytoniana
Oxalis montana
Pteridium aquilinum
Rhododendron canadense
Salix discolor
Sambucus pubens
Solidago hispida
Solidago macrophylla
Sorbus americana
Sorbus decora
Streptopus roseus
Trientalis borealis
Vaccinium angustifolium
Vaccinium caespitosum
Viburnum cassinoides
Viola incognita

(continued)

TABLE V.1—*Continued*

B. Southwestern (Alberta, southwestern Mackenzie) (Species marked with an "n" are of more northern affinity.)

Achillea millefolium
Andromeda polifolia (n)
Arctostaphylos rubra (alpina) (n)
Arnica cordifolia
Astragalus umbellatus
Carex aquatilis
Carex capillaris (n)
Carex gynocrates (n)
Carex leptalea
Carex lugens
Carex media
Carex vaginata
Dryas octopetala (n)
Elymus innovatus
Epilobium glandulosum
Equisetum arvense
Equisetum fluviatile
Equisetum pratense
Equisetum scirpoides
Empetrum nigrum (n)
Eriophorum brachyantherum
Festuca altiaca (F. scabrella)
Galium boreale
Habenaria hyperborea
Hedysarum alpinum
Juniperus communis
Ledum decumbens (n)

Listera borealis
Loiseleuria procumbens (n)
Lonicera glaucescens
Lonicera involucrata
Menyanthes trifoliata
Mertensia paniculata
Pedicularis labradorica (n)
Petasites sagittatus
Potentilla fruticosa
Potentilla palustris
Pyrola virens
Ranunculus lapponicus
Rhododendron lapponicum (n)
Rosa acicularis
Salix alaxensis (n)
Salix arbusculoides
Salix candida
Salix glauca (n)
Salix myrtillifolia
Salix pedicellaris
Salix richardsonii (n)
Salix scouleriana (n)
Shepherdia canadensis
Smilacina racemosa
Solidago multiradiata
Symphorycarpos albus

C. Southern central (western Ontario, southern Manitoba, and Saskatchewan). (Species often wide-ranging east and west.)

Actaea rubra
Alnus crispa (n)
Amelanchier alnifolia
Anemone parviflora
Aster ciliolatus
Aster puniceus
Betula glandulosa (n)
Calamagrostis canadensis
Caltha palustris
Carex brunnescens (n)
Carex canescens (n)
Carex chordorrhiza (n)
Carex disperma (n)
Carex paupercula (n)
Carex rariflora (n)
Carex rotundata (n)
Carex saxatilis (n)

Chamaedaphne calycantha
Cinna latifolia
Corallorhiza trifida
Cornus canadensis
Cornus stolonifera
Drosera rotundifolia
Dryopteris spinulosa
Empetrum nigrum (n)
Epilobium angustifolium
Equisetum sylvaticum
Eriophorum spissum
Galium triflorum
Gaultheria hispidula
Geocaulon lividum (n)
Geum macrophyllum
Goodyera repens
Habenaria dilatata

TABLE V.1—*Continued*

Habenaria obtusata
Kalmia polifolia
Ledum groenlandicum
Linnaea borealis
Listera cordata
Luzula parviflora
Lycopodium annotinum
Lycopodium complanatum
Lycopodium obscurum
Maianthemum canadense
Mitella nuda
Moneses uniflora
Myrica gale
Orchis rotundifolia
Parnassia palustris
Petasites palmatus
Pinguicula vulgaris
Polygonum amphibium
Pyrola asarifolia
Pyrola minor
Pyrola secunda
Rhamnus alnifolia
Ribes glandulosum

Ribes hudsonianum
Ribes lacustre
Ribes triste
Rubus acaulis
Rubus chamaemorus
Rubus pubescens
Rubus strigosus (idaeus)
Salix bebbiana
Salix planifolia
Scirpus caespitosus
Smilacina trifolia
Spiranthes romanzoffiana
Tofieldia pusilla
Vaccinium myrtilloides
Vaccinium oxycoccus
Vaccinium uliginosum (n)
Vaccinium vitis-idaea
Viburnum edule
Viburnum trilobum
Viola adunca
Viola pallens
Viola palustris
Viola renifolia

D. Northeastern (Labrador–Ungava and northern Quebec)

Arabis alpina
Arenaria humifusa
Arenaria sajanensis
Athyrium filix-femina
Bartsia alpina
Cardamine bellidifolia
Carex bigelowii
Carex lachenalii (biparta)
Carex terrae-novae (glacialis)
Carex williamsii
Cassiope hypnoides
Castilleja septentrionalis
Cerastium alpinum
Clintonia borealis
Cystopteris montana
Deschampsia flexuosa
Diapensia lapponica
Dryopteris disjuncta
Dryopteris phegopteris
Dryopteris spinulosa
Epilobium anagallidifolium
Gaultheria hispidula
Gnaphalium norvegicum
Gnaphalium supinum

Habenaria dilatata
Heracleum maximum
Juncus trifidus
Lonicera villosa
Luzula parviflora
Lycopodium alpinum
Oxyria digyna
Parnassia kotzebuei
Pedicularis groenlandica
Phyllodoce caerulea
Poa alpina
Prunus pensylvanica
Ranunculus allenii
Ribes glandulosum
Salix cordifolia
Salix herbacea
Salix humilis
Salix pellita
Salix vestita
Saxifraga aizoides
Stellaria crassifolia
Streptopus amplexifolius
Taraxacum lapponicum
Trientalis borealis

(continued)

TABLE V.1—Continued

Trisetum spicatum
Vaccinium angustifolium

E. Northwestern (northwestern Mackenzie)

Anemone drummondii
Aster sibiricus
Calamagrostis purpurascens
Carex aquatilis
Carex capillaris
Carex capitata
Carex gynocrates
Carex lugens
Cassiope tetragona
Castilleja raupii
Cypripedium passerinum
Dryas octopetala
Equisetum scirpoides
Festuca altiaca
Hedysarum sp.
Myrica gale

Veronica alpina

Orchis rotundifolia
Pedicularis sp.
Potentilla fruticosa
Rhododendron lapponicum
Salix alaxensis
Salix arctica
Salix arctophila
Salix glauca
Salix reticulata
Salix richardsonii
Salix scouleriana
Saussurea angustifolia
Senecio lugens
Shepherdia canadensis
Silene acaulis
Symphoricarpos albus

F. Northern central (northern Ontario, Manitoba, Saskatchewan, eastern Mackenzie, Keewatin). (Species often wide-ranging east and west.)

Achillea borealis
Alnus crispa
Agrostis borealis
Andromeda polifolia
Anemone parviflora
Arctostaphylos uva-ursi
Arctostaphylos alpina
Betula glandulosa
Calamagrostis canadensis
Carex brunnescens
Carex canescens
Carex capillaris
Carex chordorrhiza
Carex deflexa
Carex disperma
Carex leptalea
Carex limosa
Carex paupercula
Carex rariflora
Carex rotundata
Carex saxatilis
Carex scirpoidea
Carex tenuiflora
Carex vaginata
Chamaedaphne calyculata
Coptis groenlandica
Cornus canadensis

Drosera rotundifolia
Dryas integrifolia
Empetrum nigrum
Epilobium angustifolium
Equisetum pratense
Equisetum sylvaticum
Erigeron lonchophyllus
Eriophorum spissum
Galium trifidum
Geocaulon lividum
Habenaria obtusata
Hierochloe alpina
Juniperus communis
Kalmia polifolia
Ledum decumbens
Ledum groenlandicum
Linnaea borealis
Loiseleuria procumbens
Luzula confusa
Luzula parviflora
Lycopodium annotinum
Lycopodium clavatum
Lycopodium complanatum
Lycopodium selago
Menyanthes trifoliata
Mitella nuda
Moneses uniflora

TABLE V.1—*Continued*

Pedicularis labradorica	Salix bebbiana
Petasites palmatus	Salix herbacea
Pinguicula villosa	Salix myrtillifolia
Pinguicula vulgaris	Salix pedicellaris
Poa glauca	Salix planifolia
Polygonum viviparum	Saxifraga tricuspidata
Potentilla palustris	Scirpus caespitosus
Potentilla tridentata	Smilacina trifolia
Pyrola grandiflora	Solidago multiradiata
Pyrola secunda	Spiranthes romanzoffiana
Pyrola virens	Stellaria calycantha
Ranunculus lapponicus	Stellaria longipes
Ribes hudsonianum	Tofieldia pusilla
Ribes triste	Vaccinium microcarpus
Rubus acaulis	Vaccinium uliginosum
Rubus strigosus (idaeus)	Vaccinium vitis-idaea
Rubus chamaemorus	Viburnum edule
Salix arbusculoides	

[a] The species are grouped by region here; central zone listings (sections C and F) include those species that are wide-ranging east and/or west of the central zone in North America and those with circumboreal range.

[b] Key to sources (complete reference citations may be found in appropriate chapters): Southern areas—Baldwin (1958), Ontario, species listing by community; Christensen et al, (1959), northern Wisconsin, quadrat studies by community; Damman (1964), central Newfoundland, cover estimates by community; Davis (1966), Maine, species listing by community (red spruce dominant); La Roi (1967), continent-wide: divided into eastern, central, and western for the purposes here, quadrat studies by community; Horton and Lees (1961), Alberta, constancy classes in 119 stands; Hustich (1954), Knob Lake (Quebec–Labrador), species listing from sample plots by community; Hustich (1955), Ontario, species listing from sample plots by community; Hustich (1957), Ontario, species listing by community; Jurdant (1964), Quebec, species listing by community; Lambert-Day* (unpubl.), Ontario, quadrat studies by community; Larsen, 1965, 1971a,b, 1973, Ontario, Manitoba, Mackenzie, Great Slave Lake, Keewatin, quadrat studies by community; Linteau (1955), Quebec, species listing by community; Moss (1955), Alberta, species listing by community; Ritchie (1956), northern Manitoba, quadrat studies by community; Thieret (1964), Yellowknife Highway, species listing by community; Wilton (1964), Labrador, species listing by community; Zoltai* (unpubl.), Ontario, quadrat studies by community.
* Gratitude is expressed for the use of these data.

Northern areas—Argus (1966), northeastern Saskatchewan, species listing by community; Baldwin (1953), northern Manitoba, species listing by community; Cody (1960), Norman Wells, Mackenzie, species listing; Drew and Shanks (1965), Firth River valley (Alaska–Canada), species listing by community; Harper (1964), Ungava, species listing by community; Johnson and Vogel (1966), Alaska, species listing by community; Larsen (1965), Ennadai Lake, Keewatin, quadrat studies by community; Larsen (unpubl.), Inuvik, quadrat studies by community; Larsen 1971, Great Slave Lake (Ft. Reliance, Artillery Lake), quadrat studies by community; Lutz (1956), Alaska, species listing by community; Maini (1966), southeastern Mackenzie, species listing by community; Raup (1946), Athabasca, Great Slave Lake, species listing by community; Ritchie (1959), northern Manitoba, quadrat studies by community.

Appendix VI Species in Boreal Forest Literature

Listed on the following pages are those species to which reference is made with some degree of frequency in the literature on the boreal forest. Common names are given only when they can be said to apply to a single species without serious ambiguity. The nomenclature followed for the most part is that of H. J. Scoggan in "The Flora of Canada," published in 1978 by the National Museum of Natural Sciences, National Museums of Canada, Ottawa. Scoggan's four-volume work is comprehensive and provides detailed keys, although other sources must be consulted for full descriptions of the individual species. The more useful of these latter sources are listed by Scoggan (Volume I, Introduction, page 1) and include such well-known manuals as "Gray's Manual of Botany," eighth edition, as revised by M. L. Fernald, published by the American Book Company; "Manual of Vascular Plants of Northeastern United States and Adjacent Canada," Henry A. Gleason and Arthur Cronquist, D. Van Nostrand Company; "Circumpolar Arctic Flora," Nicholas Polunin, Oxford University Press, "Flora of Manitoba," H. J. Scoggan, National Museums of Canada; "Flora of Alaska and Neighboring Territories," Eric Hulten, Stanford University Press. A handy references for the trees is "Native Trees of Canada," Bulletin, 61, Canada Department of Forestry; this contains many illustrations of diagnostic features of the species, as well as a list of confusing synonyms for the species. It was noted in Chapter 5 that a number of the species have intermediate forms that result from hybridization and introgression of morphological characteristics, and these species in the listing are designated by the term "complex" following the species name. For a complete discussion of the more common of the species in which this taxonomically confusing phenomenon occurs, and its implications for field identification of species in ecological studies, the reader is referred to the discussion in Chapter 5.

TABLE VI.1

Genus/Species	Common name
Abies balsamea (L.) Mill	Balsam fir
Abies grandis (Dougl.) Lindl.	Grand fir
Abies lasiocarpa (Hook.) Nutt.	Subalpine fir, Alpine fir
Acer negundo L.	Box elder
Acer spicatum Lam.	Mountain maple
Achillea millefolium L. var. *borealis* (Bong.) Farw. (*A. borealis*)	
Achillea millefolium L.	Common yarrow
Achillea sibirica Ledeb.	
Acorus calamus L.	Sweetflag
Actaea rubra (Ait.) Willd.	Red baneberry
Agastache foeniculum (Pursh) Ktze.	Blue giant hyssop
Agropyron trachycaulum (Link) Malte	
Agrostis borealis Hartm.	
Agrostis hyemalis (Walt.) BSP. (*A. scabra*)	Hairgrass
Alnus crispa (Ait.) Pursh	Green alder
Alnus rugosa (Du Roi) Spreng.	Speckled alder
Amelanchier alnifolia Nutt. (*A. florida*)	Saskatoon berry
Amelanchier bartramiana (Tausch) Roemer	
Andromeda glaucophylla Link	Bog rosemary
Andromeda polifolia L.	
Anemone canadensis L.	Canada anemone
Anemone cylindrica Gray	Thimbleweed
Anemone drummondi Wats.	
Anemone multifida Poir.	
Anemone parviflora Michx.	
Anemone patens L. (*Pulsatilla ludoviciana*)	Pasque flower
Anemone quinquefolia L.	American wood anemone
Anemone riparia Fern.	Riverbank anemone
Antennaria neglecta Greene (*A. campestris*)	
Antennaria neglecta Greene (*A. petaloidea*)	
Apocynum androsaemifolium L.	Spreading dogbane
Aquilegia brevistyla Hook.	
Aquilegia canadensis L.	Wild columbine
Arabis alpina L.	
Aralia nudicaulis L.	Wild sarsaparilla
Aralia racemosa L.	Spikenard
Arctostaphylos alpina (L.) Spreng. (*Arctous alpina*)	Alpine bearberry
Arctostaphylos alpina ssp. *rubra* (Rehd. & Wils.) Hult (*A rubra*)	
Arctostaphylos uva-ursi (L.) Spreng.	Common bearberry
Arctous (see *Arctostaphylos*)	
Arenaria groenlandica (Retz.) Spreng.	Greenland sandwort
Arenaria humifusa Wahl.	

(continued)

TABLE VI.1—*Continued*

Genus/Species	Common name
Arenaria lateriflora L.	Grove sandwort
Arenaria sajanensis Willd.	
Armeria maritima (Miller) Willd.	
Arnica chamissonis Less.	
Arnica cordifolia Hook.	
Arnica lonchophylla Greene	
Artemisia frigida Willd.	Prairie sandwort
Aster acuminatus Michx.	Prairie sandwort
Aster borealis (T. & G.) Provancher (*A. junciformis*)	
Aster ciliolatus Lindl.	
Aster laevis L.	
Aster macrophyllus L.	
Aster puniceus L.	
Aster sibiricus L.	
Aster umbellatus Mill.	
Astragalus adsurgens Pallas (*A. striatus*)	
Astragalus alpinus L.	
Astragalus americanus (Hook.) M. E. Jones	
Astragalus umbellatus Bunge	
Athyrium filix-femina (L.) Roth	Lady fern
Bartsia alpina L.	Alpine bartsia
Betula glandulosa Michx. (*B. glandulosa* complex)	Dwarf birch
Betula lutea Michx.	Yellow birch
Betula occidentalis Hook.	Black, red, or mountain birch
Betula papyrifera Marsh.	White, paper, canoe birch
Bidens cernua L.	Sticktight
Bromus ciliatus L.	
Bromus inermis Leyss (*B. pumpellianus*)	
Calamagrostis canadensis (Michx.) Nutt.	Blue joint
Calamagrostis purpurascens R. Br.	
Caltha palustris L.	Cowslip
Calypso bulbosa (L.) Oakes	Calypso
Campanula rotundifolia L.	
Cardamine bellidifolia L.	
Carex aenea Fern.	
Carex aquatilis Wahl.	
Carex aquatilis var. *stans* (Drej.) Boott (*C. stans*)	
Carex argyrantha Tuckerm. (*C. foena*)	
Carex atherodes Spreng.	
Carex atrofusca Schkuhr.	
Carex bigelowii Torr.	
Carex brunnescens (Pers.) Poir.	
Carex canescens L.	

TABLE VI.1—*Continued*

Genus/Species	Common name
Carex capillaris L.	
Carex capitata L.	
Carex chordorrhiza L. f.	
Carex concinna R. Br.	
Carex deflexa Hornem.	
Carex diandra Schrank	
Carex disperma Dewey	
Carex eburnea Boott	
Carex foenea Willd. (*C. siccata*)	
Carex glacialis Mack.	
Carex gynocrates Wormsk.	
Carex interior Bailey	
Carex lachenalii Schkuhr.	
Carex lacustris Willd.	
Carex lasiocarpa Ehrh.	
Carex leptalea Wahl.	
Carex limosa L.	
Carex lugens Holm.	
Carex media R. Br.	
Carex membranaceae Hook.	
Carex microglochin Wahl.	
Carex misandra R. Br.	
Carex pauciflora Lightf.	
Carex paupercula Michx.	
Carex pensylvanica Lam.	
Carex pseudo-cyperus L.	
Carex rariflora (Wahl.) Sm.	
Carex rostrata Stokes	
Carex rotundata Wahl. (*C. rotundata* complex)	
Carex saxatilis L.	
Carex scirpoidea Michx.	
Carex siccata Dewey (*C. foenea, C. argyrantha*)	
Carex stans (see *C. aquatilis* var. *stans*)	
Carex tenuiflora Wahl.	
Carex trisperma Dewey	
Carex vaginata Tausch	
Carex williamsii Britt.	
Carex woodii Dewey	
Cassiope hypnoides (L.) D. Don	Moss-like mountain heather
Cassiope tetragona (L.) D. Don	White arctic bell heather
Castilleja raupii Pennell	
Castilleja pallida (L.) Spreng. ssp. *septentrionalis* (Lindl.)	Scoggan northern painted cup
Cerastium alpinum L.	

(continued)

TABLE VI.1—Continued

Genus/Species	Common name
Chamaedaphne calyculata (L.) Moench	Leatherleaf
Chimaphila umbellata (L.) Bart.	Pipsissewa
Chrysosplenium alternifolium L.	
Chrysosplenium americanum Schwein.	
Cinna latifolia (Trev.) Griseb.	
Circaea alpina L.	Enchanter's nightshade
Clintonia borealis (Ait.) Raf.	Bluebead lily
Comandra umbellata (L.) Nutt. var. *pallida* (DC.) Jones	Bastard toadflax
Comptonia peregrina (L.) Coult	Sweet fern
Coptis trifolia (L.) Salisb. (*C. groenlandica*)	Goldthread
Corallorhiza maculata Raf.	Spotted coral root
Corallorhiza striata Lindl.	Striped coral root
Corallorhiza trifida Chat.	Early coral root, pale coral root
Cornus canadensis L.	Bunchberry
Cornus stolonifera Michx.	Red osier
Corylus cornuta Marsh.	Beaked hazelnut
Cypripedium acaule Ait.	Stemless lady's slipper
Cypripedium passerinum Richards.	Sparrow's egg lady's slipper
Cypripedium reginae Walt.	Showy lady's slipper
Cystopteris bulbifera (L.) Bernh.	Bulblet fern
Cystopteris montana (Lam.) Bernh.	Mountain bladder fern
Danthonia spicata (L.) Beauv.	Poverty grass
Deschampsia flexuosa (L.) Trin.	Common hairgrass
Diapensia lapponica L.	
Diervilla lonicera Mill.	Bush honeysuckle
Drosera rotundifolia L.	Round leaved sundew
Dryas drummondii Richards.	
Dryas integrifolia Vahl	
Dryas octopetala L.	
Dryopteris (see *Thelypteris, Gymnocarpium*)	
Dryopteris austriaca (Jacq.) Woynar	Spinulose shield fern
Dryopteris cristata (L.) Gray	Crested wood fern
Dryopteris robertiana (see *Gymnocarpium*)	
Eleocharis palustris (L.) R. & S.	
Elymus innovatus Beal	
Empetrum nigrum L., s.l.	Crowberry
Epigaea repens L.	Trailing arbutus
Epilobium angustifolium L.	Fireweed
Epilobium glandulosum Lehm.	
Equisetum arvense L.	Common horsetail
Equisetum fluviatile L.	Water horsetail
Equisetum hyemale L.	Scouring rush
Equisetum palustre L.	Marsh horsetail

TABLE VI.1—Continued

Genus/Species	Common name
Equisetum pratense Ehrh.	Meadow horsetail
Equisetum scirpoides Michx.	Dwarf scouring rush
Equisetum sylvaticum L.	Wood horsetail
Erigeron compositus Pursh (*E. trifidus*)	
Erigeron glabellus Mutt.	
Erigeron lonchophyllus Hook.	
Eriophorum angustifolium Honckeny	
Eriophorum vaginatum L. (*E. spissum*)	
Eupatorium purpureum L. (*E. maculatum*)	Joe pye weed
Festuca altaica Trin.	Rough fescue
Festuca ovina L. (*F. saximontana*)	Sheep's fescue
Festuca rubra L.	Red fescue
Fragaria vesca L.	Woodland strawberry
Fragaria virginiana Dcne.	
Galium asprellum Michx.	Rough bedstraw
Galium boreale L. (*G. septentrionale*)	Northern bedstraw
Galium labradoricum Wieg.	
Galium palustre L.	Marsh bedstraw
Galium trifidum L.	
Galium triflorum Michx.	Sweet scented bedstraw
Gaultheria hispidula (L.) Muhl. (*Chiogenes hispidula*)	Creeping snowberry
Gaultheria procumbens L.	Teaberry, Checkerberry
Gentiana amarella L. (Boerner)	
Gentiana propinqua (Richards.) Gillett	
Geocaulon lividum (Richards.) Fern.	Northern comandra
Geum canadense Jacq.	
Geum macrophyllum Willd.	
Geum rivale L.	Water avens, purple avens
Glyceria borealis (Nash) Batch.	Small floating manna grass
Gnaphalium norvegicum Gunner.	
Goodyera repens (L.) R. Br.	Dwarf rattlesnake plantain
Gymnocarpium (see also *Thelypteris*)	
Gymnocarpium dryopteris (L.) Newm. (*D. disjuncta*)	Oak fern
Gymnocarpium robertianum (Hoffm.) Newm.	Northern oak fern
Habenaria dilatata (Pursh) Hook.	Leafy white orchis
Habenaria obtusata (Pursh) Richards.	Blunt leaf orchis
Habenaria orbiculata (Pursh) Torr.	Round leaved orchis
Habenaria viridis (L.) R. Br.	Frog orchis
Hedysarum alpinum L.	
Hedysarum boreale Nutt. var. *mackenzii* (Richards.) Hitchc.	
Heracleum lanatum Michx.	Cow parsnip
Heuchera richardsonii R. Br.	Alumroot

(*continued*)

TABLE VI.1—*Continued*

Genus/Species	Common name
Hieracium canadense Michx. (*Hieracium umbellatum*)	
Hierochloë alpina (Sw.) R. & S.	
Hippuris vulgaris L.	Mare's tail
Hudsonia tomentosa Nutt.	Beach heath
Hypopitys monotropa Crantz (see *Monotropa*)	Pinesap
Impatiens capensis Meerb.	Snapweed
Iris versicolor L.	Blue flag
Juncus castaneus Sm.	
Juncus trifidus L.	
Juncus triglumis L. (*J. albescens*)	
Juniperus communis L.	Common juniper
Juniperus horizontalis Moench	Creeping savin
Kalmia angustifolia L.	Sheep laurel
Kalmia polifolia Wang.	Bog laurel
Kobresia simpliciuscula (Wahlenb.) Mackenzie	
Larix laricina (Du Roi) Koch	Tamarack, American larch
Larix lyallii Parl.	Alpine, mountain, or lyall's larch
Larix occidentalis Nutt.	Western larch
Lathyrus japonicus Willd.	Beach pea
Lathyrus ochroleucus Hook.	Pale vetchling
Lathyrus palustris L.	Vetchling
Lathyrus venosus Muhl.	
Ledum palustre L. (*Ledum decumbens*)	
Ledum groenlandicum Oeder	Labrador tea
Lemna minor L.	Duckweed
Lemna trisulca L.	Star duckweed
Linnaea borealis L.	Twin flower
Linum perenne L. (*L. lewisii*)	Perennial flax
Listera borealis Morong	Northern twayblade
Listera cordata (L.) R. Br.	Heartleaf twayblade
Loiseleuria procumbens (L.) Desv.	Alpine azalea
Lonicera canadensis Bartr.	Fly honeysuckle
Lonicera dioica L. (*L. glaucescens*)	Limber honeysuckle
Lonicera involucrata (Richards.) Banks	Black twinberry
Lonicera villosa (Michx.) R. & S.	Mountain fly honeysuckle
Lupinus arcticus S. Watson	
Luzula confusa Lindeberg	
Luzula nivalis (Laest.) Beurl.	
Luzula parviflora (Ehrh.) Desv.	
Lychnis apetala L. (*Melandrium apetalum*)	
Lycopodium alpinum L.	Alpine club moss
Lycopodium annotinum L.	Stiff club moss

TABLE VI.1—*Continued*

Genus/Species	Common name
Lycopodium clavatum L.	Common club moss
Lycopodium complanatum L.	Ground cedar
Lycopodium obscurum L.	Ground pine
Lycopodium sabinaefolium Willd.	Ground fir
Lycopodium selago L.	Mountain club moss
Lycopodium tristachyum Pursh	Ground cedar
Lysimachia ciliata L.	Fringed loosestrife
Maianthemum canadense Desf.	Wild lily of the valley
Melampyrum lineare Desr.	
Melandrium apetalum (see *Lychnis*)	
Mentha arvensis L.	Mint
Menyanthes trifoliata L.	Buckbean
Mertensia paniculata (Ait.) G. Don	Lungwort
Mitella nuda L.	Mitrewort
Moneses uniflora (L.) Gray	One flowered pyrola
Monotropa hypopithys L. (*Hypopithys monotropa*)	Pinesap
Myosotis sylvatica Hoffm. var. *alpestris* (Schmidt) Koch (*M. alpestris*)	
Myrica gale L.	Sweet gale
Myriophyllum alternifolium DC.	
Myriophyllum spicatum L. (*M. verticillatum*)	
Nemopanthus mucronata (L.) Trel.	Mountain holly
Orchis rotundifolia Banks	Small round leaved orchis
Oryzopsis asperifolia Michx.	
Oryzopsis pungens (Torr.) Hitchc.	
Osmorhiza longistylis (Torr.) DC.	Anise root
Osmorhiza depauperata Phil (*O. obtusata*)	
Osmunda claytoniana L.	Interrupted fern
Oxalis montana Raf. (*O. acetosella*)	Pink wood sorrel
Oxycoccus microcarpus Turcz. (*Vaccinium oxycoccus* complex; *Oxycoccus* complex)	
Oxytropis campestris (L.) DC.	
Oxytropis leucantha (Pall.) Pers. (*O. viscida*)	
Oxytropis splendens Dougl.	
Parnassia kotzebuei Cham. & Schlecht.	Kotzebue's grass of parnassus
Parnassia palustris L. (*P. multiseta*)	
Pedicularis capitata Adams	
Pedicularis groenlandica Retz.	
Pedicularis labradorica Wirsing	
Pedicularis lanata Cham. & Schlecht.	Woolly lousewort
Pedicularis lapponica L.	
Petasites frigidus (L.) Fries	
Petasites palmatus (Ait.) Gray	

(*continued*)

TABLE VI.1—*Continued*

Genus/Species	Common name
Petasites sagittatus (Banks) Gray	
Petasites vitifolius Greene	
Phragmites australis (Cav.) Trin.	Reed
Phyllodoce caerulea (L.) Bab.	Purple mountain heather
Picea engelmannii (Parry) Engelm.	Engelmann spruce
Picea glauca (Moench) Voss	White spruce
Picea mariana (Mill.) BSP.	Black spruce
Picea rubens Sarg.	Red spruce
Picea sitchensis (Bong.) Carr.	Sitka spruce
Pinguicula villosa L.	
Pinguicula vulgaris L.	Common butterwort
Pinus albicaulis Engelm.	White bark pine
Pinus banksian Lamb.	Jack pine
Pinus contorta Dougl.	Lodgepole pine
Pinus flexilis James	Limber pine
Pinus resinosa Ait.	Red pine, Norway pine
Pinus strobus L.	Eastern white pine
Poa alpina L.	
Poa glauca Vahl.	
Poa palustris L.	
Polygala paucifolia Willd.	Fringed polygala
Polygonum amphibium L.	Water smartweed
Polygonum bistorta L.	Bistort
Polygonum viviparum L.	Alpine bistort
Populus balsamifera L.	Balsam poplar
Populus grandidentata Michx.	Large toothed aspen
Populus tremuloides Michx.	Trembling aspen
Potamogeton vaginatus Turcz.	
Potentilla fruticosa L.	Shrubby cinquefoil
Potentilla nivea L.	
Potentilla palustris (L.) Scop.	Marsh five finger
Potentilla tridentata Ait.	Three toothed cinquefoil
Prenanthes alba L.	White lettuce
Prunus pensylvanica L.f.	Pin cherry
Prunus virginiana L.	Choke cherry
Pseudotsuga menziesii (Mirb.) Franco	Douglas fir
Pteridium aquilinum (L.) Kuhn	Bracken
Pulsatilla (see *Anemone*)	
Pyrola asarifolia Michx.	Pink pyrola
Pyrola grandiflora Radius	Arctic pyrola
Pyrola minor L.	
Pyrola secunda L.	One sided pyrola
Pyrola virens Schweigger	
Ranunculus allenii Robins.	

TABLE VI.1—*Continued*

Genus/Species	Common name
Ranunculus gmelinii DC.	Small yellow water crowfoot
Ranunculus lapponicus L.	Lapland buttercup
Rhamnus alnifolia L'Her.	Alder leaved buckthorn
Rhododendron canadense (L.) Torr. (*Rhodora canadense*)	Rhodora
Rhododendron lapponicum (L.) Wahl.	Lapland rose bay
Ribes glandulosum Grauer	Skunk currant
Ribes hudsonianum Richards.	
Ribes lacustre (Pers.) Poir.	Bristly currant
Ribes oxyacanthoides L.	Canada gooseberry
Ribes triste Pallas	Red currant
Rorippa islandica (Oeder) Borbas	
Rosa acicularis Lindl. (*R. acicularis* complex)	Prickly rose
Rubus acaulis Michx.	
Rubus chamaemorus L.	Baked apple berry
Rubus idaeus L.	Red raspberry
Rubus pubescens Raf.	Dwarf raspberry
Rumex arcticus Trautv.	
Salix alaxensis (Anderss.) Cov.	
Salix arbusculoides Anderss.	
Salix arbutifolia Pallas	
Salix arctica Pallas	Arctic willow
Salix arctophila cockerell	
Salix bebbiana Sarg.	Long beaked willow
Salix brachycarpa Nutt.	
Salix candida Fluegge	Hoary willow
Salix cordata Michx.	Heartleaf willow
Salix discolor Muhl.	Pussy willow
Salix glauca L. (*S. glauca* complex)	
Salix herbacea L.	
Salix humilis Marsh.	
Salix myrtillifolia Anderss.	
Salix pedicellaris Pursh	Bog willow
Salix pellita Anderss.	
Salix petiolaris Sm.	
Salix phylicifolia L. (*S. planifolia*)	
Salix pyrifolia Anderss.	Balsam willow
Salix reticulata L.	
Salix richardsonii Hook.	
Salix scouleriana Barratt	
Salix uva-ursi Pursh	
Salix vestita Pursh	
Sambucus canadensis L.	Common elder
Sambucus racemosa L. (*S. pubens*)	Red berried elder
Sanguisorba canadensis L.	

(*continued*)

TABLE VI.1—*Continued*

Genus/Species	Common name
Sanicula marilandica L.	Black snakeroot
Sarracenia purpurea L.	Pitcher plant
Saussurea angustifolia (Willd.) DC.	
Saxifrage aizoides L.	Yellow mountain saxifrage
Saxifraga aizoön Jacq.	
Saxifraga tricuspidata Rottb.	
Scheuchzeria palustris L.	
Schizachne purpurascens (Torr.) Swallen.	
Scirpus lacustris L. (*S. acutus*)	
Scirpus caespitosus L.	
Scolochloa festucacea (Willd.) Link	
Selaginella selaginoides (L.) Link	
Selaginella sibirica (Milde) Hieron.	
Senecio integerrimus Nutt. (*S. lugens*)	
Senecio pauperculus Michx.	
Senecio tridenticulatus Rydb.	
Shepheridia canadensis (L.) Nutt.	Soapberry
Silene acaulis L.	Moss campion
Silene repens Patrin	
Sium suave Walt.	Water parsnip
Smilacina racemosa (L.) Desf.	False spikenard
Smilacina trifolia (L.) Desf.	
Solidago gigantea Ait.	
Solidago hispida Muhl.	
Solidago macrophylla Pursh	
Solidago multiradiata Ait.	
Solidago nemoralis Ait.	
Sorbus americana Marsh.	American mountain ash
Sorbus decora (Sarg.) Schneid	Mountain ash
Sorbus scopulina Greene	
Spiranthes romanzoffiana Cham.	Hooded ladies' tresses
Stachys palustris L.	Woundwort
Steironema ciliatum (L.) Raf. (see *Lysimachia*)	
Stellaria calycantha (Ledeb.) Bong.	
Stellaria crassifolia Ehrh.	
Stellaria longifolia Muhl.	
Streptopus amplexifolius (L.) DC.	
Streptopus roseus Michx.	
Symphoricarpos albus (L.) Blake	Snowberry
Symphoricarpos occidentalis Hook.	Wolfberry
Tanacetum huronense Nutt.	
Taraxacum lapponicum Kihlm.	
Taxus canadensis Marsh	Ground hemlock, American yew

TABLE VI.1—*Continued*

Genus/Species	Common name
Thalictrum dasycarpum Fisch. & Lall.	Purple meadow rue
Thalictrum venulosum Trel.	
Thelypteris (see also *Gymnocarpium*)	
Thelypteris palustris (Salisb.) Schott (*Dryopteris thelypteris*)	Marsh fern
Thelypteris phegopteris (L.) Slosson (*Dryopteris phegopteris*)	Long beech fern
Thuja occidentalis L.	White cedar
Tofieldia coccinea Richards.	
Tofieldia glutinosa (Michx.) Pers.	
Tofieldia pusilla (Michx.) Pers. (*T. palustris*)	
Trientalis borealis Raf.	Star flower
Trisetum spicatum (L.) Richter	
Tsuga canadensis (L.) Carr	Eastern hemlock
Tsuga mertensiana (Bong.) Sarg.	Mountain hemlock
Urtica dioica L.	Stinging nettle
Utricularia intermedia Hayne	
Utricularia minor L.	
Utricularia vulgaris L.	Common bladderwort
Vaccinium angustifolium Ait.	Low Sweet Blueberry
Vaccinium caespitosum Michx.	Dwarf bilberry
Vaccinium microcarpus (see *Oxycoccus* complex)	
Vaccinium myrtilloides Michx. (*V. canadense*)	Velvet leaf blueberry
Vaccinium oxycoccus complex (see *Oxycoccus* complex)	
Vaccinium uliginosum L.	Alpine bilberry
Vaccinium vitis-idaea L.	Rock cranberry
Valeriana dioica L. (*V. septentrionalis*)	Marsh valerian
Veronica peregrina L.	
Viburnum cassinoides L.	
Viburnum edule (Michx.) Raf.	Mooseberry
Viburnum opulus L. (*V. trilobum*)	Highbush cranberry
Vicia americana Muhl.	
Viola adunca Sm.	
Viola canadensis L. (*V. rugulosa*)	Canada violet
Viola incognita Brainerd	
Viola palustris L.	Northern marsh violet
Viola renifolia Gray	
Woodsia glabella R. Br.	Smooth woodsia
Zigadenus elegans Pursh	White camass

Appendix VII Special Definitions

This appendix presents terms having special definitions when employed in an ecological context. For other technical terms it is advisable to consult an appropriate textbook or technical dictionary.

Autecology The branch of ecology devoted to study of the relationships of individual species to the environment in which they are found growing. The study emphasizes life history, reproduction, nutrition, and so on.

Biomass Weight of living material existing at a given time, expressed usually on an areal basis; for example, kilograms per hectare. In some instances, it is the weight of organic material (of which all need not be considered actually living) existing in a given area at a given time. Thus, usually the weight of an entire tree is considered its biomass even though only the cambium may actually be considered living.

Catena See "Soil catena"

Climax community The stable end point of succession in which a self-perpetuating community of long duration is established. See "Succession." Here, too, the concept is a complex one both in theory and in the reality of nature (and subject to many modifications and variants, not to mention the controversies among ecologists) and textbook discussions should be consulted for further details.

Competition Competition exists when an organism attempting to obtain materials (nutrients) or to occupy sites where favorable conditions exist (light, temperature, living space) for life or reproduction meets resistance as the result of the attempts of other organisms to obtain the same materials or the same space.

Competitive exclusion A somewhat vague concept stating that two species that require very nearly the same materials and conditions for existence will not be able to survive (to "coexist") in a community together for an extended period of time; one eventually will exclude the other. While the concept may have some practical aspects in studies of communities, conditions (and species morphology and physiology) are usually more complex than can be easily understood or described, and the concept to date has had more theoretical than practical value and some ecologists have recently questioned the assumptions upon which the concept is based.

Continuum The variation in plant communities over distance is a response to gradually changing environmental conditions (see "Ecocline") and variations over short distances are slight and involve only a few species, but as distances increase the variation becomes more and more pronounced until there are few if any similarities. Thus, there is a continuity in the vegetation, but at increasing distances the similarities decline.

Ecocline A gradient in one or more ecological factors over a distance on the land surface (or beneath it in the case of soil or aquatic organisms). An example of an ecocline is a steady decline in moisture content of soil up a slope. Another is the temperature gradient from tropics to pole. A third example might be the gradient in light conditions from top to bottom of a forest canopy.

Indexes of Similarity See "Methods"

Methods The sampling procedures by which the species composition of a vegetational community is determined. The methods employed determine, for example, the numbers of individuals per given area (density) of each species, the frequency with which individuals of each species occur in quadrats of a given area (frequency), and the degree to which the various species dominate the vegetational community (dominance). In simple terms, these characteristics are measured as follows:

Density: measured by simply counting the plants present in a specified area occupied by the community. Often the specified area is 1 m^2, and density of each species would then be expressed as number of individuals per square meter. Sampling is usually conducted in a number of quadrats located randomly in the community some distance apart, and the density is obtained by averaging the values.

Frequency: measured by means of quadrats as in density above except that, instead of counting the number of individuals in each quadrat, a tabulation is made of the number of quadrats in which the species occurs (no attention being given to the numbers of individuals present). Thus, if individuals of a given species occur in 25 of 50 quadrats, the frequency is 50%. *Relative frequency* is obtained when each species frequency value for a given community is divided by the total; for example, if species A and B each have a frequency of 80 and species C has a frequency of 40, the relative frequencies, respectively, are 40, 40, and 20.

Dominance: measured directly by observing the total area of the ground surface occupied by the individuals of each species, usually expressed as a percentage (the term *cover* or *percentage cover* is often used for this value) or inferred indirectly by some arbitrary synthetic value such as density plus frequency plus basal area (for which the term *importance value* has at times been employed). Dominance values are a measure of the relative importance of each species in a community judged by whatever means may be employed.

Similarity indexes: many simple formulas are now available to measure the similarity between sets of numerical values. These can be used to measure the similarity existing between communities. One such index is described in the text.

Ordination: a technique by which similarity values for a group of communities can be arranged on graph paper, thus showing visually which communities are most similar and which are least similar. Those close together on the ordination are most similar, those farthest apart are least similar.

Principal component analysis: an analysis more mathematically sophisticated than ordination, used often to measure not only similarity between communities but their relationships to environmental factors. Ordination can be employed for this latter purpose also, but principal component analysis assigns numerical values to the relationships while ordination does not.

It must be understood that this is the briefest possible review of what is meant by *methods* or sampling methods and is meant to describe only in the most general fashion what is meant by the terms; a number of good books are available on the subject and should be consulted.

Nutrient cycling The movement of molecules (or atoms) of a given nutrient element follows a regular cycle; the nutrient elements are taken up by plants from the soil, then transferred to animals that eat the plants, and then released to the soil again when decomposition takes place. The cycle repeats when they are once again taken up by plant roots.

Productivity The weight of biological (organic) matter produced by a community over a given period of time, usually expressed as grams per meter or kilograms per hectare.

Replication Term used in the sampling of biotic communities to denote repetition of sampling procedures so as to obtain what is known as an "adequate" sample or, in other words, a sample that will truly represent the characteristics of the sampled community. The greater the number of samples (quadrat counts, for example) the greater the probability that the data represents the true characteristics of the sampled community.

Similarity See "Methods"

Soil catena Soil formation is considered to be the result of the influence of many factors, including the nature of the parent material from which it is formed, the climate, the vegetation present, the nature of the topography at a given site, and the elapsed time available for these factors to have worked their influence. A soil catena is a "slice" of soil, taken from the top to the bottom of a hillside which would be representative of the topography of the region under consideration; thus, all other factors are considered to be constant and only the topography (and the vegetation) to vary.

Statistical methods See "Methods"

Succession The tendency for the species composition of a community to change over periods of time, usually as a consequence of species adapted to pioneer open ground being replaced by species adapted to the environment found in communities that have advanced beyond the pioneer stages. Succession is complex both in theory and in the reality of the conditions found in nature, and more detailed textbook discussions should be consulted.

Synecology The branch of ecology devoted to study of the interrelationships among the species of biotic communities. The study emphasizes the communitywide influence of competition, successional trends, and environmental influences.

Trophic levels It is possible to categorize the organisms in a biotic community on the basis of the kind of nutrient material utilized. Called the food chain or energy cycle, this is usually represented by the sequence plants → herbivores → carnivores → decomposers → and back again to soil and plants. Each of these categories is termed a level. Thus herbivores are one level and carnivores another. The levels are often not clear-cut, however, since some animals are omnivorous.

Trophic model A diagrammatic representation of trophic levels in which the flow of energy (seen as arrows) indicates the direction of flow among the species of plants and animals and often the magnitude of the energy flow involved.

Subject Index*

A

Acidity, soil, 297, 303–307, *see also* Soil
Active layer, soil, 115–116
Air masses
  climatology of, 59–60
  compared to altitudinal climatic gradient, 149
  defined, 58
  frequency of specific types, 58–59
  frontal zones, 58–61, 67–68, 81, 196–197
  frontal zones in eastern and western Canada compared, 81, 83–84, 199–200
  related to vegetation, 58–59
  plant species distribution correlated with, 258–266
  storm tracks, patterns of, 59–60
Alaska, vegetational communities of, 137–139
Albedo
  bog and forest compared, 71–72
  characteristics of different vegetational types, 71–73
Algae, nitrogen-fixing, 356, 376
Allelopathy, 14
Ammonification, 120, 356
Altitudinal tree line, Labrador-Quebec, 192–193
Animals
  population cycles, 404–408
  populations
    influence of fire on, 403–404
    modeling of, 129
    trophic pyramid, 381–411

  succession, role in, 390
  wildlife, an ecosystem component, 390
Appalachian Mountains, vegetational communities, 224–225
Arctic tundra flora
  evolutionary history of, 23–30
  survival in refugia, 23–25
Artillery Lake
  vegetational communities, 154
  Pike's Portage route, 154
Arcto-Tertiary geoflora, *see* Evolutionary history
Aspen, trembling, quaking, *see Populus tremuloides*
Athabasca, Lake, regional vegetational communities, 54
Athabasca–Great Slave Lake, regional vegetational communities, 158–160
Atmospheric subsystem, 69–84
Avifauna
  biomass of, 390–392
  populations of common species, 394–395
  trophic pyramid, 382–383, 389

B

Basal area, forest trees, data on, 244–245
Bear, black, grizzly, role in ecosystem, 385–386
Beaver, role in ecosystem, 387
Biomass, birds; *see* Avifauna
Biomass
  plants and animals, compared, 382–383
  role in modeling, 129, 352
  systems ecology, role in, 129, 352
  trophic levels, 396–397

*For index to species see page 495.

483

Boreal flora
  circumpolar species, 140
  comparisons with tropical, 41
  evolution of, see Evolutionary history, Postglacial history
  northern forest–tundra, 196–198
  origins of, 21–23
  palynological record, 26–30
  postglacial history, 24–30; see also Evolutionary history, Postglacial history
  sympatric species, 140
Boreal forest, see also specific entries
  extent of, 1–5
  management and economics of, 412–439
  zonations in, analysis, 256–274
Botany, economic, boreal regions, 412–439
Brunisolic soil, 102–106
Budworm, spruce; see Spruce Budworm

C

Calcium, 299
  role in nutrient cycling and productivity, 351–380
Canada, eastern, regional vegetational communities, 176–188
Canada lynx
  population cycles in, 404–408
  role in ecosystem, 386–387
Canadian shield, vegetational communities, 161–176
Candelabra effect, in trees, 311–312
Candle Lake, Saskatchewan, vegetational communities in area, 170–171
Cape Breton–Newfoundland, regional vegetational communities, 180–182
Carbon, carbon cycle, 357
  role in nutrient cycling and productivity, 351–380
Carcajou Lake, regional vegetational communities, 145
Caribou, role in ecosystem, 384–385
Carnivores, component of trophic pyramid, 382–383
Circumpolar boreal forest, general, 133–134
Clay belt region, Ontario, vegetational communities, 170–171

Climate, see also Air masses, Temperature, and other climatic parameters
  arctic tree line, relationships with, 48–49
  classifications of, 238–239
  continentality, 63
  energy budget, 71–74
  factors, related to one another, 50, 78–79
  forest climatology, 67–69
  frontal zones, 58–61, 67–68, 196–197
  frontal zones, East and West compared, 81, 83–84
  frost-resistance in plants, 48
  gradients and vegetational communities, 130–131
  isotherms and vegetation, 47–48
  macroclimate, 49–50
    related to major biomes, 59–61
  microclimate, in northern zones, 81–83
  parameters, related; see factors, related
  permafrost, 74–76
  plant species frequency and, 240–244
  plant species
    geographical distribution, related to, 76–84, 161, 258–263
    responses to temperature, 20
  potential evapotranspiration, Labrador–Quebec, 177–178
  radiation and temperature, 50–51, 70–71
  relationships to other environmental factors, 62–63
  selected meteorological stations, data from, 52–57
  species response to, 20
  summer frosts, 311
  vegetational zones, determined by, 47–49, 258–266
  wildlife species, population cycles related to, 404–408
Climatic change, affecting forest communities, 37–40
Climatic regions
  moisture regimes of, 51
  radiation balance of, 50–51
  stations in
    grouping of, 79–80
    Canadian and Eurasian data compared, 66–67
Climatonomy, climatonomic analysis, 63

Climax
  in plant communities, 6
  Clementsian concepts, 276
  monoclimax theory, 316, 339
  subclimax concept, 416
  successional vectors, 334–335; see also Succession
Cloudiness, near large lakes, 62
Colville Lake, regional vegetational communities, 144
Communities, regional, see specific geographical areas
  changes with distance, 19–20
  climatic zones, related to, 79
  climax, subclimax communities related to environment, 7, 8, 107
  composition, structure, dynamics, 129, 285–341
  continuum in, see Continuum
  continuum controversy, 282–285
  distribution related to soil, 107
  dominant plant species in, 262–266
  environmental relationships of, 14–15, 186–187
  forest, regeneration of, in northern regions, 202–203
  forest-tundra ecotonal region, 36–40, 196–198
  lichen and moss, 208–224
  lowland black spruce forests, regional, compared, 190–191
  northern forest border, 80–83, 202–204, 418–421
  northern, in Labrador and Keewatin, compared, 203–204
  pioneer, 287–290
  postglacial, in palynological record, 26–30; see also Evolutionary history, Postglacial history
  regional relationships of, in Canada, 237–280
  response to soil moisture, 164–165, 287–295
  structure of, research needs, 422–423
  species groups in, 256–274
  transitional, 281
  tropical and boreal, compared, 41
  undisturbed, 284–285
  variation over distance, 130–131
  zonal divisions, analysis, 268–272
Competition
  adaptations to, 8, 307–313
  community relationships, role in, 257
  dynamics of, 129, 416–418
  moisture supply, related to, 296–297
  moss species, role of, 215–217
  root and shoot, 309
Continuum
  controversy regarding, 282–285
  density and frequency, species tabulations, 131
  fire, effect of, 131
  gradients in soils and plant communities, 110–111
  moss communities in, 215–216
  North American boreal continuum, 134–137
  ordination, demonstrates continuum, 286–287
  regional basis for concept of, 276
  related to climatic gradient, 6, 149
  research on, 237–238
  theory of, 111
  uniform continuum field, concept, 249
  vegetational climax, views on, 316
Cordillera, vegetational communities in, 139–140
Correlation, use of, in community studies, 286–287
Cycling, nutrient cycles, 351–380
Cycles, wildlife population
  related to fire, 406
  related to weather, 404–408

D

Decomposition, of organic matter, 91
Denitrification, 91, 356
Density, of plant species in communities, 244–245
Depauperate zone, floristic, in forest-tundra transition, 82–83
Dispersal, plant, rates of, 20–21
Dominance, relationships in major communities, 245–251, 262–266
  dominant species in major boreal communities, 269–273

E

Earth hummocks, 326
Eastern Canada, vegetational communities in, 176–188
Ecoclines, 6, 286–287
Ecological dynamics, general, 7–10
Ecological optima, 309
Ecosystems analysis, *see also* Modeling
  assumptions in, 128
  basic dynamics, 6
  basic elements of, 12–16
  economics of, 412–439
  energy flow in, 308–309
  functional aspects of, 128–130
Ecosystems
  boreal; *see also* Modeling
    general, 6
    human subsistence in, 430–433
    insects, role of, 401–403
    populations of birds and animals in, 381–411
    trophic levels in, 396–397
    wildlife, a component of, 390, 381–411
    wildlife, and palatable plant species, 430–433
  boreal model; *see also* Modeling
    forest management, role of in, 381–411
    problems in, 13–16
Economics, of boreal forest, 412–439; *see also* Forest economy
Ecotone, regional, northern forest border, 80–83, 202–204, 418–421
Ecotypic variation, 240–241
Edaphic factors, *see* Soils
Energy flow
  in ecosystems, 381–411
  through trophic levels, 396–397
Environmental factors, *see* individual entries
  acidity of soil, 303–307
  conditions, natural and greenhouse, compared, 309–310
  community response to, 286–287
  ecoclines, 6, 286–287
  field observations of, 312
  gradients in factors, gradient analysis, 285–287
  groundwater, 303–307
  hyperspace, expression of, 287

individual response of species to, 129–130
light and shade, 295–298
multifactorial in nature, 285–286
northern forest border, major factors in, 80–83, 202–204, 418–421
nutrients, 298–307
statistical treatment of, 286–287
Ericaceous shrubs, 306, 318
Eurasia, vegetational communities of, 225–226
Evapotranspiration, 73–74, 296, 329
  ecotypic adaptations related to, 74
  factors affecting rate of, 69
  potential, 78
  in Labrador–Ungava, 177–178
Evergreen bog plant species, 299–300
Evolutionary history, *see also* Postglacial history
  arctic tundra flora, 23–30
  boreal flora, 19–23
  Cretaceous period, 23
  fossil evidence of, 21–22
  migration of species, 25–27
  palynological record, 26–31
  Pleistocene epoch, 21–26
  Pliocene epoch, 21–23
  survival of species in refugia, 24–30

F

Factor analysis, employed in research, 245–251
Fire
  animal populations affected by, 403–404
  favorable conditions for, 76
  indicator species influenced by, 112
  Kalmia heath, 304–306
  litter and humus destroyed by, 138
  modeling of, 129
  role in forests of Upper Mackenzie River region, 156–159
  role in forests of northern Manitoba, 167–168
  role in vegetational dynamics, 131, 285, 317–319, 327–335
  nutrient cycling, influenced by, 366–368
  postglacial history of, 38–40

Subject Index

regeneration of plant species after, 138, 331–335
serotinous cones, role of, 137–138, 331
soil, influenced by, 104, 117
successional relationships of plants after, 314–341
wildlife, possible cause of cycles in, 406–407
Fisher, role in ecosystem, 388
Flood control, affect on moisture storage in forests, 113
Flooding, soil, 104
  related to sphagnum bogs, 158–159
Flora, boreal, *see* Communities, Evolutionary history
Florence Lake, regional vegetational communities, 145
Foliage, nutrient content of, 315
Foliar analysis, 281–307
Food chains, *see* Trophic levels
Forest, boreal, *see also* specific topic entries
  climatology of, Chapter 3
  communities, see entries for regional areas named
  communities, species in, 254–257
  economics of, 412–439
  exploitation and management, 426–430
  global boreal resource, 423–426
  human subsistence in, 431–433
  human survival and, 435, 440–443
  northern mesic species, found in boreal communities, 254–257
Forest economy, resources in, 412–439
  forest products, 433–444
  hydrocarbon fuel source, 434–435
  management of, 381–411
  native economy, 430–433
  plant nutrient management in, 376–377
  wildlife management in, 430–433
Forest growth
  rate at northern forest border, 81–82, 312–313, 412–442
  species indicators of site quality, 107–109
Forest-tundra transition zone, 80–84, 202–204, 418–421
Fort Reliance, regional vegetational communities, 145
Frequency, tabulation of in sampling, 249–250
Frontal zones, climatic, eastern and western Canada compared, 81, 83–84, 199–200
Frost, summer, 311
Fungi, role of in ecosystems, 129

G

Gaspé–Maritimes region, vegetational communities in, 178–180
Genecology, genetic variability and uniformity, 414–415
Genetic variability, see Hybridization
Geographical distribution of species, *see* Communities
Germination, 313
Glaze damage, 311–312
Gleysolic soil, 103–106
Gradient analysis
  direct and indirect, 285–286
  nutritional factors in soil, revealed by, 299–307
Great Bear Lake, regional vegetational communities, 144–145
Great Lakes Region, vegetational communities, 172–176
Great Slave Lake, regional vegetational communities, 154
Groundwater, environmental factor, 303–307
Grouse
  mortality in related to predator populations, 406
  population cycles in, 404–408
  role in ecosystem, 389
Growth rates, tree
  in northern latitudes, 49, 81–82
  nutritional environment, influence of, 298–307
  spruce, at northern limit, 312–313
  suppression, demonstrated, 308–310
Gypsy moth, 402

H

Hare, snowshoe
  population cycles in, 404–408
  role in ecosystem, 386
Hawks, population cycles in, 404–408
Heath, *Kalmia*, 304–306

Herbivores
  as component of trophic pyramid, 382–383
  succession, role of in, 390
Horizons, soil, 96–100
Hudson Bay lowlands, vegetational communities, 161–162, 169–170
Humidity, related to evapotranspiration, 69
Hummocks
  earth, 326
  vegetational cover on, 116
Hybridization, in plant species, 151–152, 201–202, 415

I

Importance value, data on, 245–251
Indicator species
  use of in soil classification, 106–110
  value of, 110–113
Insect infestation
  defoliators, 401–403
  modeling of, 129, 419
  soil, influenced by, 104
  spruce budworm, 392–393, 401–403
  succession, role of in, 314
Introgression, Hybridization
Inuvik, regional vegetational communities in area, 142–144
Invertebrates, role of in ecosystems, 129

K

Keweenaw peninsula, vegetational communities of, 173–174

L

Labrador–Quebec, altitudinal tree line in, 192–194
Labrador–Ungava, vegetational communities in, 177–188
Lesser Slave Lake, regional vegetational communities, 160
Lichens
  corticulous, 221–222
  distribution patterns of *Cladonia* species, 209–210
  distribution related to climate, 209–212, 221–222
  grazing of by caribou, 203–204
  in lichen woodlands of Labrador–Ungava, 140
  nitrogen fixation by, 375
  photosynthetic adaptations in, 209
  productivity and biomass of, 373–376
  role of in communities, 129, 318, 327, 335–336
  shade intolerance in, 210–216
  substrate preferences of, 210–211
Light
  ecological effects of, 295–298, 307–313, 323
  photoperiodicity in plants, 49
  shade tolerance
    and intolerance in plants, 287–290, 319
    shade intolerance in lichens, 210–216
    in mosses, 215–217
  species response to, modeling, 129–130
Limiting factors, 351
Luvisolic soil, 103–106
Lynx, Canada
  population cycles in, 404–408
  role in ecosystem, 386–387

M

Mackenzie, southwestern, and northern Alberta, vegetational communities in, 149–160
Mackenzie delta, vegetational communities in, 141
Mackenzie Mountains, vegetational communities in, 144–149
Mackenzie River region, upper, vegetational communities in, 155–156
Mackenzie–Yukon, vegetational communities in, 140–149
Magnesium
  in nutrient cycling and productivity, 351–380
  soil content of, 304
Mammals, as components of trophic pyramid, 382–383
Marten, role of in ecosystem, 387–388

Metabolism, selective accumulation and exclusion of mineral elements, 303–307
Methods
 continuum analysis, 282–285
 correlation, use of, 286–287
 expressions of hyperspace, 287
 gradient analysis, 285–286
 ordination, 286–287
 principal components analysis, 286–287
Mice, role in ecosystem, 388–389
Microclimate
 characteristics of in northern forest, 80–84
 forest, general, 72, 197
 leaf surface temperatures, 74–76
 species adaptation to and competition, 131
Microorganisms, soil
 aerobic and anaerobic, 356
 autochthonous, 119
 Azotobacter, 120
 breakdown of litter by, 118–119
 forms, 117–119
 iron-reducing, 356
 muskeg, microbiology of, 120
 nutrient cycling, role in, 355–357
 numbers of species and individuals, 98, 119–120
 numbers in soil horizons, 119–120
 organic decay, role in, 121–122
 Streptomyces, 120
 zymogenous, 119
Migration
 arctic species into forest zones, 83
 geographical distribution of species, role in, 131
 postglacial migration of species, 415
Mink, role of in ecosystems, 387
Model, forest management, 381–411
Modeling
 role of community structure and composition, 129
 use of computers in, 128
Moisture, soil
 critical factor in distribution of mosses, 218–219
 gradients in and community variation, 131
 Picea glauca, response of to, 151
 plants and, 285, 287–298, 311, 328
 species response to
  in modeling, 129–130
  on mountain slopes, 145–146
  regional regimes, 72
  snow on mountain slopes, 145
Moose
 plant–moose–wolf subsystem, 397–399
 role of in ecosystem, 383–384
Mosses
 continuum of in communities, 215
 ecology of, 318, 320, 323, 327, 328, 331
 paludification, role of in, 303–307
 peatlands, 306
 reproductive patterns in *Sphagnum* species, 223–224
 reproduction of tree species influenced by, 222–223
 root development in trees affected by, 222–223
 species in communities, 212–224

N

Nitrogen
 available soil, 138, 243, 299, 304
 cycling of, 357–364
 immobilized, release of, 303
 productivity, role in, 351–380
Nitrification, 91, 356
Nitrogen fixation, 91, 119, 306, 356, 375–376
Nutrient elements, 6, 7, 10, 285, 291, 327–328, 364–366
 analyses of soil for, in study areas, 242–244
 availability of, 105–106, 291
 affected by mosses, 222–223
 related to rate of organic decomposition, 315
 selective accumulation and exclusion, 303–307
Nutrient cycling
 carbon cycle, 357
 lichens, role of, 375–376

Nutrient cycling (*cont.*)
   management practices in forestry, 376–377
   methods, inadequacies of study methods, 353
   nature of cycling processes, 353–355
   nitrogen, 357–364
   organic decomposition and, 355
   productivity and, 351–380

O

Ordination, of species in study areas, 251–274
   use of in research, 286–287
Organic soil order, 103–106
Origins of the boreal flora, *see* Evolutionary history, Postglacial history
Owls, population cycles in, 404–408

P

Palatability, of plants to wildlife, 14
Palsas, structure of, 324
Paludification, in forests, 303–304, 314–315
Parent material, of soils, 241
Peatland vegetation
   communities, 110, 309, 320, 324, 356
   continuum in, 7–8
   succession in, 327–329
Peat plateaus, 324
Permafrost, 285, 291–295, 314–315, 375, 416
   characteristics of in northern central Manitoba, 167
   continuous and discontinuous, 324
   cryoturbation, 117
   effect of on growth of spruce, 202–203
   frost action and microrelief, 117
   hummocky permafrost soil, 116
   influence upon vegetation, 74–76
   mapping of, 326
   palsas and peat plateaus, 324
   soil drainage in, 115–116
   as soil factor, 113–117
   succession, role of in, 324–327
   vegetational communities, effect of on, 202–203
Pesticides, plant responses to, 14
Phosphorus, role of in nutrient cycling and productivity, 138, 243, 299–307, 315, 351–380
Photoperiod, in northern latitudes, 49
Photosynthesis
   acclimation to and species distribution, 77–78
   factors affecting rate of, 69
   insect defoliators and, 401–403
   light saturation, 295–296
   photosynthetic productivity of the earth, 434
   related to climate, 77–78
   related to respiration, 48
   related to temperature, 48
Physiological adaptations
   geographical distribution of species, role of in, 131, 257–258
   light and shade, effect of, 295–298
   response to changing climate, 24–28
Physiological optima, 309
   competitive interactions among species, 310
Picea species
   growth rates at northern limit of trees, 312–313
   introgression between *P. mariana* and *P. glauca*, speculations regarding, 201–202
   reproduction by layering, 201
   regeneration of, in northern regions, 202–203
   reproduction, sexual and nonsexual in, 201–202
Picea glauca (white spruce)
   dominant species in *P. glauca* stands, 272–273
   stand analyses, data from study areas, 257–273
Picea mariana (black spruce)
   communities in, species correlated with air masses, 261–266
   dominant species in communities, 270–271
   nutrient content of foliage, 315
   regeneration after fire, 329–335
   sapling establishment, 335–339
   serotinous cones, role of, 331
   vegetative reproduction in, 332
Pinus banksiana (jack pine)

reproduction in, 331–333
sapling establishment, 335–339
serotinous cones, role of, 331–332
stand analyses, 257–273
Podzol soil
 chelating compounds in genesis of, 92
 climate, related to, 93–97
 early research on, 98–99
 eastern and western compared, 105–106
 environmental factors and genesis of, 92–95, 99–100, 421–422
 genesis of
  chemical processes in, 92–93, 96–97
  compared to temperate deciduous forest soils, 100–101
  microorganisms in, 120–121
 geography of, 97–98
 miniature (photo), 100
 morphology on iron-rich substrate, 102
 nutrients in, 105–106
 permafrost and cryoturbation, 116–117
 site classification and, 108–110
 water storage in, role of mosses, 114
 vegetation associated with, 101–102
 vegetational ecotones, related to, 99–100
Podzolic soil order, 103–106
Pollen record, see Evolutionary history, Postglacial history
Pollutants, plant responses to, 14
Polyploidy, 415
Population densities, birds and mammals, general, 381–411
Populus tremuloides (trembling or quaking aspen)
 reproduction in, 331–332
 stand analyses, 257–273
Postglacial history, boreal forest, 24–30, see also Evolutionary history
 aspen parkland, 31–32
 community composition during, 28–41
 early postglacial communities, 28–43
 ecological significance of, 40–41
 forest-tundra communities, 36–39
 grasslands, 31–32
 human occupancy of boreal regions, 40–41
 northeastern Minnesota, 30
 southern boreal border, 31–33
 transitional boreal forest, 31

western and central interior of Canada, 30–31
Potassium, role in nutrient cycling and productivity, 243, 299, 304, 315, 351–380
Precipitation, see also Climate
 winter and summer patterns, 65–66
Productivity, 6, 329
 boreal, global, 434
 community comparisons, 368–373
 lichen productivity, 373–375
 nutrient cycling, role, 351–380
 wildlife production and management, 433

Q

Quaking aspen, see *Populus tremuloides*, also Species Index
Quebec, northern central, vegetational communities, 184–195

R

Radiation, see also Light
 model of, 71–72
 regional radiation budget, 65
 related to temperature regime, 70
 tree growth and, 71–73
 tundra, intensity over, 197
Rainfall, see Precipitation
Regional communities, see specific areas, see also Communities
 altitudinal climatic gradient, 149
 boreal vegetational zones, 256–274
 climatic classifications, 238–239
 community differences and climatic gradients, 237–240, 258–266
 compared, 147–149
 continuum, broad regional, 134–137, 276
 dominant species in, 262–266
 field sampling procedures, 240–244
 forest stands, characteristics of, 246–247
 forest-tundra transition, 147
 grasslands and broadleaved forest, 140
 northern and southern zones, 161–162
 northern forest border, 195–208, 418–421
 northern, west of Hudson Bay, 206–208
 ordinations of species data, 251–275

Regional communities (cont.)
  regional analyses, 266–274
  regional variation, 135–137, 237–280
  S/D index, 255–256
  spruce communities near waterways, 150–151
  zonal categories, 166–167
Regosolic soil order, 103–106
Reproduction, in tree species
  influence of mosses, 222–223
  vegetative at northern forest border, 48, 80–81, 312
Respiration
  rate related to climate, 77–78
  relationship to photosynthesis, 48
Roots
  adaptations to moisture conditions, 290
  competition between, 323
  development in trees, affected by mosses, 222–223
  frozen soil, affect of, 325
Ruffed grouse, population cycles in, 404–408

S

Sampling
  employed in research, 244–251
  vegetational community, 249–250
Seedling mortality, 313
Serotinous cones, 322
Similarity relationships among communities, 251–274
  associated with distance, 255–256
  coefficient of, 245, 286–287
  used in research, 251–274
Site classification, see also Soils, 108–110
Snow, accumulation on mountain slopes, 145
  abrasion by, 311
  animal habitats, affected by, 77–78
  moisture factor, 113–117, 311–312, 325
  plant community structure, affected by, 77–78
Snowshoe hare
  population cycles in, 404–408
  role in ecosystem, 386–387
Soil
  acidity, 297, 303, 315
  active layer in permafrost soil, 115–116

analyses of sampled soils, 241–244
animals, classification of, 121–122
bimodal distribution of plant species, related to, 110–111
Canadian system of classification, 91–92, 102–106
catena, 94, 107–108
chemical processes in genesis of, 96–97
classification, in boreal regions, 102–106
climate, as soil forming factor, 90–91
coniferous and deciduous, compared, 100–101
cryoturbation in permafrost soils, 116–117
description of orders, 103–104
earth hummocks, 326
ecotonal, 99–100
edaphic factors, 20
elements in, 91
fauna, 118–119
fire, affect of, 104, 106
fossil, 197–198, 332
genetic variations and plant communities, 106–112
genesis, and plant succession, 107–108
geography of podzols, 97–98
gray-brown, gray-wooded, 97–98
horizon designations, 91
hummocky permafrost type, 116, 326
litter and humus consumed by fire, 138
microrelief in permafrost soil, 116–117, 326
mineral nutrient elements, 90–91
moisture status, 113–117
mosses
  influence of, 215–216
  water storage capacity, 114
organic decomposition, 315
organisms, 117–119, see also Microorganisms
parent material, 98, 241
peat, 327–329
permafrost, 285, 291–295, see also Permafrost
pH, 291
podzol, see also Podzol soils, 92, 421–422
profile types, 91, 102–105
regions, 105–106
regional, related to vegetation, 239–242
research needs, 421–422

soil site classification, 108–110
soil subsystem, 90–127
soil system, Canadian, 102–106
water storage capacity, 113–115
weathering, role of, 93
Species
  bimodal environmental response of, 164–165
  circumpolar, 140
  climate, response to, 19–21, 76–80, 258–266
  communities, similarity of, 251–274
  depauperate zone in forest-tundra zone, 83–84
  dispersal rates, 20–21
  distribution, effect of topography, 161–162
  dominant, 266–274
  ecotypic variation, 240–241
  entomophilous, in palynology, 26–27
  environmental tolerance, 240–241
  regeneration after fire, 329–335
  responses to environment, 20–21, 130–132, 275, 282–285
  sprouting after fire, 138
  sympatric, 140
  tundra community, 265–274
  variability, 414–415
Spruce, see *Picea*
Spruce, black, see *Picea mariana*
Spruce, white, see *Picea glauca*
Spruce budworm, 392–393, 402–403
Squirrel, red
  population cycles in, 407–408
  role in ecosystem, 388
Statistics
  continuum analysis, 282–285
  correlation, use of, 286–287
  gradient analysis, 285–286
  methods of, use in vegetation research, 10–13
  methods, modeling, basis, 15–16
  ordination, 286–287
  principal components analysis, 286–287
Stratification, of seeds, 313
Subclimax, 416
Succession, 314–341, 416–418
  characteristics of, in northern zones, 82–83
  climax concept and, 316, 321–324

  cyclical, 304, 316, 334–335
  eastern and western, compared, 308
  fire, role in, 137–138, 329–335
  herbivores, as control agents, 390
  modeling, 129
  mosses, influence of, 215–216, 221
  northern, southern compared, 308–313, 317–318
  paludification, 303–307, 314–315
  patterns
    Mackenzie region, 156–159
    Canadian shield region, 163–174
  peatland, 327–329
  permafrost, role in, 324–327
  roots, in permafrost soils, 325
  sapling establishment, 335–339
  shade tolerance, effect on, 319–320
  soil moisture, 287–295
  spruce forest to bog, 138
  vectors, 334–335
Sulfur, 299
  role in nutrient cycling and productivity, 351–380
Superior region, vegetational communities, 162

T

Taiga, 1
Taxonomy, problems in, 151–152
Temperature, boreal
  extremes
    summer, 65
    winter, 63–64
  leaf surface, 75
  means, at representative stations, 52–57
  physiological responses to, 77–78
  photosynthetic optima, in trees, 313
  plant organ, 418
  related to evapotranspiration, 69
  role in modeling, 129–130
  soil, nutrient availability, 315
  soil surface, 72–76
Tertiary period, 21–25
  change in climate, 23–24
Toxic minerals, responses of plants to, 14
Trembling aspen, see *Populus tremuloides*
Trophic levels, general, 6, 396–397
  animal populations, 381–411

Trophic levels (*cont.*)
  role of insects, 401–403
  survival of breeding populations, 382–383
Tundra, vegetational communities, 265–274

V

Vectors, successional, 334–335
Voles, role in ecosystem, 388–389

W

Water relationships, *see* Moisture
Weasels, role in ecosystem, 387
White spruce, see *Picea glauca*
Wildlife
  biomass, 390–392
  in ecosystems, 390–392
  management of, 430–433
  populations, birds, 394–395
  population cycles in, 404–408
Wind
  canopy, related to temperature, 73
  related to evapotranspiration, 69, 311–312, 417

Winter kill, 311, 325
Winter Lake, vegetational communities of region, 205
Wolves
  plant–moose–wolf subsystem, 397–398
  populations of, compared to moose and wolverine, 397–400
Wood Buffalo Park, regional vegetational communities, 160

X

Xeromorphic adaptations, 299–300

Y

Yellowknife, regional vegetational communities, 158

Z

Zonation
  vegetational, designated by analysis, 268–272
  vegetational zones, boreal, 256–275

Species Index*

A

Abies balsamea, 1, 4, 30, 31, 102, 109, 150, 160, 161–163, 170–174, 176, 178–181, 192, 216, 217, 237–276, 287, 288, 296–298, 303, 320, 322, 335, 338, 402
Abies fraseri, 225
Abies lasiocarpa, 139, 160
Acer negundo, 31, 288
Acer rubrum, 164, 179, 337
Acer spicatum, 109, 164, 225, 288, 337
Achillea millefolium, 168, 289
Acorus calamus, 294
Actaea rubra, 154, 173, 179, 181
Agrostis borealis (A. scabra), 159, 293, 300
Alectoria, 75, 211, 212
Alnus crispa, 4, 37, 138, 139, 144, 153, 154, 158, 159, 168, 176, 181, 206, 214, 248, 288, 291, 306, 317, 318, 360
Alnus rugosa, 4, 160, 176, 248, 288, 292, 301, 373
Amelanchier alnifolia, 159, 168, 176, 288
Andromeda glaucophylla, 109
Andromeda polifolia, 139, 143, 144, 147, 164, 169, 174, 179, 206, 270, 289, 300
Anemone canadensis, 173, 289
Anemone multifida, 159
Anemone parviflora, 147, 159, 206
Anemone patens, 156, 159
Anemone quinquefolia, 248, 289
Apocynum androsaemifolium, 159, 176, 289
Aquilegia brevistyla, 155, 159
Aquilegia canadensis, 289

Aralia nudicaulis, 109, 153, 154, 159, 160, 168, 176, 179, 180, 216, 248, 272, 290, 296
Aralia racemosa, 173
Arctostaphylos alpina, 102, 144, 146, 147, 193–195, 206, 248, 265, 270, 273
Arctostaphylos rubra, 138, 139, 143, 153, 154, 158, 248
Arctostaphylos uva-ursi, 31, 143, 154, 156, 159, 164, 168, 175, 206, 214, 272, 289, 290
Arenaria groenlandica, 194
Arenaria humifusa, 193
Arenaria lateriflora, 155
Armeria maritima, 147
Arnica lonchophylla, 155, 159
Artemisia frigida, 159
Aster borealis (A. junciformis), 289, 293
Aster ciliolatus, 159, 160, 176, 248
Aster laevis, 289
Aster macrophyllus, 109, 173, 248
Aster puniceus, 289
Aster sibiricus, 159
Astragalus alpinus, 147
Astragalus americanus, 159
Athyrium filix-femina, 164
Aulacomnium, 143, 157, 168

B

Bartsia alpina, 190, 194
Betula glandulosa, 27, 30, 37, 38, 75, 102, 134, 143, 144, 146, 151, 152, 157–159, 168, 185–187, 189, 190, 193–195, 204, 206, 214,

*Included are significant references to common boreal plant species in the text. See also Tables and Appendix for additional listings. Lichens and mosses are referred to here by genus only; the listing of mosses is exclusive of pages 212–224 in which mosses are named with great frequency.

Betula glandulosa (cont.)
  248, 264, 265, 270, 273, 274, 289, 295, 318, 371
Betula lutea, 174, 176, 225
Betula occidentalis, 155, 156
Betula papyrifera, 27, 31, 37, 74, 75, 106, 109, 134–138, 144, 159, 162, 164, 168, 170–174, 176, 178, 179, 248, 296, 317, 320, 322, 338, 397
Bidens cernua, 293

C

Calamagrostis canadensis, 153, 160, 168, 206, 289, 292, 295, 300, 319
Calamagrostis purpurascens, 155, 156, 159
Calliergon, 109
Calliergonella, 153, 157
Caltha palustris, 160, 214, 289, 293
Calypso bulbosa, 153, 155, 159, 160
Campanula rotundifolia, 154, 159
Carex aenea, 159
Carex aquatilis (see also C. stans), 293, 301
Carex atrofusca, 147
Carex bigelowii, 37, 193, 194, 195
Carex brunnescens, 168, 248
Carex canescens, 293
Carex capillaris, 37, 158, 194
Carex capitata, 204
Carex chordorrhiza, 248, 293, 300
Carex concinna, 154
Carex deflexa, 168
Carex diandra, 293
Carex disperma, 158, 293
Carex foenea, 159
Carex glacialis, 37, 194
Carex gynocrates, 158
Carex interior, 179
Carex lacustris, 293
Carex lasiocarpa, 293, 301
Carex limosa, 169, 248, 293, 300
Carex lugens, 142
Carex media, 158
Carex membranaceae, 147
Carex microglochin, 190
Carex misandra, 147
Carex pauciflora, 190
Carex paupercula, 181, 190, 248
Carex pedunculata, 173
Carex pseudo-cyperus, 179
Carex rariflora, 204

Carex rostrata, 293, 301
Carex-rotundata, 152
Carex saxatilis, 190, 248
Carex scirpoidea, 193, 194, 206
Carex stans (see also C. aquatilis), 37, 204, 206
Carex tenuiflora, 293, 300
Carex trisperma, 176, 181, 190
Carex vaginata, 190, 194, 206
Cassiope hypnoides, 193, 194
Cassiope tetragona, 142, 143, 145, 147, 270
Cetraria, 37, 116, 139, 143, 158, 159, 211, 212, 318
Chamaedaphne calyculata, 109, 156, 158, 168, 181, 188, 190, 248, 264, 270, 289, 292, 300
Chimaphila umbellata, 154, 164, 175
Chiogenes, see Gaultheria
Cinna latifolia, 160
Circaea alpina, 164, 174
Cladonia, 37, 108, 116, 143, 144, 156–159, 186, 187, 192, 204, 205, 209–212, 291, 295, 318, 320, 333, 371, 385
Clintonia borealis, 109, 111, 173, 174, 176, 179–181, 248, 272
Coptis trifolia (C. groenlandica), 136, 176, 179–181, 184, 187, 189, 190, 194, 195, 248, 289
Corallorhiza maculata, 289
Corallorhiza trifida, 153, 154, 158–160
Cornus canadensis, 109, 136, 138, 144, 153, 154, 157, 159, 168, 173–176, 179–181, 184, 186–190, 194, 195, 215, 216, 248, 264, 265, 270, 273, 289, 296, 319, 385
Cornus stolonifera, 160, 168, 288, 292
Corydalis sempervirens, 168
Corylus cornuta, 109, 173, 216, 288, 296
Cypripedium acaule, 154
Cypripedium passerinum, 154, 155
Cystopteris montana, 154

D

Diapensia lapponica, 37, 83, 193, 194, 199
Dicranum, 77, 109, 143, 168, 186
Diervilla lonicera, 173, 176, 248, 289
Drosera rotundifolia, 143, 158, 164, 169, 174, 181, 248, 289
Dryas drummondi, 154
Dryas integrifolia, 30, 84, 139, 142–147, 154, 155, 193–195, 199, 206, 270, 273
Dryas octopetala, 84, 142–145, 199
Dryopteris austriaca, 179

Species Index

Dryopteris disjuncta, see *Gymnocarpium*
Dryopteris robertiana, 179

E

Eleocharis palustris, 294
Elymus innovatus, 159, 214
Empetrum nigrum, 37, 38, 138, 139, 143, 144, 146, 147, 154, 156, 159, 164, 168, 175, 185, 187–190, 193, 194, 204–206, 248, 264–266, 270, 273, 274
Epigaea repens, 109, 176, 179–181, 248
Epilobium angustifolium, 136, 138, 153, 155–157, 159, 160, 164, 168, 174, 179, 187, 248, 273, 327
Equisetum arvense, 138, 206, 248, 264, 289, 292, 301
Equisetum fluviatile, 294
Equisetum palustre, 190, 214
Equisetum pratense, 155, 160
Equisetum scirpoides, 138, 143, 144, 153, 155, 158, 159, 206, 264, 270, 289, 292, 300
Equisetum sylvaticum, 138, 153, 157, 158, 168, 174, 176, 181, 187–190, 214, 248, 264
Erigeron compositus, 154
Erigeron glabellus, 159
Eriophorum angustifolium, 142, 190, 295
Eriophorum vaginatum (E. spissum), 142, 147, 157, 190, 204, 206, 214, 248, 292, 295, 317, 319

F

Fragaria vesca, 153, 160, 164, 168, 174, 289
Fragaria virginiana, 109, 153, 159, 160, 176, 248, 292

G

Galium boreale (G. septentrionale), 138, 142, 144, 153, 159, 160, 176, 248
Galium labradoricum, 214
Galium trifidum, 289, 301
Galium triflorum, 173, 179, 180, 248, 289, 292
Gaultheria hispidula (Chiogenes hispidula), 109, 136, 174, 176, 179–181, 189, 190, 192, 193, 289
Gaultheria procumbens, 111, 176, 248
Gentiana amarella, 155
Geocaulon lividum, 136, 138, 145, 153–156, 158, 159, 168, 176, 181, 187, 190, 206, 248, 264, 273, 289
Geum canadense, 173, 190
Glyceria borealis, 289, 294
Glyceria grandis, 293
Glyceria striata, 301
Goodyera repens, 136, 153, 154, 159, 164, 176, 179
Gymnocarpium dryopteris (Dryopteris disjuncta), 176

H

Habenaria dilatata, 190
Habenaria hyperborea, 160, 289
Habenaria obtusata, 153, 158, 160, 289
Hedysarum alpinum, 143, 147, 154, 159, 289
Heuchera richardsonii, 173
Hieracium canadense (H. umbellatum), 159
Hierochloe alpina, 37, 147, 194
Hippuris vulgaris, 294, 301
Hudsonia tomentosa, 159, 289
Hylocomium, 109, 114, 116, 139, 153, 157, 168, 191, 291, 320, 333
Hypnum, 109, 157
Hypopitys monotropa, see *Monotropa*

J

Juncus castaneus, 37, 190
Juncus trifidus, 193, 194
Juniperus communis 143, 153, 156, 157, 159, 164, 175, 207, 289
Juniperus horizontalis, 159, 289

K

Kalmia angustifolia, 179–181, 183, 188, 304
Kalmia polifolia, 109, 168, 176, 181, 187, 189, 190, 248, 264
Kobresia simpliciuscula, 147

L

Larix laricina, 4, 31, 37, 75, 102, 106, 148, 150, 159, 161, 162, 174, 176, 181, 184, 196, 207, 216, 248, 288, 293, 298, 301, 318, 322, 338
Larix lyallii, 139
Larix occidentalis, 139
Lathyrus japonicus, 142

Lathyrus ochroleucus, 159, 248, 289
Lathyrus venosus, 289
Ledum decumbens (L. palustre ssp. decumbens), 37, 38, 139, 144, 146, 147, 154, 156, 207, 248, 265, 266, 270, 274, 295, 319
Ledum groenlandicum, 37, 102, 108, 109, 138, 139, 143, 144, 151, 154, 156, 157, 159, 164, 168, 174–176, 179, 181, 183, 185–190, 192, 194, 204, 207, 214, 248, 264, 265, 270, 273, 274, 289, 291, 292, 295, 300, 313, 318, 319
Lemna minor, 294, 301
Lemna trisulca, 293
Linnaea borealis, 109, 136, 138, 144, 153–155, 157–160, 164, 168, 174–176, 179, 181, 190, 194, 195, 207, 215, 248, 264, 270, 273, 289, 319
Listera borealis, 158, 160
Listera cordata, 176, 179, 187–190
Loiseleuria procumbens, 37, 38, 164, 175, 193–195, 204, 265
Lonicera canadensis, 109, 173
Lonicera dioica, 159, 289
Lonicera involucrata, 160
Lupinus arcticus, 145, 147
Luzula confusa, 37, 147, 193–195
Luzula nivalis, 147
Luzula parviflora, 194
Lychnis apetala, 147
Lycopodium alpinum, 187, 189
Lycopodium annotinum, 136, 138, 153, 159, 168, 175, 176, 289
Lycopodium clavatum, 175
Lycopodium complanatum, 168, 176, 289
Lycopodium obscurum, 154, 176, 289
Lycopodium selago, 187, 189, 190, 194
Lycopodium tristachyum, 154

M

Maianthemum canadense, 109, 111, 136, 153, 158–160, 173, 174, 176, 179–181, 214, 215, 248, 264, 265, 270, 272, 292
Melampyrum lineare, 176
Mentha arvensis, 176, 289
Menyanthes trifoliata, 109, 168, 214, 293
Mertensia paniculata, 138, 153, 160, 168, 176, 215, 248, 272, 289, 301
Mitella nuda, 136, 153, 154, 157, 158, 160, 168, 173–176, 179, 180, 184, 193, 194, 248, 270, 272, 289, 292, 301

Moneses uniflora, 136, 153, 154, 157, 158, 176
Monotropa hypopithys, 164
Myrica gale, 169, 360
Myriophyllum verticillatum, 294

O

Orchis rotundifolia, 153, 158, 160
Oryzopsis asperifolia, 159, 176, 248
Oryzopsis pungens, 154, 159, 215
Osmorhiza longistylis, 289
Oxalis montana, 179
Oxycoccus microcarpus, see also *Vaccinium*, 109, 143, 148, 151–152, 169, 293, 300
Oxytropis splendens, 159

P

Parmelia, 157, 158, 211
Parnassia palustris, 158, 190
Pedicularis capitata, 139, 147
Pedicularis labradorica, 143, 146, 147, 154, 156, 159, 207
Pedicularis lanata, 147
Pedicularis lapponica, 143
Peltigera, 153, 157, 211, 376
Petasites frigidus, 144
Petasites palmatus, 144, 153, 157, 168, 174–176, 187, 189, 190, 215, 248, 272, 289, 290, 292
Petasites sagittatus, 289, 293
Phyllodoce caerulea, 194
Picea engelmannii, 139
Picea glauca, 1, 4, 31, 34, 74, 75, 102, 106, 115, 134–136, 138, 139, 143, 144, 150–152, 154, 156, 157, 159–163, 168, 170–173, 176, 178, 184, 187, 188, 191, 202–205, 217, 237–276, 287, 288, 297, 298, 310, 318–320, 322, 329, 332, 334–338, 402
Picea mariana, 1, 4, 31, 34, 71, 74, 75, 80, 102, 106, 109, 113, 115, 116, 134–139, 143, 144, 146, 150–152, 154, 156, 157, 159–163, 166, 168, 170–174, 176, 178, 180, 181, 184, 187–192, 196, 198, 200–205, 207, 209, 212–217, 219, 220, 223, 237–276, 287, 288, 291, 292, 296–298, 300, 303, 306, 310, 311, 314, 315, 317, 319, 320, 322, 324, 329, 331, 332, 334–338, 340, 362, 367–371, 374, 375, 402, 419, 432
Pinguicula villosa, 143, 169

Pinguicula vulgaris, 158, 207
Pinus albicaulis, 139
Pinus banksiana 4, 31, 83, 106, 109, 113, 115, 134, 135, 150, 152, 154, 159–162, 168, 170–173, 191, 209, 219, 237–276, 297, 298, 311, 322, 331–334, 336, 337
Pinus contorta, 139, 150, 160, 331, 333, 335, 338, 363, 364, 367, 374, 375
Pinus flexilis, 139
Pinus resinosa, 163, 311
Pinus strobus, 163, 174
Pleurozium, 114, 168, 191, 291, 333
Poa alpina, 142
Poa glauca, 142, 159
Poa palustris, 142
Polygala paucifolia, 289
Polygonum bistorta, 147
Polygonum viviparum, 29, 139, 145, 147, 164 175, 194
Polytrichum, 75, 192, 320, 333
Populus balsamifera, 4, 30, 31, 109, 134, 138, 139, 150, 159–162, 164, 170, 178, 216, 287, 288, 317, 331, 332, 337, 338, 397
Populus tremuloides 4, 31, 106, 109, 134, 137, 150, 159–162, 170, 176, 178, 287, 288, 297, 298, 317, 320, 322, 330–332, 335, 337, 338, 397
Potamogeton vaginatus, 294
Potentilla fruticosa, 75, 143, 159, 207
Potentilla nivea, 147
Potentilla palustris, 214, 248, 300
Potentilla tridentata, 190, 289, 290
Prenanthes alba, 173, 289
Prunus pensylvanica 154, 168
Pseudotsuga menziesii, 139
Ptilium, 114, 191
Pulsatilla, see *Anemone*
Pyrola asarifolia 138, 154, 155, 158–160, 168, 248, 289, 292
Pyrola grandiflora, 143, 147, 154, 194, 207
Pyrola rotundifolia, 175
Pyrola secunda, 136, 138, 144, 147, 155, 158–160, 164, 173, 175, 179, 181, 207, 248, 273, 289, 319
Pyrola virens, 155, 158–160, 215, 289

R

Ranunculus gmelinii, 158, 294
Ranunculus lapponicus, 158, 190

Rhacomitrium, 193, 194, 205, 320
Rhamnus alnifolia, 109, 160
Rhododendron canadense, 181, 385
Rhododendron lapponicum, 38, 84, 139, 143, 145, 147, 156, 194, 199
Ribes glandulosum, 168, 179, 187, 189
Ribes hudsonianum, 158, 160
Ribes lacustre, 153, 159, 179
Ribes oxyacanthoides, 168
Ribes triste, 136, 138, 139, 153, 158, 160, 175, 176, 289
Rosa acicularis, 136, 138, 139, 144, 152, 153, 155, 157–160, 168, 176, 248, 273, 290, 295, 319, 397
Rubus acaulis 144, 158, 190, 207, 214, 248, 273, 289, 293
Rubus chamaemorus, 109, 143, 144, 156–158, 168, 176, 181, 188–190, 194, 204, 207, 214, 248, 264, 266, 270, 289, 292, 295, 300, 317
Rubus idaeus (*R. strigosus*), 160, 173, 176, 385
Rubus parviflorus, 173
Rubus pubescens, 136, 153, 157, 160, 174, 176, 179, 180, 248, 265, 270, 273, 289, 292, 301
Rumex arcticus, 147

S

Salix alaxensis, 139, 319
Salix arbusculoides, 138
Salix arctica, 147, 190
Salix bebbiana, 138, 156, 159, 160, 168, 248, 292
Salix discolor, 248, 293
Salix glauca 138, 139, 144, 146, 147, 151, 158, 159, 207, 274, 318, 414
Salix herbacea, 30, 193, 194, 248
Salix myrtillifolia, 138, 158, 204, 248, 292
Salix petiolaris, 288
Salix phylicifolia (*S. planifolia*), 144, 148, 207, 248
Salix pulchra, 318
Salix pyrifolia, 158
Salix reticulata, 139, 147
Salix richardsonii, 139, 143
Salix uva-ursi, 193, 194
Salix vestita, 193, 194
Sanguisorba canadensis, 187, 189
Sarracenia purpurea, 109
Saussurea angustifolia, 146, 207
Saxifraga aizoides, 147

500 Species Index

Saxifraga aizoön, 29, 147, 164, 174, 194
Saxifraga tricuspidata, 147, 156, 159
Scheuchzeria palustris, 293
Schizachne purpurascens, 154
Scirpus lacustris (S. acutus), 294, 301
Scirpus caespitosus, 181, 190, 193, 194, 204
Scolochloa festucacea, 294
Selaginella selaginoides, 158
Selaginella sibirica, 142
Senecio integerrimus (S. lugens), 139, 158
Senecio tridenticulatus, 159
Shepherdia canadensis, 138, 139, 142, 144, 153, 159, 168, 175, 207, 273, 288, 360
Silene acaulis, 139
Silene repens, 142
Smilacina trifolia, 157, 158, 169, 176, 188–190, 214, 248, 270, 289, 293, 300
Solidago hispida, 248
Solidago macrophylla, 179, 187, 189, 190, 194
Solidago multiradiata 154, 156, 190, 207
Sorbus americana, 164, 225
Sorbus decora, 176, 181
Sorbus scopulina, 154
Sphagnum, 75, 109, 116, 138, 143, 156–159, 166, 168, 204, 295, 314, 315, 317, 320, 322, 371
Spiranthes romanzoffiana, 158
Spiraea beauverdiana, 319
Stellaria crassifolia, 193, 294
Stellaria longipes, 193
Streptopus roseus, 109, 173, 176, 248
Stereocaulon, 155, 205, 211, 212, 320, 333, 375, 376

T

Tanacetum huronense, 154
Thalictrum venulosum, 160
Thelypteris, see also *Gymnocarpium*
Thelypteris palustris (Dryopteris thelypteris), 179
Thuja occidentalis, 163, 174, 179
Tofieldia pusilla (T. palustris) 143, 154, 155, 190, 207
Tomenthypnum, 114
Trientalis borealis, 109, 154, 160, 173–176, 179–181, 184, 190, 194, 195, 248, 289, 313
Trisetum spicatum, 29

U

Usnea, 157
Utricularia minor, 294
Utricularia vulgaris, 294

V

Vaccinium angustifolium, 108, 173, 174, 176, 179–181, 186, 187, 189, 190, 192, 194, 248, 385
Vaccinium microcarpus, see also *Oxycoccus*, 151–152, 181, 189, 190
Vaccinium myrtilloides (V. canadense), 108, 136, 154, 159, 168, 173, 174, 176, 179, 181, 248, 264, 289, 292, 300
Vaccinium oxycoccus, see also *Vaccinium microcarpus, Oxycoccus microcarpus*, 152, 157, 158, 164, 176, 207, 214, 248, 264, 289
Vaccinium uliginosum, 30, 37, 38, 48, 73, 74, 102, 130, 138, 139, 143, 144, 146, 147, 154, 156, 158, 164, 175, 186, 188–190, 193, 194, 204, 207, 248, 264–266, 270, 273, 274, 291, 295, 317, 319, 432
Vaccinium vitis-idaea, 37, 38, 48, 73, 75, 130, 138, 139, 143, 144, 146, 147, 153, 154, 156–159, 164, 168, 174, 179, 181, 185, 189, 190, 192, 194, 204, 205, 207, 214, 248, 264, 265, 270, 273, 274, 289, 292, 295, 300, 319, 333, 432
Viburnum edule, 136, 138, 139, 154, 158–160, 168, 176, 248, 273, 290, 319, 397
Viburnum opulus (V. trilobum), 179, 288
Vicia americana, 160, 289
Viola adunca, 190, 194
Viola canadensis (V. rugulosa), 289
Viola palustris, 160, 301
Viola renifolia, 153, 160, 248, 293

W

Woodsia alpina, 29
Woodsia glabella, 142

Z

Zigadenus elegans, 144, 157, 159